Clinical Aspects of
THE PLASMA PROTEINS

TADASHI KAWAI, M. D.

Director, Central Clinical Laboratory and Research Division
Nihon University Surugadai Hospital, Tokyo

Professor of Clinical Pathology,
Nihon University School of Medicine, Tokyo

Professor of Clinical Pathology,
Jichi Medical School, Tochigi

SPRINGER-VERLAG BERLIN HEIDELBERG GMBH

PUBLISHERS

© First edition, September 1973 by Springer-Verlag Berlin Heidelberg
Originally published by Springer-Verlag Berlin · Heidelberg · New York in 1973
Softcover reprint of the hardcover 1st edition 1973

ISBN 978-3-662-06269-2   ISBN 978-3-662-06267-8 (eBook)
DOI 10.1007/978-3-662-06267-8

Library of Congress Catalog Card Number 73-15272

Foreword

It was the year of 1969 when this monograph was originally published in Japanese by Professor TADASHI KAWAI, titled as "The Plasma Proteins, Their Fundamental and Clinical Aspects." After I read through the Japanese edition, I was impressed by its rather complete coverage of the subjects and their detailed descriptions. I have felt that this excellent monograph should be distributed not only among our Japanese scientists but also among many other colleagues throughout the world. I am happy, therefore, to know that the English edition of his monograph, partly revised, is ready to be published at this time.

Professor KAWAI received his postgraduate medical training in U.S.A. for seven years, and was certified by the American Board of Pathology in both Anatomical and Clinical Pathology in Fall, 1962. Thus, I believe, he is the most suitable fellow for publishing the English edition of this kind.

The first parts of the book are concerned by the fundamental and physiological properties of the plasma proteins. The most important part of this book deals with the pathophysiology of various plasma protein abnormalities. Among innumerable clinical cases that the author has experienced for the past 15 years, more than one hundred representative cases were selected and arranged adequately in the text with most beautiful immunoelectrophoretic patterns, each being analyzed painstakingly by his own judgement. Therefore, the readers should find it valuable to understand, through various plasma protein abnormalities, the fundamentals and pathophysiology of many important disease conditions. In addition, this book makes almost encyclopedic coverage of the diseases accompanying any plasma protein abnormality. This monograph is certainly comparable to the excellent publications by WUHRMANN and WUNDERLY, RIVA, SCHULTZE and HEREMANS, and others, and further it contains many unique opinions of his own.

By writing a foreword to this remarkable book may I praise Professor KAWAI for his hard work!

July, 1973

Hidematsu Hirai

Professor,
Department of Biochemistry,
School of Medicine,
Hokkaido University,
Sapporo, Japan

President,
The Society of Electrophoresis

Preface

The plasma proteins are indispensable for maintaining many important functions of living cells, and their main functions include the maintenance of osmotic relations between the circulating blood and the tissue spaces, the buffering action of various body fluids and the transport of various important substances necessary for the living cells and of various metabolites. The abnormalities of the plasma proteins are likely to cause more or less functional disorders of the living cells, and, on the other hand, almost every pathology of the tissues may be reflected to the plasma protein changes. Therefore, detailed studies of the plasma protein changes are extremely valuable to evaluate clinically the pathophysiological background of each patient. In addition, among the body proteins, the plasma protein components are the ones most easily purified, and thus recent development in molecular biology has been mainly met with the plasma proteins.

This book is primarily concerned with the human plasma proteins for the purpose of understanding the pathophysiologic and diagnostic aspects of their abnormalities. Since cellulose acetate electrophoretic and immunoelectrophoretic techniques have been receiving a tremendously wide application in clinical medicine, attention is focused particularly upon understanding the pathophysiological backgrounds on many different serum protein electrophoretic patterns, covering as much new knowledge in protein biology as possible. The author, as a clinical pathologist, has attempted to relate various important clinical disorders not only to plasma protein changes but also to other laboratory findings at the same time.

There has been certainly not a few comprehensive literature on human plasma proteins, including "Die Bluteiweisskörper des Menschen" (F. WHURMANN and Ch. WUNDERLY), "The Plasma Proteins" (ed. by F. W. PUTNAM), "Serum Proteins and the Dysproteinemias" (ed. by F. W. SUNDERMAN, and F. W. SUNDERMAN, Jr.) "The Plasma Proteins" (ed. by H. NEURATH and K. BAILEY), "Serum Proteins in Health and Disease" (G. SANDOR), and so on. However, no such monograph had been published originally in Japanese, and the present author attempted it in as early as 1964. In 1966, however, the first volume of "Molecular Biology of Human Proteins" by H. E. SCHULTZE and J. F. HEREMANS was published. It resembled amazingly well to the image that the author had in his mind. Therefore, the author's attempt to publish the monograph was set aside at least for the following two years when the second volume of "Molecular Biology of Human Proteins" was not published as expected.

This monograph appeared originally in 1969, in Japanese, including the following major topics: the fundamental structure of proteins, the analytical methods of plasma proteins, the physico-chemical and biological proterties of plasma proteins, the metabolism of plasma proteins, and the clinical abnormalities of plasma proteins. In this English edition, the whole section on the analytical methods and a portion of the other sections were removed, and the text was revised partly. However, the time lost for English translation has made the text of certain chapters less up to date than expected. For this the author asks the indulgence of the readers, but I sincerely hope that graduate students, medical technologists, clinical pathologists and clinical practitioners all find this book a valuable reference for understanding the pathophysiology of plasma protein diseases and their related clinical abnormalities as well.

My greatest thanks are due to all those who have contributed in studying valuable cases cited in this book, such as many clinicians and technologists in the Central Railway Hospital of the Japanese National Rialways, and the Nihon University Hospitals. I take this opportunity to express my sincere thanks also to the staff of the International Publishing Department, Igaku Shoin Ltd. for their tremendous cooperation.

Tokyo/Japan
March, 1973

T. Kawai

Laboratory Findings in Multiple Myeloma

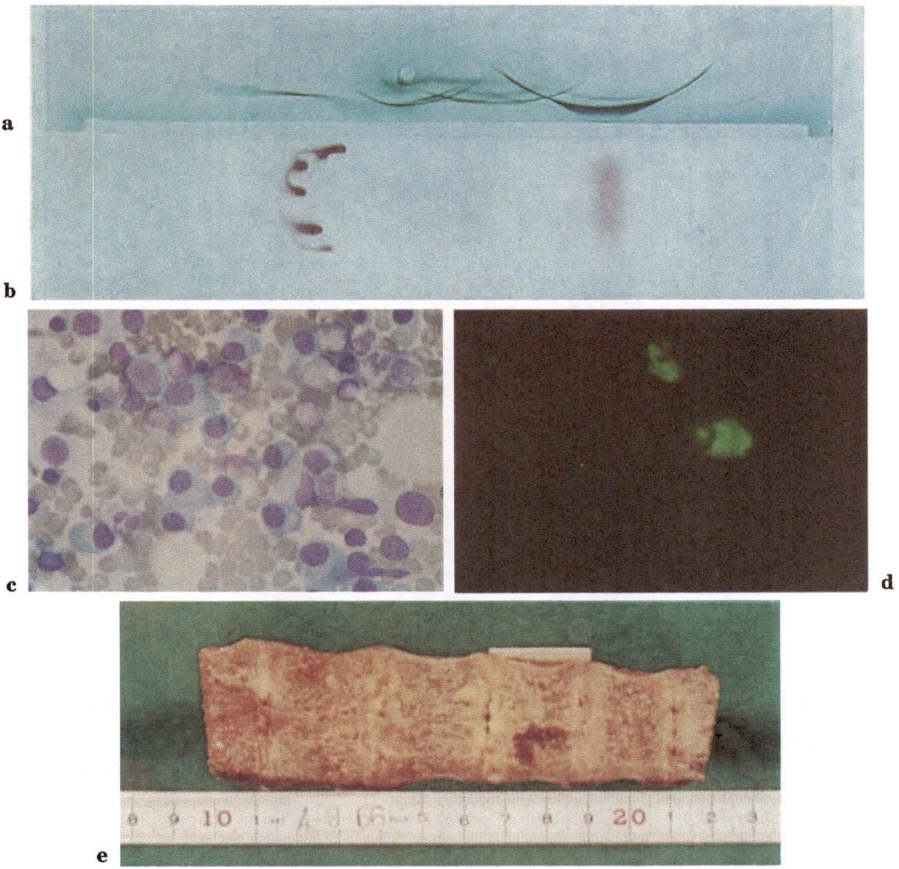

a. Agar gel immunoelectrophoretic pattern of the serum taken from the patient with IgG-K type multiple myeloma. The IgG precipitin line shows a characteristic M-bow near the point of serum application. Both the IgA and IgM lines are not recognized.

b. Cellulose acetate (Oxoid membrane) electrophoretic pattern of the same serum shows a characteristic wavy M-protein band at the cathodal end. The other protein fractions are only vaguely demonstrated because a very small quantity of the serum sample was applied.

c. The sternal bone marrow smear obtained from the same patient shows an increased number of pleomorphic plasma cells, occupying approximately 60% of the total nucleated cells.

d. One of the myeloma cells containing a large amount of IgG is shown through the indirect fluorescent antibody technique. Interestingly, the cell contains a positively stained inclusion body in the nucleus.

e. The thoracic vertebrae are involved by the neoplastic change, showing an osteolytic lesion which is recognizable macroscopically.

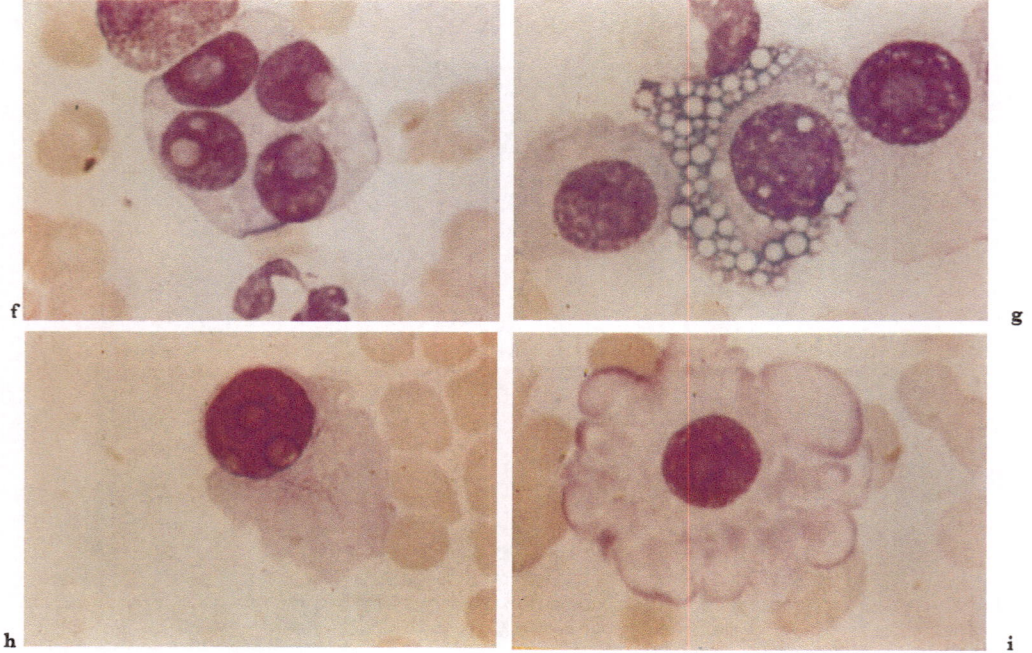

Various kinds of the abnormal plasma cells recognized in multiple myeloma:

    f. Multi-nucleated plasma cell, containing intra-nuclear inclusion bodies.

    g. Abnormal plasma cell, containing many strongly basophilic intra-cytoplasmic inclusion bodies or Russell's bodies.

    h. Abnormal plasma cell, containing several azurophilic rods in its cytoplasm.

    i. "Flame cell" or abnormal plasma cell showing distinct peripheral eosinophilia on Giemsa staining.

Contents

Section III METABOLISM OF THE PLASMA PROTEINS

Section IV DIAGNOSIS AND PATHOGENESIS OF PLASMA PROTEIN ABNORMALITIES

SECTION I

INTRODUCTION

Chapter 1

The Fundamental Structure of Proteins

The word "protein" comes from the Greek *proteios* which means "the first", that is, the most necessary material for the living body.   This material came to be called "protein" in German, with its most representative form being white albumin.   Proteins have very complicated functions and structures, and are found in all living things.   Recently protein studies in biochemistry and biology have developed remarkably.

Proteins are generally classified into three groups, according to their solubility and compositions.   These are: simple proteins, conjugated proteins, and derived proteins.   This classification is traditional but it has many drawbacks so, as research advances, rectifications are in view.

All proteins contain four elements, carbon, hydrogen, oxygen and nitrogen, with most proteins also containing sulfur.   Further, there are proteins which also contain phosphorus, iodine, iron, copper and zinc.   These elements are contained in the proportions shown below, with percentages varying in each protein.

Atom	%
C	50
H	7
O	23
N	16
S	0–3
P	0–3

Protein molecules become amino acids when they are hydrolyzed.   Moreover, all amino acids are linked together by the α-peptide bonds.   That is, in α-amino acids, the amino group ($-NH_2$) is combined with the carbon atom which is next to the carboxyl group ($-COOH$).

Amino acids are connected as in a chain by the peptide linkage and they form so-called polypeptides, or polypeptide chains.

What is called the primary structure of proteins refers to the following two factors: 1) the number of amino acids which form the polypeptide chain, and 2) the type of amino acid

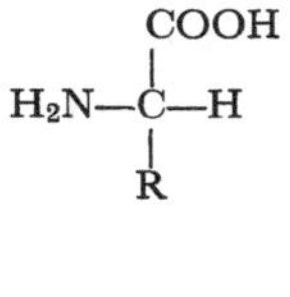

$$\begin{array}{c} COOH \\ | \\ H_2N-C-H \\ | \\ R \end{array} \quad (\alpha\text{-amino acid})$$

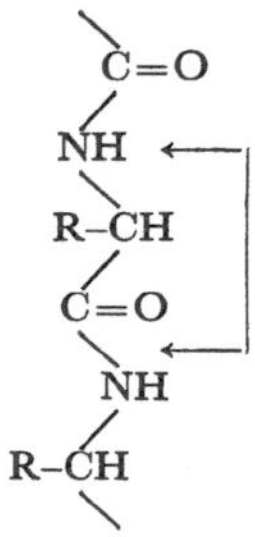

$$
\begin{array}{c}
\diagdown \\
\diagup C = O \\
NH \longleftarrow \\
\diagdown \\
R\text{--}CH \\
\diagup \\
C = O \\
\diagdown \\
NH \longleftarrow \\
\diagup \\
R\text{--}CH \\
\diagdown
\end{array}
\qquad
\begin{array}{l}
\text{one amino acid residue} \\
\text{(peptide bond)}
\end{array}
$$

composition contained therein. After the hydrolyzation of protein molecules by various methods, the primary structure of a protein may be determined by ascertaining the number and kind of amino acids through chemical reactions, electrophoresis, paper chromatography, and ion exchange column chromatography. With the invention of the automatic amino acid analyzer, the determination of primary structure has recently been greatly facilitated.

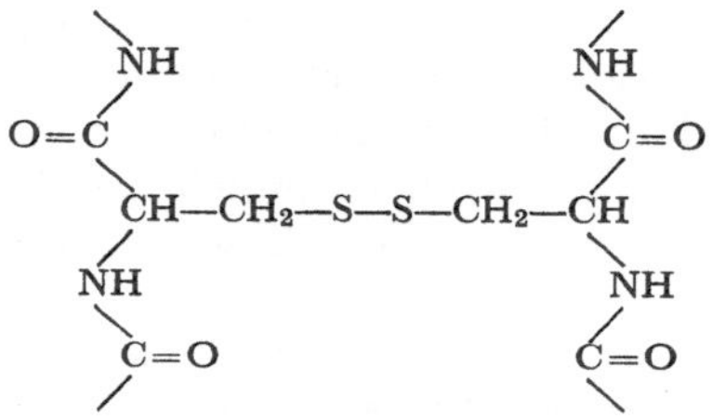

$$
\begin{array}{c}
\diagup NH \qquad\qquad\qquad \diagdown NH \\
O = C \qquad\qquad\qquad\qquad C = O \\
CH\text{--}CH_2\text{--}S\text{--}S\text{--}CH_2\text{--}CH \\
\diagup NH \qquad\qquad\qquad\qquad \diagdown NH \\
C = O \qquad\qquad\qquad\qquad C = O
\end{array}
$$

(disulfide bond)

$$
\begin{array}{c}
R \\
| \\
-C-C-N- \\
\| \quad | \quad | \\
O \quad H \quad H_* \\
\vdots \qquad\quad \vdots \\
{}^*H \quad H \quad O \\
| \quad | \quad \| \\
-N-C-C- \\
| \\
R
\end{array}
$$

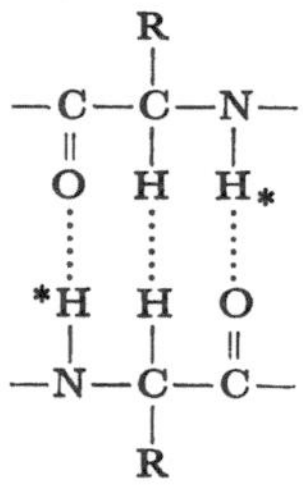

(*hydrogen bond)

Formed by several thousands of amino acids, each polypeptide chain curves in a distinctive manner, according to its amino acid sequence. It is this special manner of curving that is referred to as the secondary structure of proteins. The disulfide bond and the hydrogen bond are related factors in determining the secondary structure.

The disulfide bond is formed in the polypeptide chain which has cysteine residue. This bond is comparatively stable and usually survives through conditions which cause denaturation. However, even this disulfide bond is oxidized by the use of performic acid. On the other hand, the hydrogen bond, as shown above, is formed between NH group and oxygen atoms of the carboxyl group. This bond is comparatively unstable because it is not a covalent bond and it is easily destroyed by denaturation.

The curve of the polypeptide chain possesses not only two, but three dimensions, manifesting a structure that is spiral in nature. This is referred to as the tertiary structure of proteins and it is composed of the hydrogen bond described above or Van der Waals' force.

The shape of the protein molecule which consists of one polypeptide chain is determined by the tertiary structure. Protein molecules which consist of more than two polypeptide chains have moreover the quaternary structure. The quaternary structure is the mode of combining two or more polypeptide chains (monomeric subunits) which may or may not be identical in nature. This combination occurs during the formation of protein molecules, presumably through molecular collisions. In many cases, the quaternary structure plays an important role in enzyme activity, antibody activity, and other important biological activities. The methods which are used to ascertain the shape of protein molecules include streaming birefringence, double refraction of flow, X-ray diffraction and electron photomicrography.

Most physico-chemical characteristics of proteins derive from the characteristics of amino acids which form protein molecules, but not a few of them come from the tertiary or quaternary structures. As the physicochemical characteristics of proteins are innumerable, the classifications based on them vary accordingly. Taking plasma proteins as an example, in this group alone, there are at least 80 different constituents.

Chapter 2

General Principles of Protein Fractionation

Fractionation of plasma proteins has the following two purposes: 1) to isolate certain plasma components, clarifying the characteristics and functions of these components, and 2) to aid diagnosis and therapy in the field of clinical medicine by the knowledge attained through this research. Since the days of HIPPOCRATES and ARISTOTLE the problem of blood coagulation has occupied the attention of scientists. It is quite natural that research into the fractionation of plasma proteins has developed in relation to blood coagulation factors. Moreover, as described above, it is in response to clinical demands that attention has been focused on γ-globulin in relation to antibody reaction and that plasma protein was made for the purpose of therapy since World War II.

The problem is how to profit from the distinctive characteristics of proteins in order to fractionate plasma proteins. It may be thought that the definition of proteins is fixed, but it is difficult to find a characteristic which may be said, categorically, to be common to all proteins. Moreover, the difficulty increases when one attempts to define the purity of each component. Almost all proteins have been named according to the derivation of their components and their biological activities. They have also been classified according to solubility and electrophoretic mobility, as well as other evident physiochemical characteristics. However, as analyzing methods developed, it has become clear that most components are not only formed by single elements but also that they have many subfractions. The characteristics of proteins themselves vary greatly but those that are used for their fractionation are rather limited in number. The chief ones are: solubility, molecular size, surface charge, stability, density, biological activity, and anti-

Table 1 Plasma protein fractionation and its principles.

Techniques	Chief characteristics of proteins used for fractionation
Precipitation	Solubility against various solvents
Free electrophoresis	Surface electrical charge in certain electrolyte solutions.
Electrophoresis in various supporting media	Surface electrical charge, electroendosmosis, and adsorption to supporting media, (molecular sieve in some media).
Chromatography	Solubility against various solvents and interference with various ion exchangers.
Gel filtration	Molecular size
Ultracentrifugation	Molecular weight (molecular density)
Ultrafiltration	Molecular size
Immunological	Antigenic determinants
Hematological	Biological activities
Enzymological	,,

genicity. Some of these characteristics and representative methods of fractionation of proteins are summarized in Table 1. Of course, there are methods making use of only one characteristic, and there are others in which more than two characteristics are used at the same time. When the greatest number possible of characteristics is used, a high level of homogeneity of each fractionated component is attained. For example, in the gel immunoelectrophoretic method, surface charge, stability and antigenicity are combined, and the serum proteins are classified into more than 30 types at one time. Also, by using these fractionating methods based on differing principles in a well-ordered manner, a higher level of homogeneity of each component may be attained.

The representative fractionating methods used for clinically screening plasma protein abnormalities are the cellulose acetate electrophoretic technique and the agar gel immuno-electrophoretic technique.

SECTION II

PROPERTIES OF INDIVIDUAL
PLASMA PROTEIN COMPONENTS

Chapter 3

Plasma Proteins Included in the Albumin Fraction

The albumin fraction is the most prominent protein band observed at the extreme an-odal side when normal human serum proteins are electrophoretically separated by the standard procedure (pH 8.6). By using the filter paper or cellulose acetate electrophoresis which is used routinely, it is observed at the highest and sharpest electrophoretic peak. This fraction is composed mostly of serum albumin containing a very small amount of prealbumin.

ALBUMIN

1. Synonyms

Albumin, Serum albumin

2. Physical Characteristics

Sedimentation constant $(S_{20,w})$ 4.6S; diffusion coefficient $(D_{20,w})$ 6.1; molecular weight 69,000; frictional ratio (f/f_0) 1.28; intrinsic viscosity (η) 0.042; electrophoretic mobility 5.92; isoelectric point (pI) 4.9; extintion coefficient $(E_{280m\mu})$ 5.8

N-F transformation: Albumin is one of the most homogeneous serum proteins and in the buffer with higher pH than its isoelectric point, it forms, electrophoretically, a single protein band. However, between pH 3.5–4.5, it is divided into two or three peaks and its result is influenced by pH as well as by ionic strength and composition of an individual buffer solution used for analysis. FOSTER explains this phenomenon by the isomerization theory[45]. That is, the N form existing normally at neutral pH is transformed under the influence of acidic pH into the F form by the reversible reaction shown below.

$$N + 3H^+ \rightleftarrows F$$

The F form is of a greater mobility than the N form, and between the two, differences in the following characteristics are noted: viscosity, optical rotary dispersion, ultra violet ray absorption and binding activity with low-molecular weight substances.

3. Chemical Characteristics

Precipitation: Albumin precipitates under the following conditions[184]: 40% ethanol (pH 5.2, ionic strength 0.01, protein 2.5%), 2.6–3.0 M ammonium sulfate (pH 7.0, protein 1%), 0.0065 M Rivanol (aqueous) solution (pH 0.8, protein 1%), 0.6M perchloric acid (protein 1%), 0.15M trichloracetic acid (protein 1%), heating (pH 5.0, 0.1M acetate buffer, protein 1%).

Chemical structure: Albumin is a single, long polypeptide chain, consisting of 610 amino acids.   The N-terminal amino acid residue is aspartic acid; the C-terminal residue, leucine.   The amino acid composition of human serum albumin characterstically gives a relatively large amount of the basic amino acids (arginine, histidine and lysine) and the acidic amino acids (aspartic acid and glutamic acid), and a little amount of tryptophane.   An approximately 50% of the albumin molecule shows the α-helical configuration, and seems to form a doughnut shape, electron-microscopically, measuring approximately 100 Å in its external diameter.[129]

Its nitrogen content is 16%.   The carbohydrate content is 0.08%, containing only a minute amount of hexose and hexosamine.   The lipid content is 0.2%.[184]

Albumin is classified into two types, depending on whether or not the free sulf-hydryl group (S–H) is present.   About two thirds of blood serum albumin contain the free S-H group.   In the presence of Hg^{++} the albumin has tendency to polymerize easily and to form a dimer.   Therefore, it is called "mercaptalbumin."[79]

The remaining third, not having the free S-H group, is non-mecaptalbumin.   It has lost the polymerizability through which the free S-H groups of albumin molecule bind easily with cysteine or glutathione, which is contained in the blood.[97]   When mercaptalbumin becomes old, it tends to polymerize spontaneously, forming several protein zones on starch gel electrophoresis.[172]   Moreover, there is a report which states that a small amount of albumin in the form of a dimer exists in nephrotic or other pathological sera.[182]

With regard to albumin purification, besides COHN's ethanol fractionation, there are other methods which employ trichloracetic acid, ether, ammonium sulfate, and ion exchangers.   It has been crystallized from mercaptalbumin.[184]

4.   Hereditary Types

The serum albumins with different electrophoretic mobility have been recognized in many families.   Though it is rare in Japan, at least 10 pedigrees have been reported so far, including 5 of the authors'.   Heredity is of a non-sex-linked dominant form, depending on a pair of two co-dominant alleles.

Since KNEDEL (1957) first referred to it as double albumin, many investigators have tried naming it in various ways: WIEME called the albumin of normal mobility Al^n, of

Table 2   Various synonyms applied to hereditary types of albumin.

Alloalbuminemia (BLUMBERG et al. 1968)
    1.   Homozygous type
    2.   Heterozygous type
Paralbuminemia (DRACHMANN et al. 1965)
Doppel Albuminämie (KNEDEL, 1957)
Double albuminemia (Earle et al. 1959)
Bisalbuminemia (FRANGLEN et al. 1960)
Albumin Normal (Alb^n) (WIEME, 1960)
          = Albumin A (EARLE et al. 1959)
          = Albumin A_1 (KNEDEL, 1957)
Albumin Slow (Alb^{sl}) (WIEME, 1960)
          = Albumin B (EARLE, et al. 1959)
          = Albumin A_2 (KNEDEL, 1957)
Albumin Rapid (Alb^{ra}) (WIEME, 1960)
Albumin Reading ($Alb^{Reading}$) (TARNOKY and LESTAS, 1964)
Albumin Mexico (Alb^{Me}) (MELARTIN and BLUMBERG, 1966)
Albumin Naskapi (Alb^{Na}) (MELARTIN and BLUMBERG, 1966)

---

more rapid mobility Al^{ra} (rapid albumin), and of slower mobility Al^{sl} (slow albumin).[208] Synonyms of these terms are listed in the above Table 2.

Alb^{sl} and Alb^{ra} can easily be observed by separating them from Alb^{n} by means of the ordinary method of zone electrophoresis (Fig. 1). Alb^{ra} is less frequent than Alb^{sl}, and only 2 pedigrees of Alb^{ra} were found in Japan. Between these two there is no observable difference in immunological antigenicity and ultracentrifugal findings. In the case of double albumin, the concentration of each albumin component is, in general, the same and the total concentration of the 2 components is usually the same as that in normal serum. However, it is said that both Alb^{sl} and Alb^{ra} have a lower dye-binding capacity than Alb^{n} and when preserved, they degenerate easily.[184]

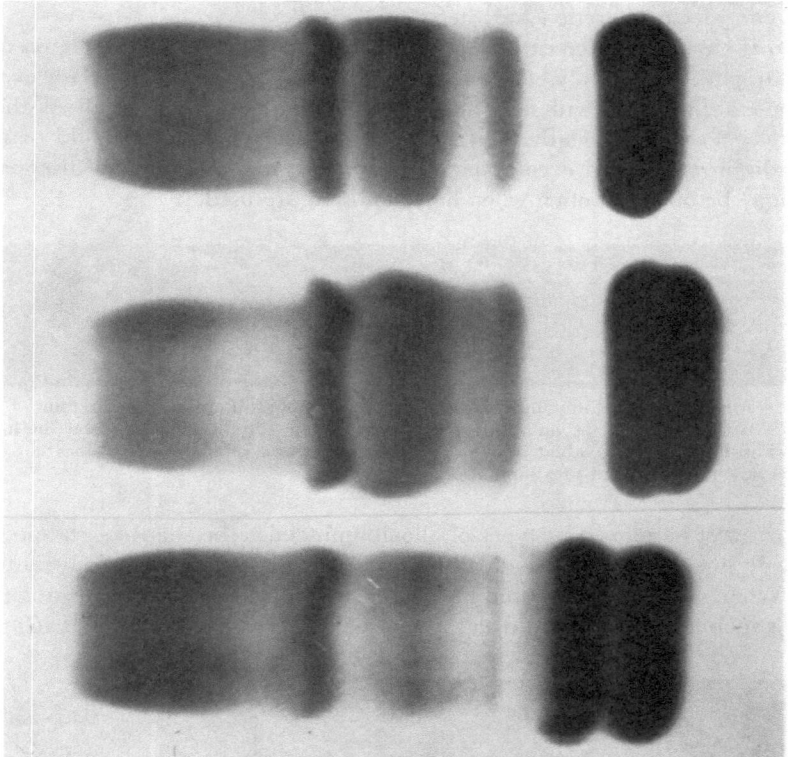

Fig. 1 Cellulose acetate electrophoretic serum protein pattern in alloalbuminemia (heterozygous type).
The upper pattern is of the normal serum, showing Alb^{n}; the middle showing Alb^{n} and Alb^{ra}; and the lower showing Alb^{n} and Alb^{sl}.

Furthermore, by means of starch-gel electrophoresis the albumin of greater electrophoretic mobility has been demonstrated recently. Even a homozygous form of Alb^{Naskapi} is known.[11] At present, 5 sub-types, as indicated in Fig. 2, are known. The hereditary type of these albumins is found also in healthy persons without being correlated with any specific disease. The difference in their molecular structure is still unknown, but it has been reported that there is a difference in their amino acid composition.[60]

MIYOSHI points out that there is fetal albumin (Alb F) which is similar to fetal hemoglobin (Hb F). In other words, compared to adult albumin (Alb A), Alb F is more

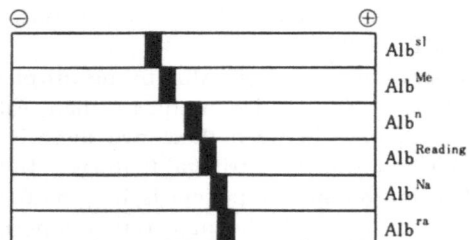

Fig. 2 Schematic diagram of starch gel electrophoretic patterns of various albumins.

resistant to alkali and proteolytic enzymes, and is different in its isomerization equation curve.   Like Hb F, Alb F transforms itself to Alb A in four months after birth.[135]

5.   Immunoelectrophoretic Characteristics

Albumin is observed at the extreme anodal side and it forms symmetric precipitin arc. Therefore, it can not be confused with other precipitin arcs (Fig. 3).   However, since it has common antigenicity with albumin of some mammalians, not all of the antisera available may possess sufficiently high antibody titer.   However, it should be noted that in the condition of antigen excess, the precipitin arc may completely disappear.   This tendency may be observed often when horse antisera are used.

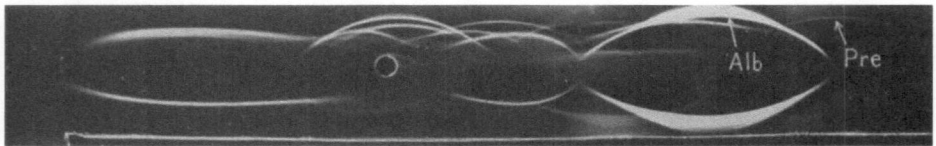

Fig. 3   Agar gel immunoelectrophoretic pattern of albumin and prealbumin.
The upper trough contains the anti-human whole serum horse immune serum, and the lower contains the anti-human serum rabbit immune serum.   The rabbit immune serum does not contain a significant amount of the anti-prealbumin.

In the cases of heterozygous types of alloalbuminemia, the shape of the precipitin arc is elongated, since none of the alloalbumins reported so far shows any antigenic difference.   When some of the albumin molecules are bound with a more negatively charged compounds such as bilirubin and drugs, it can be observed as an asymmetric arc (Fig. 4).

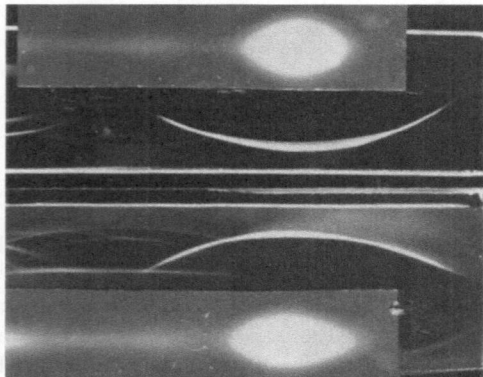

Fig. 4   Abnormal mobility of the albumin spot in icteric serum.
The upper half shows the agar gel electrophoretic and immunoelectrophoretic patterns of the normal serum albumin. The lower half shows the patterns of the icteric serum.   With the icteric serum, the albumin spot shows a definite tailing towards the anode, and the albumin line is slightly asymmetric.

6.   Biological Characteristics

Albumin concentration in serum of the normal adult is 3.5–4.5 g/100 ml.

Affinity for various ions or compounds: Albumin possesses the most remarkable non-specific binding activity among the serum proteins. That is, albumin molecules exist as negative ions in body fluids having pH higher than their iso-electric point. However, as a negative ion, it interferes non-specifically with other positive ions and at the same time, it binds also with various negative ions. For example, Hg^{++} and Cu^{++} bind with the S-H (sulfhydryl) group, and Zn^{++}, Cd^{++} and Ca^{++} are thought to combine with the imidazol group. Among various negative ions, even monovalent anions combine with albumin, but its affinity is stronger in proportion to their negative valency. Among the polyvalent anions, the greatest affinity is exhibited by surface activators such as sodium dodecyl sulfate. It is thought that such comparatively large anions combine primarily by van der WAAL's force, relying on the configurational changes of albumin molecule. It may be that the above mentioned behavior derives from the facts that the albumin molecule is prone to change the expansion of its conformation by comparatively easy isomerization and molecular expansion, and that it contains many basic or acidic amino acid residues. Table 3 lists the representative substances and drugs bound to albumin, which are thought to be physiologically or clinically important. This subject was reviewed in details by FOSTER and BENHOLD[5,45)]

Table 3   Representative compounds and drugs
binding to albumin (from PUTNAM[164)])

Ca^{++}, Cu^{++}, Zn^{++}	Aureomycin
Bilirubin	Barbiturates
Uric acid	Chloromycetin
Free vitamin C	Digitonin
Free acetyl choline	Atebrine
Cholinesterase	Neosalvalsan
Adenosine	Penicilline
Acidic dyes	Salicylates
Phenol red	Paramino salicylates
Free Congo red	Sulfa drugs
Histamine	Streptomycin
Thyroxine	Iodide contrast media
Triiodothyronine	Erythrocyte
Fatty acids	

When some of those which are listed in Table 3, especially those negatively charged, are bound to albumin, their electrophoretic mobility is changed more rapidly. As a result, they may be mistakenly identified with alloalbuminemia as mentioned previously. Brom Phenol Blue (BPB), bilirubin (in jaundice, especially obstructive), and fatty acids (in nephrosis) are good examples of this.

This type of affinity of albumin has the following two major clinical significances; namely, 1) it decreases the concentration of ionized and non-ionized free compounds in the blood, and results, for example, in buffering action such as mitigation of acidosis or neutralization of toxic substances; 2) it acts to transport various useful and harmful compounds. It transports useless or harmful compounds to kidneys or liver, through which the coumpounds are discharged from the body. Hg^{++}, bilirubin, pigments, drugs, etc. are examples of this. As to the relatively water-insoluble compounds, a great deal of it is transported into and through the body by specific binding functions of albumin. For instance, although most of the free fatty acids are bound with albumin, the fatty acid itself can be dissolved by water only to the extent of $10^{-8}M$, whereas it can be dissolved in the normal serum to the extent of $5 \times 10^{-4}M$.[63)] By making use of the binding behavior of albumin, various methods of quantitating the serum concentration of albumin have been reported. One of the most representative methods is the one using 4'-hydroxyazobenzene-caboxylic acid- (2). The pigment itself is yellow, but it becomes reddish as it is bound to albumin selectively. Thus, the reddish complex formed is now used for the colorimetric determination of serum albumin concentration.

Maintenance of blood colloidal osmotic pressure: Osmotic pressure of the solution is measured by the number of particles contained in a unit volume. The smaller the particles, the greater the resulting specific osmotic activity. The molecular weight of the albumin molecule is comparatively small, and further its serum concentration is considerably greater than other protein components. That is why the colloidal osmotic activity of the blood is determined primarily by albumin. In other words, the albumin of 1 g exhibits osmotic activity equivalent to that of 1.2 g total serum protein, and 75% of the blood colloidal osmotic pressure is dependent upon albumin.[177]

Source of amino acids: Like other tissue proteins the plasma proteins have an important role in maintaining the amino acid pool in the blood.

   7.   Clinical Abnormalities
Congenital deficiency: Congenital analbuminemia
Acquired decrease (Acquired hypoalbuminemia)
   1)   Hepatic damage
   2)   Acute phase response conditions
   3)   Protein-losing conditions
   4)   Malnutrition
   5)   Others. In almost all diseases a tendency of decreasing albumin is observable.
Acquired increase: In some cases of hemoconcentration, its serum concentration increases, but no pathological condition shows an absolute increase of the albumin concentration.
Paralbuminemia (transient)

PREALBUMIN

   1.   Synonyms
Tryptophan-rich prealbumin (TrPA) (SCHULZE et al, 1956), ϱ_1-protein (URIEL and GRABAR, 1956), Prealbumin-I (SMITHIES, 1955), Thyroxine-binding prealbumin (TBPA) (TATA, 1961), Prealbumin.

   2.   Physical Characteristics[185]
Sedimentation constant ($S_{20,w}$) 4.2S; molecular weight 61,000; electrophoretic mobility 7.6; isoelectric point (pI) 4.7; extrinction coeficient ($E_{280\,m\mu}$) 13.2.

Prealbumin cannot be observed on filter-paper electrophoresis, but by means of cellulose acetate membranes or gel media it can be observed clearly as preceding the albumin fraction.

   3.   Chemical Characteristics
Precipitation: Under the following conditions prealbumin precipitates[185]: 18% ethanol (pH 5.2, ionic strength 0.09, protein 1%) 2.6M ammonium sulfate (pH 5.0, protein 1%), 0.0065 M Rivanol aqueous solution (pH 8.0, protein 1%), 0.6M perchloric acid (protein 1%), 0.15M trichloracedic acid (protein 1%), heating (pH 5.0, 0.1M acetate buffer solution, protein 1%).

Chemical Structure: Prealbumin is composed of 538 amino acids, contains much threonine and serine, and is rich in tryptophane. Its nitrogen content is 14.9%. It contains 0.5% of carbohydrate (hexose 0.4%, hexosamine 0.1%) and no lipids.[185]

4. Immuno-electrophoretic Characteristics

Prealbumin is observed just anodal to albumin as a faint symmetric precipitin arc, partially crossing the albumin precipitin arc (Fig. 3). There are antisera not containing the anti-prealbumin, and thus it is necessary to use the antisera with high antibody titer to demonstrate the arc. When test serum becomes old, α_1-lipoprotein extends to the prealbumin zone, resembling prealbumin, but it can easily be distinguished from the latter because α_1-lipoprotein forms a straight, asymmetric precipitin line.

5. Biological Characteristics

Mono-specific antiserum against prealbumin is now commercially available, and its concentration can be measured immunochemically. Normal serum contains 28–35 mg/100 ml of prealbumin.

Prealbumin, like albumin and thyroxine-binding globulin (TBG) found in the α-fraction, combines with thyroxine and triiodothyronine, and its affinity is interfered with barbiturates.[80] (See Page 24)

6. Clinical Abnormalities

Hypoprealbuminemia: Hepatic damage, wide-spread tissue necrosis, poor protein-intake, etc.

Hyperprealbuminemia: Sometimes found in nephrotic syndrome.

Chapter **4**

Plasma Proteins Included in
the α_1-Fraction

The α_1-fraction is second to the albumin fraction in its electrophoretic migration. It contains various serum protein constituents, important among which are α_1-antitrypsin, α_1-acid glycoprotein and α_1-lipoprotein. In the standard procedure using filter-paper electrophoresis, the α_1-fraction is not clearly separated from the albumin fraction, but it is clearly recognized on cellulose acetate electrophoresis.

α_1-ACID GLYCOPROTEIN

1. Synonyms

α_1-Acid glycoprotein ($\alpha_{1_{\text{AG}_\text{D}}}$) (SCHMID, 1950), Orosomucoid (WINZLER, 1955), α_1-niedermolekulares Säureprotein (SCHULTZE et al, 1955), Seromucoide acide (DE VAUX ST. CTR et al, 1958), Seromucoide acide (MONTREUIL, 1957), α_1-Seromucoid (JAYLE and BOUSSIER, 1955), α_0-Globulin (BURTIN, 1960), X-Component (WILLIAMS and GRABAR, 1955), Prealbumin-2 (SMITHIES, 1955), Zone "b" (POULIK and SMITHIES, 1958), PS–1A (SCHULTZE et al, 1962), MP-1 (MEHL et al, 1949).

2. Physical Characteristics[185]

Sedimentation constant ($S_{20,w}$) 3.11S; diffusion coefficient ($D_{20,w}$) 5.27; molecular weight 44,100; frictional ratio (f/f_0) 1.78; intrinsic viscosity (η) 0.069; electrophoretic mobility 5.2; isoelectric point (pI) 2.7; extinction coefficient ($E_{280\,m\mu}$) 8.9.

Heterogeneity: Below its isoelectric point (pH 2.7), or above pH 2.9, α_1-acid glycoprotein is observed as a uniform protein band even on starch gel electrophoresis. However, at pH 2.9 it can be divided into as many as 7 fractions. This is because neuraminic acid then starts to dissociate partially.[181] After sialic acid has been removed from the molecule, the α_1-acid glycoprotein is separated at pH 4.8 into a major slow-moving subfraction and a minor fast-moving one.[199]

3. Chemical Characteristics

Precipitation: α_1-acid glycoprotein precipitates under the following conditions[185]: 70% ethanol (pH 6.2, ionic strength 0.05, protein 1%), 2.4–4.0M ammonium sulfate (pH 5.0, protein 1%), 0.0065M Rivanol aqueous solution (pH 8.0, protein 1%). However, under the conditions such as 0.6M perchloric acid, 0.15M tricholoracetic acid and heating, it does not precipitate like other seromucoids.

Chemical structure: α_1-acid glycoprotein is composed of 204 amino acids. It does not contain the N-terminal amino acid residue and the C-terminus is composed of serine.

Its nitrogen content is 10.1%. Lipids are not contained but carbohydrates are found as much as 41.4%. It contains 14.7% of hexose, 13.9% of hexosamine, 12.1% of sialic acid and 0.7% of fucose. Several methods of purification of this protein are reported but it is usually crystallized as its lead salt from COHN Fraction IV.[180]

4. Hereditary Types

By means of starch gel electrophoresis using acidic buffers (pH 4.8) α_1-acid glycoprotein in the normal serum can be separated into three types as shown in Fig. 5. This is thought to be under a genetic control, but the exact distribution of its homozygote and heterozygote has to be analysed in future.[199]

Fig. 5 Hereditary types of α_1-acid glycoprotein.
The schematic diagram shows the patterns obtained with the starch gel electrophoresis using the buffer of pH 4.8.

5. Immunoelectrophoretic Characteristics

This protein has relatively small molecular weight and its immunogenicity is not strong enough to produce a high antibody titer. However, high-titer antisera may be obtained from chickens, horses, and sometimes rabbits. Unlike many other serum proteins, its *in vitro* antigenicity will not be lost by boiling.

On immunoelectrophoresis, it is normally observed inside the α_1-antitrypsin line as a comparatively faint and symmetrical precipitin arc (Fig. 6). Depending on the electrophoretic conditions used for analysis the arc could be somewhat faster and it may intersect albumin and α_1-antitrypsin.

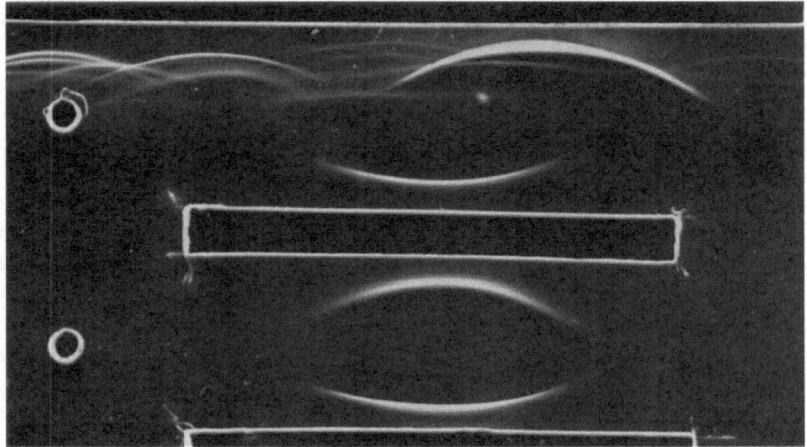

Fig. 6 Immunoelectrophoretic pattern of α_1-acid glycoprotein.
The anti-human whole serum horse immune serum (upper trough) does not contain a sufficient amount of the anti-α_1 AGP, and its precipitin line is not observed. The lower and middle troughes contain the specific antiserum against the protein, and its lines of precipitation can be seen clearly between those of albumin and α_1-antitrypsin.

Because α_1-acid glycoprotein usually forms a faint precipitin arc, it may be hidden in the albumin or α_1-antitrypsin lines which usually form prominent arcs. However, the α_1-acid glycoprotein arc can become easily recognized if an antiserum has been absorbed with purified serum albumin or if a test serum has been deproteinized with trichloracetic acid or boiled before the immunoelectrophoretic analysis is performed.

6.   Biological Characteristics

Normal serum concentration of α_1-acid glycoprotein is 75–100 mg/100 ml. Its biological function is still unkown, but it is probable that it inhibits hemagglutination of the inactivated influenza virus and it inactivates progesterone.

7.   Clinical Abnormalities

Acquired increase: As one of the acute phase reactants, it increases along with α_1-antitrypsin in the following cases:

    1)   Acute and chronic inflammatory diseases
    2)   "Stress" syndrome
    3)   Malignant tumors
    4)   Various hematological abnormalities

Acquired decrease

    1)   Hepatic damage
    2)   Nephrotic syndrome
    3)   Malnutrition and cachexia

Para-α_1 acid glycoproteinemia[181] (Decreased content of sialic acid)

α_1-ANTITRYPSIN

1.   Synonyms

α_1-Antitrypsin (α_{1AT}) (SCHULTZE et al, 1962), α_1-3.5S-Glycoprotein (SCHULTZE et al, 1962), α_1-Glycoprotein (HIRSCHFELD, 1960), α_1-Seromucoide de SCHULTZE (MONTREUIL, 1958), α_{1A}-Globulin (BURTIN, 1960), α_{1B}-Globulin (POULIK and SMITHIES, 1958), Zone "c" (POULIK and SMITHIES, 1958), PS-2C (SCHULTZE et al, 1962), α_1-Trypsin-inhibitor (BUNDY and MEHL, 1959).

2.   Physical Characteristics[185]

Sedimentation constant ($S_{20,w}$) 3.41; diffusion coefficient ($D_{20,w}$) 5.2; molecular weight 45,000; intrinsic viscosity (η) 0.068; electrophoretic mobility 5.42; isoelectric point (pI) 4.0; extinction coefficient ($E_{280\,m\mu}$) 5.3.

Heterogeneity: About 10% of α_1-antitrypsin in serum is bound with the basic low-molecular weight substances which are considered to be deprived from trypsin, and has a bit lower electrophoretic mobility.   And as will be explained later, when α_1-antitrypsin is decreased pathologically, the slow-moving fraction becomes predominant.[104]

3.   Chemical Characteristics

Precipitation: α_1-antitrypsin precipitates under the following conditions[185]:   40% ethanol (pH 5.8, ionic strength 0.09, protein 1%), 2.4–3.2M ammonium sulfate (pH 7.0, protein 1%), 0.0065 M Rivanol aqueous solution (pH 8.0, protein 1%), 0.15M trichloracetic acid, heating (pH 5.0, 0.1M acetate buffer, protein 1%).   By using perchloric acid of less concentration than 0.6M, its precipitation does not occur but it becomes slightly turbid.

Chemical structure: α_1-antitrypsin consists of 381 amino acids and contains 12.4% of carbohydrates (hexose 4.7, hexosamine 3.9, sialic acid 3.6, fucose 0.2).   Lipids are not identified.

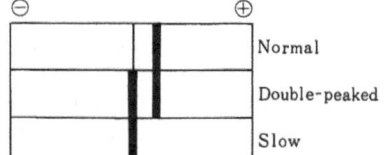

Fig. 7 Possible hereditary types of α_1-antitrypsin.
The schematic diagram shows the patterns obtained with the agarose gel electrophoresis.

Normal
Double-peaked
Slow

4. Hereditary Types

LAURELL identified subtypes of α_1-antitrypsin by using antigen-antibody crossed agarose gel electrophoresis, and classified them into three types as indicated in Fig. 7.[104]

5. Immunoelectrophoretic Characteristics

Since it is the most clearly observable precipitin arc in the α_1 zone, it is used as a guide for identifying other protein constituents. It is found very close to the antiserum trough and interesects both albumin and haptoglobin (Fig. 8). Due to the antigen-excess, central part of the arc often looks thick and broad. It is usually symmetric in its shape. Unlike α_1-acid glycoprotein, it loses *in vitro* antigenicity by boiling.

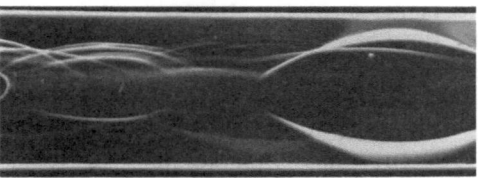

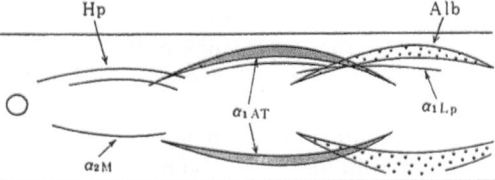

Fig. 8 Immunoelectrophoretic pattern of the α_1-globulins.
α_1-antitrypsin (α_{1AT}) yields a line of precipitation between those of albumin and α_2-macroglobulin, and it is the closest to the antiserum trough.

6. Biological Characteristics

Normal serum concentration of α_1-antitrypsin is 210–500 mg/100 ml, and its serum concentration is thought to be controlled genetically. The known biological function of α_1-antitrypsin is to inhibit the actions of trypsin and of chymotrypsin. This interaction, however, becomes inactive under high temperature and acidity of pH below 6.0.[185]

7. Clinical Abnormalities

Acquired increase: As one of the acute phase reactants it increases in the following cases:
1) Acute and chronic inflammatory diseases
2) Stress syndrome
3) Malignant tumors
4) Hematological abnormalities

Acquired decrease
1) Pulmonary diseases, especially pulmonary emphysema
2) Hepatic damage, severe
3) Nephrotic syndrome, severe
4) Malnutrition, cachexia

Congenital deficiency: Juvenile emphysema.

HORMONE-BINDING PLASMA PROTEINS

General Discussion on the Hormone-binding Plasma Proteins

Numerous substances circulating in the blood stream are bound with plasma proteins. Among these are antibiotics, vitamins, metal ions, lipids and metabolic products such as bile pigments.[167] It has become clear that there are hormones of comparatively small molecular weight which are bound with plasma proteins in the same way as the above mentioned substances. The hormones best known in connection with this phenomenon are the following three: gonadal steroids, adrenocorticoids, and thyroid hormones. Nothing about catecholamine and posterior pituitary hormones is yet known. About anterior pituitary hormones, parathyroid hormones and pancreatic hormones, a little information is available but it is not yet clear whether the phenomenon of such specific binding exists in these cases.

Gonadal steroids: In the main, gonal steroids are firmly bound to albumin while some of them are bound to COHN Fraction IV. Approximately ninety seven percent of female and male sex hormones in the serum exist bound to plasma proteins. It is thought that not only the free forms but also the conjugated forms of glucuronides and sulfates are bound with the above mentioned plasma proteins.[27,167,173]

Adrenocorticoids: These mainly bind transcortin in a specific way, and binds the albumin, also. However, binding with albumin occurs less frequently than in the case of gonadal steroids.[173] About 90–95% of the adrenocorticoids in the serum exist bound to plasma proteins.[12,211] Each albumin molecule is thought to contain two binding sites for adrenocorticoids.

Thyroid hormones: These are bound to thyroxine-binding globulin, albumin and prealbumin. As to the thyroxine-binding sites of the albumin, it is estimated that each albumin molecule contains one specific binding site of strong affinity and three other sites of relatively weak affinity. Triiodothyronine has weaker affinity than thyroxine.[173] As will be explained later, it is thyroxine-binding globulin which shows the strongest activity. Prealbumin, too, binds thyroxine in a specific manner.[80] The binding capacity of prealbumin as a whole is approximately five times more than that of thyroxine-binding globulin, but its affinity is considerably weak in the plasma having the physiological pH of 7.4, and it may be considered that physiologically there is very little activity on the part of prealbumin, even if it did exist.[21,134] Over 99.9% of thyroxine in the plasma exists bound to the above mentioned plasma proteins.[167,191]

Detection of Hormone-Protein Interaction

In order to detect hormone-protein interaction, the following two methods can be used: 1) detection of the alteration of the properties of the hormones, caused by its interaction with plasma protein; 2) detection of the alteration of the properties of certain plasma proteins caused by their interaction with hormones. Since the latter case requires that the plasma protein in question be purified and that its properties be clearly known, it is used mainly for detecting the hormone-albumin interaction.[200]

For the detection of hormone-protein interaction, the following alteration in a property of the hormone due to the presence of the binding protein will be noted: 1) increase in its solubility, 2) precipitation of the hormone in association with protein precipitation, 3) increase in the effective size of the hormone-protein complex compared with the free hormone, 4) a change in electrophoretic migration, 5) a change in chromatographic behavior on ion-exchange columns, 6) the specific shift or change in absorption, and 7) a change in *in vitro* biologic activity of the hormone.[62,98,167] In any case, it is necessary to make certain first that the property of the hormone itself will not be changed in each situation. For instance, if the hormone-protein complex is precipitated by means of the COHN fractionation, before asserting that the interaction between hormone and protein occurred, it must be established that the hormone itself was not precipitated by ethanol. In the early days of thyroxine research, the fact that a sizable amount of thyroxine is precipitated by trichloracetic acid or zinc hydroxide solution was sometimes left out of consideration.

Physiological Singificance of Hormone-Protein Interaction

In the same way that Ca^{++} and others lose their physiological activity when they are bound to plasma proteins, so hormones too are physiologically active in their free form but become inactive when bound to plasma proteins. The fact that a small amount of free hormones and a greater amount of bound hormones coexist in balance in the blood has the following physiological significances.[167]

First, association and dissociation of thyroxine with albumin, for instance, is considered to occur probably in every one thousandth second. Thus, hormone-protein binding results in buffering against the hormones secreted from the endocrine organs. In contrast, it also fulfils the role of rapidly supplying the active hormones in response to the body's demands. Second, large hormone-protein complex formed serves to prevent hormones of comparatively small molecular weight from being unnecessarily diffused into tissue spaces immediately after they have been secreted. Third, because hormone-protein complex exists in the form of a big mass, the capacity to diffuse into cells is decreased and thus the fractional rate of hormone degradation is limited. Fourth, direct degradation of hormone molecules, or unnecessary action of the glucuronate-conjugating enzymes will be limited by hormone-protein binding.

Although the aforementioned are considered to be its physiological functions, there may be no direct relation between the hormone-binding protein concentration in blood and the hormone activity. That is, even if the hormone-binding protein concentration in blood is abnormally changed, the concentration of the free hormone gains a new equilibrium and thus its normal level is maintained.[167,211] For instance, during pregnancy, hyperthyroidism does not occur even though ^{131}I-resin sponge uptake (Triosorb test) decreases.

Thyroxine-Binding Globulin (TBG)

1. Synonyms

Thyroxine-binding protein (TBP)

2. Physico-chemical Characteristics

Thyroxine-binding globulin (TBG) is a glycoprotein which has the inter-α electrophoretic mobility in pH 8.6 veronal buffer. It has the sedimentation constant $(S_{20,w})$ of 3.3 S, approximately 40,000–50,000 molecular weight, extinction coefficient $(E_{280\,m\mu})$ 4. It contains about 1% of sialic acid.[167,185] It has not yet been satisfactorily purified. It cannot be recognized on routine separation of serum proteins, but identified as a separate band on polyacrylamide or disc electrophoresis.

3. Biological Characteristics

Thyroxine-binding globulin has a tendency to bind specifically with thyroxine, and two thrids of thyroxine present in normal serum is bound with it. It also binds triiodothyronine, but thyroxine has an affinity about three times stronger than the former.

One hundred milliliter of normal serum has the capacity of binding approximately

20 μg thyroxine, and supposing it has the molecular weight of 50,000 its serum concentration is equivalent to approximately 1 mg/100 ml of TBG.[167]

4. Clinical Abnormalities

Congenital increase or decrease

Acquired increase: Pregnancy, administration of estrongens

Transcortin

1. Synonyms

Transcortin (SLAUNWHITE & SANDBERG, 1959), Corticosteroid-binding globulin (DAUGHADAY, 1958), Cortisol-binding globulin (CBG, ROBBINS, 1964).

2. Physico-chemical Characteristics

In many cases the characteristics of CBG are similar to those of TBG, and therefore in the past it was sometimes thought to be the same plasma protein. However, the two are now considered to be separate proteins.

CBG migrates electrophoretically at the α zone, but its exact mobility has not been determined. Its sedimentation constant is about 3S, while its molecular weight about 50,000. It contains sialic acid in 3.2% and is a glycoprotein containing 14.1% carbohydrate.[167]

3. Biological Characteristics

CBG specifically binds with cortisol, and approximately two thirds of cortisol present in normal serum bind to the protein. It is capable of binding with other adreno-corticosteroids, but in general, strong affinity is shown with the steroids having OH group at 17α and 21 or 11β positions. It has a relatively weak affinity with the steroids with 11β-OH, 11α-keto, 18-CHO (aldosterone) and tetrahydro-cortisol.[28,167,212]

The binding capacity of CBG depends on the temperature. At a lower temperature, more binding with cortisol occurs. It is considered that this happens probably because the structure of CBG changes at low temperatures, with the result that the binding group is exposed.[30,130] However, physiologically the binding capacity at 37°C has greater significance and, at this temperature, 100 ml of the normal serum binds 20–35 μg cortisol. According to the above-mentioned, it can be said to be equivalent to the serum concentration of 2.5–5 mg/100 ml, supposing that the molecular weight of CBG be at approximately 50,000.

4. Clinical Abnormalities

Acquired increase: Pregnancy, administration of estrogens.

OTHER α_1 GLOBULINS

Besides the clearly identified plasma proteins such as the above mentioned α_1-antitrypsin, α_1-acid glycoprotein, thyroxine-binding globulin, transcortin and α_1-lipoprotein, there are various other protein constituents in the α_1 area. Those which can be recognized by electrophoresis are α_1-easily-precipitable glycoprotein, heat labile components (HL–1, HL–2, HL–3) and esterases. An outline of those which are comparatively easy to recognize will be discussed below. In any case, their biological characteristics are not clearly known.

α_1-Easily-Precipitable Glycoprotein (α_{1PGp})

Its sedimentation constant ($S_{20,w}$) is 3.8 S, and its molecular weight of 50,000; it consists of 401 amino acids and contains 13.3% carbohydrate. By starch gel electrophoresis it is observed as postalbumin I, but by immunoelectrophoresis it is observed near, a_{1AT}.[185]

4.6 S-Postalbumin (4.6SP$_0$A)

Its sedimentation constant ($S_{20,w}$) is 4.6S, and it is observed as a postalbumin by starch gel electrophoresis. By immunoelectrophoresis it has never been observed as a precipitin arc. It contains 10.0% carbohydrate, but it precipitates in 0.6M perchloric acid.[185]

Tryptophan-Poor a_1-Glycoprotein (Trp a_1)

It is a glycoprotein which has the sedimentation constant ($S_{20,w}$) of 3.3S, and the molecular weight of 50,000–60,000. It contains 13.7% carbohydrate and does not precipitate by 0.6M perchloric acid and heating. The characteristic of this glycoprotein is to contain only a small amount of tryptophane.[185] By immunoelectrophoresis it is observed as a faint precipitin arc located inside of a_1-lipoprotein.

a_{1x}-Glycoprotein (a_{1x})

It is a glycoprotein having the sedimentation constant ($S_{20,w}$) of 3.9S, contains 22.7% carbohydrate, and does not precipitate by 0.6M perchloric acid or heating. By using rabbit antisera, its pericipitin arc can easily be observed, and it is located in the inter-a position and comparatively far away from the antiserum trough.

Chapter 5

Plasma Proteins Included in
the α_2-Fraction

α_2-fraction is the most heterogeneous and it contains many minor components. There are about 12 components detected by immunoelectrophoresis. By means of routine filter-paper electrophoresis, α_2-macroglobulin, haptoglobin and ceruloplasmin are found as the major components. However, in cellulose acetate membranes and agar gel most of low-density lipoproteins migrate also in α_2 zone.

Gc-GLOBULIN

1. Synonyms

Gc-globulin (SCHULTZE, 1962), Group-specific components (HIRSCHFELD, 1959), Gc-factor (CLEVE et al, 1964), PS-2D (SCHULTZE et al, 1962), Postalbumins 2+3 (SMITHIES, 1959), M-2 (MEHL et al. 1949).

2. Physico-chemical Characteristics

It has the sedimentation constant ($S_{20,w}$) of 3.7S and the molecular weight of 50,800, consisting of 434 amino acids. No lipid is contained, but 4.2% carbohydrate (hexose 2%, hexosamine 2%, sialic acid 0%, fucose 0.2%).

Precipitation: Gc-globulin precipitates under the following conditions[185]: 40% ethanol (pH 5.8, ionic strength 0.09, protein 1%), 2.0–2.4M ammonium sulfate (pH 5.0, protein 2%) 0.0065M Rivanol aqueous solution (pH 8.0, protein 1%), 0.15M trichloroacetic acid (protein 1%), heating (pH 5.0, 0.1M acetate buffer solution, protein 1%). It does not precipitate by 0.6M perchloric acid, but it becomes slightly turbid.

3. Hereditary Types

There are three phenotypes which are based on the two principal genes of Gc[1] and Gc[2]: namely, Gc1–1, Gc2–1 and Gc2–2. In other words, Gc 1–1 electrophoretically moves rapidly maintaining the inter-α mobility, and Gc-2-2 moves slowly and shows α_2 mobility. Gc2–1 stands midway between the two, consisting of both Gc[1] and Gc[2], and its electrophoretic mobility distributes widely from inter-α to α_2 zones.[75] Gc[2] is found less frequently than Gc[1], and even in the Kurumbas in India in whom it is found most frequently, Gc[2] is found only in a frequency of 0.356. It ranges 0.23–0.29 among Europeans, Chinese and Japanese. In the Negro, in whom it is found least frequently, it ranges 0.05–0.10.[185]

In certain races some subtypes are also found, such as Gc[Aborigine] (Gc[Ab]) and Gc[Chippewa] (Gc[Chi]), which migrate more rapidly than Gc[1], Gc[x] and Gc[Y] migrate a little slower than Gc[1].[23,76] (Fig. 9).

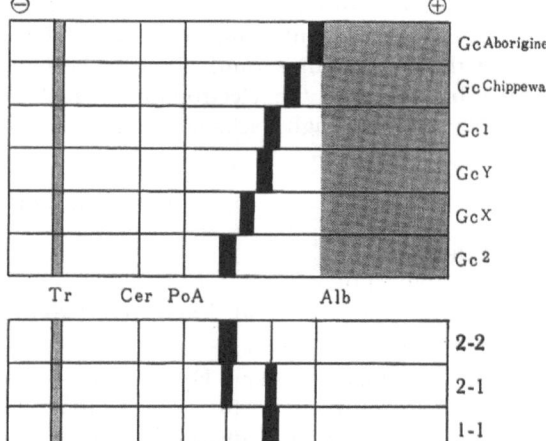

Fig. 9 Hereditary types of Gc-globulin. The upper portion shows the relative mobility of 6 different types of Gc-globulin, separated by starch gel electrophoretic techniques. The lower diagram shows the 3 representative phenotypes related to Gc-globulin.

4. Immunoelectrophoretic Characteristics

Among the various hereditary types of Gc-globulin there is no difference found in their amino acid composition and immunologic antigenicity. These hereditary types can be determined according to mobility differences detected by starch gel electrophoresis. But the three most fundamental phenotypes can be discerned more easily and quickly by immunoelectrophoresis (Fig. 10).

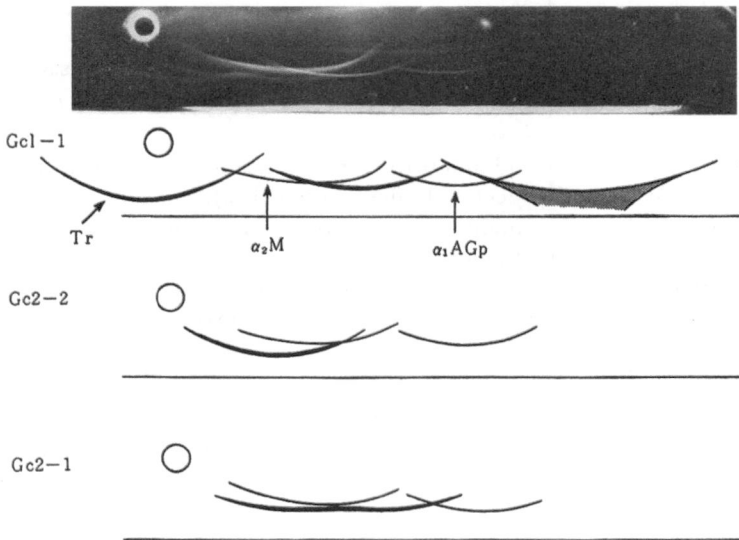

Fig. 10 Agar gel immunoelectrophoretic patterns of the different types of Gc-globulin.

Gc1–1 and Gc2–1 are observed as fine precipitin arcs close to the antiserum trough as if bridging between haptoglobin and α_1-antitrypsin. Furthermore, Gc2-1 can be distinguished with comparative ease because it appears as a biphasic precipitin arc extending to the α_2 zone. However, Gc2-2 is observed very close to the trough as almost parallel to the haptoglobin arc. When haptoglobin, ceruloplasmin or α_2-macroglobulin increase or decrease significantly, they do not necessarily remain in the same pattern.   This often makes it difficult to identify the Gc2-2 arc.   Ultimately, its identification must be done by using the specific antiserum.   It must be noted that, if the plasma sample is contaminated by bacteria, the electrophoretic mobility of Gc-globulin may become more rapid.[138]   Also, through exchange transfusion, phenotypes may become changed transiently.[77]

HAPTOGLOBIN

1. Synonyms

Haptoglobin (POLONOVSKI and JAYLE, 1939), Seromucoide a_2 (JAYLE and BOUSSIER, 1955), PS-2A (SCHULTZE et al, 1962) M-2 (MEHL et al. 1949)

2. Physical Characteristics

As will be explained below, Hp1-1, Hp2-1 and Hp2-2 are known as the three most important kinds of haptoglobin.   Among these, however, only Hp 1-1 is physico-chemically homogeneous, and the others exist as various polymers and their physical characteristics are not well known.

Hp 1-1: Sedimentation constant $(S_{20,w})$ 4.4S, diffusion coefficient $(D_{20,w})$ 4.7; molecular weight 85,000; isoelectric point (pI) 4.1; extinction coefficient $(E_{280\,m\mu})$ 12.0.

The average molecular weight of Hp 2-2 and that of Hp 2-1 is 400,000 and 200,000 respectively, and their sedimentation constants are 7.5S and 6.4S (coexists with that of 4.3S).

Any haptoglobin component has an a_2-relative mobility, but on using starch gel electrophoresis, only Hp 1-1 can be observed as a single band while Hp 2-1 and Hp 2-2 are observed as several protein bands.[164,185]

3.   Chemical Characteristics

Precipitation: Haptoglobin precipitates under the following conditions[185]: 40% ethanol (pH 5.8, ionic strength 0.09, protein 1%), 1.9-2.2M ammonium sulfate (pH 7.0, protein 1%), 0.0065M Rivanol aqueous solution (pH 8.0, pretein 1%).   However, it does not precipitate in 0.6M perchloric acid and 0.15M trichloracetic acid, and heating is to make it slightly turbid (pH 5.0, 0.1M acetate buffer, protein 1%).

Chemical structure: Hp 1-1 contains 11.2% nitrogen, is composed of 734 amino acids and has valine as the N-terminal amino acid.

It is conceivable that each haptoglobin component has a different amino acid composition.   It contains 19.3% carbohydrate (hexose 7.8%, hexosamine 5.3%, sialic acid 5.3 %, fucose 0.2%).[185]

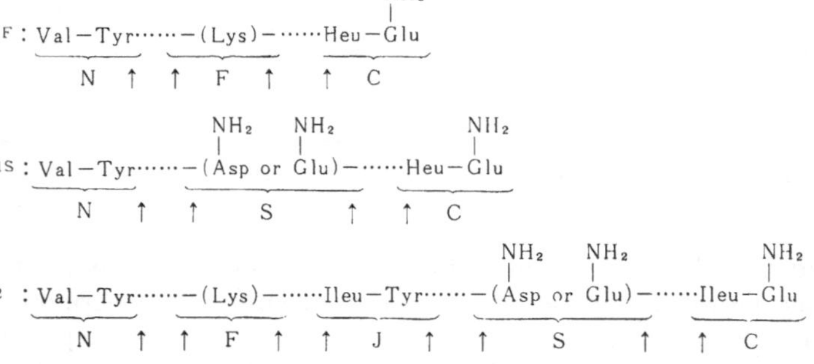

Fig. 11 Differences in the α-chain of haptoglobin. The arrows
indicate the points where chymotrypsin hydrolyzes.

When haptoglobin is treated with mercapto-ethanol in urea, the molecule is degraded into at a pair of α-chains and a β-chain. The β-chain is common to any haptoglobin component, but there are three different types of α chain; α^{1F}, α^{1S}, and α^2. Furthermore, the α-chain is divided into several fragments through digestion by chymotrypsin[190] (Fig.11). Each α-chain has the same N-terminal amino acid (glutamine) and C-terminal amino acid (valine). α^{1F} and α^{1S} have the molecular weight of 8860, while α^2 has that of 17,300. Only α^2 has the J-fragment which is not found in α^{1F} and α^{1S}.[185]

Three kinds of haptoglobin are classified into six subtypes, according to their polypeptide composition.

$$
\begin{array}{lll}
\text{Hp 1 - 1} & \begin{cases} \text{Hp 1F - 1F} & \alpha^{1F} - \beta_X - \alpha^{1F} \\ \text{Hp 1F - 1S} & \alpha^{1F} - \beta_X - \alpha^{1S} \\ \text{Hp 1S - 1S} & \alpha^{1S} - \beta_X - \alpha^{1S} \end{cases} \\[2em]
\text{Hp 2 - 1} & \begin{cases} \text{Hp 2 - 1F} & \alpha^2 - \beta_X - \alpha^{1F} \\ \text{Hp 2 - 1S} & \alpha^2 - \beta_X - \alpha^{1S} \end{cases} \\[1em]
\text{Hp 2 - 2} & \text{Hp 2 - 2} & \alpha^2 - \beta_X - \alpha^2
\end{array}
$$

These three subtypes belonging to Hp 1–1 cannot be distinguished because they themselves have no significant physico-chemical differences (starch gel electrophoretic patterns, ultra-centrifugal patterns, etc.) (Fig. 12). But the α-chain isolated can be separated by means of electrophoresis into fast-moving α^{1F} and slow-moving α^{1S}. Hp 1 shows no intermolecular aggregation. However, in a living body, a variety of combinations from dimer to tetramer can occur with Hp 2, resulting in multiple bands on starch gel electrophoretic patterns and ultra-centrifugal patterns. Since haptoglobin was first purified by JAYLE and BEUSSIER[87], various methods of purification have been employed, such as ion exchange chromatography, salting-out and so on.[185]

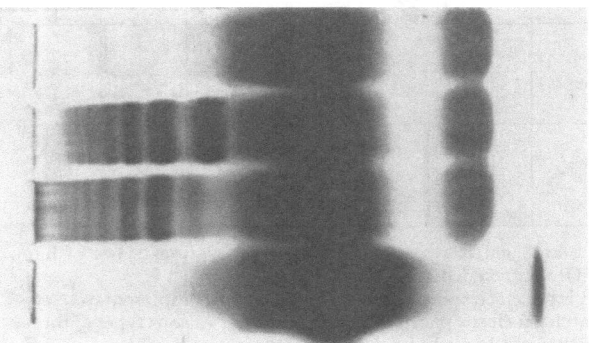

Fig. 12 Polyacrylamide gel electrophoretic patterns of the serum haptoglobins. The mixture of serum and hemolysate are electrophorized and stained with benzidine solution. Hb indicates the free hemoglobin migrating at the beta zone. The arrow indicates the Hp 1–1 which is poorly identified on the photographs, but various polymer forms of the Hp 2–2 can be clearly separated.

4. Hereditary Types

The synthesis of haptoglobin is regulated by a pair of allele in autosomes. The following three kinds of genes determine the α-chain synthesis: Hp^{1F}, Hp^{1S}, and Hp^2. As homozygous or heterozygous forms of these three, the following six phenotypes are known.

Genotype	Phenotype	Electrophoretic property
Hp^{1F}/Hp^{1F} Hp^{1F}/Hp^{1S} Hp^{1S}/Hp^{1S}	Hp1F-1F Hp1F-1S Hp1S-1S	Hp1-1
Hp^2/Hp^{1F} Hp^2/Hp^{1S}	Hp2-1F Hp2-1S	Hp2-1
Hp^2-Hp^2	Hp2-2	Hp2-2

The distribution of hereditary types of haptoglobin differs greatly according to races. GIBLETT divided into the following three groups on the basis of the frequency of Hp^1 gene: Caucasians, 0.384; Orientals, 0.261, and Negros, 0.552.[54]   This result is anthropologically very significant.   It is reported that for Japanese the distribution is as follows: Hp 1-1 8.3%, Hp 2-1 38.5% and Hp 2-2 53.2%.

In addition to the three typical hereditary types mentioned above, the haptoglobins, such as Hp 2-1 modified (Hp 2-1 M), Hp Ca, Hp JOHNSON, Hp JOHNSON modified, Hp F and Hp D were rarely found[58,100].

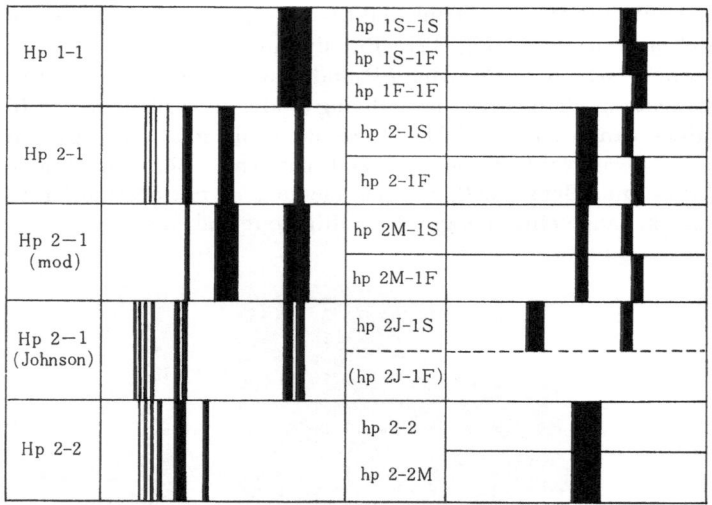

Fig. 13 Starch gel electrophoretic diagram of various types of haptoglobin (from GIBLETT and BROOKS).
Left: starch gel electrophoretic patterns showing its representative types.
Right: starch gel electrophoretic patterns showing various types of the α-chains.   Each type is further subdivided into certain subtypes.

5.   Immunoelectrophoretic Charactersitics

Haptoglobin is easily observed on immunoelectrophoretic patterns because its immunoegenicity is very strong and, in the α_2-fraction, it is the serum protein which is contained in the greatest quantity second to the α_2-macroglobulin.   Although it has been thought that any type of haptoglobin is identical in their antigenicity, it has recently been reported that Hp 2-2 could be distinguished from Hp 1-1 by using the anti- Hp 2-2 antiserum.[100]

On agar-gel electrophoresis, Hp 1-1, Hp 2-1 and Hp 2-2 are each shown to have a slightly different mobility from the others: Hp 1-1 is the most rapid and Hp 2-2 is the slowest.   However, it is probably impossible to distinguish these slight differences because in addition to the differences in hereditary types, artificial influences on mobility are not negligible. That is, even when serum is preserved for a long time, its mobility slows down, and when a significant amount of hemoglobin is present in serum, it produces the Hp–Hb complex, resulting in a decrease of its mobility and moves in the area midway between the α_2-and β-fractions.

In normal serum, haptoglobin is observed in the same position as α_2-macroglobulin within the α_2-fraction closest to the trough. It forms a symmetrical precipitin arc, the radius of which is of comparatively small curvature. Most are observed to be parallel to the α_{2M} arc, but they may sometimes be observed to be crossing the α_{2M} arc. This is especially so in the case of hemolyzed serum (Fig. 14).

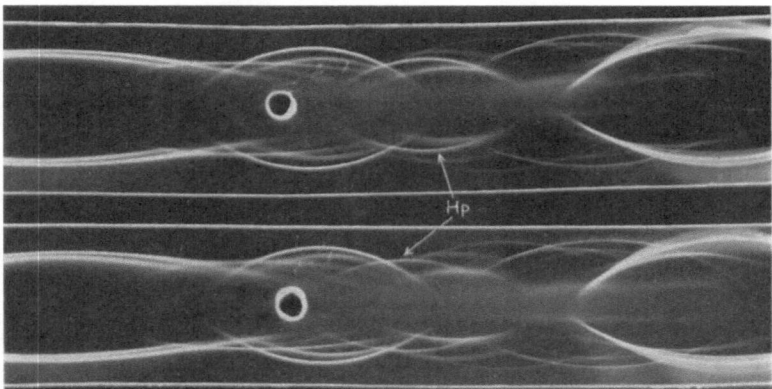

Fig. 14 Immunoelectrophoretic pattern of haptoglobin.
With the non-hemolyzed serum (upper), haptoglobin migrates between α_1-antitrypsin and transferrin, and is just outside the α_2-macroglobulin line. With the hemolyzed serum from the same blood sample (lower), the mobility of haptoglobin decreases significantly, indicating the formation of the haptoglobin-hemoglobin complex.

For its definite identification, the mono-specific antiserum should be used. Also, a decrease of its mobility after addition of hemoglobin is useful for identifying the Hp arc. Furthermore, immunoelectrophoretic patterns of the sera which have been mixed with free hemoglobin can be dyed out by peroxidase reaction (benzidine or o-dianidine stainings).

6. Biological Characteristics

The serum concentration of haptoglobin in normal adult varies greatly from individual to individual. It may be 30–290 mg/100 ml and seems to depend much on hereditary factors.[4] In fetal serum, haptoglobin cannot be detected, and in the cord blood at the time of birth, only approximately 10% of them show the presence of haptoglobin. Six months after birth, it increases so that the hereditary types of haptoglobin can be determined in almost 100 percent by starch gel electrophoresis.

Haptoglobin binds itself selectively with hemoglobin *in vivo* and *in vitro*. One molecule of Hp 1-1 binds with one molecule of hemoglobin and thus it forms the haptoglobin-hemoglobin complex (Hp–Hb complex). When Hp 1-1 is bound with hemoglobin, the molecular weight of the Hp–Hb complex becomes 155,000 and its mobility decreases. One hundred milliliters of serum has the capacity to bind 40–160 mg of hemoglobin, depending on the amount of haptoglobin. Besides immunological quantitation, the quantity of serum haptoglobin can be measured by the use of this capacity. In addition to normal hemoglobins (Hb A, Hb F), HbC, HbS, and HbD, too, bind haptoglobin, but the abnormal hemoglobins such as HbH or Barts hemoglobin, which lack the α-chain of hemoglobin, do not bind with haptoglobin.[136]

As is mentioned above, since the Hp–Hb complex has the molecular weight of at least 155,000 or more, it cannot be excreted into urine through the renal glomeruli. If hemoglobin is injected in the veins, or if hemolysis occurs excessively in the body, free hemoglobin will not be excreted into the urine until the haptoglobin in the blood be saturated. Therefore, the renal threshold of hemoglobin can be determined mainly by the quantity of haptoglobin within the blood. Accordingly, it plays an important role in preventing iron from passing out into the urine, and also in preventing the occurrence of the renal tubular damage caused by hemoglobin excretion.[57]

Binding with haptoglobin, hemoglobin increases its peroxidase-like activity but at the same time it will become phagocytized more easily in the reticulo-endothelial system. In other words, it is thought that the Hp–Hb complex is more easily taken up by phagocytosis (mainly in hepatic KUPFFER cells) than free hemoglobin, and it serves efficiently for adequate iron re-utilization. Other physiological significances still are not too clear, but it is known that, as one of the acute phase reactants, haptoglobin increases tremendously in response to various inflammatory stimuli. However, hypohaptoglobinemic state caused by *in vivo* hemolysis does not directly result in to the increased synthesis of haptoglobin. This state seems to be independent of its role as an acute phase reactant.

7. Clinical Abnormalities

Congential deficiency: Congenital anhaptoglobinemia

Acquired hypohaptoglobinemia

  1) Hemolytic anemias
  2) Acute hepatitis (poor prognosis)
  3) Severe liver damage (liver cirrhosis, others)

Acquired hyperhaptoglobinemia: As an acute phase reactant, it increases in various inflammatory diseases and tissue necrosis.

CERULOPLASMIN

1. Synonyms

Ceruloplasmin (Cer) (HOLMBERG and LAURELL, 1948), α_2–IV (HIRSCHFELD, 1960), Metalloseromucoide α (MONTREUIL, 1957)

2. Physical Characteristics[185]

Sedimentation constant ($S_{20,w}$) 7.085; diffusion coefficient ($D_{20,w}$) 3.76; molecular weight 160,000; intrinsic viscosity (η) 0.044; electrophoretic mobility 4.6; isoelectric point (pI) 4.4; extinction coefficient, $E_{280\,m\mu}$ 14.9, $E_{610\,m\mu}$ 0.68.

Ceruloplasmin is blue in color with a strong absorption at 610 mμ and a weak absorption at 320 mμ. Either by removing copper or by treatment with reducing substances, the absorption spectra at 320 mμ and 610 mμ disappear. But when it is reoxidized, the blue color returns. Consequently, it can be said that the blue color of ceruloplasmin is caused by cupric ion contained within the molecule.[104]

Ceruloplasmin deteriorates easily even on storage. Some reports stated that the heterogeneity of ceruloplasmin had been proved by electrophoresis or by chromatography. However, this was probably caused by partial alteration of the ceruloplasmin during the process of purification, resulting in appearance of the fast-moving fraction.[104]

3. Chemical Characteristics

Precipitation: Under the following conditions ceruloplasmin precipitates[185]: 18%

ethanol (pH 5.2, ionic strength 0.09, protein 1%), 1.6–2.0M ammonium sulfate (pH 7.0, protein 1%), 0.0065M Rivanol aqueous solution (pH 8.0, protein 1%), 0.6M perchloric acid (protein 1%), 0.15M trichloracetic acid (protein 1%), heating (pH 5.0, 0.1M acetate acid buffer, protein 1%). However, because ceruloplasmin is a heat-stable protein, little alteration occurs by heating approximately 30 minutes at $70° \pm 0.5°C$ in pH 5.8, 0.1M NaCl solution. This heat-stability is often utilized for ceruloplasmin purification.

Chemical structure: It contains 14.4% nitrogen, consists of approximately 1,300 amino acids, the N-terminus being valine. It contains 8.0% carbohydrate (hexose 3.0%, hexosamine 2.4%, sialic acid 2.4%, fucose 0.2%), and also contains copper in 0.34% or 8 atoms of copper in one molecule.[185]

Although it is still unknown in what form copper binds with peptide chains, at least it is clear that not all the copper atoms included in the ceruloplasmin molecule exist in the same form. However, due to the presence of Cu^{++} it is blue in color, and 40% of the copper atoms exists in the form of Cu^{++}. Where cyanides (CN^-) exist, or where there is a substance which produces chelating compounds by reacting strongly against copper, the copper is released from ceruloplasmin and thus the latter becomes apo-ceruloplasmin. MORELL and SCHEINBERG have succeeded in recombination of apo-ceruloplasmin and copper *in vitro* by adding an excess amount of ascorbic acid in pH 5.2.[127]

Diethyldithiocarbamate produces yellowish compounds by directly reacting with either free copper or the albumin-bound copper. However, copper bound with ceruloplasmin does not react with diethyl-dithiocarbamate until it is freed by the treatment of strong acids. Using this characteristic, the cerulopasmin-bound copper can be quantitatively measured and then the quantity of ceruloplasmin can be calculated from the fact that ceruloplasmin contains 0.34% of copper.[91]

It is stated that ceruloplasmin can be classified into four subunits by 8M urea and mercaptoethanol treatment.[163] Ceruloplasmin can be purified by various methods, and it can be crystallized by COHN IV-1 fraction.[185]

4. Immunoelectrophoretic Characteristics

Ceruloplasmin has strong immunogenicity and easily produces its antibody. Anti-

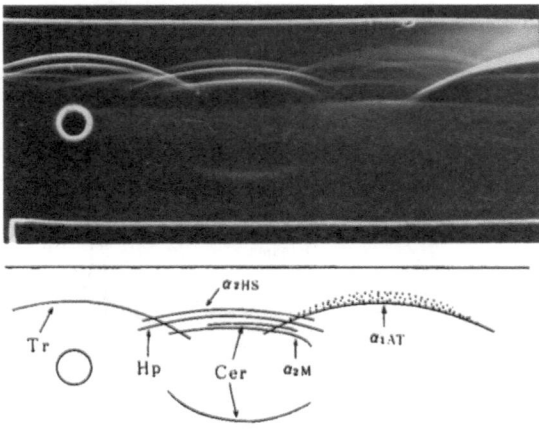

Fig. 15 Agar gel immunoelectrophoretic pattern of ceruloplasmin. Ceruloplasmin is shown as a rather faint line, between the lines of haptoglobin and a_2-macroglobulin.

ceruloplasmin reacts similarly with apo-ceruloplasmin. Therefore, immunologically, it is possible to quantitate ceruloplasmin and apo-ceruloplasmin.

In electrophoretic patterns, the ceruloplasmin arc is observed in the α_2 area, between α_2-macroglobulin and haptoglobin (Fig. 15). Because ceruloplasmin degenerates easily on storage, it is frequently difficult to be identified with certainty. In fact, old sera do not show a significant ceruloplasmin arc. However, the fact that the ceruloplasmin arc is smaller in its curvature than those of the other main α_2-globulins may serve as a guide. Specific antiserum or specific oxydase staining may be used for its definite identification.[202]

5. Biological Characteristics

The serum concentration of ceruloplasmin in normal adults is 27–63 mg/100 ml; during the neonatal period it is as low as 7 mg/100 ml and in six months after birth, it reaches almost the adult level.[100] The biological characteristics of ceruloplasmin may be summarized as below. However, its physiological role has not been definitely known.

Oxidase Activity: The oxidase activity of ceruloplasmin promotes the oxidation, in the presence of oxygen, of various compounds such as para-phenylenediamine (p-phenylenediamine, ppd), dimethyl-ppd, hydroquinone, catechol, pyrogallol, dopa, adrenalin, serotonin and ascorbid acid.[104] Especially, ppd and its derivatives are rapidly oxidized, producing a blue BANDROWSKI's complex. Since there is no other substance which has been discovered to have oxidase activity in plasma, the quantity of ceruloplasmin can be measured by comparing the blue color of BANDROWSKI's complex.[201]

The oxidase activity of ceruloplasmin is inhibited by various substances, such as KCN, NaN_3, NaOCN, KSCN, EDTA and sodium diethyldithiocarbamate[104].

Possible role in copper metabolism: In human body, a total of 100–150 mg of copper is found, 90% of it in muscles, bones and liver.[1] Normal plasma contains approximately 100 μg of copper/100 ml, over 90% of it incorporated in ceruloplasmin. Copper absorbed through the intestinal tracts binds first with albumin, flows into the blood circulation and reaches the liver. Only when apo-ceruloplasmin is synthesized in the liver is the copper incorporated into ceruloplasmin molecules. Therefore, the carrier of copper in the true sense is albumin, and the role of ceruloplasmin in copper metabolism is not clear.[91] The outline of copper metabolism is shown in Fig. 16.

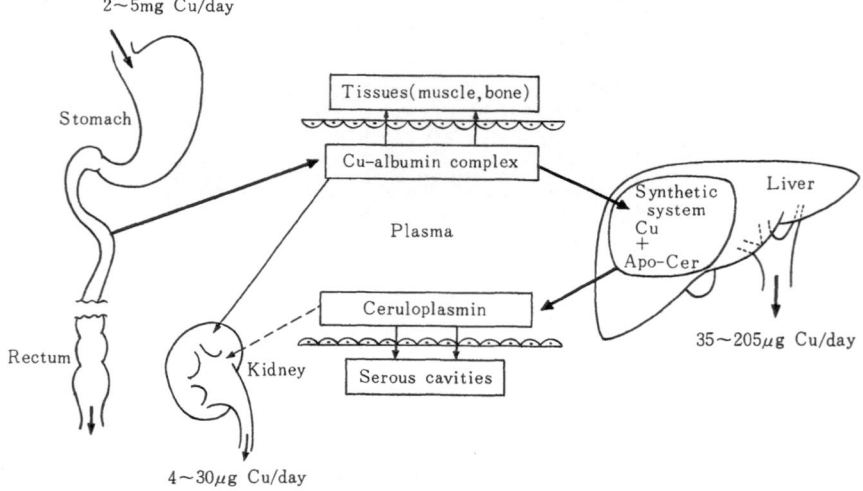

Fig. 16 Schematic diagram of the copper metabolism in body. (from KAWAI)

6. Clinical Abnormalities

Congenital hypoceruloplasminemia: WILSON's disease

Acquired hypoceruloplasminemia

1) Nephrotic syndrome, protein-losing gastroenteropathy.
2) Hypoproteinemia, anemia, malabsorption syndrome
3) Acute hepatitis, severe liver damage
4) Multiple sclerosis, and others.

Acquired hyperceruloplasminemia

1) Pregnancy, administration of estrogens
2) Acute and chronic inflammatory diseases
3) Malignant tumor, especially acute leukemia.

α₂-MACROGLOBULIN

1. Synonyms

α₂-Macroglobulin (a_{2M}) (SCHULTZE et al. 1955), Crystallizable α₂-glycoprotein (BROWN et al. 1954), Seromucoide a_2 de SCHULTZE (MONTREUIL, 1957), Heat-labile α-glycoprotein (PETERKOFSKY et al, 1956), Sa_2-Globulin (SMITHIES, 1955).

2. Physical Characteristics[185]

Sedimentation constant ($S_{20,w}$) 19.6S; diffusion coefficient ($D_{20,w}$) 2.41; molecular weight 820,000; frictional ratio (f/f_0) 1.43; intrinsic viscosity (η) 0.068; electrophoretic mobility 4.2; isoelectric point (pI) 5.4; extinction coefficient ($E_{280\,m\mu}$) 8.1.

3. Chemical Characteristics

Precipitation: Under the following conditions α₂-macroglobulin precipitates: 25% ethanol (pH 5.7, ionic strength 0.09, protein 0.4%), 1.2–1.8M ammonium sulfate (pH 7.0; protein 1%) 0.0065M Rivanol aqueous solution (pH 8.0, protein 1%), 0.6M perchloric acid (protein 1%), 0.15M trichloracetic acid (protein 1%), heating (pH 5.0, 0.1M acetate buffer, protein 1%), 0.2–0.4% calcium chloride (pH 7.0, protein 1%).

Chemical structure: It contains 14.8% nitrogen, consists of approximately 6,880 amino acids, and the N-terminus is aspartic acid. It contains 8.4% carbohydrate (hexose 3.6%, hexosamine 2.9%, sialic acid 1.8%, fucose 0.1%)[185]. α₂-macroglobulin is a large polymer consisting of several subunits, and it dissociates by 5M urea and mercaptoethanol, the subunits having faster electrophoretic mobility.[85,162]

α₂-macroglobulin was first isolated by means of ammonium sulfate precipitation by SCHÖNENBERGEN et al., and also by ultracentrifugal procedure and electrophoresis[185].

4. Immunoelectrophoretic Characteristics

It is one of the serum proteins which have the strongest immunogenicity, and its antibody can easily be produced. Therefore, by immunoelectrophoresis it is recognized as one of the clearest precipitin arcs in the α₂ area. Because its molecular weight is large, it is observed apart from the trough bridging between a_1-antitrypsin and transferrin. It has a chracteristically curved line, tracing a sort of hook toward the anodal end. The a_{2M} arc runs more or less parallel to those of the ceruloplasmin and haptoglobin. It is observable the most apart from the trough, and can be easily identified because of its asymmetric peculiar shape (Fig. 17). However, it must be noted that, if the distance of the electrophoretic movement is not long enough, the characteristics mentioned above may not be seen clearly.

5. Biological Characteristics

The serum α₂-macroglobulin concentration in the normal adult is 220–380 mg/100 ml, and it occupies approximately four fifths of the α₂-fraction.

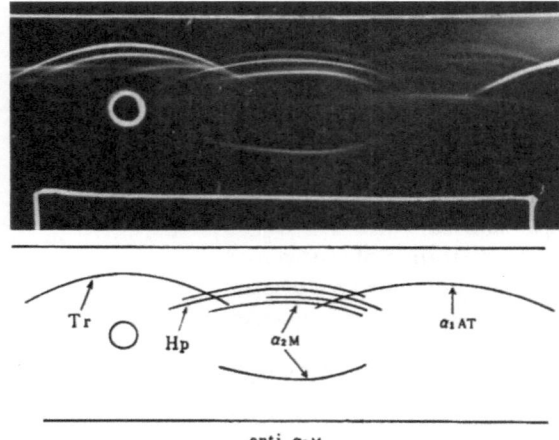

anti-α_2M

Fig. 17 Immunoelectrophoretic pattern of α_2-macroglobulin.
Its line is readily recognized since it is asymmetrically curved, appearing a sort of
hook at its anodal end, and it is always distant from the antiserum trough.

Depending on the age, the serum concentration changes considerably. During the
neonatal period it is almost the same as in adulthood, but thereafter it increases rapidly.
Around the age of 2–4 years, it reaches two to three times more than that of the adult, and
after the age of 10 years it returns to the adult level. This significant age difference in its
concentration is parallel to the body development in childhood, and thus, some people
think that α_2-macroglobulin may be closely related to the growth.[54,116] This age
difference is characteristically seen in human beings. In rats and cows, it increases
along with age; in dogs, its age difference is not clearly perceived.[53] It is therefore one
of the serum proteins which is of great intrest in comparative biochemistry.

By binding with proteolytic enzymes such as trypsin and plasmin, α_2-macroglobulin
causes them to lose their principal enzymatic activity. At the same time, the complex
formed of α_2-macroglobulin and these proteolytic enzymes gains the esterase activity and
is not interfered by inhibitors of proteolytic enzymes. The activity of the complex formed with trypsin is called the trypsin-protein esterase (TPE) activity.[53,72,103,123] TPE activity is utilized for measuring α_2-macroglobulin, also.[53] Binding activity with insulin
and some other low-molecular weight substances is also suggested.[90]

6. Clinical abnormalities
Acquired hyper-αn-macroglobulinemia
  1) Nephrotic syndrome
  2) Acute glomerulonephritis (especially in children)

OTHER α_2 GLOBULINS

In addition to the above-mentioned serum proteins in the α_2 area, it is known that
there are some enzymic proteins such as inter-α trypsin-inhibitor (HEIDE et al., 1965),
cholinesterase (or pseudo-cholinesterase), alkaline phosphatase, and lactate dehydrogenase.
There are also α_{2HS}-glycoprotein, Zn-α_2-glycoprotein, and α_2-neuramino-glycoprotein
whose biological activities are entirely unknown. Among these α_2-globulins, inter-α-
trypsin-inhibitor, α_{2HS}-glycoprotein and Zn-α_2-glycoprotein are identified by immuno-

electrophoresis. α_2-neuramino-glycoprotein is recognized by ultracentrifugal and starch gel electrophoretic procedures.[185]

$a_{2\text{HS}}$-Glycoprotein ($\alpha_{2\text{HS}}$)

This globulin has been called differently among different investigators, and some of their representative synonyms include α_{2z}-globulin (HEREMANS, 1960), Ba-α_2-glyco-protein (SCHMID and BÜRGI, 1961), $\alpha_{2\text{HS}}$-mucoid (SCHULTZE et al., 1962), PS–2B (SCHULTZE et al., 1962) and so on.

Its sedimentation constant ($S_{20,w}$) is 3.3S, and its molecular weight is approximately 49,000. It is composed of 402 amino acids and contains 13.4% nitrogen. Its electrophoretic mobility is 4.2, isoelectric point (pI) 4.1, and extinction coefficient ($E_{280\,m\mu}$) 5.6. The characteristic physicochemical property of this protein is to procipitate in the alcohol of high concentration and also in the ammonium sulfate solution of a relatively low concentration. It is precipitated under the following conditions, also: 20% ethanol (pH 5.8, protein content 2%, 0.02M barium acetate to be added), 1.0 –1.4M ammonium sulfate (pH 7.0, protein content 1%), 0.0065M Rivanol aqueous solution (pH 8.0, protein content 1%), 0.15M trichloracetic acid, heating (pH 5.0, 0.1M acetate buffer, protein content 1%). It does not precipitate in 0.6M perchloric acid.[185]

On immunoelectrophoretic analysis, the $\alpha_{2\text{HS}}$ precipitin arc is recognized in the α_2 zone, slightly faster than that of Hp 1, almost pararell to the haptoglobin arcs, being symmetric in its shape. Its serum concentration seems to be high in infancy, and its precipitin arc is frequently recognized closer to the antiserum trough than that of haptoglobin.

Zn-a_2-Glycoprotein ($\alpha_{2\text{Zn}}$)

This protein was identified by BÜRGI and SCHMID[16]. Its sedimentation constant is 3.2S, molecular weight 41,000, electrophoretic mobility 4.2, isoelectric point (pI) 3.8 and extinction coefficient ($E_{280\,m\mu}$) 18. It is composed of 287 amino acids, containing serine as the C-terminus and no N-terminal amino acid. It contains characteristically a large amount of tyrosine and tryptophane. Its carbohydrate content is 18.2%, and it does not precipitate in 0.6 M perchloric acid. It precipitates under the following conditions: 0.15 M trichloracetic acid, 20% ethanol (pH 5.8, protein content 1%, 0.02 M zinc acetate to be added). By using its specific antiserum, its precipitin arc is recognized very close to that of α_2-macroglobulin, and the normal serum may contain approximately 14 mg/100 ml.[158]

Chapter 6

Plasma Proteins Included in the β-Fraction

The most important component included in the β-fraction is transferrin. The greater part of low-density lipoprotein migrates in the β-fraction on filter paper and Tiselius electrophoreses, but it is included mostly in α_2-area on cellulose acetate and agar-gel electrophoreses. When fresh serum is run by agar-gel or cellulose acetate electrophoreses, a minor β_2-fraction is often recognized just behind the ordinary β-fraction. This comes from β_{1C}-globulin, as will be explained later. Even on filter paper electrophoresis, the β_2-fraction can also be demonstrated when Ca^{++} is added to the buffer.

TRANSFERRIN

1. Synonyms

Transferrin (Tr) (HOLMBERG and LAURELL, 1947), Siderophilin (SCHADE et al, 1949), β_1-Metal combining globulin (SURGENOR et al, 1949), β_{1S}-globulin (GRABAR and BURTIN, 1955), Metalloseromucoide β_1 (MONTREUIL, 1957).

2. Physical Characteristics[185]

Sedimentation constant ($S_{20,w}$) 5.5S; diffusion coefficient ($D_{20,w}$) 5.0; molecular weight 90,000; frictional ratio (f/f_0) 1.37; intrinsic viscosity (η) 0.055; electrophoretic mobility 3.1; isoelectric point (pI) 5.9; extinction coefficient ($E_{280\,m\mu}$) 11.2.

Transferrin which does not contain iron is colorless, but when it binds with iron, it becomes reddish and shows the maximum absorption at 470 mμ and 410 mμ.[164]

3. Chemical Characteristics

Precipitation: Transferrin precipitates under the following conditions; 40% ethanol (pH 5.8, ionic strength 0.09, protein 1%), 2.4–2.8M ammonium sulfate (pH 7.0, protein 1%), 0.6M perchloric acid (protein 1%), 0.15M trichloracetic acid (protein 1%), heating (pH 5.0, 0.1M acetate buffer, protein 1%). However, it does not precipitate in 0.0065 M Rivanol aqueous solution. This characteristic is often used for its purification.

Chemical structure: Transferrin contains a nitrogen content of 15.4%, and it is a glycoprotein containing 5.8% carbohydrate (hexose 2.4%, hexosamine 2.0%, sialic acid 1.4%, fucose 0.07%). Transferrin can be crystallized, and is composed with 827 amino acids with its N-terminus being valine.[185]

One transferrin molecule is able to bind with two ferric-ions (Fe^{+++}). Besides, this iron-binding phenomenon is pH–dependent; that is, it shows very stable binding between pH 7.5–10.0, but it dissociates in pH of less than 4. In addition, Fe^{+++} may be removed from the molecule through dialysis. Purified transferrin binds more rapidly with Fe^{+++}

than with Fe^{++} (ferrous-ion). However, the transferrin aerobically included in normal plasma binds more rapidly with Fe^{++} than with Fe^{+++}. Transferrin has strong affinity with iron, but it is known that transferrin also binds, though weakly, with copper, magnanese, zinc and other metals.[164]

In chickens, the serum transferrin is physicochemically and immunologically the same as conalbumin, which is the iron-binding protein in the egg-albumin. It is said that the difference between them is only in the carbohydrate portion.[178,205]

However, with regard to amino acid and carbohydrate compositions in the lactotransferrin of human milk, it is greatly different from the serum transferrin. It is also affirmed that there is no difference in the amino acid composition between adult serum transferrin and the tranferrin contained in cord blood.[151]

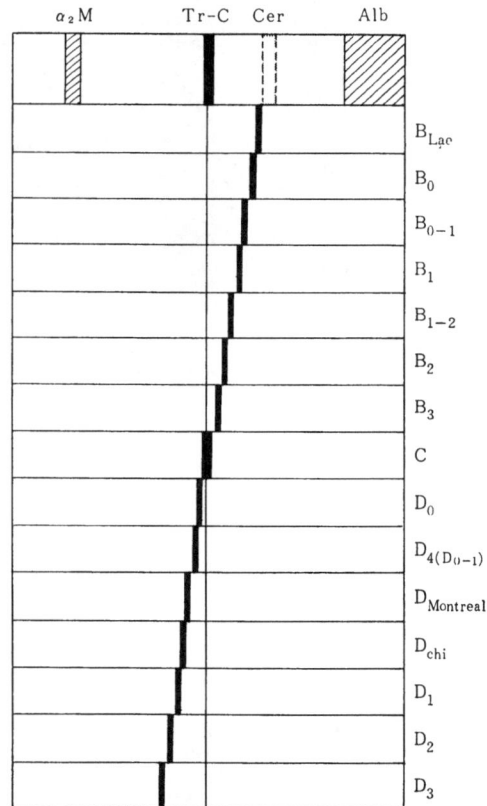

Fig. 18 Hereditary types of transferrin. The diagram shows relative mobilities of different transferrin variants on starch gel electrophoresis (from PARKER). Transferrin-Atalanti (migrating between TrB_{0-1} and TrB_1), transferrin-Wigan (migrating between $Tr\ D_0$ and $Tr\ D_{0-1}$) and transferrin Adelaide (migrating between $Tr\ D_0$ and $Tr\ D_{Wigan}$) are not included.

4. Hereditary Types

Since SMITHIES' discovery through starch-gel electrophoresis, 18 hereditary types of transferrin have been reported (Fig. 18). SMITHIES named the most fundamental one, transferrin-C (Tr-C). He called those with more rapid mobility, the transferrin-B group, and those having slower mobility, the transferrin-D group. Tr-C appears with overwhelming frequency, while Tr-B and Tr-D are rarely found. It is noted that 98.4% of the Japanese have only Tr-C. It is very rare that other transferrins are found in humans in a homozygous form, and many are observed with Tr-C in a heterozygous form. Transferrins other than Tr-C are rarely observed, so there are many points about their hereditary types which remain unclear. However, it may be assumed that transferrin is controlled by a pair of co-dominant non-sex linked alleles.[185]

The differences among hereditary types of transferrin seem to depend on their minor structural changes in the polypeptide. But amino acid analysis of their molecules yields no significant difference among Tr–C, B_2, D_1 and D_3.[150]

5. Immunoelectrophoretic Characteristics

Transferrin is clearly recognized on immunoelectrophoretic patterns because it forms the most distinctive precipitin arc in the β area. Being observed nearest to the trough, a symmetric arc with comparatively small curvature is easily identified in normal serum (Fig. 19). However, it is difficult to distinguish transferrin from hemopexin when trans-

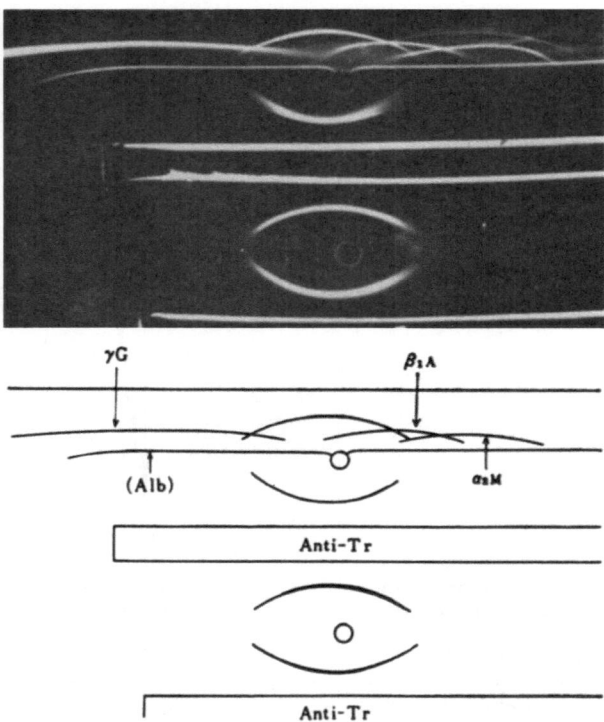

Fig. 19 Agar gel immunoelectrophoretic patterns of transferrin.
Transferrin is easily recognized as a symmetric line of the precipitation, bridging between the α_2-macroglobulin and the IgG-globulin lines. (Alb) indicates an extra precipitin line due to the excessive amount of albumin added to the immune serum during the absorption procedure.

ferrin decreases considerably. Therefore, in such cases, it is necessary to use specific antiserum or autoradiography with ^{59}Fe. Among the various hedreditary types of transferrin, there exist no antigenic distinctions. Thus, it is difficult to clearly differentiate various hereditary types by means of immunoelectrophoretic techniques.

6. Biological Characteristics

In normal serum, transferrin is contained in a concentration of 200–300 mg/100 ml. Among the β-globulins, it is present in the greatest quantity, occupying approximately 50%.

Transferrin binds easily with iron under the physiological pH, and thus it forms an iron-transferrin complex (Fe-Tr complex). Under various biological conditions the Fe–Tr complex releases iron from the molecule. Because such a reaction is also observed *in vitro*, transferrin is also measured as serum total iron-binding capacity (TIBC), being 250–400 μg/100 ml in normal serum. One third of the serum transferrin is bound with iron, normally. More than 99% of the serum iron, about 110 μg/100 ml in serum, binds with transferrin.

Physiological role of transferrin through iron-binding can be summarized as follows[104]:

1) It neutralizes an excess amount of ionic iron. When there is an excess of ionic iron in plasma, such toxic symptoms as nausea, vomiting and flush appear. However,

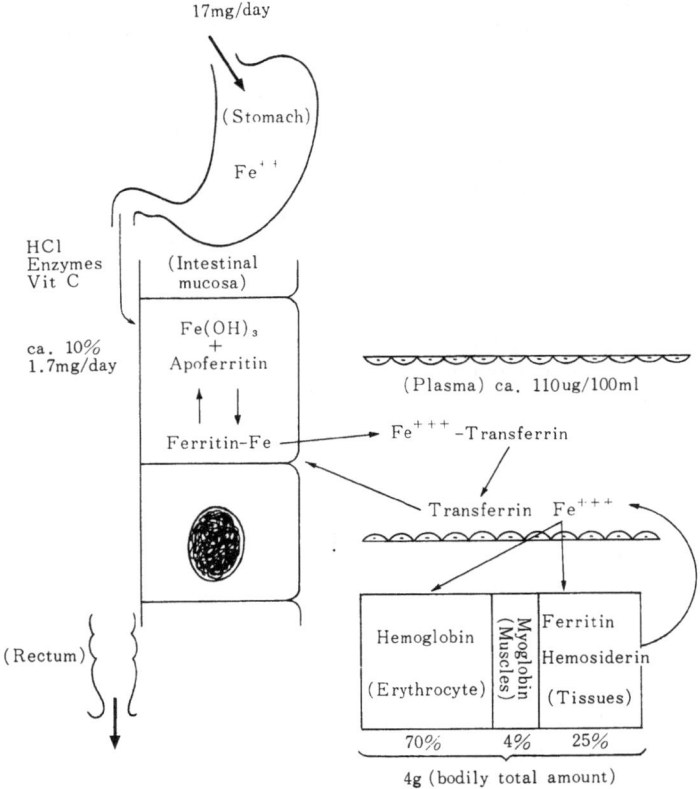

Fig. 20 Schematic diagram of the iron metabolism in normal body. The iron bound with transferrin in the plasma amounts to approximately 110 μg/100 ml, and the daily turnover of iron is about 38 mg. Iron is lost only minimally, amounting to 0.5–1.5 mg/day totally. Its urinary loss is less than 0.1 mg/day.

transferrin binds immediately with the ionic iron, when pathologically increased. In this way, it works to prevent the appearance of toxic symptoms. For example, a normal human being can bear an intravenous injection of 8–12 mg Fe^{+++} when one third of the serum transferrin normally has been bound with iron. But if the transferrin has been saturated by iron, toxic symptoms appear even when only a few milligrams of iron are injected.

2) It has a buffering action to ionic iron. In human body, the greatest buffering capa-

city for ionic iron is met with the apoferritin-ferritin system and the hemosiderin system present in cells or tissues. However, since these systems operate gradually they cannot work effectively in emergency. Transferrin is the one which shows an effective buffering action in acute toxic state. This reaction is similar to that of the carbonic acid-bicarbonate buffer system against hydrogen ions. In other words, transferrin has a capacity for immediately counteracting acute iron intoxication.

3) It keeps to minimize ionic iron concentration in plasma, and it stores iron reserves. Free Fe ions, like other metal ions, can inactivate various biological activities present in the blood. Therefore, it seems necessary to keep the ion concentration as least as possible. On the other hand, 25–30 mg of iron daily is necessary to synthesize hemoglobin in the bone marrow, so it is necessary to supply a great deal of iron into the body. To satisfy those two conditions, transferrin, as a harmless Fe–Tr complex, must circulate in the body in such a way as to be able to immediately release iron into various tissues. Transferrin is thus an important carrier for iron.

4) It prevents iron to be excreted into urine. Ionic iron itself can pass easily through the renal glomeruli. Since iron is scarcely reabsorbed in the renal tubules, the presence of transferrin as an Fe–Tr complex prevents iron to be excreted into urine. This function is closely related to iron metabolism[125] (Fig. 20). Further, it is said that transferrin not only has an important role in iron metabolism, but also it inhibits bacterial and viral growth. This anti-microbial activity is kept even when heated at 56 °C.[120,121]

7. Clinical Abnormalities

Congenital decrease or deficiency: Congenital atransferrinemia.

Acquired decrease: Usually follows to hypoalbuminemia.
  1) Acute and chronic active inflammatory diseases.
  2) Nephrotic syndrome
  3) Protein-losing conditions
  4) Hypoproteinemia

Acquired increase
  1) Pregnancy
  2) Chronic ion-deficiency anemia
  3) Acute hepatitis (early stage)

HEMOPEXIN

1. Synonyms

Hemopexin (Hpx) (GRABAR et al, 1960), Heme-binding α-globulin (NYMAN, 1960), Seromucoide α_{1A} (BISERTE and LATURAZE, 1960), Seromucoide α_{1B} (BISERTE et al, 1960), α_{1B}-Globulin (BURTIN, 1960), Cytochromophilin (BISERTE et al, 1960), α_1-Haptoglobin (α_{1H}) (KORINEK and MACH, 1961), PS–3A (SCHULTZE et al, 1962), MP–3 (MEHL et al, 1949).

2. Physical Characteristics[185]

Sedimentation constant ($S_{20, \omega}$) 4.8S; molecular weight 80,000; electrophoretic mobility 3.1; extinction coefficient ($E_{280 m\mu}$) 16.9.

3. Chemical Characteristics

Precipitation: Hemopexin precipitates under the following conditions:[185] 18% ethanol (pH 5.2, ionic strength 0.09, protein 1%), 2.4–3.2 M ammonium sulfate (pH 5.0, protein 1%). However, it does not precipitate in 0.0065 M Rivanol, 0.6 M perchloric acid, 0.15 M trichloracetic acid. Hemopexin is one of the most soluble proteins in serum.

Chemical structure: Hemopexin is composed of 554 amino acids, and it is a glycoprotein (hexose 9.0%, hexosamine 7.4%, sialic acid 5.8%, fucose 0.4%) containing 22.6% carbohydrate.[185]

4.   Immunoelectrophoretic Characteristics

Hemopexin has rather strong immunogenicity, producing easily antihemopexin in immunized animals.   It is easily observed on immunoelectrophoretic patterns by using anti-human whole serum.

The electrophoretic mobility of hemopexin is almost identical to that of transferrin, but it may sometimes be located slightly anodal to the transferrin precipitin arc, probably depending on a variation of the electrophoretic conditions used (Fig. 21).   The

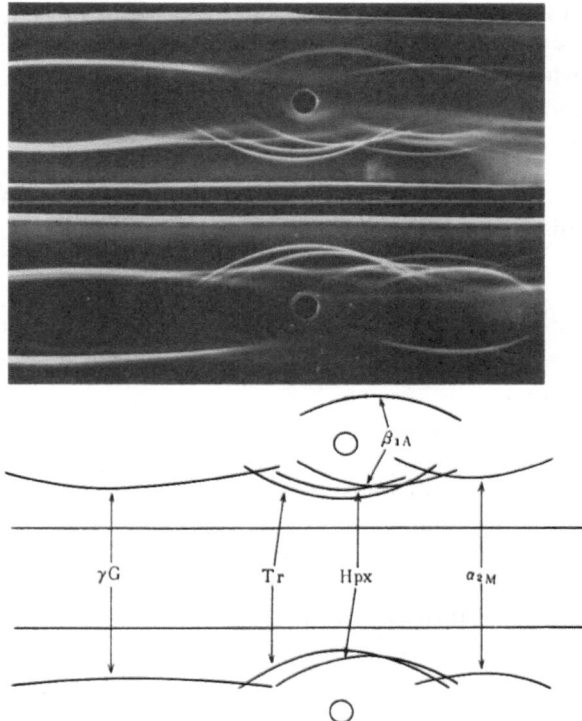

Fig. 21   Immunoelectrophoretic patterns of hemopexin.
The hemopexin line is easily identified just inside the transferrin line. It migrates at the same position as transferrin, but sometimes it may be slightly faster.   In many pathological sera, it may show a difficulty to be differentiated from β_{1A}-globulin.   However, their differentiation is usually easy when both the two are demonstrated clearly.

shape of the hemopexin arc is quite similar to that of transferrin, and it usually is recognized just inside the transferrin arc.   Of course, when transferrin decreases markedly, the hemopexin arc is observed outside the transferrin arc, so it has to be identified by using specific anti-serum.   Some of the anti-whole serum contain only a relatively low titer of hemopexin antibody, and it often forms an indistinct precipitin arc.

5.   Biological Characteristics

In normal serum, hemopexin exists in a concentration of 80–100 mg/100 ml.   Hemopexin binds with heme; one hemopexin molecule binds with 15–20 molecules of heme.

However, it does not bind with hemoglobin or cytochrome.[185]   It has long been recognized that heme binds with serum albumin, forming methemalbumin.[38]   But in recent years, this heme-binding capacity of hemopexin has drawn much attentions.[137]

The physiological role of the heme-binding capacity of hemopexin is as yet unknown. However, it can at least be affirmed that when abnormally large quantities of hemoglobin are released in plasma, as in the case of hemolytic anemia, heme-binding activity of hemopexin becomes important.   Homoglobin binds first with haptoglobin, and after haptoglobin be saturated, heme is released probably from hemoglobin, resulting in binding with hemopexin.   Finally the hemopexin-heme complex is rapidly eliminated from the blood circulation.   In many cases where the serum hemopexin concentration is decreased, haptoglobin is also decreased significantly.   Nevertheless, in cases such as thalassemia where there is no noticeable descrease of haptoglobin, a decrease of hemopexin is sometimes significant.   This is because heme is released from other sources than hemoglobin and hemopexin binds with it.[42]

6.   Clinical Abnormalities
Acquired hypohemopexinemia
1)   Hemolytic anemia
2)   Internal bleeding into body cavities and organs (especially, hemorrhagic pancreatitis)

$\beta_{1C/A}$-GLOBULIN AND THE COMPLEMENT COMPONENTS

Complement System

Complements(C) exist in normal serum and have the capacity to bind non-specifically with various antigen-antibody complexes.   The complements lose their activity when they bind with antigen-antibody complexes.   Besides, they do not increase even by immunization, unlike various specific antibodies.

Complements are heat-labile, and their activity is lost if it is submitted to heating at 56° for 30 minutes.   At ordinary room temperature, their activity weakens rapidly.   Even in 0°C, they lose nearly all their activity in three to four days.   Exposure to ultra-violet rays, mechanical shacking, and various chemicals also cause loss of their activity.

A complement system is a rather complex one, consisting of at least nine components. Also, Ca^{++} and Mg^{++} are included in the system.[107]

Physicochemical Characteristics of the Complements

The complements were physico-chemically identified for the first time by PILLEMER and others, and proved to be protein in nature.[156]   At the present, DEAE-cellulose column-chromatography is most widely used for their separation.

It was first supposed to consist of four components: C'1, C'2, C'3, and C'4, but further research by NISHIOKA and LINSCOTT, followed by INOUE and NELSON, showed clearly that the C'3 in guinea pig was further fractionated into six components, named as C'3a, C'3b, C'3c, C'3d, C'3e and C'3f in the sequence of their discovery.[81,141]   The similar complement system has been reported also in human beings.   In recent years, it has been suggested that each component be named C3, C5, C6, C7, C8 and C9 respecitively in the sequence of binding (Table 4).

1.   C3 (β_{1C}-globulin)
MÜLLER-EBERHARD and NILSSON highly purified β_{1C}-globulin and reported that it had

Table 4 Various complement components and their synonyms.

	Synonyms
Complement (C)	Komplement (EHRLICH et al. 1900)
	Alexin (BUCHNER, 1889; BORDET, 1895)
Complement components	
C1	First component (BRAND, 1907)
C2	Second component (BRAND, 1907)
C3	Third component (RITZ, 1912)
	C'3c (LINSCOTT and NISHIOKA, 1963)
	β_{1C}-Globulin (MÜLLER-EBERHARD et al. 1960)
	β_{2C}-Globulin (WIEME, 1965)
	β_{2B}-Globulin (SCHEIDEGGER, 1957)
	β_{1B}-Globulin (HINTZIG, 1961)
C4	Fourth component (GORDON et al. 1926)
	β_{1E}-Globulin (MÜLLER-EBERHARD et al. 1963)
C5	C'3b (LINSCOTT and NISHIOKA, 1963)
	β_{1F}-Globulin (NILSSON and MÜLLER EBERHARD, 1964)
C6	C'3e (INOUE and NELSON, 1965)
C7	C'3f (INOUE and NELSON, 1965)
C8	C'3a (LINSCOTT and NISHIOKA, 1963)
C9	C'3d (LINSCOTT and NISHIOKA, 1963)

the activity of the third complement component (C3).[133] They thought that it was equivalent to the C'3a of LEON and the C'3c+C'3b of LINSCOTT and NISHIOKA. However, they later found that purified β_{1C}-globulin was bound to the sensitized red cells, R · Ab · C' 1,4,2 and they made it clear that β_{1C}-globulin was equivalent to C'3 or C'3c.[119]

The sedimentation constant ($S_{20,w}$) of β_{1C}-globulin is 9.5S, and its extinction coefficient ($E_{280m\mu}$) is 10.0. It is one of the glycoproteins containing 3.03% carbohydrate. It is present in normal serum at a concentration of about 35 mg/100 ml, and precipitates under the following conditions:[185] 25% ethanol (6.9 pH, 0.09 ionic strength, 3% protein), 1.42–1.8M ammonium sulfate (pH 7.0, protein 1%), 0.0065M Rivanol aqueous solution (pH 8.0, protein 1%), 0.6M perchloric acid, 0.15M trichloracetic acid, and heating (pH 5.0, 0.1M acetate buffer, protein 1%).

β_{1C}-globulin exists only in fresh serum. As the serum becomes old, β_{1C}-globulin gradually transforms to inactive β_{1A}-globulin. Immunologically, β_{1A}-globulin has nearly the same antigenicity as β_{1C}-globulin, and it has the sedimentation constant ($S_{20,w}$) of 6.9S. The electrophoretic mobility of β_{1A}-globulin is faster than that of β_{1C}-globulin. β_{1C}-globulin has strong immunogenicity and can be easily detected by routine immunoelectrophoresis. In fresh serum, β_{1C}-globulin is seen as a symmetric precipitin arc bridging between transferrin and IgG-immunoglobulin. In stored serum, β_{1A}-globulin has a higher mobility than β_{1C}-globulin, and its symmetric precipitin arc is located slightly anodal to transferrin and hemopexin. Intermediate biphasic patterns sometimes appear in stored serum, depending on varying proportions between β_{1C} and β_{1A}-globulins (Fig. 22). β_{1C}-β_{1A} conversion is promoted when the globulins are heated or when antigen-antibody precipitates, zymosan, C1-esterase or proteolytic enzymes are present. EDTA inhibits the β_{1C}-β_{1A} conversion.[185]

On the other hand, inactive β_{1G}-globulin is formed when R· Ab· C1,4,2 and β_{1C}-globulin are mixed *in vitro*. Electrophoretically, β_{1G}-globulin has the same mobility as β_{1A}, while the same sedimentation constant by means of ultracentrifugation. It is thought that β_{1C}-β_{1G} conversion depends probably on the catalytic action of C2.[132]

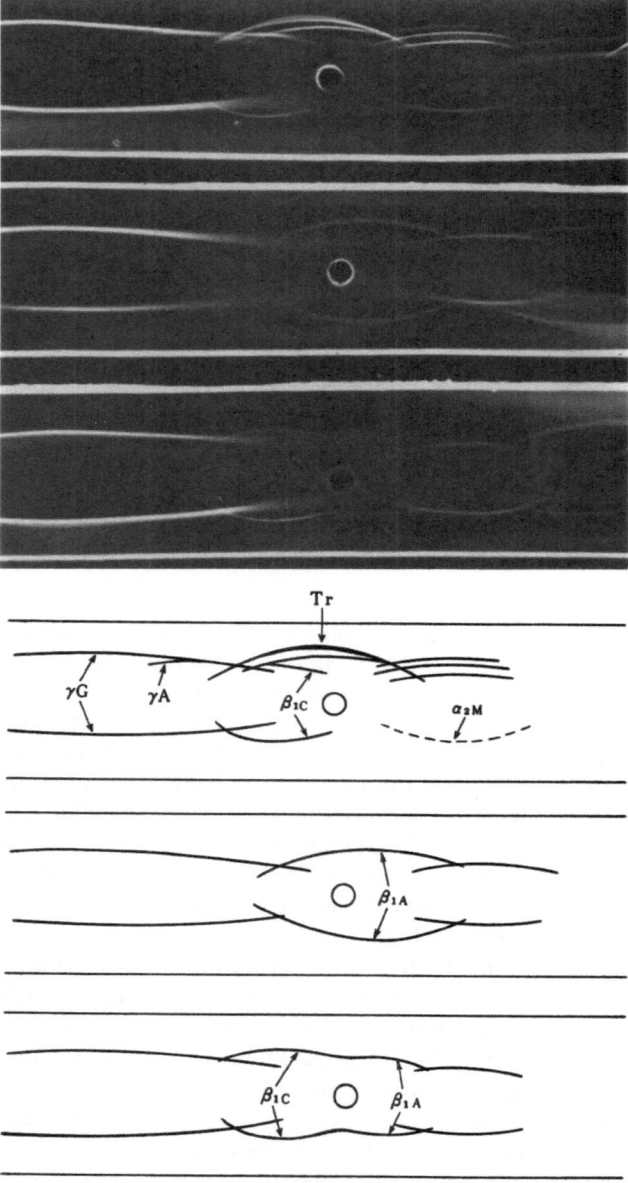

Fig. 22 Immunoelectrophoretic patterns of β_{1A}-globulin and β_{1C}-globulin.
The upper pattern of a fresh serum shows only the β_{1C}-globulin line, and the middle
pattern of a stored serum shows only the β_{1A}-globulin line. The lower pattern,
however, shows a double-humped line.

2.   C4 (β_{1E}-globulin)

β_{1E}-globulin has the sedimentation constant ($S_{20,w}$) of 10.0S and its activity is not lost
by heating at 55 °C. β_{1E}-globulin has immunogenicity, and shows the same electrophoretic mobility as transferrin.[131] It is observed as a fine symmetric precipitin arc with
rather large curvature and is located apart from the antiserum trough. β_{1E}-globulin is

converted into inactive β_{1E1} (a_2-mobility) and β_{1E2} (β-mobility) through the action of hydrazine, antigen-antibody complex or Cl-esterase.[153]

3. C 5 (β_{1F}-globulin)

It has the gulobulin nature and is heat-labile.   It loses the complement activity by heating at 56°C.   Its sedimentation constant is 9.4S.[140]

As for other complement components, they are not sufficiently purified to determine with certainty their physico-chemical and immunological characteristics.

Biological Characteristics of the Complement System

Complements themselves have no serological specificities, but they are related to the following specific antigen-antibody reactions.

 1) Lytic reactions
   1. Hemolytic reaction   2. Bacteriolytic reaction   3. Cytolytic reaction
 2) Complement fixation reactions
 3) Conglutination reaction
 4) Immune adherence reaction
 5) Phagocytosis
 6) Bactericidal reaction

It is not known how the complements are bound to the antigen-antibody complex, but they are fixed in the following order.

 1) Red blood cells (R)+hemolysin (Ab) $\longrightarrow$ Sensitized red blood cells (R·Ab)
 2) R·Ab+Cl $\xrightarrow{Ca^{++}}$ R·Ab·Cl
   At this phase Cl-esterase is formed from Cl, and it inactivates C4 and C2.
 3) R·Ab·Cl+C4 $\longrightarrow$ R·Ab·Cl, 4
   R·Ab·Cl, 4 is an unstable intermediate product.   C4 is also strongly bound to the other cell surface than the binding site of Cl.   However, its significance is not known.
 4) R·Ab·Cl, 4+C2 $\longrightarrow$ R·Ab·Cl, 4, 2
   R·Ab·Cl, 4, 2 is extremely unstable.   On standing, it returns back to R·Ab·Cl,4 again, but R·Ab·C'1,4,2 are re-formed by adding fresh C2.
 5) R·Ab·Cl,4,2+C3 $\longrightarrow$ R Ab Cl,4,2,3
   The sensitized cell, R·Ab·Cl,4,2,3 can produce specific immune adherence and also becomes prone to be involved by leukocytic phagocytosis.
 6) R·Ab·Cl,4,2,3+other components $\longrightarrow$ R·Ab·Cl,2,4,3,5,6,7,8,9

In hemolysis, it is necessary that all the nine components be fixed.   Besides, it seems that C5 is related to the formation of anaphylatoxin and that C6 is related to producing the substances necessary for chemotaxis of white blood cells.

It is not known what sort of role serologically non-specific complements play in various immunological phenomena of the body.   However, LEPOW[107] considers the occurrence of specific antigen-antibody reaction as the primary phenomenon resulting in subsequent complement fixation.   Cell injuries resulting from complement fixation can be divided roughly into two types.   Direct cell destruction refers to that caused by hemolytic, cytolytic and bacteriolytic reactions.

However, it sometimes happens that phagocytosis, immune adherence reaction, and others occur before complement fixation is completed.   In those cases, complement fixation itself does not directly cause cell destruction, but it does so indirectly.   Moreover, it may be further affirmed that some kind of substances supposedly releasing from complement fixation may lead indirectly to cell injury, provoking inflammatory reaction.

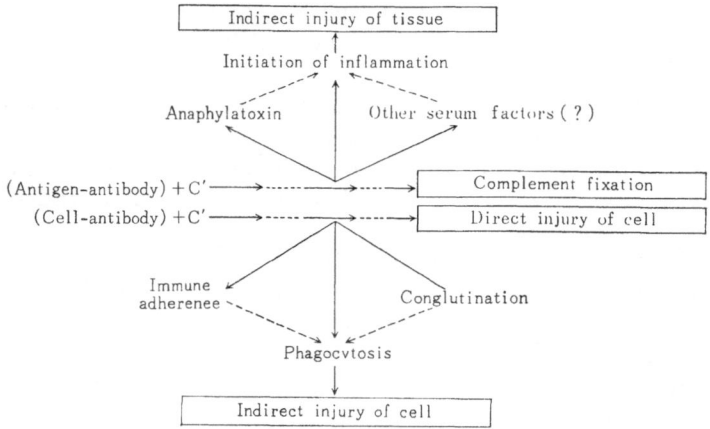

Fig. 23 Several possible pathways of cellular and tissue injury mediated by antibody and complement (from LEPOW)

In Fig. 23 shown schematically is the significance of the complement in immunological phenomena.

Two methods are employed for measuring the amount of complements in serum. One is the immunological quantitation of $\beta_{1C/A}$, β_{1E} and β_{1F}-globulins, and the other is the quantitation of the serum complement titer measured from the complement activity using the hemolytic system.

Clinical Abnormalities

Acquired increase: As "acute phase reactant", its increase is seen in acute inflammatory diseases, acute stress syndromes, connective tissue diseases, excluding SLE.

Acquired decrease
1) Serum sickness
2) Acute glomerulonephritis
3) Systemic lupus erythematosus and other autoimmune diseases

PROPERDIN

1. Physico-chemical Characteristics

Since properdin was first reported by PILLEMER et al.[155] there have been many doubts about its existence. However, since it was purified by PENSKY et al, its existence has been confirmed.[154]

In normal serum, it is contained at a concentration of approximately 0.2 mg/100 ml. It is a serum protein with about 230,000 molecular weight. Its sedimentation constant $(S_{20,w})$ is 5.2S, and its diffusion coefficient (D_{20}) is 2.2. It has β_2-mobility.[154] In the earlier days of research, it was misunderstood due to poor purification procedure, that properdin belonged to the macroglobulin group.

2. Immunoelectrophoretic Characteristics

Human properdin has an immunogenicity in rabbits and produces its specific antibodies. By OUCHTERLONY technique or immunoelectrophoresis using rabbit antiserum, a single symmetric precipitin arc is formed in β zone.[157]

3. Biological Characteristics

Properdin was first reported as a serum factor necessary for zymosan (insoluble polysaccharides of yeast) to inactivate the third complement component (C3). However, it was made clear later that the "properdin system" also requires Mg^{++} and other complement-like components other than properdin, zymosan and C3. Formerly, it was considered to be a nonspecific reaction, but recently it is thought that the "properdin system" consists of various components present in normal serum, including properdin, Mg^{++}, complement-like components, and specific antibodies. Serologically, properdin itself reacts in a nonspecific way. However, each "properdin system" is of specific antigen-antibody reaction, and the function of properdin is similar to the action of complements in the hemolytic system. Moreover, not only does the "properdin system" have an intimate role in the bactericidal and virus-neutralizing reactions of normal serum, but it is also related to paroxysmal nocturnal hemoglobinuria.[107] This kind of physiological role can be used to measure the properdin activity, but usually bactericidal and zymosan reactions are used for its direct quantitation.

4. Clinical Abnormalities

Acquired decrease

1) Gram-negative bacterial infections, pneumonia
2) Infectious mononucleosis, aplastic anemia, paroxysmal nocturnal hemoglobinuria, leukemia, myeloma
3) Advanced carcinomas

OTHER β-GLOBULINS

In addition to the above-mentioned β-globulins, several precipitin arcs are sometimes seen in the β-fraction. Besides, the β_2-glycoprotein identified by SCHULTZE et al,[184] there are β_{2S}-globulin and β_{2B}-globulin. The biological and physico-chemical characteristics of these β-globulins are unknown.

β_2-Glycoprotein

β_2-glycoprotein is also called β_{2X}-globulin (HEREMANS, 1959), β_2-mucoid (SCHULTZE et al, 1961), and PS–4A (SCHULTZE et al, 1962). It was identified by SCHULTZE et al[184]. It is a glycoprotein with sedimentation constant of 2.9S, containing 17.7% carbohydrate.

Normal serum contains 20–25 mg/100 ml of β_2-glycoprotein. By immunoelectrophoresis, its precipitin arc crosses with transferrin, close to the trough, just outside the IgA arc.

β_{2S}-Glycoprotein

β_{2S}-globulin, named by ZIMMER, is a minor component in the β_2 zone. It is an extremely unstable globulin present in fresh serum, disappearing on heating at 56 °C for 10 minutes.[213] Immunoelectrophoretically, it is found just inside the IgG arc in nearly the same position as β_{1C}-globulin. Its relationship to β_{1A} and other complement components is not clear yet.

Chapter 7

Fibrinogen and its Degradation Products

There are many protein constituents contained in plasma in minute amounts, having various biological activities, such as enzymes, hormones, antibodies and factors involving blood coagulation and fibrinolysis. However, only fibrinogen and its degradation products are discussed in this chapter, for they are sometimes tested both electrophoretically and immunochemically as other plasma proteins.

Blood flows constantly in the blood vessels normally, and its liquid state appears to be kept in a balance between blood coagulation and fibrinolysis. Blood coagulation is apparently a transformation of soluble fibrinogen molecules into a solid clot through a series of catalytic activity of various blood coagulation factors. Among the blood coagulation factors contained in plasma, fibrinogen is found in the largest amount, and it migrates electrophoretically as the so-called "ϕ fraction" between the β and γ fractions. Other coagulation factors of protein nature are contained only in a minute amount, and their activity is usually tested by various hematological procedures.

Fibrinolysis is the phenomenon in which fibrinogen or fibrin is cleaved enzymatically with plasmin activity. During the fibrinolytic process, fibrin and fibrinogen are degraded into various small peptides, called fibrinogen degradation products, which are identified immunochemically.

FIBRINOGEN

1. Synonyms

Fibrinogen, Blood coagulation factor I

2. Physical Characteristics[185]

Sedimentation constant ($S_{20,w}$) 7.63S; molecular weight 341,000; intrinsic viscosity (η) 0.23; electrophoretic mobility 2.1 (β_2); isoelectric point (pI) 5.8

3. Chemical Characteristics

Precipitation: Fibrinogen precipitates under the following conditions: 8% ethanol (pH 7.2, ionic strength 0.14, protein 5%), 0.6M ammonium sulfate (pH 7.0, protein 1%), 0.0065 M Rivanol aqueous solution (pH 8.0, protein 1%), and 0.6M perchloric acid (protein 1%). Fibrinogen shows irreversible precipitation on heating at 56°C, and is included in the euglobulin fraction of serum. It is adsorbed to $Al(OH)_3$ and $Mg(OH)_2$, but not to $BaSO_4$. Although fibrinogen is difficult to be separated from factor VIII, it may be separated by bentonite adsorption. That is, fibrinogen is adsorbed to bentonite, while factor VIII is not.

Chemical Structure: Fibrinogen consists of about 2,900 amino acids, the N-terminal amino acids being alanine and tyrosine.[15] Its nitrogen content is 16.7% and its carbohydrate content is 2.5% (hexose 1.0%, hexosamine 0.9%, sialic acid 0.6%, fucose 0%).[186]

Fibrinogen is a long nodular molecule (Fig. 24) with a length of 600–700Å.[69] It is composed of three pairs of different polypeptide chains, the molecular weight of each chain being as follows: γ-chain, 47,000; β-chain, 56,000; α-chain, 63,000. The γ-polypeptide chain, the smallest of the three, does not change during the coagulation process.[122] The N-terminal peptides of the three pairs of polypeptide chains are said to be combined, forming "disulphide knots" composed of –S–S– bonds (Fig. 25). These "disulphide knots" become separated by adding CNBr.[8]

Heterogeneity: According to DEAE-cellulose chromatography, fibrinogen is divided into two fractions, but no significant immunological and biological differences are noted between the two fractions.[43]

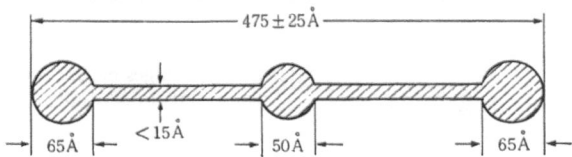

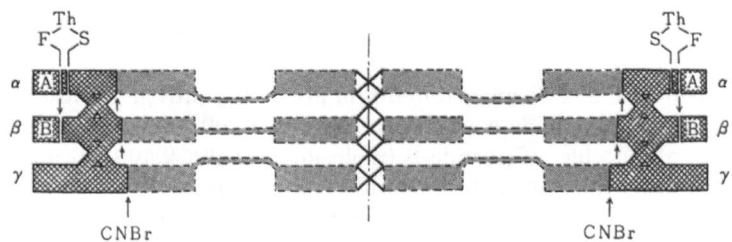

Fig. 25 A schematic diagram of the fibrinogen molecule (from BLOMBÄCK et al.). The fibrinogen molecule consists of 6 polypeptide chains, or 3 pairs. Th-F indicates the site of fast thrombin action, releasing "peptide A". Th-S indicates the site of slow thrombin action, releasing the tripeptide (Gly-Pro-Arg) after the peptide A is cut off. CNBr indicates the site of the methionyl bond opened when CNBr is added. The N-terminal ends of α, β and γ chains seem to form the "disulphide knots".

4. Immunoelectrophoretic Characteristics

Human fibrinogen is immunologically species-specific, and does not cross-react with bovine or other animal fibrinogens. On immunoelectrophoresis, since the molecular weight of fibrinogen is considerably large, its precipitin arc is located away from the antiserum trough. Using agar gel electrophoresis, the fibrinogen arc is frequently asymmetric, showing slight tailing towards the anode. Using cellulose acetate membranes, however, the fibrinogen arc becomes symmetric. It is not known how this immunoelectrophoretic finding corresponds either the cryo-precipitable fraction or the heterogeneity observed in chromatographic analysis.

5. Biological Characteristics

Fibrinogen is present in normal plasma at concentrations between 200 and 600 mg/ 100 ml (290 mg/100 ml on the average), and is indispensable for blood coagulation.

By the enzyme thrombin, fibrinogen is converted to insoluble fibrin and this reaction proceeds in four stages,[111,179] as shown in Fig. 26. Stage 1 is partial hydrolysis; Stage 2, polymerization of fibrin monomers; Stage 3, formation of urea-soluble fibrin-s; and Stage 4, formation of urea-insoluble fibrin-i. In the fourth stage, the action of Factor XIII (fibrin-stabilizing factor, FSF) is required. SATO et al, reporting on the electron-microscopical aspects of fibrin formation, stated that fibrin is formed by passing through a pro-fibrin, then a pre-fibrin stages (Fig. 27).[194]   However, it is not known how this report is related to the above described chemical or immunological findings.

Fibrinogen

(1) ↓ thrombin

Fibrin monomer+2A Peptides+2 B Peptides

(2) ↓

Fibrin polymers(intermed.)

(3) ↓

Fibrin clot(sol.)

(4) ↓ Factor XIII
     Ca++

Fibrin clot (insol.)+Carbohydrate+NH₃

Fig. 26   Stages of fibrin clot formation from fibrinogen.

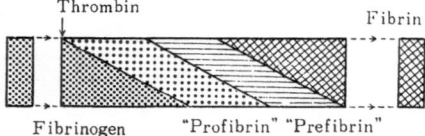

Fibrinogen   "Profibrin" "Prefibrin"

Fig. 27   Schematic diagram of fibrin clot formation from fibrinogen (from SATO, et al.).
It displays stages inferred from electron microscopic findings, but its relation with chemical findings is not known.

During the conversion of fibrinogen to fibrin, about 3% nitrogen is lost.   Through this process two molecules of peptide A (molecular weight, 1,900) are released from one pair of α-chains, and two molecules of peptide B (molecular weight, 2,400) from one pair of β-chains.   The amino acid sequences of peptide A and peptide B which are cleaved from bovine fibrinogen are shown in Fig. 28.   Peptide A is cleaved faster than peptide B.[29]

Peptide A
  Glu-Asp-Gly-Ser-Asp-Pro-Pro-Ser-Gly-Asp-Phe-Leu-Thr ⌉
                               Arg-Val-Gly-Gly-Gly-Glu ⌋

Peptide B   SO₄
                             |
  N-acetyl-Thr-Glu-Phe-Pro-Asp-Tyr-Asp-Glu-Gly-Glu-Asp-Asp ⌉
                      Arg-Ala-Gly-Leu-Gly-Val-Lys-Pro-Arg ⌋

Fig. 28   Amino acid sequences of peptide A and peptide B which are cleaved from bovine fibrinogen.

In the case of human fibrinogen, acidic peptide chains are cleaved by cutting the bond between arginine and glycine in the same manner as in the case of bovine fibrinogen. In the case of human fibrinogen, phosphorus is not necessary for fibrin clotting, but human fibrinogen not containing phosphorus is clotted rapidly by thrombin.[9,146]

Peptides A and B not only prevent the polymerization of fibrinogen molecules but also have strong biological activity.   That is, the peptides are effective in provoking strong activity of contracting smooth muscles by bradykinin.[102]   When two pairs of peptides A and B are combined with fibrin monomers, the molecular polymerization of fibrin monomers is prevented effectively.   In this way, fibrinogen is present physiologically in soluble form.   It is thought that the end-to-end binding of fibrin monomers is caused by the cleavage of peptide A, and also that the side-to-side binding is caused by the cleavage of peptide B. Thus the net structure of fibrin polymers is completed finally.[102]

  6.   Clinical Abnormalities

Congenital a-or hypofibrinogenemia

Acquired hypofibribrinogenemia

  1)   Severe hepatocellular damage and malnutrition
  2)   Defibrinating syndrome
  3)   Accentuated fibrinolysis

Acquired hyperfibrinogenemia: It is increased as an acute phase reactant in cases of inflammatory and necrotizing processes.

Parafibrinogenemia (BECK):   Parafibrinogenemia is mainly recognized in a congenital form, referring to the case in which abnormal fibrinogen appears in blood.

FIBRINOGEN DEGRADATION PRODUCTS

  1.   Immunochemical and Physico-chemical Characteristics

Activated plasmin acts on fibrin and digests it into metafibrin and fibrinolysopeptides. Metafibrin specifically cross-reacts against the antifibrinogen.

Plasmin not only reacts to fibrin but also reacts on fibrinogen as a proteoloytic enzyme resulting in fibrinogenolysis.   Fibrinogenolysis by plasmin is divided into the following three stages, by MARDER as shown in Fig. 29.[115]   In the first stage, fragment X is mainly produced; in the second, in addition to fragment X, fragment Y is produced; third, both

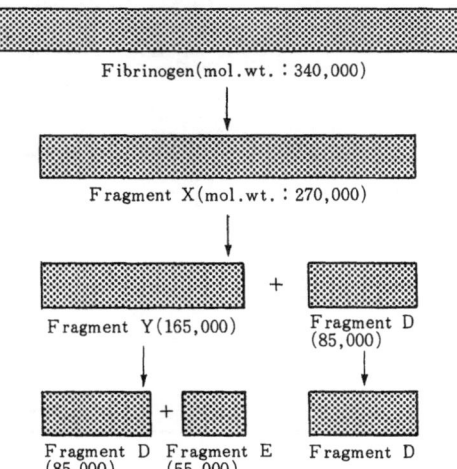

Fig. 29 Suggested scheme for plasmin fragmentation of fibrinogen (from MARDER).
Fragment D occupies approximately 55%, and fragment E occupies 15–20% among the "final products" of digestion. All the fragments shown above react specifically against the anti-fibrinogen.

Fibrinogen(mol.wt. : 340,000)

Fragment X(mol.wt. : 270,000)

Fragment Y(165,000)   +   Fragment D (85,000)

Fragment D (85,000) + Fragment E (55,000)   Fragment D

fragment X and fragment Y disappear, and the end products, fragment D and fragment E are formed.   It is only fragment E among those degradation products which does not precipitate on heating at 56°C for 30 minutes.   Fragment X and fragment Y have strong anti-coagulant activity, and are considered to be a cause of marked bleeding in the patients with accentuated fibrinolysis.   Fragment X, Y, D, and E react specifically to the anti-fibrinogen (Fig. 30, 31).   Thus, the detection of fibrinogen degradation products in serum can be done immunologically[41,124,189] (Fig. 32).

For detection of fibrinogen degradation products in serum, it is important to collect serum samples in the presence of a plasmin inhibitor.   Usually ε-aminocaproic acid will be added to avoid false-positive result due to *in vitro* hydrolysis.

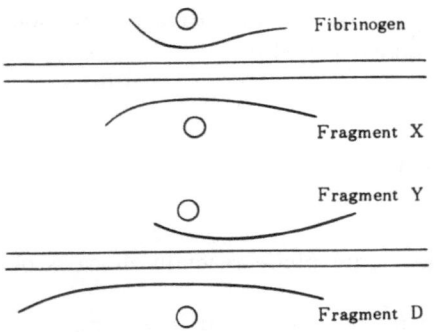

Fibrinogen

Fragment X

Fragment Y

Fragment D

Fig. 30 Schematic diagram of fibrinogen and its degradation products on agar gel electrophoresis. (from MARDER)

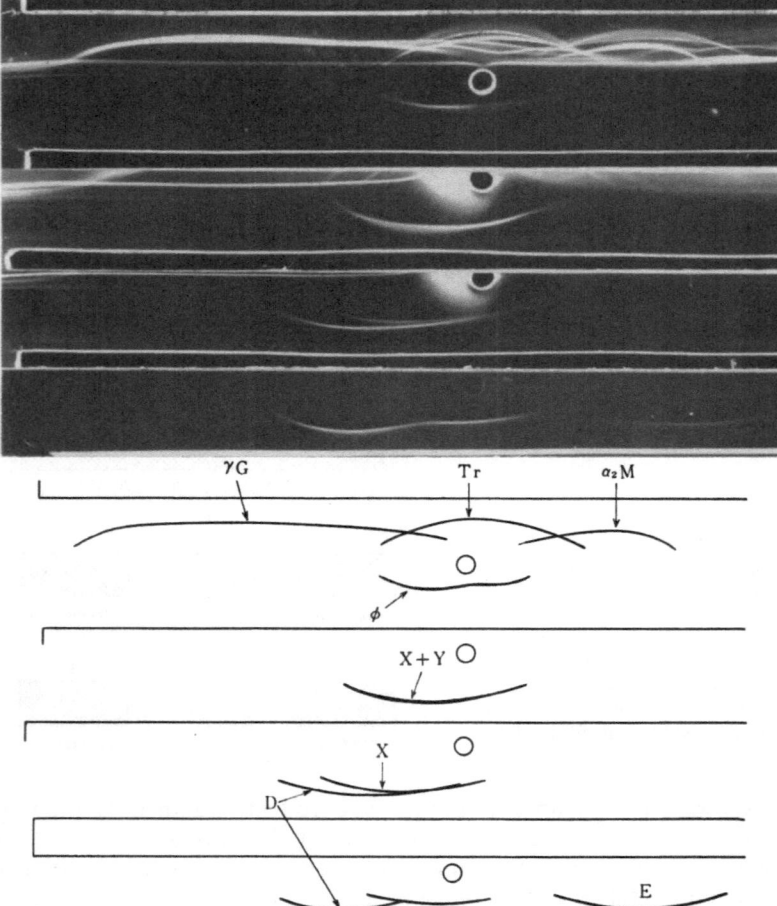

Fig. 31 Agar gel immunoelectrophoretic patterns of fibrinogen and its degradation products. The uppermost pattern is of normal plasma, while the other patterns are of the sera obtained from the patients with accentuated fibrinolysis.

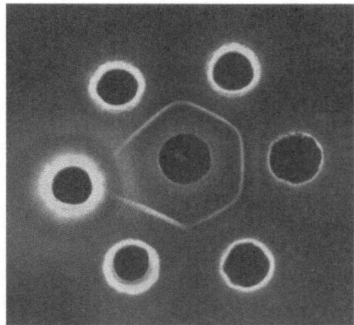 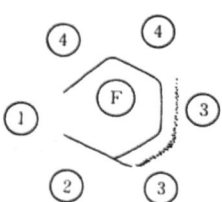

Fig. 32 Identification of the fibrinogen degradation products in patients' sera.
F: anti-human fibrinogen, 1: serum with completed fibrinolysis, 2: normal plasma, 3: the patient Y.Y.,
4: the patient Y. K. The patients' sera are obtained by adding ε-ACA.

2. Increase in Fibrinogen Degradation Products

When fibrinolysis is accentuated pathologically in the body, markedly increased plasmin activity destroys fibrinogen, and hypofibrinogenemia usually occurs. In addition, various amounts of fibrinogen degradation products are found in serum. It has been only about 20 years since abnormal fibrinolysis was clinically noted. Therefore, it is natural that abnormalities in fibrinolysis have not yet been systematized though there has been remarkable progress in this field of research.

In Table 5 listed are some of the conditions in which accentuated fibrinolysis is frequently encountered.

Table 5 Representative clinical conditions associated
with accentuated fibrinolysis (from MIALE).

1. Liver cirrhosis
2. Acute bleeding
3. Shock
4. Severe burn
5. Barbiturate intoxication
6. Methanol intoxication
7. Major surgical operation (especially after hysterectomy and prostatectomy)
8. Stress (fear, emotional upset, convulsive seizure)
9. Thrombolytic purpura
10. Administration of streptokinase, plasminogen and urokinase.

Chapter 8

Immunoglobulins

GENERAL DISCUSSION ON THE IMMUNOGLOBULINS

The term γ-globulin was used by TISELIUS in 1936 referring to the slowest moving serum globulin. The following year, TISELIUS demonstrated that antibodies are to be found in the γ-globulin.[198] However, soon after that, it was discovered that the serum proteins with antibody activity not only belong to the γ-globulin group but is also found in the fast-moving regions.[122] This discovery was followed by further discovery of γ-macroglobulin and other similar proteins. Although for a number of years the proteins having antibody activity were included within the term γ-globulins, this electrophoretic designation ultimately became very much confusing.

In 1960, HEREMANS[74] proposed the term "immunoglobulin", and later in 1964, the World Health Organization (WHO) published "Nomenclature for Human Immuno-globulins". Since then it has been in general use, and the Ig is frequently used to indicate the immunoglobulin.

As described above, γ-globulin designates only electrophoretically to the slowest moving serum globulin fraction in the buffer of pH 8.6, as used for the first time by TISELIUS. The author, fearing confusion with the former use of the term γ-globulin, has named simply the γ-fraction for the slowest moving serum protein group separated by electrophoretic methods. When analysis is made by substantial electrophoretic methods which are widely used in routine analysis (filter-paper, cellulose acetate, agar gel, etc.), almost all the proteins included in the γ fraction are IgG-globulins. A portion of the immuno-glubulins is also migrated in γ_1-zone. However, there are some minor serum proteins which migrate to the γ fraction and at the same time they have nothing to do with the immunoglobulins physicochemically. These serum globulins exist in very small quantity and are simply classified into the following three groups: 3S γ_1-globulins, 2S γ_2-globulins and 0.5S γ_2-globulins.[59]

Immunoglobulins constitute a functionally and structurally diverse group of proteins, and are the general term applied for a certain group of the serum globulins which are disguished by the following four properties.

First, they are synthesized in the lympho-reticular tissues including plasma cells and related lymphoid cells. Secondly, they are biologically active, having a characteristic antibody activity. Not all of the immunoglobulin molecules have been proved by now to have antibody activity. It is probably because we do not have sufficient numbers of different antigenic substances to detect all the antibody activities. However, this bio-logical activity has been proved only in the immunoglobulin molecules. Thirdly, they have a very similar chemical structure (Fig. 33).[25,106,147] The unit molecule of immunoglobulins is composed of four polypeptide chains held together by inter-chain di-sulphide and noncovalent bonds, including one pair of the heavy polypeptide chains (H chains) and one pair of the light polypeptide chains (L chains).   The heavy chains, having a molecular weight of approximately 60,000, are divided into five major classes, γ, a, μ, δ, ε, depending on their antigenic specificity.   Some of the heavy chain classes are further divided into different subclasses.   The light chains, having a molecular weight of appro-ximately 23,000, are the same in all classes of immunoglobulins, and they are further divided into two types; $\varkappa$ (kappa) chain or K type, and λ (lambda) chain or L type.   The L chain consists of two connected segments each about 100 amino acids long, the variable portion and the constant portion.   The C-terminal segment, called the constant portion, has an essentially constant amino acid sequence for a given type of L chains, whereas the

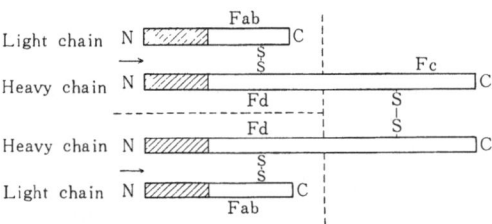

Fig. 33 Suggested fundamental structure of the immunoglobulin
molecule.
N: N-terminus, C: C-terminus
The arrows indicate the possible sites for antigen binding. The
shaded areas of the H and L chains are the variable portions, while
the others are the constant portions.

N-terminal segment, the variable portion, is variable in amino acid composition and sequence from one L chain to another.   The H chains appear to consist of four segments of approximately 100 amino acid residues each.   The N-terminal segment is known as the variable portion, because it shows variability in its amino acid composition and se-quence.   The C-terminal three quarters of the H chains, or the constant portion, appear to be essentially constant for a given class or subclass of the H chain.   The variable por-tions of the H and L chains have been assumed to contribute to the binding affinity of an antibody globulin for its specific antigen.   Forthly, each class of immunoglobulins shows polyclonicity, characterized electrophoretically by a broad band.   This continuous variability is principally based on chemical heterogeneity present in the variable portions of both H and L chains.   A tremendous progress in immunoglobulin structure and functions has been made and is still going day by day.

IgG-GOLBULIN

1.   Synonyms
γG-immunoglobulin (WHO, 1964), γ-globulin, γ_2-globulin, 7Sγ-globulin (MÜLLER-EBERHARD et al. 1956), Kohlenhydratarme γ-Globulin (SCHULTZE and HEIDE, 1960), γ_{ss}-globulin (WALDENSTRÖM, 1962).

2.   Physical Characteristics
Sedimentation constant ($S_{20,w}$) 6.6–7.2S; diffusion coefficient ($D_{20,w}$) 4.0; molecular weight 156,000–161,000; frictional coefficient (f/f_0) 1.38; intrinsic viscosity (η) 0.060; iso-electric point (pI) 5.8–7.2; extinction coefficient, ($E_{280m\mu}$) 13.8; mobility, 1.2 (IgG moves

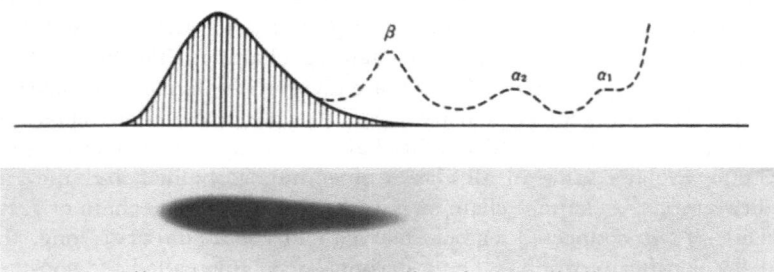

The dotted line indicates the densitometric pattern of the normal serum protein fractions, while the solid line is of the normal IgG-globulin. The lower spot indicates the agar gel electrophoretic pattern of the normal IgG, showing a broad distribution rather than a monoclonal spot.

within a very wide range from γ_1 to slow-γ on electrophoresis, and the modal mobility is found in the middle of γ_2 (Fig. 34).

3. Chemical Characteristics

Precipitation: IgG is precipitated under the following conditions:[185] 25% ethanol (pH 6.9, ionic strength 0.005, protein content 1%), 1.2–1.8M ammonium sulfate (pH 7.0, protein content 1%), 0.6M perchloric acid, 0.15M trichlor-acetic acid, heating (pH 5.0, 0.1M acetate buffer, protein content 1%). IgG is soluble in 0.0065M Rivanol aqueous solution (pH 8.0).

Chemical structure: IgG-globulins are composed of about 1,555 amino acids. The unit molecule is made up of two γ chains and two L chains ($\varkappa$ chains or λ chains). Four polypeptide chains are combined with each other by inter-chain disulfide bonds. As N-terminal amino acids, there are aspartic and glutamic acids, while serine and glycine are known as the C-terminal amino acids.[185] On electron microscopy, IgG molecule appears as a triangular pillar, 80Å in width and 35Å in height, with a length of 105–120Å[40]. Nitrogen content is 16.0%, carbohydrate content, 2.9% (hexose 1.10%, hexosamine 1.30%, sialic acid, 0.30%, fucose, 0.20%). It is the enzymatic cleavage (hydrolysis) of PORTER that has paved the way for the discovery of the IgG structure.[159] PORTER reported that three fragments of approximately the same size are formed by the papain digestion of the rabbit IgG. Later, the same facts were observed in regard to the IgG molecules of other animals. The three fragments include two "antigen-binding" fragments or Fab (also called, piece I or II, fragment S, fragments A and C) and one "crystallizable" fragment or Fc (also called piece III, fragment F, fragment B). When, instead of papain, pepsin is used in pH 4.0, IgG is cleaved into one big fragment, F(ab')2 or 5S, and several small peptide chains. F(ab')2 has a molecular weight of 90,000 and is able to combine with antigen, having 2 antigen-binding sites. It is further disrupted into two Fab' by the use of thiol.[142] The Fab' fragment resembles very much the Fab fragment gained through papain digestion, but they are not identical. It may be inferred from the research employing enzymatic cleavage (hydrolysis) that IgG molecule is composed of three parts as shown in Fig. 35.[143] Moreover, these three parts are independent of each other, each forming a three-dimensional structure that is not influenced by the other two. Two Fab contain antibody-active sites which bind with antigens, and Fc has various kinds of other biological properties than antibody activities.

Further research has proved that IgG is cleaved into two kinds of polypeptide chains by the following three methods. They are: 1) reduction in urea solution by EDELMAN and POULIK,[34] 2) reduction and alkylation by FLEISCHMAN,[44] and 3) S-sulfonation by

FRANEK.[46)] The two kinds of polypeptide chains are γ chain (formerly called H chain or A-chain) and L chain (light chain, formerly known also as B chain). The molecular weight of γ chain is 50,000; that of L chain is 25,000. It has been found that 67% of the molecule is occupied by γ chains while 33% by L chains. Therefore, it is certain now that an IgG molecule is composed of four polypeptide chains (Fig. 35). Also, by immuno-chemical analysis, it has been found that Fab is composed of L chain and the N-terminal half of γ chain (Fd piece) and that Fc is a dimer of the C-terminal half of γ chains. Therefore, the structure of IgG molecules may be shown as $(\gamma_2 \varkappa_2)$ or $(\gamma_2 \lambda_2)$.[160)]

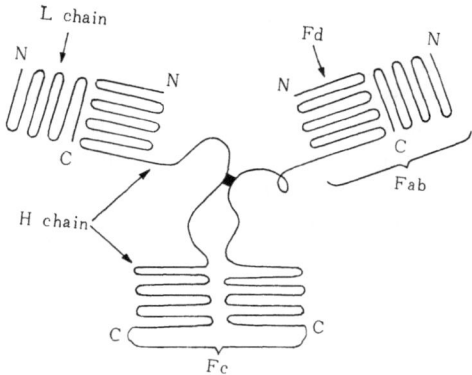

Fig. 35 A structural model of the IgG-globulin molecule (from NOELKEN et al.).
The black dot in the central portion of the molecule carries the disulfide bond connecting the two heavy chains, and it represents the portion most susceptible to enzymatic digestion. Other disulfide bonds are not shown in the scheme.

L-chains and Fd are bound by the disulfide bonds, and even after reduction and alkylation, a rather high proportion of L chains remains to be combined. Therefore, a strong affinity is thought to be between the two, without the disulfide bonds. Also, C-terminal halves of γ chains are bound together by the disulfide bond to form Fc fragment. Between these two, there exists a stong affinity. In fact, under adequate *in vitro* conditions, it is possible to reproduce IgG molecules from L chains and γ chains, Fab from L chains and Fd, and also F(ab')2 from Fab'. Therefore, it is not necessary to rely solely on the disulfide bonds, but on non-covalent bonds.[266)]

Heterogeneity: IgG shows very remarkable heterogeneity in electrophoretic mobility and in solubility (euglobulin and pseudoglobulin), and the term IgG family would seem to be more appropriate. The widespread mobility of IgG is based on variable mobility of γ chains themselves, as well as on the characteristics of Fab.[47,148)] On the other hand, L chains are divided into about 10 components of remarkable heterogeneity by starch gel electrophoresis, and the difference of their mobility does not seem to affect the mobility of IgG molecules.[26)] Therefore, the difference of the mobility of IgG is determined mainly by the characteristics of Fd.

According to the difference in antigenity, the IgG is further subdivided. L-chains are classified into $\varkappa$ chains and λ chains, depending on the difference in their antigenicity. Each IgG molecule has one pair of L-chains, both of which are identical in nature. The molecules containing $\varkappa$ chains are called K-type, while the molecules with λ chains are called **L-type**.[101)] In the western countries, it has been reported that the percentage of K-type and L-type in normal IgG molecules is 60% and 30%, respectively.[114]   In Japan, however, the L-type is present relatively in larger proportion.

The γ chains are further divided into at least four subclasses, γ_1, γ_2, γ_3 and γ_4, depending on their antigenic differences. Similarly, the IgG can be subdivided into the following four subclasses, IgG 1 (b or We), IgG 2 (a or Ne), IgG 3 (c or Vi) and IgG 4 (d or Ge).[32,65,144]   In normal sera, the foregoing are contained in the following percentages, respectively: 77%, 11%, 9% and 3%.[51]

Primary structure of the polypeptide chains in IgG molecules has become apparent to some extent by analyzing many individual myeloma proteins and some specific antibody globulins isolated. The C-terminal half of the L chain has an essentially constant amino acid sequence for a given type, and it is called "constant portion." In contrast, the N-terminal half is variable in amino acid composition and sequence from one L chain to another, and it is called "variable portion". The γ chains also consists of four different segments. The N-terminal segment is called "variable portion", corresponding to that of the L chain. The C-terminal three quarters of the γ chains are called "constant portion", showing essentially constant amino acid composition for a given class or subclass of the γ chains. Among various antigenic specificities related to the IgG molecules, only idiotype specificity is reflected by the primary structure of the variable portions in the γ and L chains.[209]

4. Hereditary Types

Hereditary types of the IgG molecules involve two systems which are controlled independently; the InV system related to L chains and the Gm system related to γ chains.

It was GRUBB and LAURELL[66] who first recognized the Gm-phenotypes by using Rh-positive Group O erythrocytes sensitized with an incomplete anti-D serum of human origin. The sensitized erythrocytes are agglutinated with a corresponding RA-positive human serum. This agglutination may be inhibited by some of human sera. According to their study among the Swedish population, they could classify into two groups with the use of their particular system: Gm (a+) or inhibitors in about 60%, and Gm (−) or non-inhibitors in about 40%. In Japan, Gm (a+) is 100%, which is the same for other Mongolians. Later, by using various combinations of human anti-D and RA-positive human sera, many other Gm factors have been reported: Gm (b1), (b2), (b4), (b^x), (b$^\beta$), (b^r), (c), (r), (e), (p), (f), (s), (t), (m), (g) and others[119]. At least, 23 different Gm factors have been reported, and it is proposed that they are to be called in numbers; for example, Gm 1, . . . Gm 23, etc.[210] The hereditary pattern of the Gm-system has not been completely clarified yet. It varies among different races, but it seems to be controlled by three codominant allelic genes. The Gm factors are restricted to γ chains, and are closely related to the subclasses of γ chains. Gm (a), (x) and (f) are found in IgG1, while Gm(b), (g), (s), (t) and (c) are found in IgG3. Also, the Gm(a) factor is located in the Fc fragment, and the Gm(f) is in the Fab framgent.[119,128,197]

The InV-system was found first by ROPARTZ et al.[169] by using the same principle of agglutination inhibition reactions as with the Gm-system. Besides InV(a) factor, InV (b), (1) and other factors have been reported. The InV factors are restricted to the L chains, and are found not only in IgG but also other classes of immunoglobulins.[37,70118,197]

5. Immunoelectrophoretic Characteristics

IgG exists in a large quantity in normal serum, and its precipitin arc is very distinct, having a characteristic shape (Fig. 36). It shows a very long line which not only runs through the entire γ-zone but also reaches even to the α_2 zone. The anodal three quarters of the line is almost straight and the cathodal end shows asymmetical curve. The shape of the IgG line differs somewhat in different electrophoretic conditions. With the use of the veronal buffer having the ionic strength of 0.07–0.1, the usual pattern of the precipitation can be recognized. However, on agar gel electrophoresis in the veronal buffer having the ionic strength of 0.03, the IgG line appears somewhat biphasic, as seen in Fig. 37. In addition, the IgG line tends to extend farther toward the anode, and

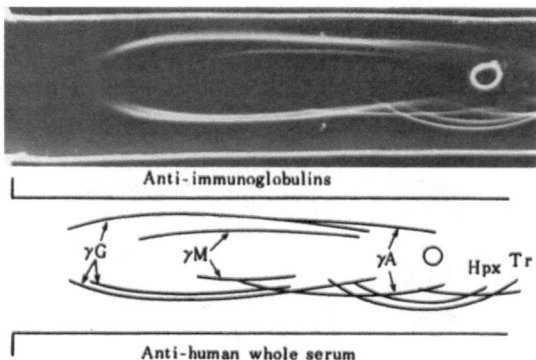

Fig. 36 Agar gel immunoelectrophoretic patterns of the immunoglobulins.

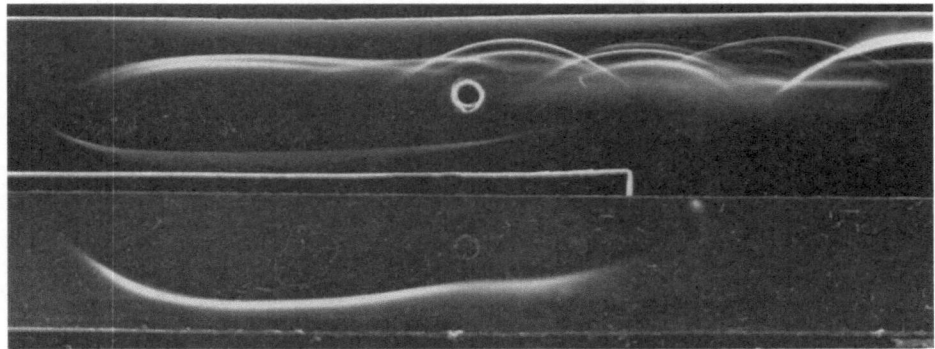

Fig. 37 The precipitin lines of IgG in the buffers having different ionic strength. The upper pattern is obtained in the Veronal buffer (pH 8.6) of the ionic stregth 0.06, while the lower pattern is obtained in the buffer with the ionic strength of 0.03. Even with the normal serum, the IgG line extends more towards the anode in the latter.

the anodal end of the IgG line sometimes shows forking. These phenomena seem to occur due to the interaction between some IgG molecules and agar gel. Agar gel seems to act as an ion exchanger in the buffer having a low ionic strength, and interact with some IgG molecules, particularly slow-migrating molecules. Therefore, it is not advisable for routine analysis to use the buffer having the ionic strength of less than 0.05 on agar gel electrophoresis.

With rabbit and goat immune sera, IgG shows a single line of precipitation. With certain horse immune sera, however, two lines of precipitation are recognized consistently when normal human sera are studied. The horse immune serum which has been prepared in our laboratory demonstrates always two IgG lines (Fig. 37). The line of precipitation nearest to the antibody trough, conveniently named as IgG_{II}, extends farther

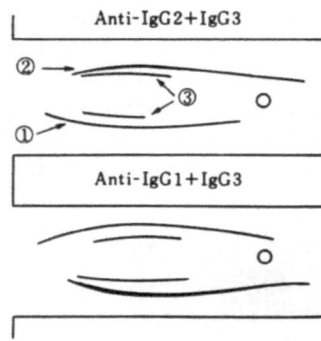

Fig. 38　Immunoelectrophoretic diagrams of the IgG subclasses.

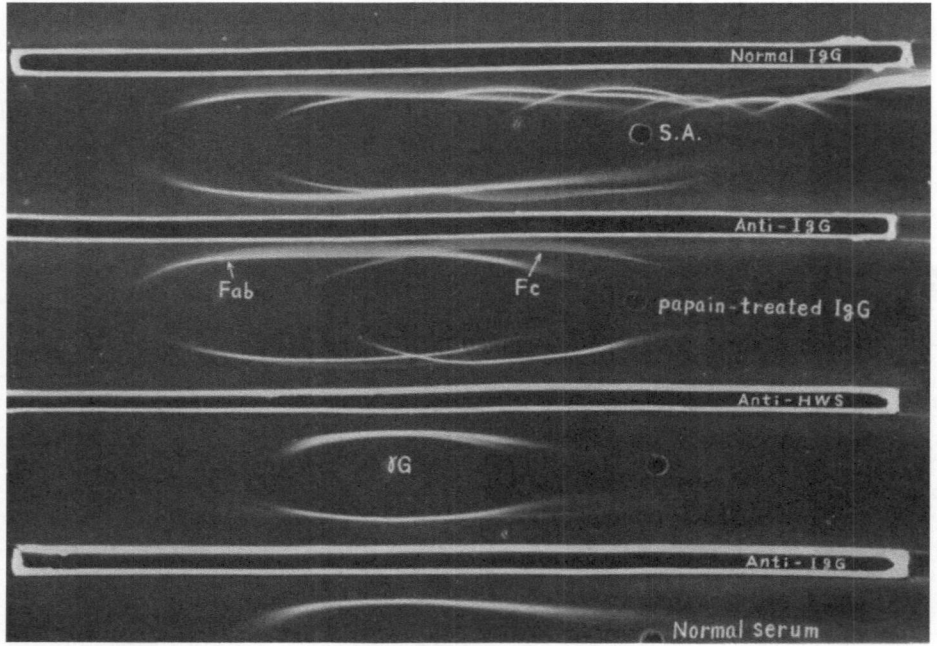

Fig. 39　Agar gel immunoelectrophoretic patterns of IgG and its papain digests. Anti-HWS: anti-human whole serum horse immune serum.　Fab and Fc indicate Fab fragment and Fc fragment of the IgG molecules.　The serum S. A. is taken from the patient with polyclonal hyperimmunoglobulinemia, and its stored sample contains both the Fab and Fc fragments.

toward both ends than the other, IgG_I.　These patterns have not compared to any of the IgG subclasses.　However, the immunoelectrophoretic patterns on different IgG subclasses are reproduced in Fig. 38.　When human serum is preserved for a long time or at 37 °C for a considerable length of time, the anodal end of the IgG line sometimes splits. This is similar to the pattern of enzymatic split-products of IgG (Fig. 39).

IgM-GLOBULIN

1. Synonyms

γ_M-immunoglobulin (WHO, 1964) γ_{1M}-globulin, γ_1-macroglobulin, 19Sγ-globulin (MÜLLER-EBERHARD et al. 1956), β_{2M}-globulin (BURTIN et al. 1957), β_{2C}-(β_C)-globulin (SCHEIDEGGER et al. 1956), Iota (SOOTHILL, 1962).

2. Physical Characteristics[185]

Sedimentation constant ($S_{20,w}$) 18–20S; molecular weight 1,000,000; mobility 2.1; extinction coefficient ($E_{180\,m\mu}$) 13.3.

3. Chemical Characteristics

Precipitation: IgM precipitates under the following conditions:[185] 25% ethanol (pH 6.9, protein content 0.2%), 1.2–1.8M ammonium sulfate (pH 6.9, protein content 1%), 0.0065M Rivanol aqueous solution (pH 8.0, protein content 1%), 0.6M perchloric acid, 0.15M trichlor-acetic acid, heating (pH 5.0, 0.1M acetate buffer, protein content 1%). In contrast to IgG, IgM precipitates very easily when Rivanol is used.

Chemical structure: IgM is composed of about 8,500 amino acids and its amino acid composition closely resembles that of IgG. Just as in the case of IgG, the N-terminal amino acids of IgM are aspartic and glutamic acids. Among the immunoglobulins, IgM is the one that contains carbohydrate in the greatest amout, 11.8% (hexose, 5.4%; hexosamine, 4.4%; sialic acid, 1.3%; fucose, 0.7%).

The IgM molecule is cleaved by mercapto-ethanol treatment into subunits (γMs) of 6-7S, each having a molecular weight of 160,000[31]. As the unit molecule, γMs is composed of one pair of μ chains (that is, the H chain peculiar to IgM) and one pair of L chains. L chains have exactly the same antigenity as those of the other immunoglobulins, being divided into $\varkappa$ chains and λ chains.[25,35] The molecular weight of the μ chain is about 70,000, carrying the carbohydrate moiety. An IgM molecule consists of five γMs which are combined together with five disulfide bonds.[126] An IgM molecule has either the $\varkappa$ chains or the λ chains. Therefore, the structure of IgM present in a normal human serum can be expressed as $(\mu_2\varkappa_2)_5$ or $(\mu_2\lambda_2)_5$. IgM fragments such as Fab, Fc and F(ab')2 may be obtained through various enzymatic cleavages, similarly as in IgG. On electron microscopy, the center of the IgM molecule appears as a ring structure, measuring approximately 100Å, and the entire IgM molecule looks like a star fish.[64] In certain pathological conditions, 7S IgM, which is very similar but not identical to γMs, are recognized in serum. It is quite interesting to note that this 7S IgM is observed even under normal conditions in certain kinds of animals such as lemon shark.[22,170]

4. Hereditary Types

Just as in the case of IgG, the InV-system is known in relation to the L chains. However, with the μ chain, no definite genetic polymorphism is known.

5. Immunoelectrophoretic Characteristics

The electrophoretic mobility of IgM is faster than that of IgG, migrating around the γ_1-zone. IgM is a macromolecular component and is difficult to diffuse through agar gel. Therefore, the IgM line appears at a distance farthest from the antibody trough, and shows an almost straight form. Generally, the IgM line becomes observable slowly, frequently necessitating about 3 days to be completed. This tendency is particularly so with the use of horse immune serum. On electrophoresis with the buffer having a low ionic strength, the IgM line tends to extend farther toward the anode, and IgM is easily precipitated at the point of application when the serum concentration of IgM is increased pathologically.

Specific antisera against IgM or μ chain are used to identify the IgM line of precipitation, if necessary (Fig. 40). When treated with mercaptoethanol or penicillamine, the precipitation line, apparently of γMs, is located more toward the anode, closer to the antibody trough, and shows a rather symmetric arc.

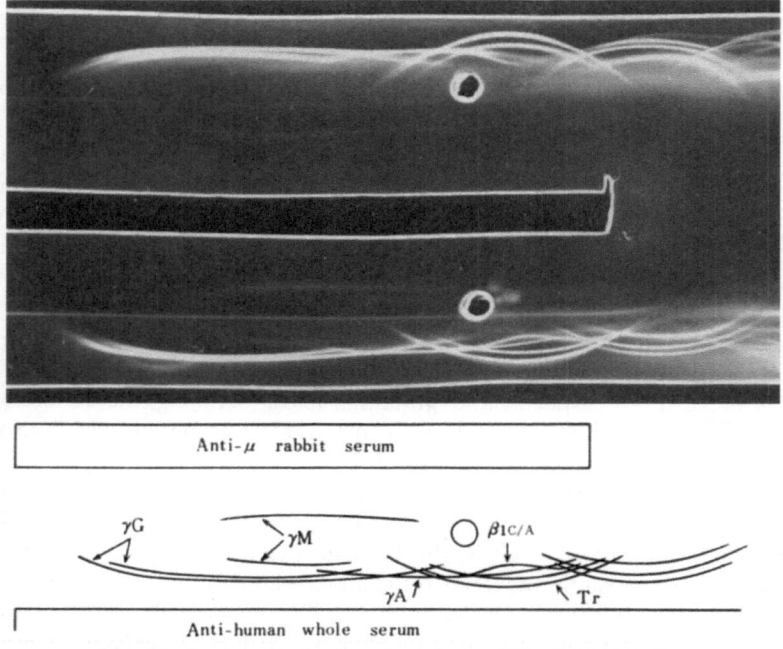

Fig. 40 Immunoelectrophoretic patterns of the IgM-globulin. An almost straight line of precipitation is seen farther from the antibody trough, migrating at the γ_1-zone.

IgA-GLOBULIN

1. Synonyms
γ_A-immunoglobulin (WHO, 1964), γ_{1A}-globulin (HEREMANS, 1960), β_{2A}-globulin (BURTIN et al. 1957), β_X-globulin (WILLIAMS, 1954).

2. Physical Characterstics.[185]
Sedimentation constant ($S_{20,w}$) 7, 10, 13, 15, 17S; molecular weight, about 160,000 (7S); extinction co-efficient ($E_{280\,m\mu}$) 13.4; mobility, 2.1.

3. Chemical Characteristics
Precipitation: IgA precipitates under the following conditions:[185] 18% ethanol (pH 5.2, ionic strength 0.015, protein content, 1%); 1.2–1.8M ammonium sulfate (pH 7.0, protein content 1%); 0.0065M Rivanol aqueous solution (pH 8.0, protein content 1%); 0.6M perchloric acid, 0.15M trichlor-acetic acid, heating (pH 5.0, 0.1M acetate buffer, protein content 1%).

Chemical structure: IgA is observed as various kinds of polymers having 7S monomer as a unit molecule. In normal serum, IgA consists mainly of 7S monomer and 5–10% of all IgA exists as polymers of about 10–11S.

Dimer form of the serum IgA is disrupted into 7S IgA by mercaptoethanol treatment. 7S IgA which has a molecular weight of about 160,000 consists of one pair of a chains (H chains peculiar to IgA) and one pair of L chains.[18,25,36] Since the L chains can be divided into $\varkappa$ chains and λ chains, as in other immunoglobulins, the structure of IgA may be expressed as $(a_2\varkappa_2)_n$ or $(a_2\lambda_2)_n$.

7S IgA consists of about 1,500 amino acids and shows an amino acid composition similar to IgG or IgM. N-terminal amino acids of IgA are aspartic and glutamic acids. IgA contains approximately 8–10% carbohydrate (hexose, 3.2%; hexosamine, 2.3%; sialic acid, 1.8%; fucose, 0.22%).[185]

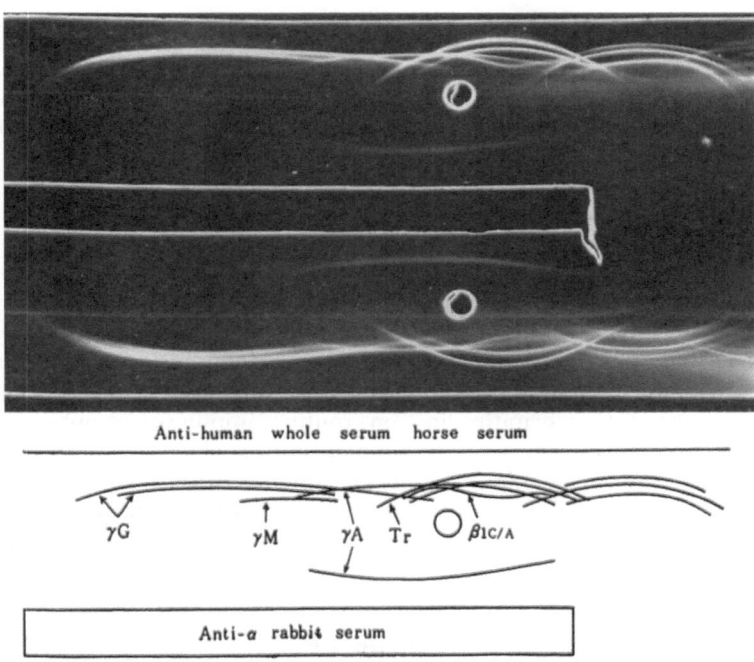

Fig. 41 Agar gel immunoelectrophoretic patterns of IgA.
IgA is recognized as a symmetric line of precipitation, bridging between the IgG and the transferrin lines.

IgA is further divided into at least two subclasses. The major component, designated as IgA 1, occupies approximately 80% of all normal serum IgA, while the minor component, IgA 2, occupies the rest.[93,203] IgA 2 is thought to have a peculiar chemical structure, in which non-covalent bonds, instead of the disulfide bonds, contribute for binding the a chains and the L chains.

4. Hereditary Types

The InVsystem is known in relation to the L chain. But, nothing is yet known about the a chain.

5. Immunoelectrophoretic Characteristics

IgA migrates around γ_1-zone, and shows a symmetric arc of precipitation, located between those of transferrin and IgG (Fig. 41). With the use of certain horse immune sera, IgA appears occasionally as two lines of precipitation.

The IgA lines vary in their shape, depending on the types of immune sera used.

With most of horse immune sera which hardly contain the antibodies against the L chain components, the IgA line crosses completely with the IgG line. However, with many of rabbit and goat immune sera, the anodal portion of the IgA line completely fuses into the IgG line, and the IgA line appears to branch off from the IgG line. When identification of IgA is necessary, the specific antiserum against IgA or a chain has to be used. However, care must be given because many of its specific antisera are not completely monospecific, although labeled as such.

In normal infants, the IgA line is not recognized frequently and the immunochemical quantitation is necessary in order to rule out its complete deficiency state.

IgD-GLOBULIN

IgD or γD-immunoglobulin was first identified by ROWE and FAHEY by using the rabbit immune serum against the myeloma globulin which was isolated from a patient (S. J.)[171]

IgD has a sedimentation constant ($S_{20,w}$) of 7S, and migrates around the γ_1-zone. IgD molecules are composed basically of one pair of δ chains (the H chains peculiar to IgD) and one pair of the L chains, represented as $(\delta_2\varkappa_2)$ and $(\delta_2\lambda_2)$. The molecular weight of the IgD molecule is approximately 200,000 and it contains carbohydrates in approximately 14%.[73]

On immunoelectrophoretic analysis (Fig. 42), the IgD line of precipitation appears quite similar to that of IgA. The most of the commercially available anti-human whole serum do not contain the antibody against IgD or a very little, if any. Therefore, one cannot identify the IgD precipitin line on routine immunoelectrophoretic analysis. The horse immune serum against pooled human serum, which has been used routinely in the author's laboratory, contains a significant amount of the IgD antibody, and IgD can be recognized as a precipitin arc showing a partial identity with the IgG line, as show in Fig. 210. The specific immune serum against IgD has to be used to identify the IgD abnormality.

IgE-GLOBULIN

IgE or γE-immunoglobulin was first identified by ISHIZAKA et al.[83] as the fifth class of of immunoglobulins having reaginic activity. LATER, JOHANSSON et al.[88] reported the most unusual myeloma protein designated as IgND, and it was found to belong to the IgE class.

IgE molecules have the sedimentation constant ($S_{20,w}$) of 8.0 S and the molecular weight of approximately 200,000. It migrates electrophoretically around γ_1-zone, almost identical to IgA. IgE molecules are composed basically of one pair of ε chains (the H chains peculiar to IgE) and one pair of the L chains, represented as $(\varepsilon_2\varkappa_2)$ and $(\varepsilon_2\lambda_2)$. The carbohydrate content of IgE is approximately 11%. ε chain has the molecular weight of approximately 75,500, and is significantly larger than the other heavy chains. ε chains characteristically contain a large amount of intrachain disulfide bonds and methionine.[6,83]

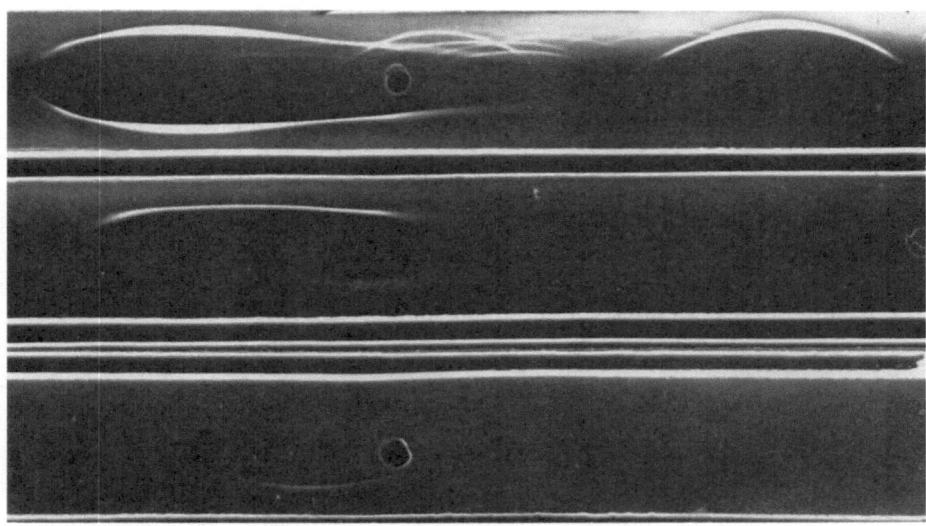

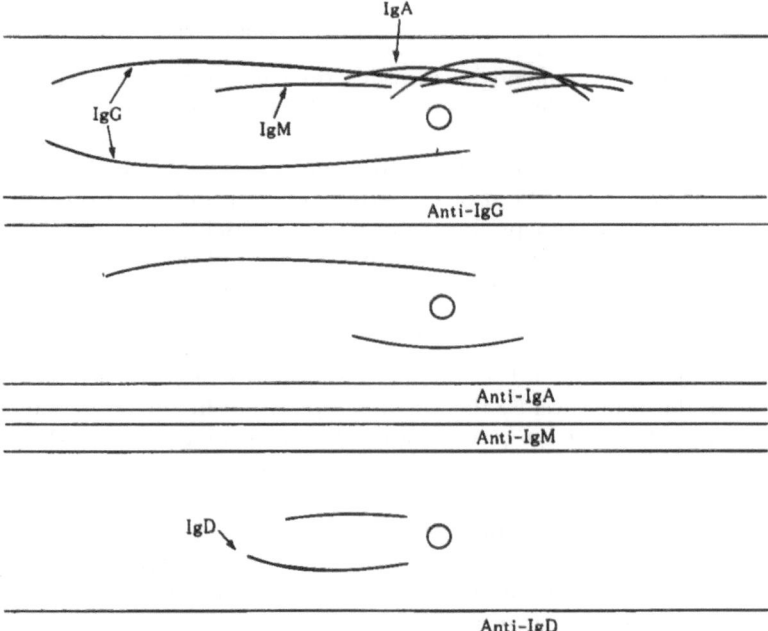

Fig. 42 Agar gel immunoelectrophoretic patterns of IgD.

BIOLOGICAL CHARACTERISTICS
OF THE IMMUNOGLOBULINS

Normal Serum Concentration of the Immunoglobulins

Serum concentrations of immunoglobulins in normal individuals vary among different laboratories. It is mainly due to the present lack of uniformity of the numerous immunoglobulin standards used in different laboratories. Therefore, World Health Organization has recently recommended to use the International Reference Preparation which has the values expressed in terms of International Units per ml. Two representative reports on the normal values of serum immunoglobulins are sited in Table 6. [2,185].

Table 6 Serum concentrations of immunoglobulins in normal adults.

	SCHULTZE & HEREMANS	ALLANSMITH et al.
IgG	900 – 1,500	710 – 1,540 (m = 1,045)
IgM	39 – 117 (m = 75)	♂ 37 – 204 (m = 90)
		♀ 42 – 261 (m = 104)
IgA	110 – 180	60 – 490 (m = 170)
IgD	< 0.3 – 30	

Unit: mg/100 ml.

The concentrations of immunoglobulins in serum show remarkable change after birth. Fig. 43 shows the physiological variations of IgG, IgM, and IgA concentrations. The synthesis of immunoglobulins is thought to begin in more than 20 week-old fetuses.[204] The newborn baby with uneventful prenatal course has normally IgG and IgM in its serum at the time of birth. Most of the serum IgG is transfered from the mother, but a very small quantity is also produced by the baby itself (Fig. 44)[2,52] The fetus is potentially capable of producing IgA, but in the serum of normal newborn infants IgA is not observed, though in cases of fetal *in utero* infection IgA may be demonstrated in its serum.

One week after birth, the serum IgM begins to increase rapidly, but after from two to four weeks, its increase diminishes as the serum IgG increases[2]. In males, the serum IgM reaches to the adult level by the age of one year, but in females it comes around the age of two years. After the age of seven or eight years, the serum IgM is higher in females (Fig. 43).

The synthesis of IgG begins to increase in two to four weeks after birth, but the IgG of maternal origin is quickly eliminated. Therefore, the serum IgG concentration of the newborn baby becomes the lowest around four months after birth. After this, the synthesis of IgG becomes active and reaches the adult level when the child becomes about eight years of age. Moreover, of interest to note that when the serum IgG reaches the adult level, the sex difference in the serum IgM concentration becomes prominent.[2]

The production of IgA usually increases very slowy in two to three weeks after birth. It reaches the adult level at puberty (about 12 years of age).

As to IgD, it is occasionally detected in a very small quantity even at the time of birth. Within six weeks after birth, it may be detected in almost all newborns. It seems to reach the adult level in 2–5 years of age[113].

Antibody Activities of the Immunoglobulins

1. Specific Antigen Binding

The function which is specific to and common to immunoglobulins is antibody activity.

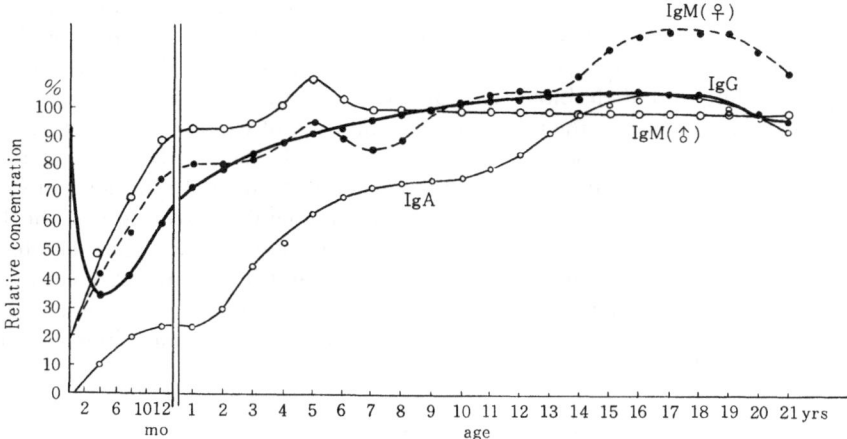

Fig. 43 The changes of the serum immunoglobulin levels (from ALLANSMITH et al.).

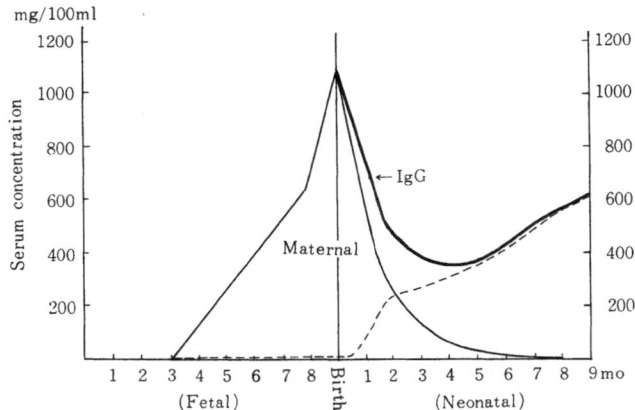

Fig. 44 IgG change during the fetal and neonatal periods
(from ALLANSMITH et al.).
The dotted line indicates the change of the serum concentration
of the IgG produced by infants.

Antibodies may be classified from different points of view. One way of classifying them is as follows:

Humoral antibodies

1) Antitoxin
2) Precipitin
3) Agglutinin
4) Lysin
5) Opsonin
6) Neutralizing antibody
7) Compement-fixing antibody
8) Blocking antibody

Cellular antibodies

Antibodies are combined specifically with corresponding antigens, being observed

as various phenomena described above.   However, the activity common to all antibodies is specific binding with antigens.   Despite a tremendous progress on immunoglobulin chemistry, it has to be admitted that the relation between the chemical structure
of immunoglobulins and their biological functions as antibodies is a subject to be investigated further.   However, the following is a summary of what is known at the present.

Generally speaking, an antibody molecule has two active sites.   It is, therefore, called
a bivalent antibody which is thought to have two subunits located symmetrically, as in a
mirror.[78,152]   There are many studies regarding the chemical structure of the binding
site.[161]   As far as IgG molecules are concerned, it may be said that H chains are
absolutely necessary for binding with antigens and that L chains also contribute to
binding with antigens.   That is, the antigen binding site is composed of one L chain
and the Fd part of an H chain.   It is thought to correspond to depressions at both ends
of the EDELMAN and GALLY model (Fig. 45).[33]

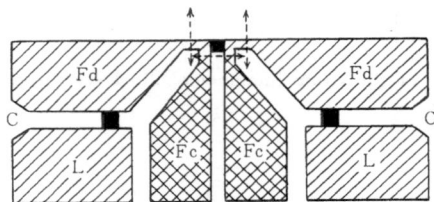

Fig. 45   A structural model of the IgG-globulin (from HAUROWITZ).
The diagram shows a longitudinal section through a rod-like IgG
molecule.   L = light chains, Fc and Fd = fragments of the heavy
chains.   The letters C indicate the probable location of the two
specific combining sites.   The vertical arrows indicate the sites of
cleavage by papain, while the horizontal arrows indicate the sites of
cleavage by pepsin.   The black areas show the three interchain disulfide bonds.

The ability to bind with a specific antigen must be related to those amino acid sequences
which are unique to those antibody molecules showing the specific activity.   The binding
site is apparently composed of the variable portions of the L-chain and the Fd fragment,
representing so-called idiotypic specificity.[68,207]

    2.   Complement Fixation

Complements are not only adsorbed or activated by the antigen-antibody complex which
is formed through specific antigen-antibody reactions, but also, without any antigen, they
are adsorbed onto aggregated immunoglobulin molecules and the immunoglobulins chemically combined with BDB (bis-diazotized benzidine).   The complement fixation is
thought to occur when IgG molecules change sterically under the various conditions
described above.[206]   However, when there is an excessive supply of antigen resulting
the formation of $(Ag)_2Ab$ complex, the complement fixation is not observed.   So it may
be said that at least two IgG molecules are necessary for complement fixation.[13,82]   Moreover, the mono-valent IgG antibodies cannot fix complements even they can passively
sensitize the skin in guinea pigs.[206]

Fc fragments themselves fix complements, and also $F(ab')_2$ or Fab fragments may be
able to fix complements weakly.[82,165,187]   Therefore, it is chiefly Fc fragments that are
connected with complement fixation, although other parts may contribute to it.

IgG and IgM are capable of fixing complements, but this capability has not yet been
observed with IgA.[84]   As described above, IgG does not fix complements unless at least
two molecules are close to each other, but as in the case of IgM, one IgM molecule combined with antigen is sufficient to adsorb one complement molecule.[13]

3. Hemagglutination

Agglutinin, which is concerned with hemagglutination, is divided into saline agglutinin (complete antibody, typical antibody, or bivalent antibody) and albumin agglutinin (incomplete antibody, atypical antibody, or univalent antibody).   However, in ordinary serological conditions, the antibody must be bivalent for hemagglutination to take place. Until now, there has been no evidence to prove that the type of hemagglutination by the so-called 7S incomplete antibody is due to univalency.[206]

Among the immunoglobulins which are concerned with hemagglutination, the IgM antibody has a much stronger activity than the IgG antibody.   With IgA, its molecular aggregate shows stronger agglutinability.   Moreover, it is said that five IgM molecules combined with one erythrocyte sufficient to cause hemagglutination[84].   WEBB and GOODMAN assert that the reason why IgG antibodies have lower agglutinability than IgM antibodies is that when the antigenic determinants exist close to each other on the surface of the erythrocyte, a single bivalent IgG molecule is apt to react by both of its combining sites on the same erythrocytes.[206]

It is generally stated that the activity of IgM antibodies is destroyed by mercaptoethanol treatment.   But it has been confirmed that 7S subunits of IgM also have the antigen-binding site(s).[20,86,145]   It is not definitely known whether the 7S subunits of IgM molecules have bivalent antibody activity or not, although they are composed of two H chains and two L chains.

Based on the facts stated above, STERZL has suggested that multivalent and larger IgM antibodies may serve the biological function of being more effective in dealing with large or particulate antigens, such as bacteria and erythrocytes, whereas the 7S bivalent antibody might be more efficient for neutralizing toxins or other relatively smaller antigens with fewer antigenic determinants.[192]   Actually, it has been proved that the IgM antibodies include isohemagglutinin, cold hemagglutinin, rheumatoid factor, O-agglutinin of Gram-negative bacilli, etc.[92]   Therefore, the differences of biological activities between IgG and IgM antibodies may be related mainly to the differences in the size of the molecule (the numbers of combining sites available) as well as the differences in the structure of the H chain.

4. Neutralization of Virus and Toxin

Regarding the activity of virus-neutralizing antibodies and antitoxin, it may be said that IgG antibodies are more efficient than IgM antibodies.[168,149]   The activity of IgG as a virus-neutralizing antibody diminishes as it is degraded into 5S $F(ab')_2$ and Fab' through enzymatic treatment, and a whole IgG molecule is thought to contribute for the neutralizing reaction[96].   On the other hand, in toxin-antitoxin reaction, the antitoxin active sites seem to exist only in Fab fragments.[149]

5. Opsonization

Opsonins and complements are necessary for phagocytosis.   It is generally accepted that an important role of opsonins is to prepare bacteria and other foreign particles for ingestion by phagocytes.   Two kinds of opsonins are known: normal (or natural) opsonins which exist in normal serum, and immune opsonins which are produced through immunization.   Normal opsonin is chiefly IgM.   Both IgM and IgG may contain immune opsonins.   As for opsonin antibodies against *S. typhimurium*, IgM antibodies are 500–1,000 times more efficient than IgG antibodies.[168]   Opsonin activity is reduced remarkably by reduction-alkylation of IgM antibodies, but their 7S subunits also possess opsonin activity, though it is very weak.   It is still unknown how phagocytes recognize the particles combined with opsonin.

Antibodies similar to these are cytophilic antibodies.   BOYDEN and SORKIN[14] discovered that a special class of immunoglobulins in the serum of animals immunized with human albumin combine with normal spleen cells, and called them "cytophilic antibodies." They thought that these belonged to probably IgG-immunoglobulins.

On the other hand, it has been recognized that human monocytes, macrophages, and certain kinds of lymphocytes are combined specifically with IgG-sensitized red cells, and the red cells combined with macrophages are noticeably deformed.[110] This phenomenon is specifically inhibited by the IgG molecules used for sensitization and also by their Fc fragments. Whether this phenomenon is a certain kind of antigen-antibody reaction or not is unknown, but it is thought that there may be the receptors for IgG on the surface of the mononuclear cells.

6. Skin Sensitization and Anaphylaxis

Allergic tissue reactions are classified into four types: Type I— anaphylactoid reaction, Type II— cytotoxic reaction, Type III— ARTHUS phenomenon and serum sickness, and Type IV— delayed type or tuberculin type reaction.[55] Type IV allergic tissue reaction is well known to occur in relation to the cell-mediated immune reaction. Those antibodies which are related to so-called immediate type tissue reactions including Types I–III belong to humoral antibodies, generally known as anaphylactic antibodies. In man, the following two anaphylactic antibodies are known: 1) reagin, homologous anaphylactic antibody, or homocytotropic antibody, and 2) heterologous anaphylactic antibody, or heterotropic antibody.

At the present, research in this field is being carried on chiefly with passive cutaneous anaphylaxis (PCA) and by P–K reaction (PRAUSNITZ–KÜSTNER reaction). Future development of new research techniques may result in finding additional kinds of anaphylactic antibodies.

Reagin (homologous anaphylactic antibody): These antibodies sensitize tissue cells (probably mast cell) of the same species or of closely-related animals and cause generalized or localized allergic reaction. In man, this procedure is passively observed by PRAUSNITZ–KÜSTNER reaction. It can also be observed by using a certain kind of experimental animals. It was once thought that it was IgA that had this antibody activity, but recently ISHIZAKA et al. have proved that newly discovered IgE is related with the reaginic antibody activity[83]. IgE characteristically binds to histamine-containing cells of the homologous or closely-related heterologous tissues. The skin-sensitizing activity is apparently present in the Fc fragment of the ε (epsilon) chain, or H chain of IgE, and it disappears after heating at 56 °C for 2 hours. Histamine is assumed to be released when cell-bound IgE anti-

Table 7 Some physiological characteristics of immunoglobulins related to heavy chain classes. (modified from METZER)

	γ(IgG)				α(IgA)	μ(IgM)	δ(IgD)	ε(IgE)
	γ1	γ2	γ3	γ4				
Serum level (mg/ml)	9	2.5	1	0.5	4	1	<0.03	0.0003
Half-life (days)		23			6	5	?	?
Rate of synthesis (g/day)		2.3			2.7	0.4	?	?
Traverses placenta	+	+	+	?	−	−	?	−
Fixes complements	+	+	+	−	−	+	?	−
Gives Arthus reaction		+			−	+	...	...
Agglutination activity		"1"			variable	10–1000	?	+
In vitro opsonization		500–1000			...	"1"	...	...
Fixes to guinea-pig skin	+	−	+	+	−	−	−	−
Reaginic activity		−			−	−	−	+

bodies combine with specific allergens.[206] Interestingly, IgE-producing plasma cells are predominantly detected in the tonsils, the gastrointestinal and bronchopulmonary mucosa, and the mesenteric and bronchopulmonary lymph nodes.[196] They are the sites where allergic reactions occur predominantly.

The anaphylactic antibodies, which have similar but not identical biological activity to human reaginic skin-sensitizing antibody, has been described in the guinea-pig and mouse. However, they have different physicochemical properties from reagin-IgE.

Heterologous anaphylactic antibodies: They are the electrophoretically slow IgG antibodies from heterologous species which are capable of eliciting passive cutaneous anaphylaxis in guinea-pigs. The sensitizing activity resides in Fc fragments of IgGl, IgG3 and IgG4, but not of IgG2.[197]

For detection of the antibody activity, the antibodies are injected into guinea-pig skin, and at least three hours are allowed for the skin sensitization. Local anaphylaxis is detected by transudation of dye into the wheal produced by an increased vascular permeability, when antigen with dye is injected intravascularly.

Some of the important physiological characteristics of different classes of immunoglobulins are summarized in Table 7.

CLINICAL ABNORMALITIES OF THE IMMUNOGLOBULINS

1. Congenital Deficiencies
Certain types of primary immunodeficiency syndromes
2. Acquired Deficiencies
Some types of primary immunodeficiency syndromes
Secondary hypogammaglobulinemias:
   1) Protein-losing conditions
   2) Malignant lymphomas
   3) Leukemias
   4) Multiple myeloma
3. Polyclonal Hyperimmunoglobulinemias
   1) Acute and chronic liver diseases
   2) Chronic infections
   3) Malignancies
   4) Autoimmune diseases
   5) Miscellaneous conditions
4. Monoclonal Immunoglobulinopathies
   1) Multiple myeloma
   2) Primary macroglobulinemia
   3) Essential monoclonal immunoglobulinopathy
   4) Miscellaneous conditions

Chapter 9

Glycoproteins and Lipoproteins

GLYCOPROTEINS

1. Synonyms and Nomenclature

As far as glycoproteins are concerned, at present there is no common international nomenclature. Many investigators have examined them by using various methods of analysis and used more or less different terms independently. Therefore, it is necessary to understand sufficiently the methods of analysis used by each investigator. In this monograph, the following definitions are proposed according mainly with those of SHETLAR.[188]

Glycoprotein: This refers to a protein component which contains more than 1% of carbohydrate other than nucleic acids, and in which the carbohydrate is bound to the polypeptide moiety by covalent bonds. Moreover, some investigators advance the definition of glycoprotein as the protein which contains at least one molecule of carbohydrate in one protein molecule.

Mucopolysaccharide-protein complex: Those belonging to this complex easily free carbohydrate in strong salt solutions or weak alkali at ordinary room temperature. As a synonym, there is mucoprotein. However, the term mucoprotein is not used in this book as it has often been used with other meanings.

Protein-bound carbohydrate: This is the general term for carbohydrates bound to proteins obtained from blood, body fluids, or tissues through chemical precipitation or dialysis.
Both glycoporteins and mucopolysaccharide-protein complexes are included.

Seromucoid: It is the general term for the serum proteins which are soluble against sulfosalicylic acid and, in addition, insoluble against phosphotungstic acid or alcohol. The term mucoprotein has often been used iterchangeably with seromucoid.

Neuraminic acid: This is a carbonic acid consisting of nine carbons formed by the condensation of pyruvic acid and mannosamine.

Sialic acid: This is used for acetyl or glycolyl derivatives of neuraminic acid. In man, it is present in the form of N-acetyl-neuraminic acid.

Hexosamine: This is the term for derivatives of hexoses where the hydroxyl group has been replaced by the amino group.

2.   Historical Backgrounds[56]

It has been long known that proteins have reducing activity. However, as for mucins in animals, E. EICHWALD proved the existence of protein-bound carbohydrate for the first time in 1865. In 1897, C. ZANETTI obtained a similar finding in human serum and named it seromucoid. Then, at a joint committee meeting held in 1908, the American Society of Biological Chemists and the American Physiological Society defined glycoprotein in terms which have nearly the same meaning as current terminology. Since about 1945, individual protein components included in glycoprotein began to be identified, and monographs or review articles about protein-bound carbohydrates began to appear. Since GRABAR introduced immunoelectrophoresis in 1953, the identification of individual glycoproteins has advanced rapidly (Table 8). As for seromucoid, it began to be separated into various fractions by applying such means as chromatography and electrophoresis.

Table 8 Various glycoproteins and their carbohydrate contents (from SCHULTZE and HEREMANS).

	Plasma glycoproteins	Carbohydrate (%)
1.	α_1-Neuramino-glycoprotein	42.6
2.	α_1-Acid glycoprotein	41.4
3.	Human chorionic gonadotropin	31.4*
4.	α_1x-Glycoprotein	22.7
5.	Hemopexin	22.6
6.	Haptoglobin	19.3
7.	Zn-α_2-Glycoprotein	18.2
8.	β_2-Glycoprotein	17.1
9.	α_2-Mucoprotein of SCHMID	15.5*
10.	Transcortin	14.1
11.	Tryptophan-poor-α_1-glycoprotein	13.7
12.	α_{2HS}-Glycoprotein	13.4
13.	α_1-Easily-precipitable glycoprotein	13.3
14.	α_1-Antitrypsin	12.4
15.	Prothrombin (beef)	12.1*
16.	IgM-immunoglobulin	11.8
17.	Thyroxine-binding globulin	10.6*
18.	4.6 S-Postalbumin	10.0
19.	Inter-α-trypsin -inhibitor	9.1
20.	α_1-Macroglobulin	8.4
21.	Ceruloplasmin	8.0
22.	IgA-immunoglobulin	7.52
23.	Transferrin	5.87
24.	Gc-globulin	4.2
25.	β_{1A}-and β_{1C}-Globulin	3.03
26.	IgG-immunoglobulin	2.9
27.	Fibrinogen	2.5
28.	β-Lipoprotein	1.8
29.	α_2-Lipoprotein	1.68
30.	α_1-Lipoprotein	1.50

* from GERBARG

Table 9 Uncertain glycoproteins (from GERBARG).

1. Inhibitor of virus hemagglutination
2. Plasma cholinesterase
3. Complement components
4. Lipemia clearing factor inhibitor
5. Platelet cofactor 1
6. Hyaluronidase inhibitor
7. Plasminogen
8. Rheumatoid factor
9. Erythropoietin
10. Thyrotropic hormone
11. Interstitial cell-stimulating hormone
12. Follicle-stimulating hormone

3. Principle of Analytical Methods

The details of the analytical methods were treated in a review article by SHETLAR,[188] and in this section, only its general outline will be given.

When protein-bound carbohydrate is measured, it is necessary first of all to hydrolyze glycoprotein. Following three analytical principles are used.

1) The quantity of carbohydrate is measured before and after hydrolysis. This principle is used infrequently today, but it was formerly used for serum samples.

2) After proteins have been precipitated out first, they are hydrolyzed and their carbohydrate is measured. Ordinarily, this is used for measuring the quantity of individual components of carbohydrate contained in serum.

3) First, individual glycoprotein or glycoprotein group is isolated, and then, after hydrolyzed, the quantity of carbohydrate is measured. For isolation of glycoprotein used are electrophoresis and chromatography. This principle is frequently applied to measure the fractional quantitation of the seromucoid. Furthermore, recently each glycoprotein component has been purified and its carbohydrate content is known (Tables 8, 9).

Many quantitative methods of measuring hexose content are known. All of them use color reactions related to orcinol, tryptophane, carbazole, anthrone, and thymol by making use of the reducing action of hexoses, which have been released as a result of protein hydrolysis in strong sulfuric acid solution.

As far as hexosamines are concerned, quantitation is met by application of the ELSON-MORGAN reaction after they are released as a result of hydrolysis. This reaction is characteristic of glycoprotein, because carbohydrate in glycogen and nucleic acid do not react.

As sialic acid is decomposed by strong hydrolysis, its quantitation is carried out by applying resorcinol-HCl reaction or thio-barbital reaction after it has been selectively hydrolyzed by weak hydrolysis or neuraminidase.

With regard to fucose, its reaction with cysteine is applied after hydrolysis by heating in strong acids.

Protein-bound carbohydrate contents in normal serum are listed in Table 10.[188]

Table 10 Normal serum concentration of protein-bound carbohydrates (from SHETLAR).

Carbohydrate	mg/100 ml	g/100 g protein
Hexose	114±3	1.6
Hexosamine	71±4	1.01
Sialic acid	62±2	0.84
Fucose	8.9±0.6	
Total	256	3.46

4. Chemical Characteristics

Each glycoprotein differs greatly from other glycoproteins in its physico-chemical properties, so they will be described separately. In this section, the points common to all glycoproteins will be summarized.

Low-molecular weight glycoproteins, such as acid glycoproteins moving in the a zone, generally contain a large amount of carbohydrate and it is characteristic that they are not precipitated by boiling.

Carbohydrates included in glycoproteins are divided into the following four groups: 1) hexose (mannose, galactose, and rarely glucose), 2) methylpentose (fucose), 3) hexosamine (glucosamine, galactosamine), 4) neuraminic acid derivatives (sialic acid). Their chemical formulae are shown in Fig. 46.

Many points of how these carbohydrates are bound to polypeptide moiety remain unknown. However, in glycoproteins, they are strongly bound with each other and may be separated only by means of acid hydrolysis, alkali hydrolysis, and enzymatic hydrolysis. On the other hand, mucopolysaccharide-protein complexes show comparatively loose binding between protein and carbohydrate, and carbohydrates are easily freed by strong salt solutions or by alkali at ordinary room temperature.

The chemical structures of glycoproteins are not clear in many points. Until very recently, only carbohydrate content and amino acid analysis were done. However, recently, analysis of carbohydrate moiety has finally begun. Review articles on this subject by WINZLER and PUTNAM are available.[164]

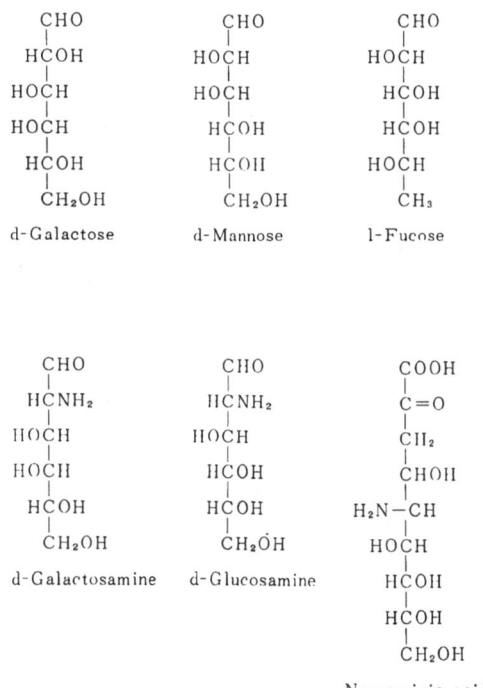

Fig. 46 The structural formulae of carbohydrates contained in glycoproteins.

5. Biological Characteristics

There are a large number of plasma protein components included in glycoproteins, and each has a rather characteristic biological function. The details of each glycoprotein component, therefore, will be described in its own section, while here the general outlines will be discussed.

As glycoproteins fulfilling the function of carriers or binders, transcortin (CBG), thyroxine-binding globulin (TBG) and haptoglobin are known. Ceruloplasmin and transferrin are known as those which are bound especially to metals.

Those serving as blood coagulation factors are prothrombin, fibrinogen, platelet cofactor 1, and, in addition, plasminogen participating in fibrinolysis.

As for the glycoproteins which have some relation to immunity, there are various im-

munoglobulins.   Also, besides virus hemagglutination inhibitor, there are various complements.   Furthermore, the rheumatoid factor is also considered a glycoprotein.

Typical glycoproteins having enzymatic activity are a_1-antitrypsin and cholinesterase.
Hyaluronidase inhibitor and lipemia-clearing factors may perhaps belong to this group.

The one that has hormonal activity is chorionic gonadotropin.   Moreover, erythropoietin is also thought to be a glycoprotein.

There are many glycoproteins whose biological functions are still unknown.   Among
them are the low-molecular weight glycoproteins which, along with haptoglobin and C-
reactive protein, are the so-called acute phase reactants.

6.   Seromucoids

The greater part of serum protein components contain more or less carbohydrate, and
glycoproteins as listed in Table 8 are clearly identified.   Moreover, electrophoretically,
they are widely distributed in all fractions, but many of them have a_1, a_2 and β-mobility.
They are fractionated by chromatography or isoelectric technique into various components, but seromucoid is the best known.[17,166]

For separating seromucoid, major protein components are first precipitated out by
adding 0.6 N-perchloric acid to serum.   Then, phosphotungstic acid is added to the
supernatant, and the precipitate is collected.   The seromucoid fractioned in this way is
not physicochemically homogeneous.   In the main, it has a–β mobility on filter-paper
electrophoresis (Tris buffer, pH 8.60), and is divided into 4–5 subfractions.   Also, it is
known to contain albumin, a_1-antitrypsin, a_1-acid glycoprotein, haptoglobin, and hemopexin by immunoelectrophoretic analysis.[139]

Seromucoid occupies approximately 10–15% of the total glycoproteins found in serum.

7.   Clinical Abnormalities

The serum glycoprotein level shows significant age difference.   In fetus it is approximately 80% of the adult level.   Its concentration in aged people is about 10% higher
than that of an average adult, while that of a pregnant woman is about 20–50% higher.
(Table 11)   In pregnancy, however, there is an increase of total glycoprotein, while
there is no notable increase of seromucoid.

Acquired hypoglycoproteinemia: 1) Hepatocellular damage, 2) Nephrotic syndrome, 3)
Malnutrition, cachexia.

Acquired . hyperglycoproteinemia: 1) Acute or chronic inflammatory diseases, 2) Stress
syndrome, 3) Malignancy, 4) Various hematological diseases.

Table 11   Serum glycoprotein concentration in healthy individuals (from SHETLAR).

	Hexose in glycoprotein		Hexose in seromucoid mg/100 ml
	mg/100 ml	g/100 g protein	
Fetus (15)	80(62–103)	1.41(1.05–1.77)	6 (3–9)
3–8 years (8)	105(94–118)	1.60(1.47–1.82)	
21–40 years, male (18)	110(93–126)	1.58(1.26–1.90)	
21–40 years, female (10)	111(100–125)	1.58(1.42–1.81)	12 (8–18)
61–85 years (15)	129(104–138)	1.79(1.62–2.06)	
Pregnancy (8–10 mo.)			
Primipara (13)	152(120–172)	2.46(2.26–2.63)	
Multipara (16)	150(123–191)	2.42(1.96–2.82)	12 (8–14)
Pregnancy (4 mo.) (6)	122(110–136)	1.87(1.70–2.10)	

(   ) indicates the number of the cases studied.

LIPOPROTEINS

1. Synonyms and Nomenclature

As in the case of glycoproteins, there is no generally accepted nomenclature on lipoproteins. Different investigators have used the same term for different preparations and, conversely, different terms have been used for the same preparation. Therefore, it is necessary to keep in mind the procedure used for analysis. The following nomenclature will be used here mainly from clinical standpoints.

Lipid-protein complex: All of the lipids in plasma are themselves water-insoluble, and they are bound to proteins in order to be dissolved in plasma water. These are generally called as "lipid-protein complex", which is divided into 3 groups: 1) free fatty acids bound to serum albumin, 2) lipoproteins and 3) chylomicrons.

Lipoproteins: The term has been used interchangeably with lipid-protein complex. However, the author prefers to use the term lipoprotein to mean any water-soluble macromolecule or micromicelle binding plasma lipids. It does not include the albumin bound with free fatty acids. It is divided into two major groups: high-density lipoproteins (HDL) and low-density lipoproteins (LDL).

High-density lipoproteins (HDL): They include the lipoproteins which do not float in the solvent having the density of 1.063, but which do float in the solvent density of 1.21. Electrophoretically, they are sometimes called either α-lipoproteins or α_1-lipoproteins. They are further subdivided ultracentrifugally into HDL_2 and HDL_3.

Low-density lipoproteins (LDL): They include the lipoproteins which float in the solvent having the density of 1.063. Electrophoretically, they may be called by the terms such as lipoproteins, β-lipoprotein, pre-β lipoprotein and α_2-lipoprotein. These terms are very confusing, because low-density lipoproteins migrate differently on different electrophoretic techniques. They will be treated in detail, later.

Apolipoproteins: FREDRICKSON et al. used the term lipoprotein apoproteins. The author prefers to use the term apolipoproteins. The term apolipoprotein refers to the protein moiety of lipoproteins. At least two different apolipoproteins, possibly three, have been identified: Apolipoprotein A and apolipoprotein B, possibly apolipoprotein C.

2. Historical Backgrounds

It was already known in the latter part of the 19th century that some lipid-protein complex was present in living tissues. About 1877, EDMUNDS observed highly refractile minute particles in serum, and they were later called "Haemokonien" by MÜLLER in 1896. GAGE and FISH (1924) assumed that they derived from chyle, and named them "chylomicron", which is still currently used. This is how the existence of lipids in serum came to be known. It was MACHEBOEUF (1929) who clarified the fact that hydrophilic serum proteins combined with lipids in the body.

As these modern hypotheses became dominant, one after the other, various kinds of analytical methods regarding serum proteins were introduced for analyzing the lipid-protein complex. MUTZENBACHER applied ultracentrifugation to analyze serum. MC-FARLANE (1935) noticed a new fraction which was named "X-protein". It came to be studied by PEDERSEN (1939) working in the SVEDBERG's laboratory, leading to the establishment of the floatation method for lipoprotein analysis.

On the other hand, by the introduction of TISELIUS' electrophoretic method (1937), the β fraction was noted to be particularly lactescent. Later, BLIX et al. (1941) reported that serum lipids were migrated in the α and β fractions.

In this way, lipoproteins had been studied physicochemically. After 1946, COHN et al.

has introduced chemical analytic methods, and studied with both COHN's ethanol fractionation and ultracentrifugation. About 1952, electrophoretic techniques using various supporting media such as filter-paper opened a new era of study. In 1955, GRABAR et al. introduced immunoelectrophoretic techniques, which promised a vast advance in this field.

3.   Principle of Analytical Methods

Several properties of the lipoproteins have been exploited for studying their changes in the serum. There seems to be two approaches: (1) studying the lipoprotein molecules as a whole or the apolipoproteins, and (2) studying the lipid moiety of the lipoproteins.

In all lipoprotein studies, the storage of serum samples and purified lipoprotein fractions is an important problem. As described later, lipoproteins, particularly the low-density lipoproteins, are easily denatured. Therefore, it is desirable that they be preserved at $4\,^{\circ}$C, and that all routine analysis of lipoproteins be finished within two weeks. Storage at usual freezing temperatures gives rise to marked degradation within a few days. If they have to be stored for a long time, it is necessary to add inhibitory agents against lipoprotein lipase and fatty acid transferase. Also, since degradation of lipoproteins is likely to proceed with the presence of metal ions, lipoprotein analyses are preferably done with the addition of some chelating agents such as EDTA. Practically, storage is possible at room temperature for several days provided bacterial growth is suppressed.

Analysis of lipoproteins   Various methods of isolating and studying the serum lipoproteins have been introduced, including salting-out, COHN ethanol precipitation, immunochemical precipitation using specific antisera, electrophoresis, ultracentrifugation, chromatography, non-specific precipitation by polyanions (such as dextran sulfate, amylopectin, etc.).[49,109,174] Among these, ultracentrifugation and electrophoresis are the two most widely applied for clinical purposes.

Ultracentrifugal analysis: Lipoproteins have comparatively low density, and they float in high-density solvents upon ultracentrifugation. By using this principle, lipoproteins are separated into different classes. The ultracentrifugal procedures have been most widely applied for studying the serum lipoproteins, since they would allow isolation in as closely the native state as possible. The modern system of analyzing the serum lipoproteins ultracentrifugally has been established by GOFMAN et al.,[61] and it has been improved by many workers.[176]

Usually the serum lipoproteins are separated into two major classes by using sodium chloride solution of different solvent densities. The low-density lipoproteins float in the solvent density of 1.063. In contrast, the high-density lipoproteins float in the solvent density of 1.21, but not in the density of 1.063. In connection with the low-density lipoproteins, the term S_f or the flotation constant is used. It refers to a particular flotation rate expressed by SVEDBERG unit (10^{-13} cm/sec/dyne/g) in a sodium chloride solution of density 1.063 g/ml at $26\,^{\circ}$C. If different analytical conditions are used, a more accurate description of the flotation media should be added; for example,

$$S_{f1.3920}(D_2O\text{–}NaNO_3)$$

Electrophoretic analysis: From the practical viewpoint, ultracentrifugal methods seem to be the best. However, the equipment is expensive and the procedures are cumbersome. Therefore, they have not applied routinely for screening the patients with various lipid abnormalities. In contrast, the electrophoretic methods, particularly using various supporting media, are simpler and easily applicable for screening purposes.

Filter paper, agar gel, starch gel, polyacrylamide gel and cellulose acetate membrane are the ones which have been used for electrophoretic separation of lipoproteins. None of these supporting media is ideal, and they have some interferrence with lipoproteins. One of the practical methods is the new filter-paper electrophoretic procedure developed by LEES and HATCH.[105] In this procedure, the electrophoretic separation of lipoproteins is done in the buffer containing 1% serum albumin, and as shown in Fig. 47 the serum lipoproteins are separated into 4 fractions: α_1-lipoprotein fraction, pre-β lipoprotein fraction, β-lipoprotein fraction and the origin (chylomicron).

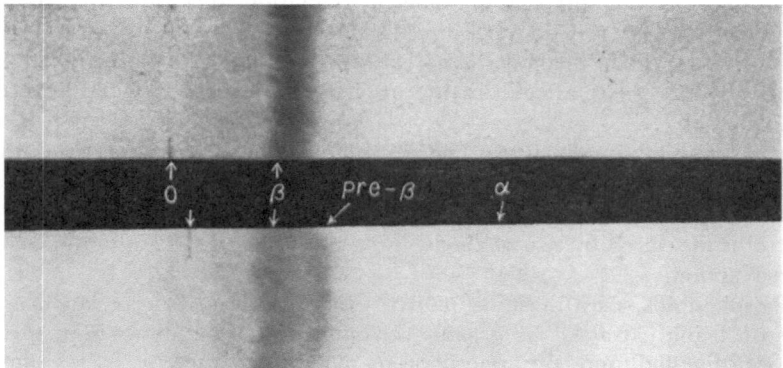

Fig. 47 Filter-paper electrophoretic patterns of the serum lipoproteins. The upper strip demonstrated a marked increase of the β-lipoprotein fraction, and the lower strip shows clearly separated the β-lipoprotein and the pre-β lipoprotein fractions.

Agaropectin, which is contained in agar, binds with low-density lipoproteins to form precipitate. Therefore, agar gel electrophoresis cannot be used for separating the serum lipoproteins. Purified agar gel or agarose gel devoid of most of the interfering substances do not develop precipitation with low-density lipoproteins. Agarose gel electrophoresis results in fairly well resolution of the serum lipoprotein fractions.

Cellulose acetate electrophoresis also gives satisfactory separation of the serum lipoproteins, and it can be easily applied for clinical screening purpose. However, one of the disadvantages on using cellulose acetate membranes is the fact that the membrane itself is heavily stained by lipophilic dyes. Also with some of the cellulose acetate membranes β-lipoproteins migrate in different positions, ranging from α_1 zone to β_2 zone. Therefore, the kind of cellulose acetate membrane should be carefully selected for serum lipoprotein separation. In our experience, Cellogel, Sepraphor III and Selecta membranes seem to be satisfactory, because β-lipoproteins migrate around β zone.

In order to detect the lipid moiety on electrophoretic strips, there are two major staining procedures available for clinical purpose. First, the lipid moiety is stained with lipophilic dyes, such as Oil Red O, Sudan Black, etc. Since the dyes are dissolved in alcoholic or other organic solvents, some of the lipids in lipoproteins may be eluted out into the solvent during staining.[89,95] This tendency is particularly notable in so-called slow-migrating or obstructive lipoprotein fractions. On cellulose acetate electrophoresis, both pre-staining and chlorination procedures have to be used for lipophilic staining.[95] Secondly ozone-Schiff staining procedure[99] is used mainly for cellulose acetate electrophoresis. With this staining, however, only unsaturated fatty acids can be detected.

Analysis of apolipoproteins Apolipoproteins are obtained through delipidation. Namely, lipids are gradually eluted from lipoprotein molecules in ethanol at low temperature.[176]   Apolipoproteins are water-soluble and comparatively stable.   Therefore, they can be used for studying the peptide moiety of lipoproteins.

Analysis of lipid moiety   Since lipids in serum always exist as a complex bound with serum proteins, there is a close relation between serum lipid composition and lipoproteins. Therefore, analysis of the serum lipids is also important.   Total lipid, neutral fat (tri-glyceride), cholesterol, and cholesterol ester, phospolipids, and free fatty acids are usually analysed in serum samples.

The total lipids are measured by gravimetric, oxidation, turbidimetric, volumetric, or dye-binding methods after extracting lipids from the serum by using various organic solvents.   However, these method do not always instigate the same reaction to lipid components.   Besides, there are still many problems to be solved on the lipid extraction process.

There are two methods of measuring neutral fat, the direct and the indirect.   At present, the direct method is more widely applied, and is based on colorimetric quantitation of glycerol contained in neutral fat.

Cholesterol is measured by two methods, using the LIBERMANN-BURCHARD reaction and the KILIANI reaction.

As to phospholipids, care must be paid to exclude impurities as much as possible, when they are being extracted by organic solvents.   After the extraction, the phospholipids are wet-digested and the phosphorous contained therein is measured colorimetrically.

Free fatty acids are titrated by alkali (NaOH) after they are extracted in an acidic condition.

4.   Physical Characteristics.

Lipid-protein complexes are classified as in Fig. 48, according to differences of molecular density by using ultracentrifugation.   Free fatty acids combined with albumin form the FFA-albumin complex.   The very high-density lipoprotein (VHDL) is the lipoprotein fraction which does not float in the density >1.21 g/ml, and it contains phospholipids (about 10% of total serum phospholipids) and a small amount of neutral fat.[108,175]   The high-density lipoprotein class (HDL) is separated at the solvent density from 1.063 to 1.21 g/ml.   It is further subdivided into two groups: HDL_2, floating between the solvent density of 1.063 and 1.125, and HDL_3, floating between the solvent density of 1.125 and 1.21.   The differentiation of both subclasses is not necessarily clear-cut, and there are still some ambiguity on various points.   The low-density lipoproteins (LDL) sometimes refer to the lipoproteins floating in a solvent density of 1.063, and are also called as β lipoproteins. This fraction is also further subdivided into two major classes: low-density lipoprotein class (LDL), having a flotation rate between 0 to 20 at a solvent density of 1.063 g/ml and very low-density lipoprotein class (VLDL), having a flotation rate between 20 to 400.   Chylomicrons have a flotation rate of more than 400 in a solvent density of 1.063 g/ml, but they are usually defined in physiological terms as particles of alimentary origin.[50]   There is no definite chemical evidence for being different from VLDL.   In fact, GUFSTAFSON et al. have included chylomicrons in the VLDL class, which were subdivided into five sub-classes A, B, C, D and E.[67]   BIERMAN has subdivided into α-chylomicron and β-chylo-micron.

The HDL average about 200,000 in molecular weight, while the LDL have a very high molecular weight of from 2,000,000 to more than 20,000,000.   But the diameter of the molecule increases in even greater proportion.   Therefore, the density becomes remarkably low; that is, the density of the HDL is 1.05–1.20, that of the LDL is 1.02–1.05, that of the VLDL is 0.96–1.02, and that of chylomicrons is 0.93–0.95.

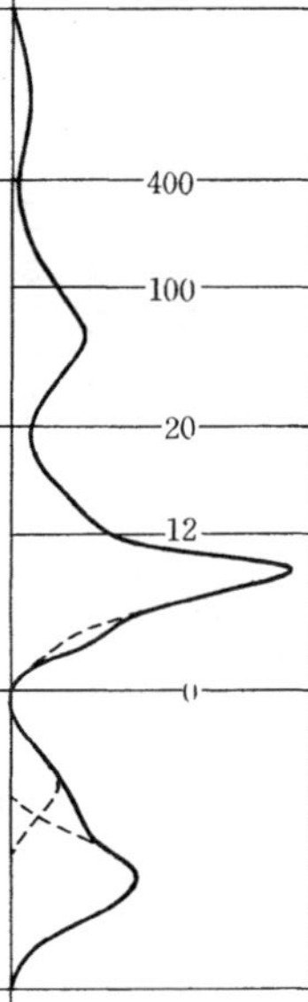

Class	Subfraction or Abbreviation	Free electro. mobility	Solvent density (g/ml)	S_f	Filter Paper* electro. mobility	(Class)*	Diameter* (mμ)	Molecular* weight	Appearance*
Chylomicrons	α-Chylo. / β-Chylo.	α₂–Alb / β	1.006		Origin	Chylomicrons	>200	10^{10}	turbid
Very low density lipoproteins	VLDL	α₂	1.006	400 / 100	Trail	(Small particles)	70~200	10^9	opalescent
			(1.006) ~ 1.006 ~ 1.019	20	pre-β	(VLDL)	30~70	2×10^7	
Low density lipoproteins	LDL₁	β₁	1.006 ~ 1.019	12					
	LDL₂	β₁	1.019 ~ 1.063	0	β	β-lipoprotein	15~30	2×10^6	
High density lipoproteins	HDL₂	α₁	(1.063) ~ 1.063 ~ 1.12		α₁	α₁-lipoprotein	5~15	2×10^5	transparent
	HDL₃	α₁	1.12 ~ 1.21						
Very high density lipoproteins	VHDL	α₂–β	(1.21) ~ 1.21						
	FFA–Alb	Alb			Alb	FFA–Alb			

Fig. 48 Classification of the serum lipid-protein complexes and their properties. (*from HATCH, 1964)

Electrophoretic mobility of HDL is α_1, that of chylomicrons and VLDL is α_2, and that of LDL is β zone.   However, through the improved filter-paper electrophoretic method by Lees and Hatch, chylomicrons are separated at the origin and VLDL at pre-β zone.[105]

5.   Chemical Characteristics

It is not known how lipids and proteins combine in lipoprotein molecules.   It is thought that in most cases, they are combined by covalent bonds.   Also, regarding their tertiary structure, it is unknown whether peptide chains surround lipids which are the core, or whether peptide chains face lipids like two slices of bread as in a sandwich.   However, speaking of at least HDL, the presence of lipids may not directly influence their tertiary structure.[174]

The lipid composition differs among different classes of lipoproteins, as shown in Fig. 49.   Both chylomicrons and VLDL are mostly composed of neutral fat, while cholesterol is mainly contained in the LDL class.

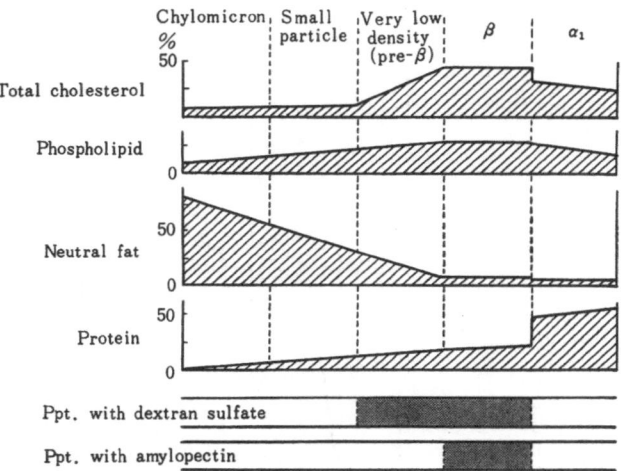

Fig. 49   Diagram showing chemical properties of the serum lipid-protein complexes.

There are still many points to be clarified with regard to the chemical structure of lipid-protein complexes, but what has been known about major classes of the serum lipoproteins is summarized here.

α_1 *lipoproteins (HDL):* α_1-lipoproteins precipitate under the following conditions[185]: 18% ethanol (pH 5.2, ionic strength 0.09, protein content, 1.6%), 2.4–2.8M ammonium sulfate (pH 7.0, protein content, 1%), 0.0065 M Rivanol aqueous solution (pH 8.0, protein content, 1%), 0.6 M perchloric acid, 0.15 M trichloroacetic acid, heating (pH 5.0, 0.1 M acetate buffer, protein content, 1%).   But, being different from LDL, this class does not precipitate by polyanions (dextran sulfate, heparin, etc.).

There are various methods of purifying α_1-lipoproteins, the most efficient being ultra-centrifugation.   The fraction floating in a solvent density of 1.21 g/ml is collected and then LDL is removed by precipitation using heparin and Mn.[174]

α_1-apolipoprotein (A-protein) is apt to cause molecular aggregation and easily forms 4.5S polymers.   It consists of 2 to 6 subunits which have the molecular weight of 23,000–36,000 (about 2.3S).   These subunits contain 1.5% carbohydrate (hexose, 0.9%; hexosamine, 0.25%; sialic acid, 0.3%; fucose, 0.05%).   They are rich in leucine and glutamic acid, and contain only a small amount of isoleucine, methionine and cystine. Also, the N-terminal amino acid is aspartic acid, and the C-terminal amino acid is threonine.[50,185]

It is impossible to distinguish between HDL and VHDL as far as their immunological antigenicity is concerned, but A-protein is clearly different antigenically from B-protein.

a_2 or pre-β lipoproteins (VLDL): a_2-lipoproteins precipitate under the following conditions[185]: 18% ethyl ether (pH 5.2; ionic strength, 0.09; protein content, 1.6%), 2.0–2.4 M ammonium sulfate (pH 7.0, protein content, 1%), 0.0065 M Rivanol aqueous solution (pH 8.0), 0.6 M perchloric acid, 0.15 M trichloroacetic acid. Also, it precipitates in reaction to dextran sulfate.

Not only is the amino acid composition of its peptide moiety unknown but also the relation to B-protein is not clear. However, a specific apoprotein of the a_2-lipoproteins is often called C-protein, and it is said to contain 1.68% carbohydrate (hexose, 1.2%; hexosamine, 0.2%, sialic acid, 0.2% fucose, 0.08). It contains serine and threonine in the N-terminus and alanine in the C-terminus.

β-lipoproteins (LDL): β-lipoproteins precipitate under the following conditions[185]: 20% ethanol (pH 6.9; ionic strength, 0.09; protein content, 3%), 1.6–2.4 M ammonium sulfate (pH 7.0; protein content, 1%), 0.0065 M Rivanol aqueous solution (pH 8.0; protein content, 1%), 0.6 M perchloric acid, 0.15 M trichlor-acetic acid. They are precipitated also with various polyanions such as dextran sulfate, amylopectin, etc.

B-apolipoprotein (B-protein) is not as well known as A-protein. It has 1.8% carbohydrate content (hexose, 1.1%; hexosamine, 0.4%; sialic acid, 0.3%). B-protein is thought to consist of several subunits of the molecular weight of 100,000. It has glutamic acid in the N-terminus and serine in the C-terminus. The amino acid composition of B-protein clearly differs from that of A-protein.[50]

6. Hereditary Types

The hereditary polymorphisms of the serum lipoproteins are recognized in two independent systems: Ag-system and Lp-system. The hereditary types of the serum β lipoproteins were first reported by ALLISON and BLUMBERG.[3] Incidentally they found the presence of isoprecipitating antibody against human β lipoproteins in serum of the patient (C. de B.) who had received multiple blood transfusion. Ag(a+) is the one showing a precipitation reaction with the patient's serum, while Ag(−) is the one showing no precipitation reaction. In Orientals, approximately 70–91% are classified as Ag(a+). Since then, more complex systems have been reported such as Ag(r), Ag(x), Ag(y), and Ag (z).[10]

Also BERG reported on Lp-system of the serum lipoproteins.[7] Lp-system is tested by using the rabbit antiserum against human β-lipoproteins; namely, Lp(a+) and Lp(−). The Lp-system is apparently independent from the Ag-system.

7. Immunoelectrophoretic Characteristics

a_1-lipoproteins or A-protein are completely different immunologically from β-lipoproteins or B-protein. Immunogenicity (*in vivo* antigenicity) is definitely weaker in a_1-lipoproteins. A-protein is not completely identical to a_1-lipoproteins. Cross reaction is observed between the anti-A-protein and the anti-a_1-lipoproteins. In fresh serum, only one a_1-lipoprotein (a LP$_A$) exists immunologically. But after ultracentrifugation, storage, freezing and thawing, two kinds of a_1-lipoproteins (a LP$_A$, a LP$_B$) are observed. a LP$_A$ is fast-moving and a LP$_B$ is slow-moving. There is usually only partial identity immunologically between the two. Only a LP$_A$ is detected in HDL$_2$, while both a LP$_A$ and a LP$_B$ are detected in HDL$_3$.[50]

Since some lots of anti-β-lipoproteins do not react with a phospholipid-B protein complex, neutral fat in β-lipoproteins may play an important role as haptene. At the present, it is thought that, considered from the immunological standpoint, β-lipoproteins are all of one kind.[50] Under normal conditions, there is a rather large amount of A-protein and B-protein in pre-$\beta(a_2$-$)$ lipoproteins. However, it is said that C-apoprotein isolated from pre-β lipoproteins does not react with anti-a_1-lipoproteins and anti-β-lipoproteins. There are many points which are not clear with regard to the immunological characteristics of C-protein and pre- β lipoproteins.[50]

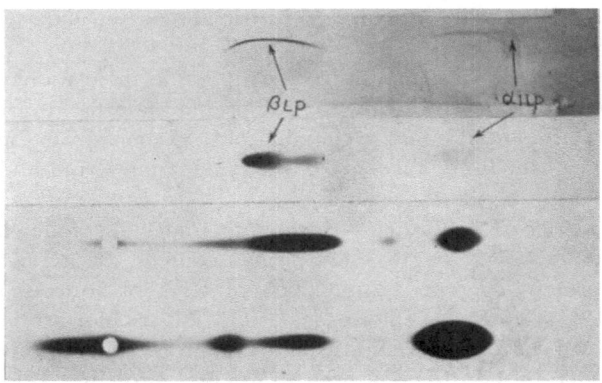

Fig. 50 Agarose gel electrophoretic patterns of the serum lipoproteins.
The upper three patterns are of nephrotic serum, and the lowermost pattern is of normal serum. The upper two are the immunoelectrophoretic and agarose gel electrophoretic patterns stained with Oil Red O, showing clearly separated a_1-lipoprotein and β-lipoproteins. The lower two agarose gel electrophoretic patterns are stained with Amido Black 10B.

When immunoelectrophoretic methods are used, it is necessary to employ agarose gel as a supporting medium. If agar gel is used instead of agarose gel, the reaction of LDL to agaropectin causes precipitation. It then becomes difficult to distinguish this physicochemical precipitation from immunological reaction.

a_1-lipoproteins become fast-moving in agarose gel (or in agar gel), and are usually observed in the albumin zone. However, when a_1-lipoproteins are contained in large amounts, their mobility becomes relatively slow. On storage, their mobility increases remarkably and then may be migrated at the pre-albumin zone. Thus, the a_1-lipoproteins vary greatly in their electrophoretic mobility. When fresh serum is used, their precipitation line is observed as a rather symmetric arc, but it is frequently observed as a straight line extending from the albumin zone to the pre-albumin zone.

β-lipoproteins become slow-moving when they increase in abnormal quantity, as is the case of a_1-lipoproteins. However, they usually are observed over a wide range, from the a_2 to β zones. Because their molecular weight is large, they hardly diffuse out, and are observed as an almost straight line, mostly within the a_2 zone. The precipitation line of β-lipoproteins usually appear characteristically milky white, because of their high lipid content. Therefore, they are easily distinguished from other protein lines. When agar gel is used, the agar gel itself becomes turbid, and the precipitation lines are difficult to be observed.

When it is difficult to identify them, lipoproteins may be made clearer by using specific anti-sera or by lipid staining (Fig. 50).

8. Biological Characteristics
The normal values of plasma lipids and lipoproteins differ greatly among many reports.

Generally, their normal values may vary depending on the differences of age, sex, environment, food and race.   Therefore, it is necessary to know the normal values characteristic for the population to be studied with a certain analytical method to be used.   As reference, data relevant to this subject is shown in Tables 12, 13, 14, 15.   The plasma lipids consist of free fatty acids, neutral fat (triglyceride), phospholipids, and cholesterol. They are water-insoluble and so are combined with strongly polar proteins, forming a lipid-protein complex.   Lipid-protein complexes are necessary for transporting lipids and fat-soluble vitamins, especially carcinoid and vitamin E.

Free fatty acids and neutral fat are the two which change greatly in various physiological conditions.   In contrast, cholesterol and phospholipid concentration in serum do not vary significantly.   Phospholipids have strong surface activity and are thought to increase colloidal stability of the lipid-protein complexes.   On the other hand, free fatty acids make up only 2 to 5 % of the total lipid content in the serum.   But they have a great turnover and as much as 25 grams of free fatty acids are exchanged in one hour.

Table 12   Normal values of plasma lipids and lipoproteins by age group (from FREDERICKSON et al.).

Age	Total Cholesterol (mg/dl)	Neutral fat (mg/dl)	Pre-β Cholesterol (mg/dl)	β-Cholesterol (mg/dl)	α-Cholesterol (mg/dl) male	female
0–19	120–230	10–140	5–25	50–170	30–65	30–70
20–29	120–240	10–140	5–25	60–170	35–70	35–75
30–39	140–270	10–150	5–35	70–190	30–65	35–80
40–49	150–310	10–160	5–35	80–190	30–65	40–85
50–59	160–330	10–190	10–40	80–210	30–65	35–85

Table 13   Normal values of plasma lipids in Japanese (from YAMAGATA et al.).

Total lipids	500–700
Neural fat	40–110
Total cholesterol	150–200
Free cholesterol	30– 70
Cholesterol ester	100–150
Phospholipid	10– 15
Free fatty acids	(400–600 μmEq/L)

Unit: mg/dl

Table 14   Normal values of plasma lipoproteins (from LINDGREN and NICHOLS).

Age	Sex	S_f 100–400	S_f 20–100	S_f 12–20	S_f 0–12	HDL$_2$	HDL$_3$
17–29	male	37±43	75±41	40±21	322±86	37±28	217±40
	female	9±14	44±29	30±16	283±68	80±41	228±38
30–39	male	51±64	91±54	51±23	355±84	36±28	219±42
	female	13±17	51±36	41±22	324±86	81±45	235±38
40–49	male	66±91	107±66	57±23	380±84	37±28	226±50
	female	18±24	65±51	42±21	346±67	89±53	241±43
50–65	male	58±70	103±58	56±24	383±75	42±32	224±51
	female	32±37	77±48	93±36	437±40	117±66	270±54

Unit: mg/dl

Table 15   Changes of serum lipoproteins (from YASUGI et al.).

Age	Sex	S_f 0–12	S_f 12–20	S_f 20–100	S_f 100–400
20–30	male	239±44	29± 7	31±19	9±14
	female	206±26	13± 6	23±22	5± 8
30–40	male	255±57	28±10	47±25	7± 9
	female	233±34	21±5	30±17	5± 5
40–50	male	271±45	35±20	70±14	18±14
	female	287±57	30± 8	62±26	11± 9
50–60	male	291±51	41±12	88±38	18±16
	female	300±73	35±16	90±29	28±26
60–70	male	251±50	27±14	44±25	7± 9
	female	295±71	44±17	58±38	13±20

Unit: mg/dl

It is the serum albumin that carries this largeam ount of free fatty acids. When they reach the liver, muscles, heart and other organs, they are supplied to tissues, leaving the albumin behind. Postprandially, 50 to 90% of body energy source depends on free fatty acids coming from fatty tissues, so they are important also as a source of energy.

There are two kinds of neutral fat in plasma, that is, exogenous (dietary) and endogenous. Since the lipids in foods are supplied mostly in the form of neutral fat, it may be said that neutral fat has a primary importance in lipid metabolism. Neutral fat metabolism is summarized in Fig. 51. Fatty acids derived from food digestion are absorbed, then synthesized into neutral fat in the intestinal epithelial cells. They become chylomicrons that pass through the thoracic duct and flow into the major circulatory system. Chylomicrons are degraded by the lipoprotein lipase contained in the blood capillary walls. As a result, the separated fatty acids are added into the plasma fatty acid pool, and immediately combine with albumin. Chylomicrons reach the fatty tissue, liver, heart and other body organs, where they are hydrolyzed and stored as esterified fatty acids. On the other hand, the neutral fat mainly synthesized in the liver ("endogenous" triglyceride) is circulated in a form of pre-β lipoproteins. Neutral fat is not only synthesized and stored in fatty tissues but also hydrolyzed by hormone-sensitive lipase, supplying free fatty acids to the body pool.[39,50,71]

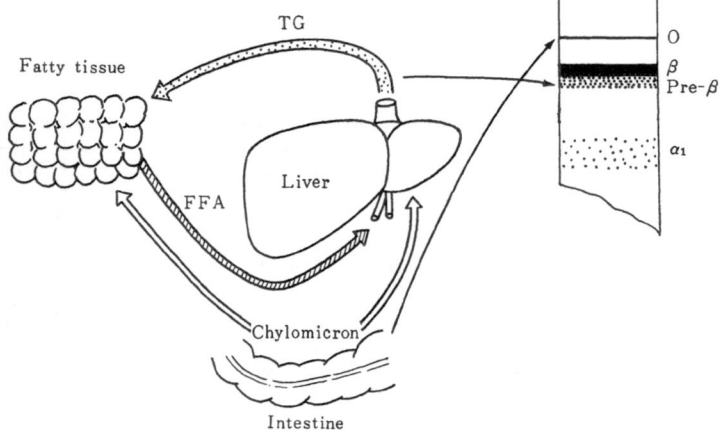

Fig. 51   Metabolism of the plasma neutral fat.

9.   Clinical Abnormalities

Congenital hypo- or a-lipoproteinemia
1)   Analpha-lipoproteinemia, Tangier disease
2)   A- or hypo-β-lipoproteinemia

Familial hyperlipoproteinemia
1)   Type I (Hyperchylomicronemia)
2)   Type II (Hyperbeta-lipoproteinemia)
3)   Type III (Hyperbeta-and pre-beta-lipoproteinemia)
4)   Type IV (Hyperpre-beta-lipoproteinemia)
5)   Type V (Hyperpre-beta-lipoproteinemia and hyperchylomicronemia)

Acquired hypolipoproteinemia
1)   Severe hepatocellular damage
2)   Obstructive jaunce (especially the decrease of α_1-lipoporteins)
3)   Malignancies

Acquired hyperlipoproteinemia: As seen in the familial abnormalities, this one is also divided into five types, but among the following diseases, Types II, III, and IV occur most frequently.

1)   Diabetes mellitus
2)   Nephrotic syndrome
3)   Pancreatitis
4)   Acute alcohol intoxication
5)   Hypothyroidism
6)   Glycogen-storage disease
7)   Immunoglobulinopathies
8)   Pregnancy
9)   Miscellaneous

REFERENCES

1) ADELSTEIN, S. L. and VALEE, B. L.: New Engl. J. Med., **265**, 892, 1961.
2) ALLANSMITH, M., McCLELLAN, B. H., BUTTERWORTH, M. and MALONEY, J. R.: J. Ped., **72**, 276, 1968.
3) ALLISON, A. C. and BLUMBERG, B. S.: Lancet I, 634, 1961.
4) BAYANI-SIOSON, P. S., LOUCH, J., SUTTON, H. E., NEEL, J. V., HORNE, S. L. and GERSCHOWITZ, H.: Am. J. HUM, Genet., **14**, 210, 1962.
5) BENNHOLD, H.: in "Protides of the Biological Fluids" (edited by PEETERS, H.), Elsevier, Amsterdam, 1962.
6) BENNICH, H. and JOHANSSON, S. G. O., Novel Symposium III, pp. 199, ALMQUIST and WISKSELL, Stockholm, 1967.
7) BERG, K.: Acta path. microbiol. scand, **59**, 369, 1963.
8) BLOMBÄCK, B., BLOMBÄCK, M., HENSCHEN, A., HESSEL, B., IWANAGA, S., and WOODS, K. R.: Nature, **218**, 130, 1968.
9) BLOMBÄCK, B., BLOMBÄCK, M. and SEARLE, J.: Biochem. et Biophys. Acta, **74**, 148, 1963.
10) BLUMBERG, B. S., ALTER, H. J., RIDDELL, N. M. and ERLANDSON, M.: Vox Sang., **9**, 128, 1964.
11) BLUMBERG, B. S., and MARTIN, J. R.: J. A. M. A., **203**, 180, 1968.
12) BOOTH, M., DIXON, P. F., GRAY, C. H., GREENWAY, J. M. and HOLNESS, N. J.: J. ENDOCRIN. **23**, 25, 1961.
13) BOROS, T. and RAPP, H. J.: Science, **150**, 505, 1965.
14) BOYCLEN, S. V. and SORKIN, E.: Immunology, **4**, 244, 1961.
15) BRAND, E., KASSEL, B. and SAIDEL, L. J.: J. Clin. Invest., **23**, 439, 1944.
16) BÜRGI, W. and SCHMID, K.: J. Biol. Chem., **236**, 1066, 1961.
17) BYWATERS, W. W.: Biochem. Z., **15**, 322, 1909.
18) CARBONARA, A. O. and HEREMANS, J. F.: Arch. Biochem. Biophys., **102**, 137, 1963.
19) CEPPELLINI, R.: Atti. Assoc. genet. ital., **12**, 3, 1967.
20) CHAN, P. C. Y. and DEUTSCH, H. F.: J. Immun., **85**, 37, 1961.
21) CHRISTENSEN, L. K. and LITONJUA, A. D.: J. Clin. Endocr., **21**, 104, 1961.
22) CLEM, L. W. and SMALL, P. A.: Federation Proc., **25**, 437, 1966.
23) CLEVE, H., KIRK, R. L., PARKER, W. C., BEARN, A. G., SCHACHT, L. E., KLEINMAN, H. and HOLSFALL, W. R.: Am. J. Human Genet., **15**, 368, 1963.
24) COHEN, S.: Biochem. J., **89**, 334, 1963.
25) COHEN, S. and MILSTEIN, C.: Advanc. Immunol., **7**, 1, 1967.
26) COHEN, S. and PORTER, R. R.: Adv. Immunol., **4**, 287, 1964.
27) DAUGHADAY, W. H.: J. Clin. Invest., **37**, 511, 1958.
28) DAUGHDAY, W. H., ADLER, R. E., MANZ, I. K. and ROSINSKI, D. C.: J. Clin. Endocr., **22**, 704, 1962.
29) DAVIE, E. W. and RATNOFF, O. D.: The Proteins, Vol. III. (ed. by H. NEURATH), Academic Press, New York, 1965.
30) DeMOOR, P., HEIRWEGH, K., HEREMANS, J. F. and DeCLERK-RASKIN, M.: J. Clin. Invest., **41**, 816, 1962.
31) DEUTSCH, H. F. and MORTON, J. I.: J. Biol. Chem., **231**, 1107, 1958.
32) DRAY, S.: Science, **132**, 1313, 1960.
33) EDELMAN, G. M. and GALLY, E.: Proc. Nat. Acad. Sci., **51**, 846 1964.
34) EDELMAN, G. M. and POULIK, M. D.: J. Exp. Med., **11**, 861, 1961.
35) FAHEY, J. L.: J. Immunol., **91**, 448, 1963.
36) FAHEY, J. L.: J. Immunol., **91**, 438, 1963.
37) FAHEY, J. L. and LAWLER, S. D.: J. Natn. Cancer Inst., **27**, 973, 1961.
39) FARBER, E.: Advance in Lipid Research **5**, 119, 1967.
40) FEINSTEIN, A. and ROWE, A.: Nature, **205**, 147, London, 1965.

41) FERREIRA, H. C. and MURAT, L. G.: Brit. J. Haemat, **9**, 299, 1963.
42) FESSAS, P., LEUKOPOULOS, D. and KALTSOYA, A.: Biochem. et Biophys. Acta., **124**, 430 1966.
43) FINLAYSON, J. S.: Fibrinogen, (ed. by LAKI), M. Dekker, New York, 1968.
44) FLEISCHMAN, J. B., PAIN, R. H. and PORTER, R. R.: Arch. Biochem. Biophys. Suppl. **1**, 174, 1962.
45) FOSTER, J. F.: in "The Plasma Proteins", (ed. by F. W. PUTNAM), Academic Press, New York 1960.
46) FRANEK, F.: Biochem. Biophys. Res. Common., **4**, 28, 1961.
47) FRANKLIN, E. C.: J. Clin. Invest., **39**, 1933, 1960.
48) FRANKLIN, E. C.: Progr. Allergy, **8**, 58, 1964.
49) FREDERICKSON, D. S. and GORDON, R. S.: Physiol. Rev., **38**, 585, 1958.
50) FREDERICKSON, D. S., LEVY, R. I., and LEES, R. S.: New Engl. J. Med., **276**, 34, 94, 148, 215, 273, 1967.
51) FUNDENBERG, H. H.: Rev. Microbiol., **19**, 301, 1965.
52) FUDENBERG, H. H. and FUDENBERG, B. R.: Science, **145**, 170, 1964.
53) GANROT, P. O.: Scand. J. Clin. Lab. Invest., **21**, 177, 1968.
54) GANROT, P. O. and SCHERSTEN. B.: Clin. Chem. Acta., **15**, 113, 1967.
55) GELL, P. G. H. and COOMBS, R. R. A.: Clinical Aspects of Immunology, 2nd ed., DAVIS Co. Philadelphia, 1964.
56) GERBARG, D. S.: in "Serum Proteins and The Dysproteinemias", (ed. by SUNDERMAN), Lippincott, Co., Philadelphia, 1964.
57) GIBLETT, E. R.: Vox Sang., **6**, 513, 1961.
58) GIBLETT, E. R. and BROOKS, L. E.: Nature, **197**, 676, 1963.
59) GITLIN, D., BOESMAN, M., SCHMID, K. and VUOPIO, P.: Science, **160**, 538, 1968.
60) GITLIN, D., SCHMID, K., EARLE, D. P. and GIVELBER, H..: J. Clin. Invest., **40**, 820, 1961.
61) GOFMAN, J. W., DELALLA, O. F. GLAZIER, F., FREEMAN, N. K., LINDGREN, F. T. and NICHOLS, A. V.: Plasma, 2, 413, Milano, 1954.
62) GOLDSTEIN, A.: Pharmacol. Rev., **1**, 102, 1949.
63) GOODMAN, D. S.: J. Am. Chem. Soc., **80**, 3892, 1958.
64) GREEN, N. M.: Advanc. Immunol., **11**, 1, 1969.
65) GREY, H. M. and KUNKEL, H. G.: J. Exp. Med., **120**, 253, 1964.
66) GRUBB, R. and LAURELL, A. B.: Acta Pathol. Microbiol. Scand., **39**, 390. 1956.
67) GUSTAFSON, A., ALAUPOVIC, P. and FURMAN, R. H.: Biochem. Biophys. Acta, **84**, 767, 1964.
68) HABER, E.: Proc. Nat. Acad. Sci., U.S.A., **52**, 1099, 1964.
69) HALL, C. E. and SLAYTER, H. S.: J. Biophys. Biophyem. Cytol., **5**, 11, 1959.
70) HARBOE, M., MANNIK, M. and KUNKEL, H. G.,: J. Exp. Med.: **116**, 719, 1962.
71) HAVEL, R. J.: Advances in Internal Medicine **15**, 117, 1969.
72) HAVERBACK, B. J., DYCE, B., BUNDY, H. F., WIRTSCHAFTER, S. K. and EDMONDSON, H.: J. Clin. Invest., **41**, 972, 1962.
73) HEINER, D. C., SAHA, A. and ROSE, B.: Fed. Proc., **28**, 766, 1969.
74) HEREMANS, J. F.: Les Globulines Sérique de Systéme Gramma Brussels, Arsica, 1960.
75) HIRSCHFELD, J.: Acta Path. Microb. Scand., **47**, 160, 1959.
76) HIRSCHFELD, J.: Progress of Allergy, 6, 155, 1962.
77) HIRSCHFELD, J. and NILSSON, B. A.: Acta Pathol. Microbiol. Scand., **56**, 471, 1962.
78) HONG, R. and NISONOFF, A.: J. Biol. Chem., **240**, 3883, 1965.
79) HUGHES, W. L. Jr.: J. Am. Chem. Soc., **69**, 1836, 1947.
80) INGBAR, S. H.: Endocrinology, **63**, 256, 1958.
81) INOUE, K. and NELSON, R. A. Jr.: J. Immunol., **96**, 386, 1965.
82) ISHIZAKA, K.: Progr. Allergy, **7**, 32, 1963.
83) ISHIZAKA, K., ISHIZAKA, T. and HORNBECK, M.: J. Immunol., **97**, 75, 1966.
84) ISHIZAKA, K., ISHIZAKA, T. and LEE, E. H.: J. Immun., **95**, 771, 1965.
85) ISLIKER, H.: Helv. Med. Acta, **25**, 41, 1958.
86) JACOT-GUILLARMOD, H. and ISLIKER, H.: Vox Sang., **7**, 675, 1962.
86) JACOT-GUILARMOD, H. and ISLIKER, H.: Vox Sang., **7**, 675, 1962.
87) JAYLE, M. F. and BOUSSIER, G.: Bull Soc. Chem. Biol., **36**, 959 1954.

88) JOHANSSON, S. G. O. and BENNICH, H.: Immunology 13, 381, 1967.
89) JENCKS, W. P., DURRUM, E. L. and JETTON, M. R.: J. Clin. Invest., 34, 1437, 1955.
90) KALLEE, E.: Helv. Med. Acta, 30, 150, 1963.
91) KAWAI, T.: Bull. Univ. Miami Sch. Med., 16, 89, 1962.
92) KAWAI, T., HASUNUMA, S., AOKI, T., SATO, K. YOSHIDA, N., and OHSHIMA, S.: Physico-Chemical Biology (Jap.), 13, 24, 1968.
93) KAWAI, T. and HASUNUMA, S.: Protides of the Biological Fluids, vol. 19, pp. 35, Pergamon Press, Oxford and New York, 1972.
94) KELLER, R.: J. Immun., 94, 532, 1966.
95) KING, T. P.: J. Biol. Chem., 236, PC 5, 1961.
96) KLOTZ, I. M.: in "The Proteins" Vol. 1, ed. by NEEURATH and BAILEY, Acad. Press, New York 1953.
97) KOHN, J.: Nature (London), 189, 312, 1961.
98) KORNGOLD, L.: Progress in Clinical Pathology, Vol. 1, Chapter 9, Grune & Stratton, New York, 1966.
99) KORNGOLD, L. and LIPARI, R.: Cancer, 9, 262, 1956.
100) KUNKEL. H. G.: in "Plasma Proteins" (ed. by F. W. PUTNAM), Academic Press, New York, 1960.
101) KUNKEL, H. G. and PRENDERGAST, R. A.: Proc. Soc. Exp. Biol. Med., 122, 910, 1966.
102) LAKI, K.: Fibrinogen, M. Dekker, New York, 1968.
103) LANCHANTIN, G. F., FRIEDMAN, J. A., De GROOT, J. and MEHL, J. W: Biol. Chem., 238, 238, 1963.
104) LAURELL, C. B.: "The Plasma Proteins" (ed. by F. W. PUTMAN), Acad. Press, New York, 1960, Scand. J. Clin. Lab. Invest., 17, 271, 1965.
105) LEES, R. S. and HATCH, F. T.: J. Lab. Clin. Med., 61, 518, 1963.
106) LENOX, E. S. and COHN, M.: Ann. Rev. Biochem., 36, 365, 1967.
107) LEPOW, I. H.: in "Immunological Diseases" (ed. by M. SAMTER), Little Brown, Co., Boston, 1965.
108) LEVY, R. I. and FREDERICKSON, D. S.: J. Clin. Invest., 44, 426, 1965.
109) LINDGREN, F. T. and NICHOLS, A. V.: in "Plasma Proteins", (ed. by F. W. PUTNAM), Academic Press, New York, 1960.
110) LoBUGLIO, A. F., CORTAN, R. S. and JANDL, J. H.: Science, 158, 1582, 1968.
111) LOEWY, A. G., DUNATHAN, K., KRIEL, R. and WOLFINGER, H. L. Jr.: J. Biol. Chem., 236, 2625, 1961.
112) LUNDGREN, H. P. PAPPENHEIMER, A. M. and WILLIAMS, J. W.: J. Am. Chem. Soc., 61, 533, 1939.
113) MANGALIK, A., ALDOWS, E. W., WAGASTAFF, B. and HEINER, D.: American Pediatric Society Meeting, Atlantic City, 1966.
114) MANNIK, M. and KUNKEL, H. G.: J. Exp. Med., 116, 859, 1962.
115) MARDER, V. J.: Fibrinogen, (ed. by LAKI), M. Dekker, New York, 1968.
116) MARR, A. G. M., OWEN, J. A. and WILSON, G. S.: Biochem. Biophys. Acta, 63, 276, 1962.
118) MARTENSSON, L. and KUNKEL, H. G.: J. Exp. Med., 122, 799, 1965.
119) MARTENSSON, L., van LOGHEM, E., MATSUMOTO, H. and NIELSON, J.: Vox Sang. 11, 393, 1966.
120) MARTIN, C. M.: Am. J. Med. Sci., 244, 334, 1962.
121) MARTIN, C. M. and JANDL, J. H.: Clin. Invest., 38, 1024, 1959.
122) MCKEE, P. A., ROGERS, L. A., MARLER, E. and HILL, R.: Arch. Biochem. Biophys., 116, 271, 1966.
123) MEHL, J. W., O'CONNELL, W. and DeGROOT, J.: Science, 145, 821, 1964.
124) MERSKEY, C., KLEINER, G. J., and JOHNSON, A. J.: Blood, 28, 1, 1966.
125) MIALE, J. B.: Laboratory medicine—HEMATOLOGY, 3rd Ed. Mosby Co., St. Louis, 1967.
126) MILLER, F. and METZGER, H.: J. Biol. Chem., 240, 3325, 1965.
127) MORELL, A. G. and SCHEINBERG, I. H.: Science, 127, 588, 1958.
128) MUIR, W. A. and STEINBERG, A. G.: Hematology 4, 156, 1967.
129) MURAYAMA, M.: in "Serum Proteins and Dysproteinemias" (ed. by SUNDERMAN), Lippincott, Philadelphia, 1964.

130) Murphy, B. P. and Pattee, J.: J. Clin. Endocr., 23, 459, 1963.
131) Müller-Eberhard, H. J. and Biro, C. E.: J. Exp. Med., 118, 447, 1963.
132) Müller-Eberhard, H. J., Calcott, M. A. and Mardiney, M. R.: Fed. Proc., 23, 506, 1964.
133) Müller-Eberhard, H. J. and Nilsson, U. J. Exp. Med., 111, 217, 1960.
134) Myant, N. B. and Osorio, C.: J. Physiol., 152, 601, 1960.
135) Miyoshi, K.: J. Jap. Soc. Intern. Med. 57, 179, 1968.
136) Nagel, R. L. and Ranney, H. M.: Science, 144, 1014, 1964.
137) Neale, F. C., Aber, G. M. and Northam, B. E.: J. Clin. Path., 11, 206, 1958.
138) Nerstrøm, B. & Frederikson, W.: Acta. Pathol. Microbiol. Scand., 61, 474, 1964.
139) Neuhaus, O. W. and Liu, A.: Proc. Soc. Exp. Biol. Med., 117, 244. 1964.
140) Nilsson, U. and Müller-Eberhard, H. J.: Fed. Proc., 23, 506, 1964
141) Nishioka, K. and Linscott, W. D.: J. Exp. Med., 118, 767, 1963.
142) Nisonoff, A., Wissler, F. C., Lipman. L. N. and Woernley, D. L.: Arch. Biochem. Biophys., 89, 230, 1960.
143) Noelken, M. E., Nelson, C. A., Buckley, C. E. and Tanford, C.: J. Biol. Chem., 240, 218, 1965.
144) Notation for Human Immunoglobulin Subclasses, Bull. World Health Organ. 35, 953, 1966.
145) Onoue, K., Yagi, Y., Grossberg, A. L. and Pressman, D.: Immunochemistry, 2, 401, 1965.
146) Osbahr, A. J. Jr., Glandner, J. A. and Laki, K.: Biochem. Biophs. Research Communs., 13, 462, 1963.
147) Oudin, J.: J. cell. comp. Physiol., 67, Suppl. 1, 77, 1966.
148) Palmer, J. H., Mandy, W. J. & Nisonoff, A.: Proc. Natn. Acad. Sci. U.S.A., 48, 49, 1962.
149) Parke, J. A. C. and Aris, P. J.: Immunology, 7, 248, 1964.
150) Parker, W. C.: Thesis, Rockefeller Institute, New York, 1963.
151) Parker, W. C. and Bearn, A. G.: Science, 137, 854,: J. Exp. Med., 115, 83, 1962.
152) Pauling, L., Pressman, D. and Ikeda, C.: J. Am. Chem. Soc., 64, 3010, 1942.
153) Peetom, F. and Pondman, K. W., Vox Sang, 8, 605, 1063.
154) Pensky, J., Hinz, C. F. Jr., Todd, E. W., Wedgwood, R. J. and Lepow, I. H.: Fed. Proc. 23, 505, 1964.
155) Pillemer, L., Blum, L., Lepow, I. H., Ross, O. A., Todd, E. W. and Wardlaw, A. C.: Science, 120, 279, 1954.
156) Pillemer, L. and Ecker, E. E.: J. Biol Chem., 137, 139, 1941.
157) Pillemer, L., Hinz, C. F. Jr. and Wurz L.: Science, 125, 1244, 1957.
158) Poortmans, J. R. and Schmid, K.: J. Lab. Clin. Med., 71, 807, 1668.
159) Porter, R. R.: Biochem. J., 73, 119, 1959.
160) Porter, R. R.: in "The Structure of γ-Globulin & Antibody", (ed. by Gellhorn and Hirschberg), Columbia University Press, 1962.
161) Porter, R. R. and Weir, R. C.: J. Cell. Physiol., 67 (Suppl. 1), 51, 1966.
162) Poulik, M. D.: Biochem. Biphys. Acta, 44, 390, 1960.
163) Poulik, M. D.: in "Protides of the Biological Fluids" (ed, by H. Peeters), p. 170, Elsevier, Amsterdam, 1964.
164) Putnam, F. W.: in "The Proteins", (ed. by H. Neurath), Vol. 111, Academic Press, New York, 1965.
165) Reiss, A. M. and Plescia, O. J.: Science, 141, 812, 1963.
166) Remington, C.: Biochem. J. 34, 931, 1940.
167) Robbins, J.: in "Serum Proteins and the Dysproteinemias" (ed. by Sunderman and Sunderman), Lippincott Co., Philadelphia, 1965.
168) Robbins, J. B.: in, Molecular and Cellular Basis of Antibody Formation", (ed. by Sterzl.), Czechoslovak Acad. of Science, Prague, 1965.
169) Ropartz, C., Lenoir, J. and Rirat, J.: Nature, 189, 586, 1961.
170) Rotnfield, N. F., Frangione, B. and Franklin, E. C.: J. Clin. Invest., 44, 62, 1965.
171) Rowe, D. S. and Fahey, J. L.: J. Exp. Med., 121, 171, 185, 1965.
172) Saifer, A., Robin, M. and Ventrice, M.: Arch. Biochem. Biophys., 92, 409, 1961.
173) Sandberg, A. A., Slaunwhite, W. R. and Antoniades, H. N.: Recent Progr. Horm. Res., 13, 209, 1957.

174) SCANU, A.: Advances in lipid research, (ed. by KRITCHEVSKY and PAOLETTI), Academic Press, New York, 1965.
175) SCANU, A. and GRANDA, J. L.: J. Lab. Clin. Med., 64, 1002, 1964.
176) SCANU, A. and GRANDA, J. L.: Progress in Clinical Pathology, I, 398, 1965.
177) SCATCHARD, G., BATCHELDER, A. and BROWN, A.: J. Clin. Invest., 23, 458, 1944.
178) SCHADE, A.: REINHART, R. and LEVY, H.: Arch. Biochem., 20, 170, 1949.
179) SCHERAGA, H. A.: Protein Structure, Academic Press, New York, 1961.
180) SCHMID, K.: J. Am. Chem. Soc., 75, 60, 1953.
181) SCHMID, K.: Nature, 204; 75, 1964.
182) SCHMID, K. BINETTE, J. P. and KAMIYAMA, S.: Biochemistry, 1, 959, 1962.
183) SCHMID, K., POLIS, A. and TAKAHASHI, S.: Biochem. Biophys. Acta, 57, 48, 1962.
184) SCHULTZE, H. E., HEIDE, K. and HAUPT, H.: Naturwissenschaften, 48, 719, 1961.
185) SCHULTZE, H. E. and HEREMANS, J. F.: Molecular Biology of Human Proteins, Vol. 1, Elsevier Pub. Co., Amsterdam, 1966.
186) SCHULTZE, H. E., SCHMIDTBERGER, R. and HOUPT, H.: Biochem. Z. 329, 490, 1958.
187) SCHUR, P. H. and BEEKER, E. L.: J. Exp. Med., 118, 891, 1963.
188) SHETLAR, M. R.: Glycoproteins, Progress in Clinical Pathology, Vol. 1, 419, Grune & Stratton, New York, 1966.
189) SINGER, J. M., ALTMAN, G., GOLDENBERG, A. and PLOTZ, C. M.: Arthritis Rheum, 3, 515, 1960.
190) SMITHIES, O., CONNELL, G. E. and DIXON, G. H.: Am. J. Human Genet, 14, 14, 1962.
191) STERLING, K. and HEGEDUS, A.: J. Clin. Invest., 41, 1031, 1962.
192) STERZL, J.: In "Molecular and Cellular Basis of Antibody Formation, Czechoslovak Academy of Sciences, Prague, 1965.
193) SVEHAG, SEVEN-ERIC: J. Exp. Med., 119, 517, 1964.
194) SATO, T.: J. Jap. Soc. Legal Med. (Jap.); 8, 220, 1954.
195) SOFUE, S. and OBAYASHI, H.: Physico-Chemical Biology (Jap.), 12, 211, 1967.
196) TADA, T. and ISHIZAKA, K.: J. Immunol., 104, 377, 1970.
197) TERRY, W. D. and STEINBERG, A. G.: J. Exp. Med. 122, 1087, 1965.
198) TISELIUS, A.: Biochem. J., 31, 1464, 1937.
199) TOKITA, K. and SCHMID, K.: Nature, 200, 266, 1963.
200) TRITSCH, G. L., RATHK, C. E., TRITSCH, N. E. and WEIS, C. M.: J. Biol. Chem., 236, 3163, 1961.
201) URIEL, J.: Nature, 181, 999, 1958.
202) URIEL, J.: Ann. N. Y. Acad. Sci., 103, 956, 1963.
203) VAERMANN, J. P. and HEREMANS, J. F.: Science, 158, 647, 1966.
204) VAN FURTH, R., SCHUIT, H. anp HIJMANS, W.: J. Exp. Med., 122, 1173, 1965.
205) WARNER, R. C. and WEBER, I.: J. Biol. Chem., 191, 173, 1951.
206) WEBB, T. and GOODMAN, H. C.: Modern Trends in Immunology, vol. 2, (ed. by CRUICKSHANK and WEIR), Butterworths, London, 1967.
207) WHITNEY, P. L. and TANFORD, C.: Proc. Nat. Acad. Sci., 53, 524, 1965.
208) WIEME, R. J.: Lancet, 1, 830, 1960.
208) WIEME, R. J.: Lancet, 1, 830, 1960.
209) WHO: Genetics of the Immune Response, WHO techn. Rep. Ser. No. 402, 1968.
210) WHO: Bull. W. H. O. 33: 721, 1965.
211) YATES, F. E. and URQUHART, J.: Physiol., 42, 359, 1962.
212) YOUTH, W. J., HONG, R., SELIGMANN, M. Good, R. and KUNKEL, H. G.: J. Clin. Invest., 49, 1957, 1970.
213) ZIMMER, J.: Bull. soc. franç. dermatol. syph., 67, 616, 1960.

SECTION III

METABOLISM OF THE PLASMA PROTEINS

Chapter 10

General Survey of the Plasma Protein Metabolism

DYNAMIC EQUILIBRIUM OF THE PLASMA PROTEIN METABOLISM

The plasma proteins, like other constituents of the body, are continuously being lost and replaced by newly synthesized molecules, resulting in their characteristic homeostasis. Therefore, when plasma proteins included in the blood are being examined, only a part of their dynamic equilibrium may be momentarily apprehended. In order to understand the pathophysiology related to the plasma proteins, an effort should be made to see pathological alterations of each component dynamically.

To understand the metabolism of plasma proteins, the first requirement is to know about amino acids, the fundamental structural components of proteins. The metabolism of body proteins, including amino acids, is summarized in Fig. 52. Food protein constituents are absorbed into the blood circulation after digested into amino acids in the gastrointestinal tract under the action of digestive enzymes. The amino acids absorbed through the intestinal tract are mostly carried to the liver by way of the portal vein. In the liver, some of the amino acids transported from the intestinal tract are used for the synthesis of its own proteins. On the other hand, the amino acids synthesized in the

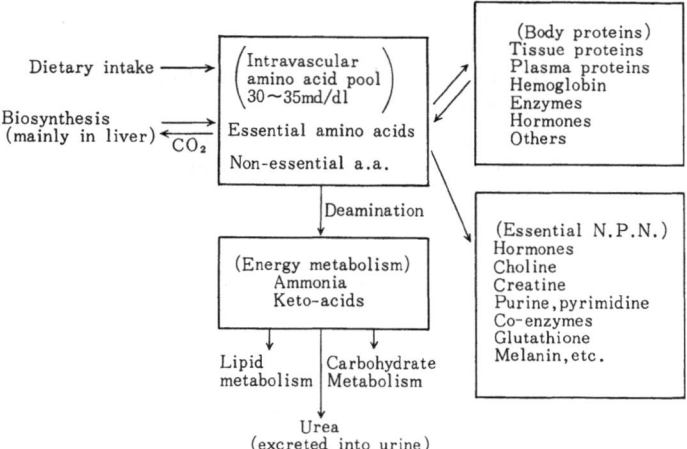

Fig. 52 Protein metabolism in the body.

liver are also added into the blood. Besides, the amino acids which are formed through the spontaneous destruction of the tissue proteins are also added into the blood. Thus, the blood amino acid pool comes from the following: 1) intestinal absorption, 2) biosynthesis, and 3) tissue destruction. In normal individuals this amounts to 30–35 mg/100 ml or 4–4.5 mg/100 ml as amino acid-nitrogen content. Amino acids constituting the blood amino acid pool are shown in Table 16.[143)]

Table 16   Amino acid concentration in human blood (from STEIN).

Amino acid	Concentration
Alanine	3.4–4.2
α-Aminobutyric acid	0.2–0.3
Arginine	1.5–2.5
Asparagine	1.0–1.4
Aspartic acid	0.01–0.07
Citrulline	0.5
Cystine	1.0–1.4
Glutamic acid	0.7–4.4
Glutamine	5.8–9.7
Glycine	1.1–2.8
Histidine	1.0–2.1
Isoleucine	0.9–1.8
Leucine	1.7–2.4
Lysine	2.2–3.0
Methionine	0.3–0.9
Ornicine	0.6–0.8
Phynylalanine	0.8–1.9
Proline	2.4–2.9
Serine	1.0–1.3
Threonine	1.3–3.1
Tryptophan	1.1–1.7
Tyrosine	1.0–1.5
Valine	2.7–3.4
Total	32.3–55.1

All the body proteins including plasma proteins are synthesized into the ones which are adequate for an individual need by obtaining necessary amino acids from the blood amino acid pool. Morever, in addition to body proteins, many non-protein nitrogenous substances essential for maintaining tissue activities, such as creatine, choline, and glutathion are also synthesized from the blood amino acid pool. The excess amino acids are degraded into ammonia and keto acids through deamination in the liver. The ammonia is excreted into urine after it has become urea, while the keto acids are used for lipid metabolism or carbohydrate metabolism after being converted into ketone bodies or glucose. Keto acids, eventually oxidized into carbon dioxide and water, yield energy in the form of ATP.

Such sulfur-containing amino acids as cystine, methionine and cysteine discharge sulfur in their metabolic process. They are eventually excreted into urine in the form of sulfate.

The greater part of nucleic acids which come from the decomposition of nucleoproteins are excreted into urine in the form of uric acid, after they have been destroyed into phosphoric acid and purines. Thus, most of the final metabolic products are excreted into urine.

Plasma proteins, like other body proteins, are widely distributed throughout the body by means of the blood stream, after they have been synthesized from the blood amino acid pool. Individual proteins have their characteristic life spans. They are destroyed spontaneously and some of them return to the amino acid pool, again. In addition, some of the plasma proteins are lost externally into various secretory fluids. In this way, the plasma proteins maintain a dynamic equilibrium. This dynamic state is kept under the control of four fundamental mechanisms: synthesis, degradation, internal bodily distribution, and external loss. The relation with these four mechanisms resembles the distillation process, as shown in Fig. 53.[76)]

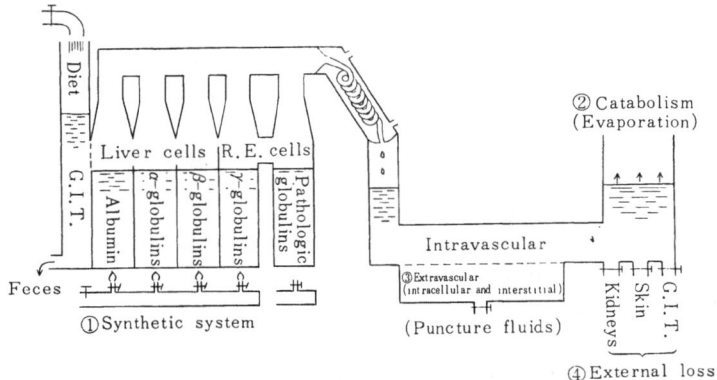

Fig. 53 A metaphoric representation of the metabolism of the plasma proteins (from KAWAI).
The metabolism of the plasma proteins is metaphorically represented by a distillation apparatus, indicating the four major factors; synthesis, catabolism, distribution and external loss.

The organs that are playing a chief role in the synthesis of plasma proteins are the liver and reticulo-endothelial cells. It is mainly the immunoglobulins that are synthesized in the reticulo-endothelial cells. These cells compare to the distillation tank in the figure. The absolute quantity of the individual plasma protein component synthesized daily is determined by the number of the cells contributing to its synthesis and the functional activity of the individual synthetic cell. This fact resembles that the amount of distilled water is determined by the number of distillation tanks and the size of the flame of the burners. The size of the flame of the burner is controlled by the proportion of oxygen used and the extent to which the burner cock is opened. These two are probably comparable to the hereditary factor and the feedback mechanism in the synthesis of proteins. If there is a smaller number of distillation tanks, as is the case in liver cirrhosis which causes a decrease of the synthetic cells, hypoproteinemia naturally occurs. Conversely, in myeloma where synthetic cells are markedly increased in number, hyperproteinemia may result. Moreover, even if synthetic activity of the cells is normal, hypoproteinemia may result when amino acids run short as in the case of starvation and malnutrition. This may resemble the fact in the Figure, that if the water to be distilled is not obtained from the faucet, no distilled water is obtainable. The catabolism (degradation) of plasma proteins in various body tissues or organs is equivalent to evaporation of distilled water, and if catabolism increases, hypoproteinemia may also occur. It seems that when the catabolism of tissue proteins is accelerated, the catabolism of plasma proteins is generally increased, too. Particularly, the catabolism of albumin is influenced by various hormones such as thyroid hormone and adrenocorticoids. If the synthesis does not overtake the catabolism thus accelerated, hypoproteinemia results.

Plasma proteins are present not only in the intravascular space but also, more widely, in the extravascular space. For example, in the case of the plasma proteins of comparatively small molecular weight such as albumin, transferrin, and a_1-glycoproteins, only about 40% of their total quantity in the body is present in the blood vessels. When edema suddenly occurs in a wide area, or when a great quantity of pleural fluids and/or ascites have accumulated, this causes a significant increase in the extravascular portion of the proteins accordingly and then hypoproteinemia results.

A small quantity of plasma proteins is continuously being lost normally into digestive fluids and urine. However, with the exception of excessive bleeding, the kidneys, gastrointestinal tract, and skin are the three sites through which abnormally great quantities of plasma proteins are lost externally. This external protein loss can be compared to ④ in the figure.

Thus, part of the plasma proteins are constatly being degraded and an equal quantity of plasma proteins is simultaneously being synthesized anew. In a healthy individual, degradation and synthesis of plasma proteins are balanced and a steady state is thus obtained. In order that this be maintained, there is a daily turnover of plasma proteins which may reach as much as 20–25 g a day. This amounts to approximately one tenth of the total quantity of plasma proteins in the body.

PRINCIPLE OF THE ANALYTICAL METHODS

The idea of dynamic equilibrium on plasma proteins, or "turnover" was introduced already by BORSOOK and KEIGHLEY, in 1935.[18] However, it was after the procedure of SCHOEHEIMER et al. using radioactive isotopes that this idea was experimentally verified.[132] Even at the present, the procedure of injecting radioactively labeled plasma proteins is used.

Labelling of plasma proteins by radioactive iodine: There are two methods of labelling plasma proteins by using radioactive isotopes. They are, 1) *in vivo* labelling and 2) *in vitro* labelling. With *in vivo* labelling, amino acids are first labelled. When the labelled amino acids are injected into the body of animals, they are trapped into plasma proteins which are synthesized in the body. Isotopes such as ^{15}N, ^{14}C, ^{35}S, and ^{3}H are used.[4,48] In the earlier days of research, this procedure was widely used. However, reproduceable results were not obtained by this method because, with *in vivo* labelling, the labelled amino acids are used again for synthesizing new proteins after the degradation of the labelled plasma proteins. On the other hand, *in vitro* labelling has come to be applied exclusively since the *in vitro* method for direct labelling radioactive isotopes (^{131}I) to plasma proteins was established.

Outline of the metabolic study using radioactive iodine-labelled plasma proteins: LUGOL's solution is administered orally to the subjects to be studied just prior to and throughout the period of observation in order to block the thyroid gland. Thus, ^{131}I freed by metabolism of plasma proteins is easily excreted into urine. A certain amount of the labelled proteins are injected intravenously, and serial blood samples are drawn to determine the plasma volume and to follow the daily plasma radioactivity and protein concentration over a period of about 3 to 4 weeks. Serial 24-hour urine samples are also collected. Moreover, other body fluids such as pleural fluid, ascites or fecal material are collected, if necessary.

Analysis of data and definition of various terms: When the metabolism of a particular protein is examined by the methods described, analysis is made on the following four assumptions: 1) radioactive iodine-labelled proteins are metabolized in the same manner as unlabelled native proteins; 2) [131]I which has been freed from the protein molecules is excreted rapidly and also, quantitatively, out of the body; 3) during the studying period, the subjects are in a steady state; 4) [131]I freed by metabolism is not reutilized for synthesizing new protein molecules *in vivo*. Up to the present, these four assumptions have been validated, if proceeded carefully. However, it is usually true that the subject under study metabolizes his own labelled albumin more slowly than that prepared from other individuals.[92]

Soon after injected intravenously, the labelled albumin (a) is mixed evenly with unlabelled native albumin (A) in circulating blood. Supposed that the labelled albumin does not leak out extravascularly in the mixing phase, the total circulating albumin and circulating plasma volume are calculated with the following formulae by measuring "specific radioactivity" (a/A) for the serum sample drawn in 10–15 minutes after its intravenous injection.[13,27,149]

1) Plasma volume (ml) $=\dfrac{\text{Total amount of radioactivity injected}}{\text{Radioactivity per ml of plasma}}$

2) Total circulating albumin (g) $=\dfrac{\text{Total amount of radioactivity injected}}{\text{Specific radioactivity of albumin}}$

   or $=$ Plasma volume (ml)
   $\times$ plasma albumin concentration (g/ml)

Later, plasma radioactivity gradually decreases because the labelled albumin diffuses extravascularly and is catabolized. Serial plasma and 24-hour urine samples are analyzed for their radioactivity, and their data are plotted on a semi-logarithmic graphic paper.[8,157] Other data are calculated as follows:

3) Total body radioactivity (cpm)
   $=$ total radioactivity injected (cpm)
   $-$ cumulative urinary excretion of radioactivity (cmp).

4) T 1/2 albumin (half-time survival in days) is calculated from the slope of the third portion of the plasma specific activity curve.

5) Fractional degradation rate $=\dfrac{0.693}{\text{T1/2 albumin}}$

6) Per cent intravascular radioactive albumin
   $=\dfrac{\text{plasma volume} \times \text{plasma radioactivity/ml} \times 100}{\text{total body radioactivity}}$

7) Total exchangeable albumin (g)
   $=\dfrac{\text{total circulating albumin}}{\text{percent intravascular radioactive albumin}}$

8) Albumin turnover (g per day) (Catabolic rate)
   $=$ total body albumin $\times$ fractional degradation rate

9) Daily degradation rate
   $=\dfrac{\text{urinary radioactivity on a given day}}{\text{radioactivity remaining in the body (midpoint of that day)}}$

Definition of the terms and expressions with regard to the above mentioned calculation are not absolutely uniform. The procedure for calculating the distribution rate of albumin between intra- and extravascular spaces tends to differ markedly among investigators. As

to procedures, BEEKEN et al. have considered the following four fundamental methods[9]:
1) the extrapolation method of STERLING[145], 2) the equilibration time method, 3) the
radioactivity distribution rate proposed by CAMPBELL et al,[19] and 4) the multicompart-
mental analysis of LEWALLEN et al.[82]   These differences come from the fact that calcula-
tions are made on various assumptions, by simplifying the models of albumin metabolism.

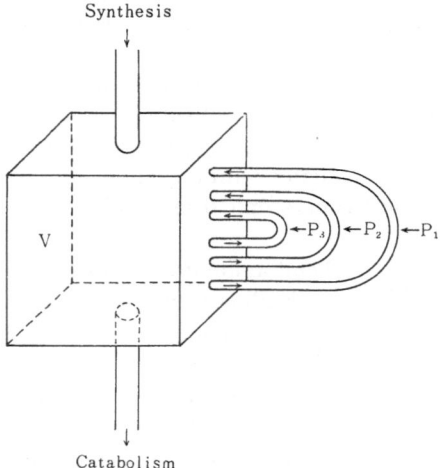

Fig. 54 "Multi-pipe model" for the analysis of plasma protein
metabolism.
Plasma proteins synthesized at the speed of Ks are released into
plasma (V... capacity) and catabolized at the speed of km, while
the plasma proteins remaining in the vessels are distributed outside
the vessels and again return into the vessels at different speeds passing
through various stages, P_1, P_2, P_3 .... P_z. Therefore, unlike the
classical two-compartment model, complicated elements are
further added.

Actually, they are more complicated.   Detailed descriptonis of these catabolism models
are found in SCHULTE and HEREMANS' monograph[133].   The actual albumin metabolism
is not so simple as the classical two-compartment model employed by CAMPBELL et al.[19]
It has been thought that the multi-pipe model of REEVE and BAILEY is clinically better
adapted[118], and, in fact, it is currently applied in many studies (Fig. 54).

Chapter 11

Synthesis of the Plasma Proteins

SITES OF SYNTHESIS OF THE PLASMA PROTEINS

1. Principle of the Analytical Methods

Various methods of demonstrating the sites of plasma protein synthesis have been reported. Fundamentally, after the amino acids labelled with a radioactive isotope are added to a particular system and incubated for a certain period, then radioactivity is detected in particular proteins which increase by synthesis in the system studied. This procedure was first introduced by PETERS and ANFINSEN.[110] That is, $^{14}CO_2$ was added to the liver section of chickens, and albumin was separated after a certain time in which radioactive carbon was found to exist. Thus, it was proved that albumin is synthesized in the liver.

Following a similar procedure, studies were made concerning various plasma protein components. However, in the early days of research, quite a number of experiments were carried out in regard to such electrophoretically-separated fractions as α-globulin, β-globulin, and others. Consequently, however, reliable results have become available through the progress of the procedures in purifying plasma proteins.

The methods for demonstrating the sites of plasma protein synthesis are classified roughly into *in vitro* and *in vivo* procedures.

In vitro procedures: There are four *in vitro* procedures, 1) the incubation and perfusion of organs which have been removed from the body; 2) the use of fresh tissues or organ sections for the same purpose; 3) the use of cell or tissue cultures; and 4) the use of microsomal fractions of cells.

The problem in the use of these procedures is that the protein components which are already present in cells before the experiment may be freed and be mistaken as newly synthesized protein. There is also a possibility of some proteins to form a complex with the protein to be studied. Therefore, it is necessary for sure to prove that a radioactive isotope has been incorporated into the polypeptide chain.[122]

In vivo procedures: The most direct procedure for determining the site of protein synthesis is the one reported by GRABAR et al.[52] It involves two species of animals having plasma proteins of different antigenicity. In this method, the organs of one animal are first transplanted to the other whose immunity has been depressed by strong irradiation, and then the plasma proteins of the latter animal are studied immunochemically.

In men, this procedure was adapted coincidentally on liver transplantation between two indivduals having haptoglobins of different hereditary types. The haptoglobin type of the donor was subsequently detected in the blood of the recipient. Thus, it was verified *in vivo* that haptoglobins are synthesized in the liver.[91] The sites of synthesis may be indirectly demonstrated from the fact that plasma proteins decrease after a particular organ is experimentally or surgically removed.   Also, the sites of protein synthesis may be assumed by a significant decrease of the particular proteins followed by diffuse necrosis or injury of a certain organ.   By clinical observations such as these, it had been supposed that many plasma proteins other than immunoglobulins were synthesized in the liver cells but *in vitro* procedures have served to confirm this assumption.

In addition to these procedures, fluorescent antibody techniques may be mentioned though they are not direct ways of demonstrating the sites of plasma protein synthesis. That is, the localization of particular plasma proteins may be determined by direct or indirect methods of fluoresent antibody techniques.[55,99]   Even if particular plasma proteins are demonstrated to be present with this method, it does not necessarily indicate the site of synthesis.   When a particular organ is known through other analytical methods to be the site of synthesis, fluorescent antibody techniques are useful to confirm the result at the cell level.

2.   Synthetic Organs

It has long been thought that the liver plays an important role in plasma protein synthesis, and it has been cofirmed gradually by the above mentioned procedures, especially in *in vitro* experiments.   In Table 17, the results on various plasma proteins are summarized.[133]   It may be thus concluded that the liver and reticuloendothelial system has a major role in plasma protein synthesis but the nerves, the heart, the muscles, the renal parenchyma and the endocrine organs are not involved.   As shown in Table 17, the liver mainly synthesizes the following proteins: albumin, α_1-acid glycoprotein, α_1-antitrypsin, haptoglobin, α_2-macroglobulin, Gc-globulin, transferrin, various lipoproteins, chylomicron, fibrinogen, prothrombin, factor VII, and C-reactive protein.   Moreover, ceruloplasmin and hemopexin are also thought to be synthesized in the liver.   Complement components and immunoglobulins are not synthesized at least in the healthy liver.[66,133]   Therefore, when the greater part of the liver cells are removed or destroyed as in hepatectomy or liver cirrhosis, there is a marked decrease of albumin, fibrinogen and lipoproteins in plasma. In severe liver damage, a further decrease of α- and β-globulins may result.

The second major site of plasma protein synthesis is the reticuloendothelial system. The reticuloendothelial cells are widely distributed throughout the system and involved chiefly in the synthesis of immunoglobulins.   Not all immunoglobulins are produced from reticuloendothelial cells.   Rather, a variety of cells are involved in their production.   For instance, IgG-immunoglobulin is synthesized mainly in mature plasma cells.[29,89]   It is also synthesized in the cells which are present in germinal centers of the lymph node. Moreover, most investigators believe that one plasma cell synthesizes the immunoglobulin of either the K or the L type, one or the other, while there are reports saying that the light chain of both the K and the L type is contained in one cell.[109]   There are some discrepancy among investigators with regard to the cells that synthesize IgM-immunoglobulin. Though it is produced in plasma cells, they may not be exactly the same as IgG-producing mature plasma cells.[160]   In fact, in WALDENSTRÖM's macroglobulinemia, the cells with atypical morphology, known as lymphocytoid plasma cells or plasmacytoid lymphocytes produce IgM-immunoglobulin.[37,138]   IgA-immunoglobulin is synthesized mainly in mature plasma cells.[21,89]   Plasma cells and similar immunoglobulin-producing cells are distributed widely in nearly all the tissues and organs.   However, the following organs are particularly rich in plasma cell: the spleen, the lymph nodes, the bone marrow, the lungs, the intestinal mucosa, the appendix, the tonsils, the interstitial tissues of various exocrine glands, and the hepatic sinusoids.   In normal conditions, immunoglobulins are produced chiefly in these organs.   The activity of individual organs may depend on the way how certain antigenic stimuli enter. For example, when the influenza virus enters through the tracheobronchial route, local production of immunoglobulins occurs very actively at the tracheobronchial mucosa, as compared to that in the case of percutaneous vaccination. Even in the organs, such as nervous tissues, which hardly produce any antibody globulin normally, immunoglobulins may be produced in pathological conditions.[94]

As shown in Table 17, $\beta_{1C/A}$-globulin, which is the third component of the complements, is produced in the reticulo-endothelial system, excluding the liver. However, it is not known which cells in the RE system are involved. With regard to lipoproteins, transferrin, and α_1-glycoproteins, the liver is thought to play a role but other organs may contribute for their synthesis.

Table 17 Synthetic sites for plasma proteins (modified from SCHULTZE and HEREMANS).

| | Organs including many R. E. cells | | | | | | Liver | Brain | Heart | Kidneys | Adrenals |
	Spleen	Lymph-node	Bone marrow	G. I. mucosa	Lung	Thymus					
Albumin	—	—	—	—	—	—	◎	—	—	—	—
α_1-Acid glycoprotein							◎				
α_1-Antitrypsin	—	—	—	—	—	—	○	—	—	—	
Ceruloplasmin	—						○				
Haptoglobin	—?	—	?	—	—	—	◎	—	—	—	—
α_2-Macroglobulin	—						○	—			
Gc-globulin		—					◎				
Transferrin	?—		?	○—	?	—	◎	—	—	—	—
$\beta_{1C/A}$-Globulins	○	○	○	—	○	—	—	—	—	—	—
Hemopexin	—?						?	—?			
Immunoglobulins (IgG, IgA, IgM)	◎	◎	◎	◎	○	○	—(○)	—	—	—	—
α_1-Lipoprotein							◎			?	
β-Lipoprotein and chylomicron	?	?	?	○	?		◎			?	
Fibrinogen							◎				
Prothrombin							◎				
Factor VII							◎				
C-reactive protein	—	—	—			—	◎			—	

◎ = Important and convincingly demonstrated synthesis
○ = Definite but probably quantitatively unimportant synthesis
? = Suggestive evidence but insufficient proof of synthesis
— = No demonstrable synthesis

MECHANISMS OF SYNTHESIS OF THE PLASMA PROTEINS

1. The Ultrastructure of the Cells Synthesizing the Plasma Proteins

Plasma proteins are synthesized mainly in the liver cells and plasma cells. When the ultrastructure is examined on electron microscopy, these cells are seen to have the same basic structures as the ones shown in Fig. 55. Detailed descriptions of these are found elsewhere,[133] and only outline of the subject is discussed in the following.

It is known that all cells synthesize some proteins and that ribosomes are the sites of synthesis. Therefore, every cell has ribosomes which are observed on light microscopy as cytoplasmic basophilia. However, not every cell has ribosomes in the same form.

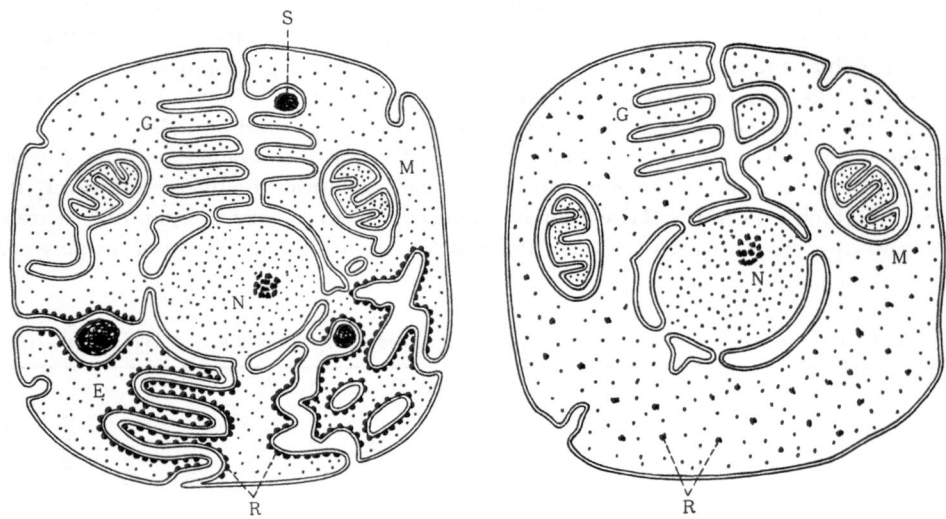

Fig. 55　The electron-microscopical diagrams of a secretory cell (left) and a non-secretory cell (right). N = nucleolus, M = mitochondria, G = GOLGI apparatus, S = secretory granules, R = ribosomes, E = endoplasmic reticulum.　In secretory cells, the ribosomes are attached to the endoplasmic reticulum, but they are free in nonsecretory cells.

Functionally and morphologically, they are divided into two types, secretory and non-secretory cells.

In secretory cells, the proteins synthesized in ribosomes are finally secreted out of the cells, while in non-secretory cells the proteins are retained mainly in their own cytoplasm. For example, in non-secretory cells included are normoblasts and reticulocytes which synthesize hemoglobin; epidermal cells which synthesize keratin and keratohyaline; and myocytes which synthesize actomyosin.　Moreover, rapidly proliferating cells usually have the similar morphological aspects to non-secretory cells as they require great quantities of cellular proteins for their own growth; for example, various embryonal cells, neoplastic cells and various blast cells.　These cells synthesize proteins for themselves, and the cell organels excreting proteins are poorly developed.　As a result, the ribosomes appear freely scattered in protoplasm on electron microscopy, and the endoplasmic reticulum is hardly seen.　Conversely, in secretory cells included are exocrine cells such as in the pancreas and salivary glands, thyroidal epithelial cells, and fibroblasts.　Plasma cells and liver cells also fall in this category.　In these cells, the cell organels necessary for transporting proteins are well developed and rich in endoplasmic reticulum.　This is due to the fact that the proteins synthesized in the ribosomes need to be excreted out of the cells rapidly.　Moreover, in many cases, the ribosomes are attached to the membranes of the endoplasmic reticulum, and form a so-called rough-surfaced endoplasmic reticulum. The proteins synthesized in the ribosomes are transported directly into the endoplasmic reticulum, and passing through it, are excreted either directly out of the cells or by way of the GOLGI apparatus.

Plasma cells and liver cells belong fundamentally to the secretory cell group. Neverthe-less, there are great differences between two.　As shown in Fig. 56, plasma cells have most markedly developed rough-surfaced endoplasmic reticulum.　By the fluorescent antibody

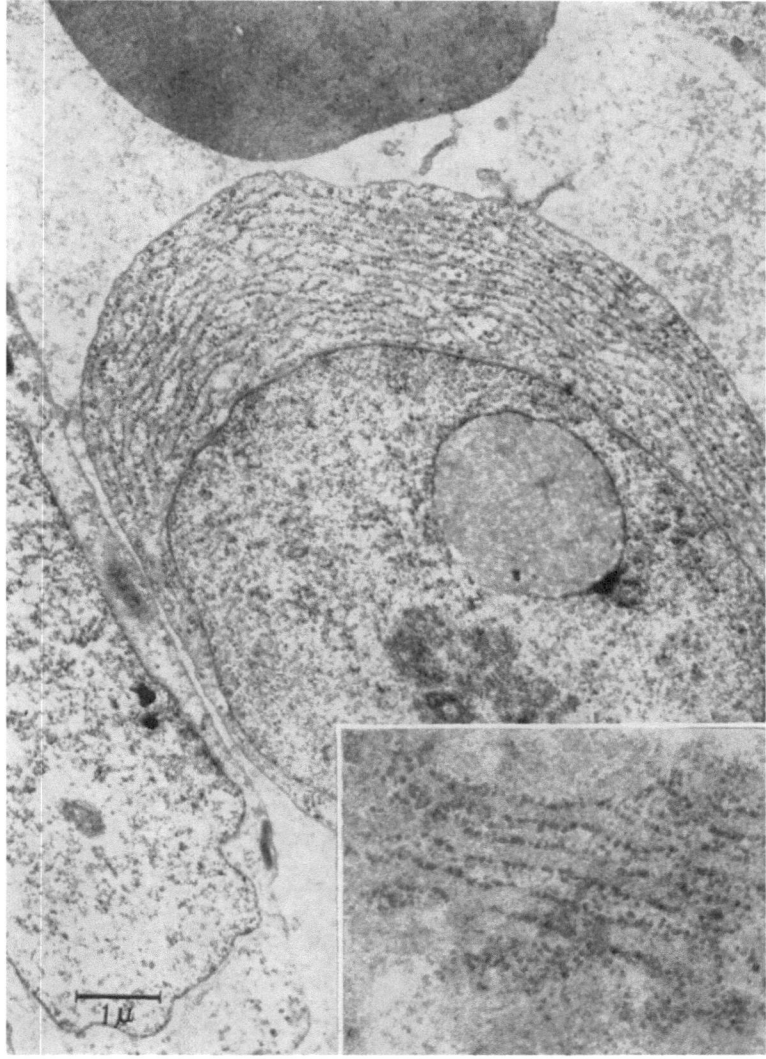

Fig. 56 Electron-microscopic photographs of a plasma cell. In the cytoplasm seen are the well-developed rough-surfaced endoplasmic reticulum showing a characteristic lamellar structure. In the nucleus, an inclusion body having a single limiting membrane is recognized. The right lower picture shows the close-up view of the rough-surfaced endoplasmic reticulum, containing a homogeneous material.

technique and the ferritin-antibody technique, demonstrated are immunoglobulins stored in the endoplasmic reticulum. On the other hand, liver cells show well developed mitochondria (Fig. 57). As has been described, the majority of plasma proteins (with the exception of complements and immunoglobulins) is synthesized in liver cells, but these plasma proteins are hardly detectable in their cytoplasm with fluorescent antibody techniques. Because, unlike immunoglobulins made in plasma cells, the proteins are likely to be released rapidly without being retained in the cytoplasm.

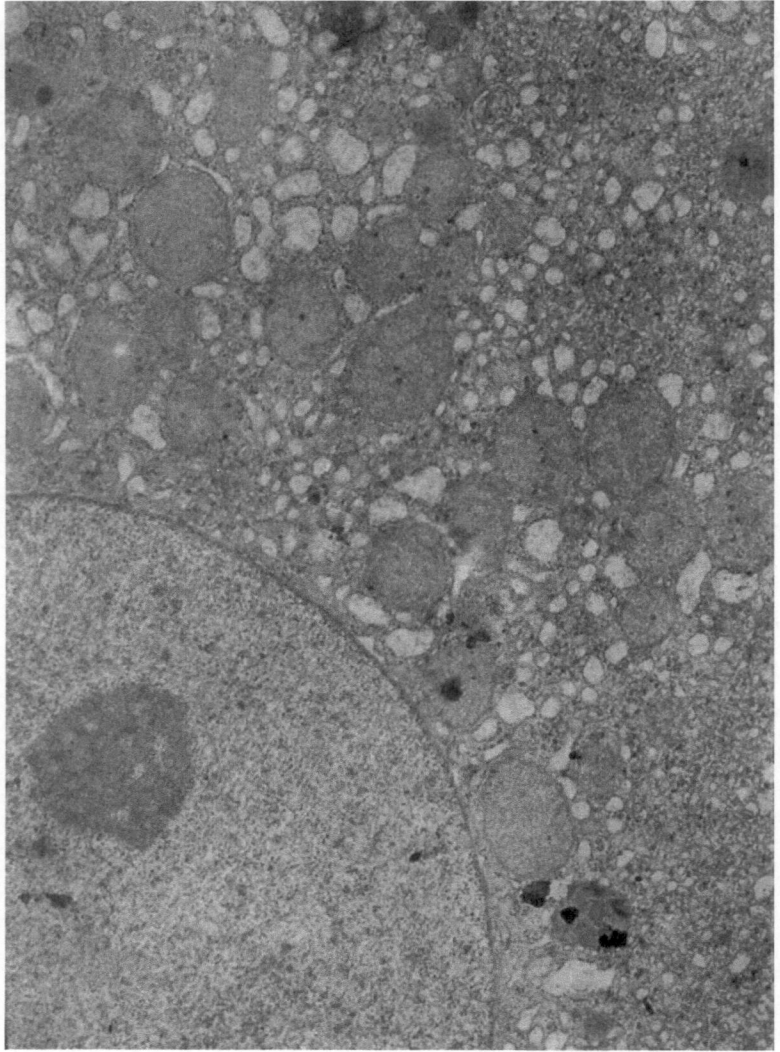

Fig. 57 Electron-microscopic photographs of a liver cell.
In the cytoplasm seen are many mitochondria and vacuole-like
structures. However, the rough-surfaced endoplasmic reticulum
is not apparent.

2.   The Synthetic Mechanisms of Proteins in the Cells

Ribosomes are the fundamental sites of protein synthesis, and of course, other cell
components are also closely related to the synthesis.   The four kinds of nucleic acid
present in the nucleus and the cytoplasm are involved, and various enzymes contribute to
their mechanisms.[60,133]

Nucleic acids are divided, according to their chemical structures, into DNA (deoxyri-
bonucleic acid) and RNA (ribonucleic acid).   DNA exists only in the nuclei and maintains hereditary information concerning protein synthesis, while RNA is present in the
nucleus and cytoplasm.   RNA is contained in cell nuclei by 10–20%, being distributed

chiefly in the nucleoli and chromatins.   The greater part of RNA (80–90%) is present
in the cytoplasm, and moreover, more than 50% exists in ribosomes.   Functionally, RNA
is divided into mRNA (messenger RNA), tRNA (transfer RNA, also called sRNA, soluble
RNA, adaptor RNA) and mitochrondrial RNA.   mRNA exists either in a free form or a
ribosome-bound form.

Fig. 58 is a model of the mechanisms of protein synthesis in cytoplasm.   As is shown in
WATSON–CRICK's model, DNA is thought to have a double-helical molecular structure.
On a nuclear division, DNA having the same structure (that is, the same genetic code)
as a mother nucleus, is transferred to daughter nuclei through the catalytic action of
RNA polymerase (WEISS).   In this way, mRNA produced in cell nuclei passes through
the endoplasmic reticulum and is distributed in the cytoplasm, particularly in the ribosomes.   In order for amino acids to be used in protein synthesis, it is necessary for them
first to be activated by reacting with ATP through the catalytic action of the amino acid
activating enzyme (aminoacyl-RNA synthetase), and to assume an aminoacyl-AMP-
enzyme complex.   Each enzyme is needed in the activation of individual amino acid, and
the activating enzymes for tryptophane, tyrosine, arginine, threonine and the like have
been isolated.

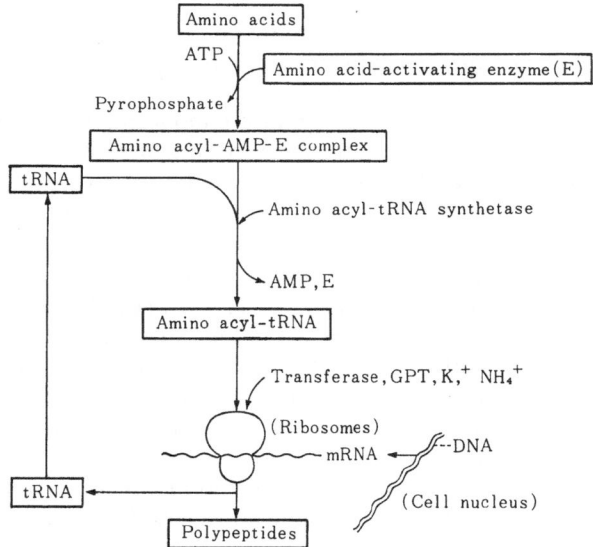

Fig. 58 A schematic representation of the protein biosynthesis in cell.

At the second stage of protein synthesis, the activated amino acids are bound to each
corresponding tRNA.   At this stage, also, the catalytic action of an enzyme, amino acyl-
tRNA synthetase, is required.   After the activated amino acids have been transferred to
each corresponding tRNA, the amino acid activating enzymes and AMP are released out
of the complex.

At the third stage, aminoacyl-tRNA forms peptide bonds one after the other in the ribosomes.   This is done in a fixed order according to the mRNA code.   It is thought that,
in reality, this occurs in polysomes which are the complex of several ribosomes and mRNA,
and that ribosomes translate genetic information while moving on a long chain of mRNA.

However, in the case of ribosomes attached to the endoplasmic reticulum, it is not known how this movement takes place. Thus, mRNA, which plays a fundamental role as a template in protein synthesis does not maintain its function permanently. It is presumed that mRNA is degraded within 10 to 20 times, but that new mRNA is continuously being synthesized in cell nuclei.

Once the polypeptide chain is synthesized in ribosomes, it will simply assume the three-dimensional configuration (such as the secondary or tertiary structure), depending on its amino acid sequence. Its three-dimentional configuration seems to have the lowest configurational free energy in order to keep its stable shape. It is thought that no biological mechanism are needed in this process. With regard to glycoproteins, carbohydrate seems to be bound inside the microsomes after the polypeptide chain has been completed. The mechanisms for incorporation of carbohydrate residues into glycoprotein seem to work independently from those of polypeptide chains.

One monomeric subunit or one polypeptide chain synthesized in the polysomes is generally thought to be controlled by one structural gene, or the principle of one structural gene-one polypeptide chain. In the case of proteins composed of many different subunits (as is the case with many plasma proteins), each subunit seems to be synthesized separately, followed by polymerization of subunits which may occur simply by molecular movements.

3. The Genetic Control Mechanism on Protein Synthesis

Plasma proteins, like all other constituents of the body, are kept in a steady state. It has long been thought that plasma protein synthesis was regulated by some control mechanism. However, it was in the 1950's that the modern idea on control mechanisms was discussed at the gene level. The ultrastructure of genes has been elucidated by advanced techniques in molecular biology, and genes began to be described in greater detail at the molecular level. Finally, the operon concept of control mechanisms was discussed by JACOB and MONOD.[70]

Operon theory: The two kinds of control genes are directly involved in protein synthesis, that is, structural genes and operator genes. The structural gene is situated on chromosomes in close proximity to the operator gene (Fig. 59). DNA of the structural gene promote protein synthesis directly through the aid of the mRNA template, while the operator gene controls the adjacent structural genes. The structural genes cannot trigger protein synthesis by themselves, and instead the operator gene initiates the transcription of the structural gene into mRNA. One or more structural genes exist side by side with one operator gene on the same chromosome. These form a functional unit which is called an operon. That is, if we compare the operon to an electric fan, the structural gene would be the motor, but it would not be sufficient to start the fan operating. The switch (the operator gene) which turns it off and on is required. As the fan begins to move only when the switch is turned on, so the operon begins to function only with the action of the operator gene.

Moreover, there is the regulator gene as the third factor. It controls the activity of operator gene. This function is mediated through the repressor which is synthesized by the regulator gene. The operator gene loses its activity or is repressed if it is bound to the repressor, and it will not be able to stimulate the sturctural genes any longer.

Protein synthesis in cells, the synthesis of adaptive enzymes in particular, may be induced only when it is given a particular consitutent, an inducer. In this case, the operon is repressed as a result of the combining of the repressor, and the operon is derepressed by an inducer, resulting in activation of the operator gene.

On the other hand, the synthesis of proteins which are constantly being produced, especially constitutive enzymes, may be repressed by various metabolic products. In the repressible system, the repressor produced by the regulator gene is itself not active, and it is called an aporepressor. However, only when reacted with co-repressor, the aporepressor is converted to the complete represser which represses the operon.

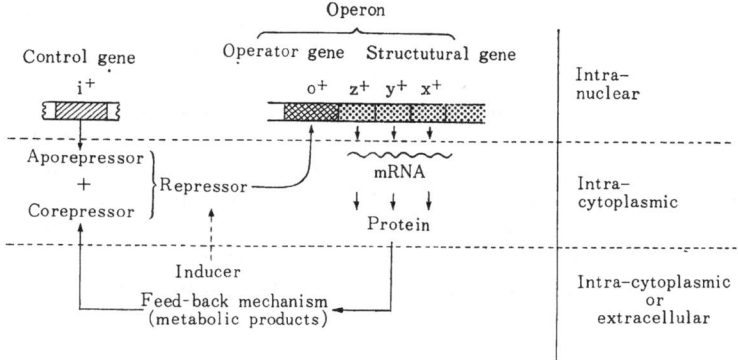

Fig. 59 The operon theory for the induction and inhibition of protein synthesis.

MECHANISMS OF SYNTHETIC ABNORMALITIES
OF THE PLASMA PROTEINS

The synthetic process of plasma proteins which is described in the previous section will be summarized as in Fig. 60. As this schematic diagram indicates, synthetic abnormalities of plasma proteins, whether congenital or acquired, may occur roughly in the following three stages: 1) abnormalities in blood amino acid pool, 2) abnormalities in synthesizing cells, and 3) abnormalities in feedback mechanisms.

Fig. 60 A schematic representation of the synthetic process of proteins.
The abnormalities of protein synthesis may appear under the following three mechanisms; an abnormality of the amino acid pool, of the synthetic cells, and of the feed-back regulation.

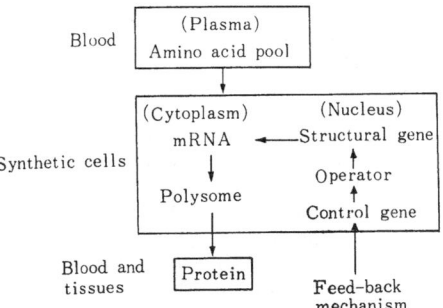

If the blood amino acid pool is deficient, there occurs a decrease of protein synthesis even if synthesizing cells have their normal functions; e.g. in malnutrition, cachexia, and starvation. However, the deficiency of amino acids is not the only element involved. It may be assumed that where there are abnormalities in the blood amino acid pool, the synthesizing cells themselves are already suffering from nutritional disturbance. Therefore, it is clinically inconceivable that malnutrition alone cause hypoproteinemia. With experimental animals, it is extremely difficult to produce hypoproteinemic state on low-protein diets, and the animals usually die soon when hypoproteinemia appears. Therefore, it seems very difficult for hypoproteinemia to occur clinically until an extremely sever condition is met.

Abnormalities of synthesizing cells will be divided into increased protein synthesis and decreased protein synthesis.

a) Increased synthesis can be encountered in (1) the increase in the total number of synthesizing cells, and (2) the increase of the activity of each synthesizing cell.

The former mechanism may arise in a neoplastic or reactive proliferation of synthesizing cells, such hyperimmunoglobulinemia as in multiple myeloma and liver cirrhosis. With the latter mechanism, the increase of "acute phase reactants" in inflammatory diseases can occur. Of course, the exact mechanism by which acute phase reactants increase in inflammatory processes is still unknown. However, many investigators assume that inflammation causes a change in the feedback mechanism in plasma protein synthesis. On the other hand, a-fetoprotein, like the adaptive enzymes mentioned previously, is not found in normal conditions. However, it may also be thought that with a repressor acting asan inducing factor, in some way or other, an active hepato-cellular proliferation begins to produce a-fetoprotein.

The normal levels of various plasma proteins differ according to nationality and geo-

Table 18 Possible genetic causes of protein deficiencies (from SCHULTZE and HEREMAN).

Pseudo-deficiency (the dummy-protein condition)
   The protein is produced in normal amounts but is not-functional.
True deficiency
   A. At the level of the cell
       Absence or non-function of the tissue responsible for the synthesis
   B. At the level of the genes
    I. Regulator genes
       (a) Mutation making repressor unable to combine with inducer
       (b) Mutation making repressor unable to undergo allosteric transition upon combination with inducer
       (c) Mutation producing a repressor already in the allosteric transition state without the aid of a co-repressor
    II. Operator genes
       (a) Deletion of the operator
       (b) Inactive operator gene
    III. Structural genes
       (a) Complete deletion of structural gene
       (b) Mutation making the gene unreadable for transcription into mRNA
       (c) Mutation transcribled into mRNA
           (i) mRNA unreadable, block of synthesis
           (ii) mRNA incompletely readable owing to block of synthesis at mutation site: production of peptide interfering with feedback control?
           (iii) mRNA completely readable: production of a full protein molecule
               (1) Protein insoluble, unable to leave the cell
               (2) Protein non-functional: the dummy-protein condition
               (3) Protein incomplete owing to partial deletion of the structural gene: production of a peptide interfering with feedback control?
   C. At the level of the ribosomes
       Ribosomal defect preventing reading of mRNA
   D. Interference with feedback control

---

graphic location. A comparatively great individual difference is also noted. This is presumably controlled genetically in some way or other.

b)   Decreased synthesis can be encountered in (1) the decrease of tht total number or complete lack of synthesizing cells, (2) the decrease or lack of activity of each synthesizing cell, and (3) abnormal activity of synthesizing cells. As for hereditary protein deficiency, SCHULTZE and HEREMAN have mentioned theoretical possibilities (Table 18).[133]

Abnormalities at the cell level include a congenital lack of tissues or organs containing synthesizing cells. The lack of plasma cells and lymphocytes in agammaglobulinemia is the most typical example of this condition. The destruction or removal of synthesizing cells as a result of hepatectomy, irradiation, drug administration or inflammatory necrosis is also in this category.

Abnormalities at the gene level are caused by various gene mutations. They are observed in congenital deficiency. However, they may be brought about by some acquired inducing factors (e.g. irradiation, drug administration). As to dummy-protein, there are many points to be studied in future.

Abnormalities at the ribosome level may occur congenitally, but they may also be caused by acquired cell injury due to drug administration. It is conceivable that deficient DNA or RNA synthesis often accompany this condition. Where abnormalities in the feedback mechanism occur, some incomplete molecules are synthesized congenitally and their metabolic products may not enter the feedback mechanism. Also, the feedback mechanism may be blocked acquiredly by drugs or pathological metabolic products.

The mechanisms mentioned here are only theoretical. They are not clear in many clinical cases. Some of the abnormalities may occur due to two or more mechanisms combined.

Chapter 12

Bodily Distribution of the Plasma Proteins

EXTRAVASCULAR CIRCULATION AND DISTRIBUTION OF THE PLASMA PROTEINS

Plasma proteins are present not only in the blood vessels, but under normal conditions, they also penetrate the vascular wall and are distributed extravascularly such as in the interstitial tissue fluid, the synovial fluid, the cerebrospinal fluid, the fluid of serous cavities, and the intraocular fluid. Moreover, the extravascular portion of plasma proteins are not stationary but are actively recycled both in and out the blood vessels through the lymphatic vascular system.[13,35,88,104,114,133] Extravascular circulation is remarkably slow as compared with intravascular circulation, but still, an exchange of a considerable amount of plasma proteins is constantly getting under way. Albumin, for instance, exchanged daily between the tissues and plasma amounts probably to equal the total amount of intravascular albumin (110–140 g, on the average).[133] Plasma proteins, albumin in particular, play an important role in maintaining the colloid osmotic pressure not only in the blood but also in various extravascular body fluids. This makes it possible for the tissues to maintain the appropriate lymphatic circulation.

Each plasma protein component contributes in various degrees to maintain the osmotic pressure. The lower the molecular weight of a component, the higher its osmotic pressure obtained by a unit mass (Fig. 61).[105] In particular, albumin, which is of comparatively low molecular weight and is contained in a large amount, plays a major role in maintaining colloid osmotic pressure of various body fluids.

It is likely that the plasma protein compositions of various extravascular body fluids are influenced by the structure of the capillary walls. BENNET et al. have classified the

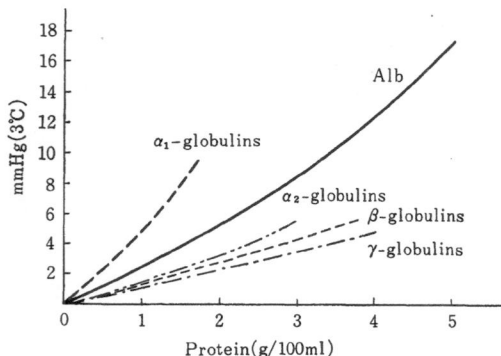

Fig. 61 Effects of the plasma protein fractions to the colloid osmotic pressure (from OTT).
Each protein fraction is obtained through filter-paper electrophoretic separations.

capillary vessels according to their fine structural difference as is shown in Table 19.[10] Type A–1 is the most resistant to the passage of plasma proteins. The "blood-brain barrier" observed in the central nervous system is a good example of this.[152] On the other hand, Type B–3 has the greatest permeability. The hepatic lymph, for example, contains nearly the same components as the plasma. That is, with the exception of such particular capillary vessels as sinusoids existing in the liver, the spleen, and the bone marrow, the capillary vessels more or less show a molecular sieve effect; namely, low-molecular weight proteins are apt to be passed more easily than high-molecular weight proteins.

Table 19 Structural features of different kinds of blood capillaries (from BENNETT et al.).

	Type A Capillaries ensheathed by a complete and continuous basement membrane	Type B Capillaries without a complete, continuous basement membrane
Type 1 Endothelium without fenestrations	Muscle Central nervous system Lung Skin	
Type 2 Endothelium with intracellular fenestrations	Kidney (mammalian) Intestinal villi Some endocrine organs	
Type 3 Discontinuous endothelium (= intercellular gaps)		Liver Spleen Bone marrow

Macroglobulins such as IgM-immunoglobulin and fibrinogen are present mostly (approximately 80% of the total amount in the body) in the plasma, while less than 20% is distributed outside the blood vessels.[133] On the other hand, albumin, prealbumin, IgG-immunoglobulin, IgA-immunoglobulin, transferrin, and β-lipoprotein exist more in the extravascular pool, with 40–50% of the total existing in the plasma.[133]

MECHANISMS OF DISTRIBUTIONAL ABNORMALITIES OF THE PLASMA PROTEINS

The extravascular portion of plasma proteins maintains a certain dynamic equilibrium, and normally it is in a "steady state". However, under various abnormal conditions, this equilibrium is disturbed. Consequently, the extravascular portion of plasma proteins is increased, and then edema, as well as abnormal retention of peritoneal, pleural and pericardial fluids appear. These distributional abnormalities of plasma proteins, edema in particular, may occur with various mechanisms, as shown in Fig. 62.

Abnormalities in the Tissue Interstitium

An example of this is edema due to increased amount of ground substance in the interstitial tissue. It also arises as a result of the increased water content or hydrophilia of the interstitial ground substance. As a result, the distribution of plasma proteins may alter. Clinically, myxedema is thought to be included in this category (① in the Figure). Ordinarily, lymphatic circulation decreases significantly.

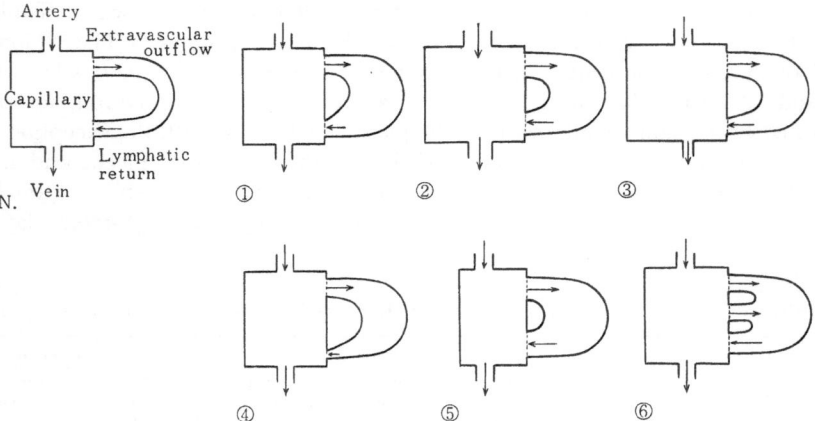

Fig. 62 A schematic diagram showing different mechanisms of edema formation. N = normal, 1 = increased hydration of the ground substance, 2 = increased arterial pressure, 3 = increased venous pressure, 4 = lymphatic obstruction, 5 = hypopro- teinemia, 6 = increased capillary permeability. The arrows indicate the extent and direction of water circulation. The rectagular portion indicates the size of the intra- vascular protein pool, while the U-shaped portion indicates the size of the extravascular protein pool.

For abnormal retention of serous fluids in body cavities, the second to sixth mechanisms in Fig. 62 are applicable. However, when dialysis with peritoneal perfusion or artificial kidney is performed, the first mechanism may be applied, supposing that the tissue inter- stitium has been infinitely expanded.

Abnormalities Due to Hydrostatic Changes

These may be classified in the following three groups: 1) increased arterial capillary pressure (② in the Figure) causes increased capillary filtration. As a result, the plasma proteins as well as water leaving the blood vessels increases. This is clinically met with hyperemia in inflammatory lesions and hypervolemia resulting from water and salt retention. 2) increased venous pressure (③ in the Figure) causes increased venous capillary pressure thus decreasing venous capillary circulation and increasing the interstitial fluid. Clinically, this condition occurs in local venous obstruction (e.g. thrombosis) and also in sys- temic water and salt retention (e.g. congestive heart failure, and nephrotic edema). 3) lymphatic obstruction (④ in the Figure) causes decreased lymphatic return, and clinically observed are congenital and acquired lymph edema.

Abnormalities Due to Osmotic Changes (⑤ in the Figure)

The colloid osmotic pressure in the blood becomes low due to hypoproteinemia (due particularly to the decrease of albumin). This lowers the gradient of the osmotic pressure between the plasma and tissue fluid. Thus, a portion of the plasma water moves into the tissues. Clinically, liver cirrhosis, nephrotic syndrome, malnutrition and analbuminemia may be associated with edema in this way.

Abnormalities Due to Increased Capillary Permeability (⑥ in the Figure)

Clinically, they are associated with hyperemia due to inflammation or inflammatory destruction of the vascular walls. In the latter, a large amount of plasma proteins are lost through extravasation.

Edema occurs with various mechanisms and the plasma protein composition in edema fluids differs. It is generally determined by interaction of the following four factors; namely, 1) the plasma protein composition in the plasma, 2) the molecular size of plasma proteins, 3) permeability of the capillary walls and the adjacent tissue, and 4) local production of plasma proteins, especially immunoglobulins. Extravascular retention fluids vary in their protein content. Edema fluid or transudate is produced through the capillaries with almost normal permeability, and it contains the least amount of proteins. Exudate, in contrast, is produced through the severely damaged capillaries and its protein content is similar to that of the plasma.

INTERSTITIAL TISSUE FLUIDS AND EDEMA FLUID

Under normal conditions, the interstitial tissue fluid, which is contained in loose connective tissue fluid, is the largest extravascular pool. The organs having the loose connective tissues in large quantity play a major role in the extravascular pool of plasma proteins. According to ROTHSCHILD et al., 40% of albumin is distributed in the plasma, 18% in the skin, 15% in the muscles, and the remaining 4% is found in the heart, the lungs, the liver, the kidneys, and the spleen.[124] It has been shown anatomically that the skin and muscles in particular are rich in loose connective tissue.

If the extracts from the skin and muscles, that is, the tissue fluids are electrophoretically fractionated, their protein patterns are quite similar to those of serum. By immunoelectrophoresis, the presence of albumin, α_1-antitrypsin, α_1-acid glycoprotein, transferrin, hemopexin, IgG-immunoglobulin and IgM-immunoglobulin is demonstrated.[97,164] Of these, albumin is contained in greatest quantity and amounts to approximately 40% of the serum albumin concentration.[30] The composition of edema fluid depends upon the total protein concentration, disease, and the degree of vascular injury. The total protein concentration measured in various diseases ranges from 0.02 to 6.8 g/100 ml (Fig. 63). When normal permeability of local capillaries is maintained, as in hypoproteinemia, heart failure or constriction pericarditis, the total protein concentration is below 1 g/ 100 ml. On

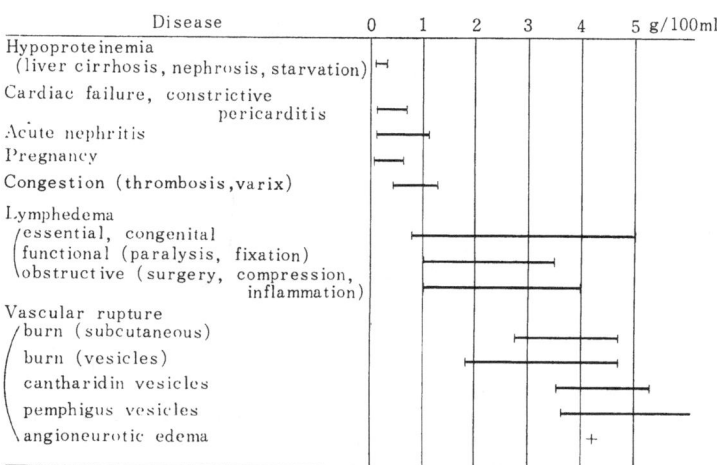

Fig. 63 Total protein concentration of various edema fluids (modified from SCHULTZE and HEREMANS).
The highest value experienced in the author's laboratory is 6.8 g/ 100 ml in the vesicular fluid obtained from pemphigus vulgaris.

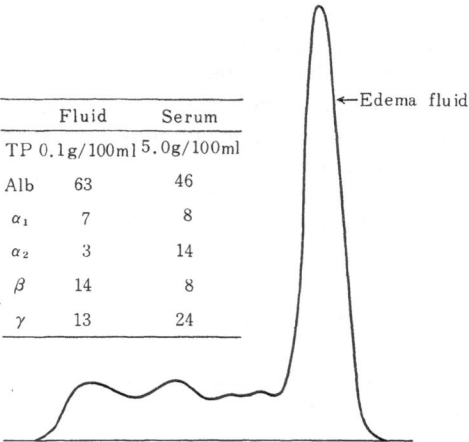

	Fluid	Serum
TP	0.1g/100ml	5.0g/100ml
Alb	63	46
α_1	7	8
α_2	3	14
β	14	8
γ	13	24

←Edema fluid

Fig. 64 Cellulose acetate electrophoretic pattern of the edema fluid obtained from the lower leg. 55 y.o., male, diabetic glomerulosclerosis and myocardial infarction.

Suffered from diabetes mellitus complicated with diabetic glomerulosclerosis. Died of uremia (urea N 68.4 mg/100 ml, creatinine 8.9 mg/100 ml) and myocardial infarction. Moderate anemia (Hb 6.4 g/100 ml) and hypoproteinemia (5.0 g/100 ml). Marked proteinuria, amounting to 1.6 g/day. The serum protein electrophoretic pattern is of protein-losing type.

Marked anasarca noted. The edema fluid obtained through a subcutaneous puncture is colorless and transparent, the specific gravity of 1.009, Rivalta reaction (−), the total protein 0.1 g/100 ml, urea N 67.8 mg/100 ml, creatinine 9.6 mg/100 ml, Na 140 mEq/L, Cl 116 mEq/L, and K 5.9 mEq/L. The electrophoretic pattern of the edema fluid shows a relative increase in the albumin and β fractions, and a relative decrease in the α_2 and γ fractions, in comparison to the serum pattern.

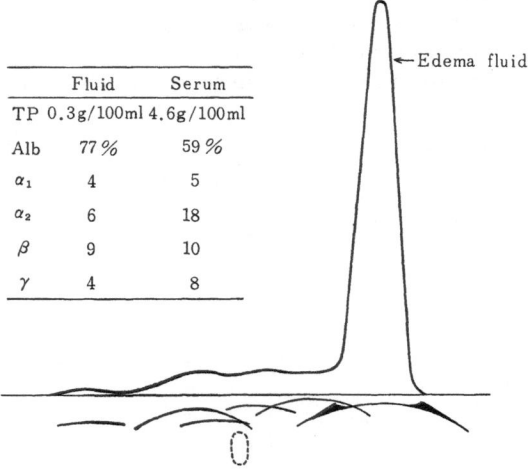

	Fluid	Serum
TP	0.3g/100ml	4.6g/100ml
Alb	77 %	59 %
α_1	4	5
α_2	6	18
β	9	10
γ	4	8

←Edema fluid

Fig. 65 Cellulose acetate electrophoretic pattern of the edema fluid.
3 m. o., male, congenital lymphedema.

Compained of painless swelling of the left upper arm and the lower extremities since the time of birth. No other clinical abnormality except for hypoproteinemia (4.6 g/ 100 ml). Lymphangiography is unsuccessful.

The cellulose acetate electrophoretic pattern of the edema fluid shows a relative increase in the albumin fraction and a relative decrease in the α_2 and γ fractions in comparison to the serum pattern. The total protein concentration is 0.3 g/100 ml, and its immuno-electrophoretic analysis shows albumin, α_1-antitrypsin and transferrin; also, haptoglobin, Gc-globulin, hemopexin and IgG in minute quantities.

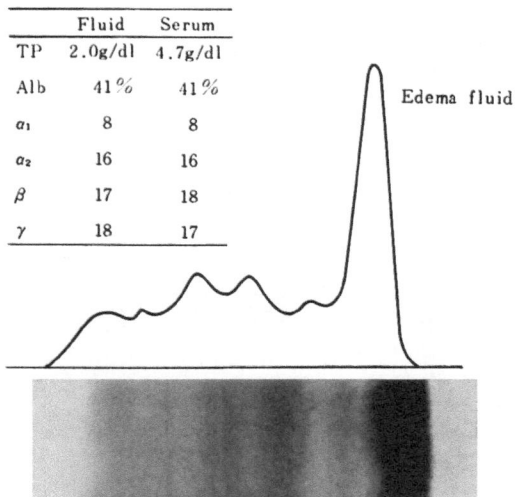

	Fluid	Serum
TP	2.0g/dl	4.7g/dl
Alb	41%	41%
a_1	8	8
a_2	16	16
β	17	18
γ	18	17

Edema fluid

Fig. 66 Filter paper electrophoretic pattern of the edema fluid. 51 y.o., male, elephantiasis of the right lower extremity.
Chronic long-standing lymphedema and elephantiasis developed due to obstruction of the retroperitoneal lymphatics (etiology unknown, but filariasis suggested). After surgical removal of the fibrous tissue in the retroperitoneum, the edema fluid was accumulated extraperitoneally in the small pelvic cavity, and it was obtained for laboratory examinations. The serum total protein concentration is 4.7 g/100 ml, while the edema fluid shows the total protein of 2.0 g/100 ml.

electrophoretic patterns, the so-called molecular sieve effect is obvious, showing a relative increase of the albumin fraction and a relative decrease of the a_2 and γ fractions as compared to those of serum (Fig. 64, 65). Besides, protein components almost similar to those of normal tissue fluids are also demonstrated on immunoelectrophoretic patterns.

On the other hand, in the inflammatory edema fluid, or the exudate, the more serious the vessels injury, the larger the total protein concentration. Its protein pattern resembles that of plasma (Fig. 66). This tendency is especially marked in the vesicular content of the skin with pemphigus vulgaris. Its immunoelectrophoretic patterns are hardly distinguishable from those of serum. Low-density lipoproteins are also demonstrated (Fig. 67).

It should be noticed here that, despite a very wide variation of the total protein concentration of edema fluid seen in various diseases, the ratio of the percentage of the albumin faction in edema fluid (Alb_E %) as compared to that of serum (Alb_S %) is fairly constant in most cases being about 1.10–1.20.[129] This fact shows that as far as there is no true hemorrhage, the selective permeability of capillary vessels can be fairly well preserved. The edema fluid seen in lipoid nephrosis is an exception. It shows remarkably high Alb_E/Alb_S ratio, being 1.59 and 1.79. This tendency is stronger as hypoproteinemia increases in severity. It is possible that there is some mechanism at work preserving the level of the extravascular colloid osmotic pressure as at constant a point as possible. Besides, with the pleural effusions in cases of pyothorax, the ratio is remarkably low, 0.84. This is thought to be due to the destruction by proteases coming from the leukocytes in effusions.

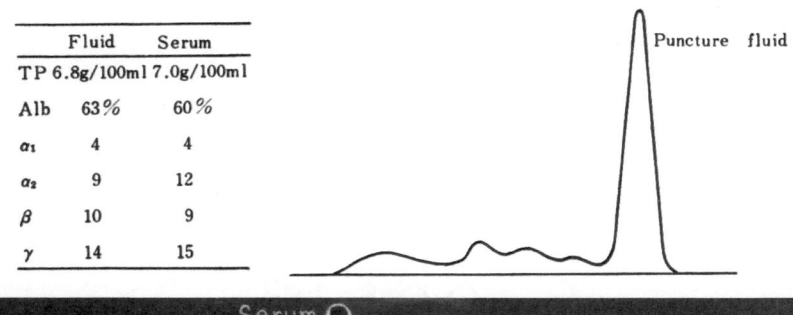

	Fluid	Serum
TP	6.8g/100ml	7.0g/100ml
Alb	63%	60%
α_1	4	4
α_2	9	12
β	10	9
γ	14	15

Puncture fluid

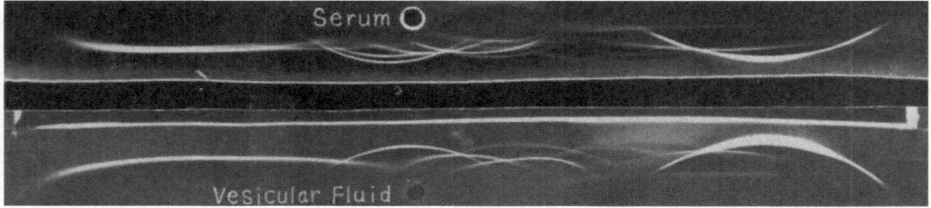

Fig. 67 Cellulose acetate electrophoretic and agar gel immunoelectrophoretic patterns of the vesicular content.
49 y.o., male, pemphigus vulgaris.
The skin over the trunk is entirely involved by the exudative lesion, and the vesicular content is obtained through puncture of vesicles. The total protein concentration in serum is 7.0 g/100 ml, while 6.8 g/100 ml in the vesicular fluid. On immunoelectrophoretic analysis, also, the fluid pattern is entirely the same as the serum pattern.

EFFUSIONS IN SEROUS CAVITIES

Little is known about the characteristics of normal serous fluids in the pleural cavities, the pericardial cavity, the peritoneal cavity and testicular sac, because they are present in extremely small amounts. In this section, pleural fluid, ascites, pericardial fluid, and the effusion of hydrocele which are found pathologically will be discussed. These effusions have fundamentally the same characteristics. But when they are compared to each other in the same patient, a slight difference is seen in their total protein concentration; to list them in order of their total protein concentration, in general, ascites, pericardial fluid, pleural fluid, and hydrocele effusion.[155]

Ascites: Its total protein concentration differs significantly among different disease conditions, and is similar to that of edema fluids mentioned previously. In cases where hepatic venous congestion is apparent, such as congestive heart failure, liver cirrhosis, hepatoma, and metastatic carcinoma of the liver, the total protein concentration may reach 2.0–4.0 g/100 ml even without significant vascular damage of the peritoneal tissue. This is thought to be due to the transudation through the hepatic surface of the plasma proteins which have been forced out from the highly permeable sinusoids as a result of hepatic congestion.[133]

In any case, ascites of various character are encountered from the transudate seen in nephrosis and liver cirrhosis to the exudate seen in peritonitis, depending on the degree of injury to the peritoneal capillary walls.[71] As in Fig. 68, a significant molecular sieve effect is seen in liver cirrhosis; albumin, α_1-glycoproteins, transferrin and IgG-immunoglobulin are contained chiefly, while traces of Gc-globulin, α_2-macroglobulin, haptoglobin, prealbumin, hemopexin, IgA, IgM-immunoglobulins and C-reactive protein are found, but none of low-density lipoprotein. As in Fig. 69, the ascites accompanying carcinomatous

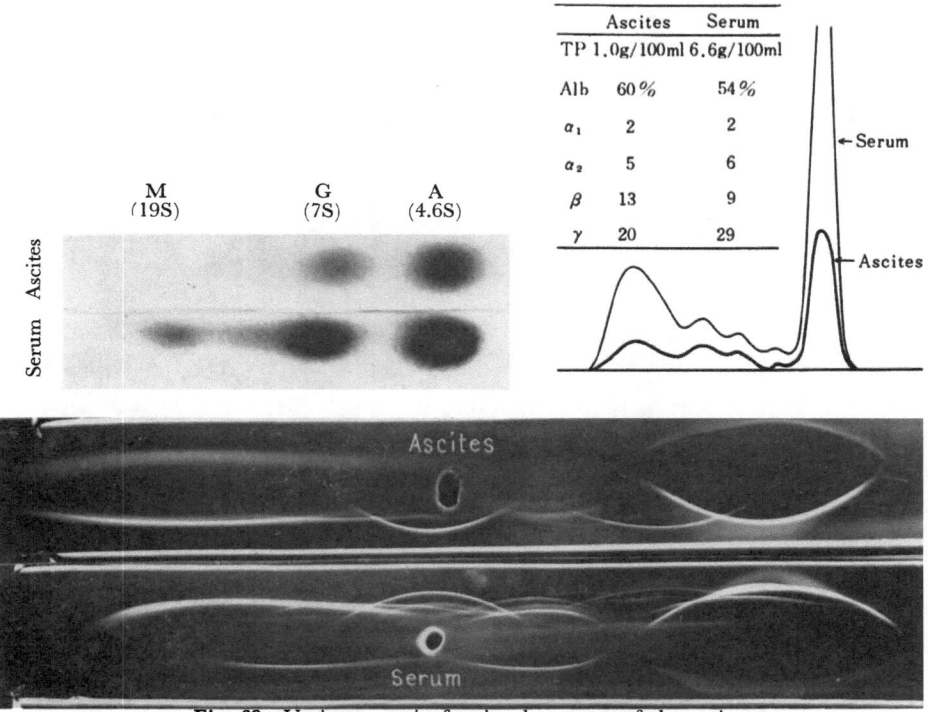

	Ascites	Serum
TP	1.0g/100ml	6.6g/100ml
Alb	60%	54%
α_1	2	2
α_2	5	6
β	13	9
γ	20	29

Fig. 68 Various protein fractional patterns of the ascites.
48 y.o., female, liver cirrhosis.
The serum total protein is 6.6 g/100 ml, and the cellulose acetate pattern shows a decrease in the albumin fraction and a marked increase in the γ fraction. The immunoelectrophoretic pattern shows markedly increased IgG, IgA and IgM.
The ascites is yellowish and transparent, the total protein being 1 g/100 ml with negative Rivalta reaction. Its electrophoretic pattern shows, in comparison to the serum pattern, a relative increase in the albumin and γ fractions, and a relative decrease in the α_2 and β fractions. The thin-layer gel filtration pattern (upper left) of the ascites clearly shows absence of the M spot, indicating the absence of the macromolecular components, such as the LDL, α_{2M} and IgM.

peritonitis are hardly distinguishable from serum. As is obvious in this case, when the M-component is present in serum, it is demonstrable also in ascites though their percentage differs with different molecular size of the M-component.

Pleural and pericardial fluids: These have nearly the same composition as ascites. Those of various compositions, from the transudate to the exudate, are found in abnormal conditions. As in the case of the ascites, the molecular sieve effect is observed more or less. When the total protein concentration is great, low-density lipoprotein is demonstrated and fibrin formation may also be observed (Fig. 70, 71).

As a specific case, the content of cystic tumors may be cited. However, as shown in Fig. 72, in the case of serous content, there are the same characteristics as in the transudate of serous cavities. That is, the serous content shows a molecular sieve effect and its major components are albumin and transferrin. It may contain tissue protein components, depending on the kinds of the tumor present.

The Rivalta *reaction:* Clinically speaking, the Rivalta reaction is used to differentiate transude from exudate or *vice versa*. When an exudate is dripped into acetic acid which has been sufficiently diluted by water, it shows white turbidity and may form white precipitate. According to Caputo,[20] the positive phenomenon of the Rivalta reaction

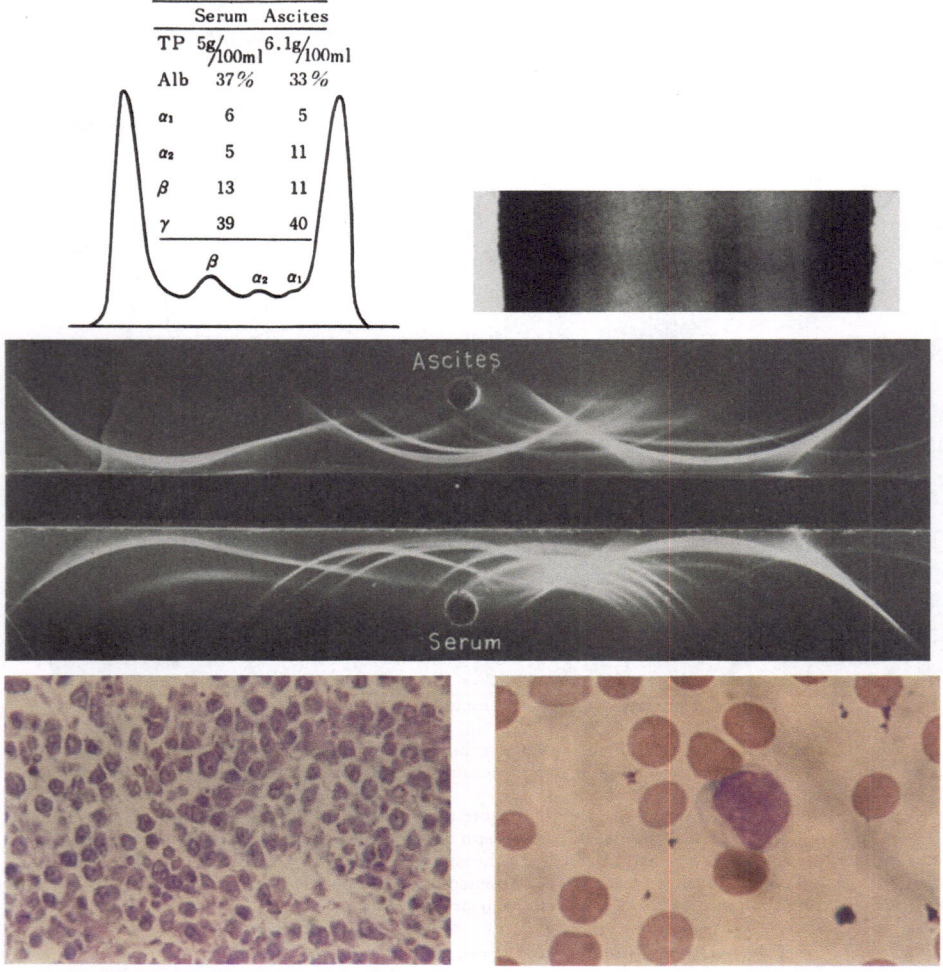

	Serum	Ascites
TP	5g/100ml	6.1g/100ml
Alb	37%	33%
α_1	6	5
α_2	5	11
β	13	11
γ	39	40

Fig. 69 Filter paper electrophoretic and immunoelectrophoretic patterns of the ascites.
75 y.o., female, reticulum cell sarcoma of the retroperitoneum.
Admitted with moderate anemia and ascites, and diagnosed to have reticulum cell sarcoma in the retroperitoneum.
Filter paper electrophoresis and immunoelectrophoresis of the serum proteins showed the IgG-type M-protein at the slow-γ zone. No skeletal lesion nor myeloma cells in the bone marrow. The ascites was yellow and bloody, the specific gravity of 1.026 and the positive Rivalta reaction. The electrophoretic analysis of the ascites demonstrated the IgG type M-protein which was identical to that of the serum.
The sections from the retroperitoneal lymph node (lower left) shows a typical histological pattern of reticulum cell sarcoma. The peripheral blood smear (lower right) also demonstrates occasional abnormal cells.

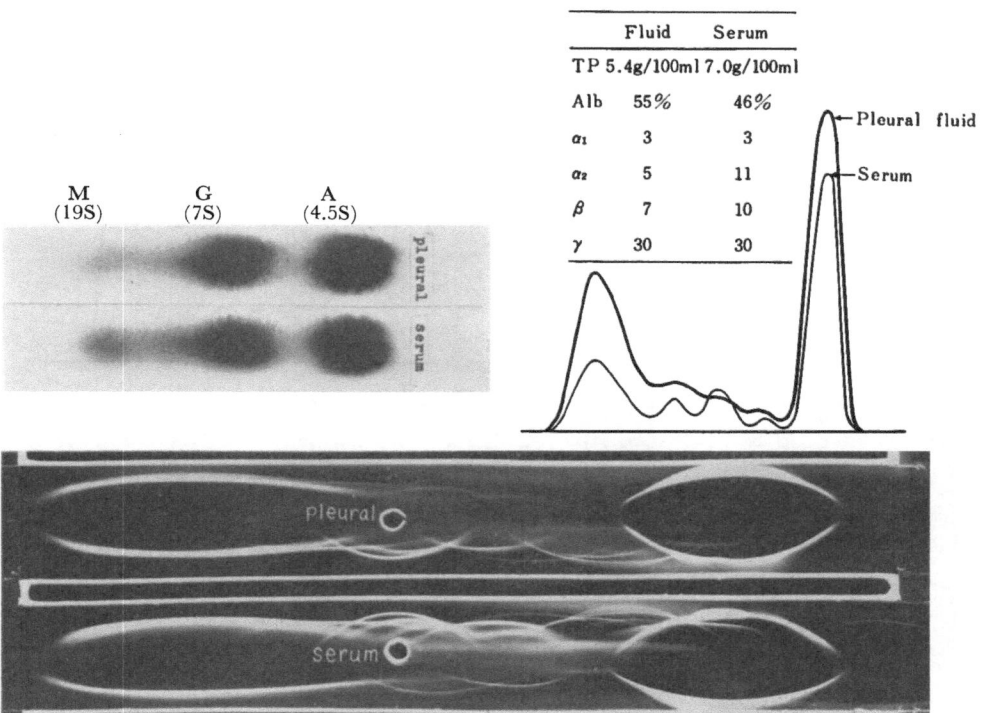

	Fluid	Serum
TP	5.4g/100ml	7.0g/100ml
Alb	55%	46%
α_1	3	3
α_2	5	11
β	7	10
γ	30	30

M (19S)   G (7S)   A (4.5S)

Fig. 70 Various protein fractional patterns of the pleural fluid.
60 y.o., male, carcinoma of the pancreas and pleuritis carcinomatosa.
Upon admission, moderate anemia (Hb 9.0 g/100 ml), normal liver functions and normal serum alkaline phosphatase activity.
The pleural fluid aspirated from the left pleural cavity forms jelly-like floccula. The total protein is 5.0 g/100 ml, the specific gravity 1.039, and negative Rivalta reaction. The protein electrophoretic fractionation pattern of the pleural fluid is quite similar to that of the serum, but the thin-layer gel filtration pattern of the pleural fluid (upper left) shows a significantly lower value of the M fraction than that of the serum, indicating the molecular sieve effect. On immunoelectrophoresis, also, the macro-molecular components like LDL, α_{2M} and IgM are not distinctly recognized in the pleural fluid.
In this case, the pleura has been infiltrated by the metastatic carcinoma quite extensively, leaving almost no normal serous surface. Therefore, the discrepancy between the protein content and the Rivalta reaction seems to be due to lack of mucopolysaccharides secreted from the serous surface.

comes from the formation of a polysaccharide-protein complex which is insoluble in acetic acid. Thus, important are also hyaluronic acid and acid polysaccharides which are increased by the inflammation of the covering mesothelium. Therefore, in the exudate effused by inflammatory change, the RIVALTA reaction becomes positive with the increase of protein content. However, when the mesothelial lining of the serous cavities are severely damaged, the RIVALTA reaction may become negative, even if the exudation of plasma proteins is remarkable (Fig. 70).

SYNOVIAL FLUID

The synovial cavity, considered embryologically, has developed in the connective tissues in order to fulfil a special function. Thus, the synovial fluid is not like the effusion in serous cavities, but bears a close resemblance to the interstitial tissue fluid. However, in accordance with the specially developed function of the synovial membrane, hyaluronic acid

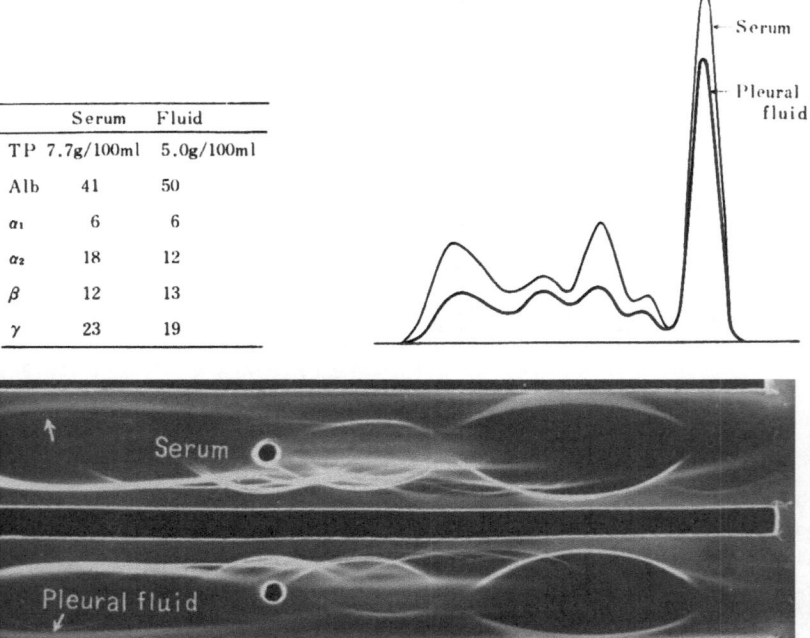

	Serum	Fluid
TP	7.7g/100ml	5.0g/100ml
Alb	41	50
α_1	6	6
α_2	18	12
β	12	13
γ	23	19

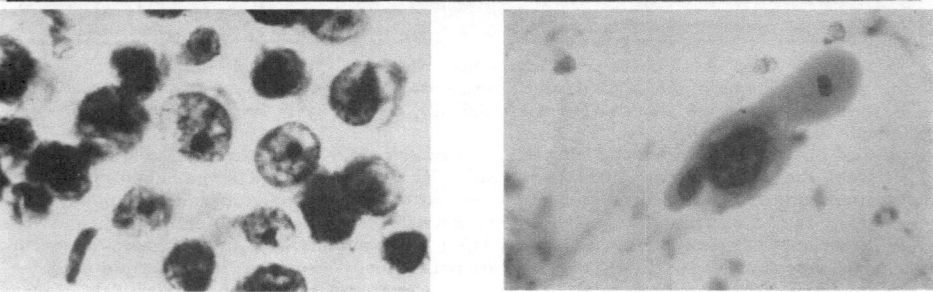

Fig. 71 Various protein fractional patterns of the pleural fluid.
38 y.o., male, epidermoid carcinoma of the lung and pleuritis carcinomatosa.
The pleural fluid is bloody and turbid, its specific gravity 1.041, its protein content 5.3 g/100 ml, and positive Rivalta reaction. The sediment of the pleural fluid contains 20–30 red blood cells in every high-power field showing marked rouleaux formation, 8–10 white blood cells and many cancer cells.
The electrophoretic patterns of the pleural fluid are similar to those shown in Fig. 70, but the C-reactive protein (arrows) is clearly demonstrated with the particular antiserum used here.

is contained in large quantities in the synovial fluid. Because of this, the viscosity of the synovial fluid is high, and the electrophoretic separation of protein using various supporting media is not clear. Therefore, it is necessary for hyaluronic acid to be destroyed by adding hyaluronidase before analysis of the synovial fluid proteins (Fig. 73).

It may be affirmed that almost all proteins in the synovial fluid, at least in normal condition, come from plasma. After having permeated the capillary vessels, the plasma proteins move into the joint lumen by diffusing through the hyaluronic gelous supporting tissue in the synovial membrane. Therefore, a molecular sieve effect by the synovial membrane must be great. It is thought that the reabsorption of plasma proteins occur mainly through lymphatic vessels.[6,133] Accordingly, it is said that the molecular weight

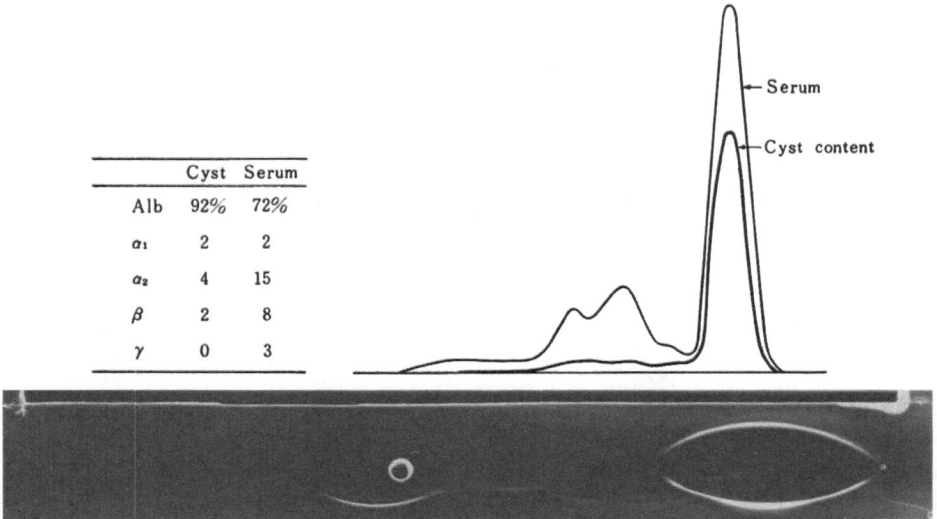

	Cyst	Serum
Alb	92%	72%
α_1	2	2
α_2	4	15
β	2	8
γ	0	3

Fig. 72 Various protein fractional patterns of the cyst content.
5 m.o., male, benign teratoma of the mediastinum.
Admitted with suspected pericarditis. A mediastinal puncture yielded yellow-brown, slightly turbid fluid which was thought to be the pericardial fluid. However, on surgery, the pericardium was intact, and the fluid was confirmed to have been the cyst content removed from a benign teratoma, $10 \times 7 \times 5$ cm.
The protein content of the fluid is 0.8 g/100 ml, the specific gravity 1.011, neutral, and negative Rivalta reaction. In the sediment seen are small numbers of the erythrocytes, leukocytes and large epithelial cells. No malignant cells seen.

of each protein component affects their transfer into synovial cavities, but that on re-absorption all protein components enter the lymphatic vessels in the same way.[123]

The total protein concentration of synovial fluid is approximately 3 g/100 ml and its electrophoretic patterns resemble those of serum. However, relative increase of the albumin fraction (Alb_E/Alb_S is about 1.10) and a relative decrease of α_2 fraction are observed in comparison to those of serum. The α_1 fraction/α_2 fraction ratio is also higher than that of serum.[133] The following components are demonstrated on immunoelectrophoretic patterns: prealbumin, albumin, α_1-antitrypsin, α_1-acid glycoprotein, α_1-lipoprotein, thyroxine-binding globulin, ceruloplasmin, haptoglobin, α_2-macroglobulin, low-density lipoprotein, transferrin, hemopexin, immunoglobulins, fibrinogen.[134] However, the transfer of plasma proteins into the synovial fluid is presumably not so great.

In the case of arthritis, a relative decrease of the albumin fraction and a relative increase of the γ fraction are noted. Further, with regard to IgG-immunoglobulin, the ratio of its concentrations in the synovial fluid and serum is usually higher than 0.60. This was thought to be due to the increase of albumin reabsorption when arthritis followed the chronic process[129]. However, recently, many investigators are apt to believe that this is due rather to a local production of the immunoglobulins, especially in the case of rheumatoid arthritis.[65,137]

CEREBROSPINAL FLUID

1. The Proteins in Normal Cerebrospinal Fluid

In normal adult, an approximately 100–150 ml of the cerebrospinal fluid can be drained,

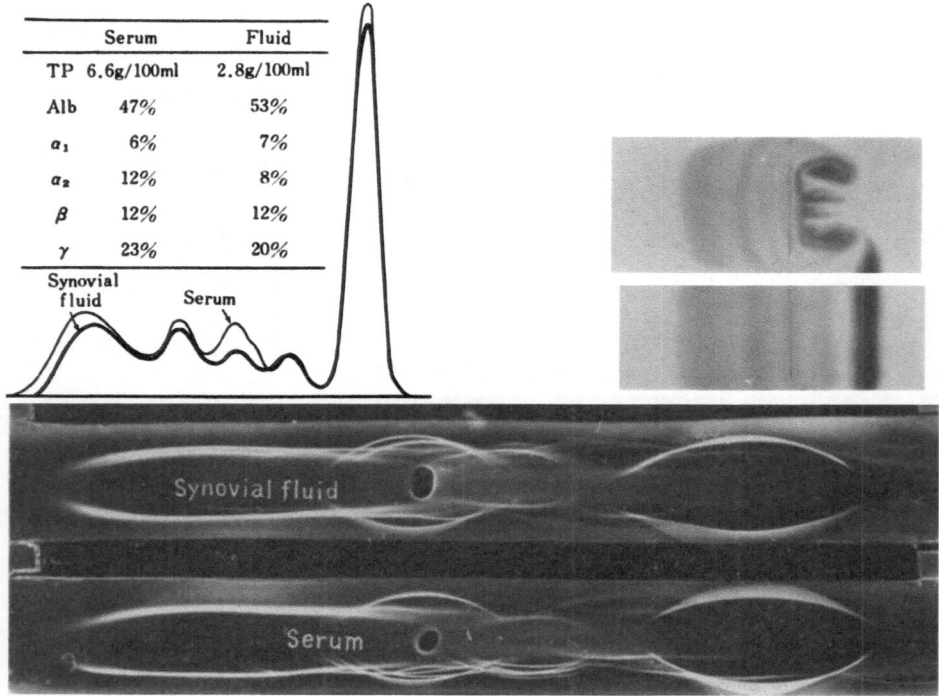

	Serum	Fluid
TP	6.6g/100ml	2.8g/100ml
Alb	47%	53%
a_1	6%	7%
a_2	12%	8%
β	12%	12%
γ	23%	20%

Fig. 73 Various protein fractional patterns of the synovial fluid.
K. U. 53 y.o., female, rheumatoid arthritis.
RA-test 1+, CRP 5+, sed rate 93 mm/l hr. The synovial fluid is viscous, and it shows a poor separation of the protein fractions on cellulose acetate electrophoresis (upper right). However, after treated with hyaluronidase, the cellulose acetate electrophoretic separation of the protein in the synovial fluid is almost identical to that of the serum, although the low density lipoprotein is barely seen. IgM is more prominent in the synovial fluid, but it is not recognized in the serum which has been reacted with RA-test reagent (see Fig. 190).

and its total protein concentration is 15–65 mg/100 ml.   Thus, the total amount of plasma proteins in the cerebrospinal fluid constitutes only a small part of the extravascular pool. In the case of hydrocephalus, cerebrospinal fluid amounts to 2–3 l, but a significant decrease of the total protein concentration in the plasma is rare. It is thought that cerebrospinal fluid comes from the choroid plexus in the lateral ventricles, as well as the third and fourth ventricles.   It flows through the foramen of MAGENDIE, the foramen of LUSCHKA, and is distributed in the subarachnoidal space, draining over the surface of the brain and the spinal cord.   It is eventually reabsorbed through the venous system.   Accordingly, it is possible that the brain-blood barrier against plasma proteins exists in the ventricular choroid plexus.[133)]

The total protein concentration of normal cerebrospinal fluid is remarkably low, in comparison to other body fluids.   Ordinarily, it is 15–65 mg/100 ml, while in neonates and aged people it is a little higher.   Its total protein concentration depends on the site of puncture and tends to increase as the site goes down, such as is the case of ventricular fluid, cysternal fluid, and lumbar fluid.   Because of their low total protein concentration, electrophoretic analysis is performed after cerebrospinal fluid has been concentrated.   However, care should be taken as analytical values vary according to different concentration procedures.   In general, the electrophoretic patterns of normal cerebrospinal fluid resemble

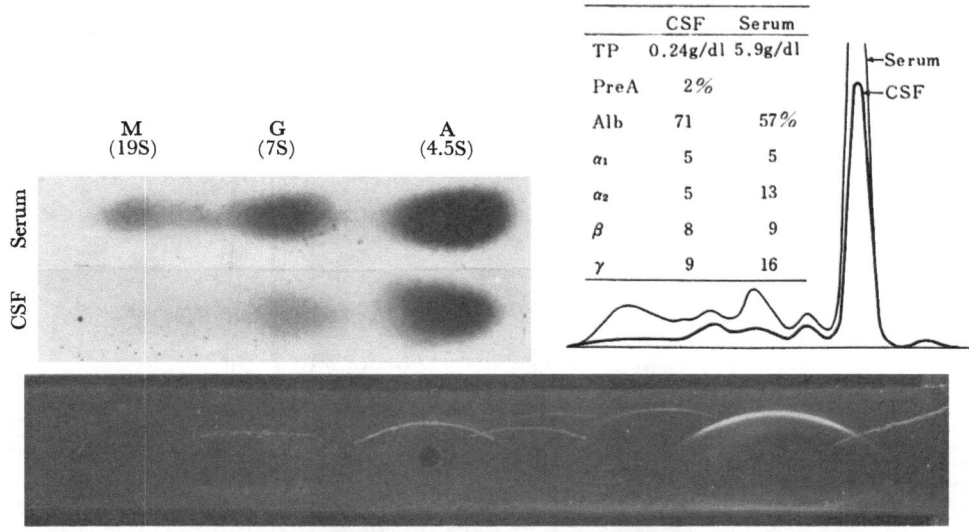

	CSF	Serum
TP	0.24g/dl	5.9g/dl
PreA	2%	
Alb	71	57%
α_1	5	5
α_2	5	13
β	8	9
γ	9	16

Fig. 74 Protein fractional patterns of the cerebrospinal fluid.

42 y.o., male, solitary myeloma.

About one year and a half ago, admitted complaining of paresthesia and increasing muscular weakness. On X-ray examination, the XII thoracic vertebra was noted to be destroyed, and the biopsy showed plasmacytoma.   No other skeletal lesion, and no abnormal cells in the sternal marrow.   The serum proteins and other blood chemistry works showed nothing remarkable.

The CSF obtained through a lumbar puncture was colorless and transparent, its total protein being 0.24 g/100 ml.   The cellulose acetate electrophoretic pattern of the CSF proteins is somewhat similar to that of the serum proteins, showing a relative increase of the albumin fraction and a relative decrease of the α_2 and γ fractions.   On immunoelectrophoresis, identified are prealbumin, albumin, α_1-antitrypsin, Gc-globulin, transferrin and IgG.   On thin-layer gel filtration (upper left), the protein components of relatively large molecular weight seem to be absent, showing a lack of the M fraction and a significant decrease of the G fraction.

Table 20 Protein fractions of normal cerebrospinal fluid obtained by cellulose acetate (Oxoid) electrophoresis.

Cerebrospinal fluids were obtained by lumbar puncture from the patients with acute appendicitis.

Fractions	Average values (%)
Prealbumin	1.8
Albumin	57.7
α_1	5.1
α_2	7.0
β	15.6
γ	12.8

those of serum, but they are frequently separated into seven fractions including the prealbumin fraction and the β_2 fraction.   Moreover, a relative increase of the albumin fraction and a relative decrease of the α_2 and γ fractions occur (Fig. 74).   The average values of each fraction examined in the author's laboratory by cellulose acetate electrophoresis are shown in Table 20.

Appearance of the prealbumin fraction: Prealbumin was first described by KABAT et al.[75] In normal adults, it is contained at an average of 1.8%.   The amount of the prealbumin fraction varies in different places.   In the ventricular fluid, it amounts to as much as

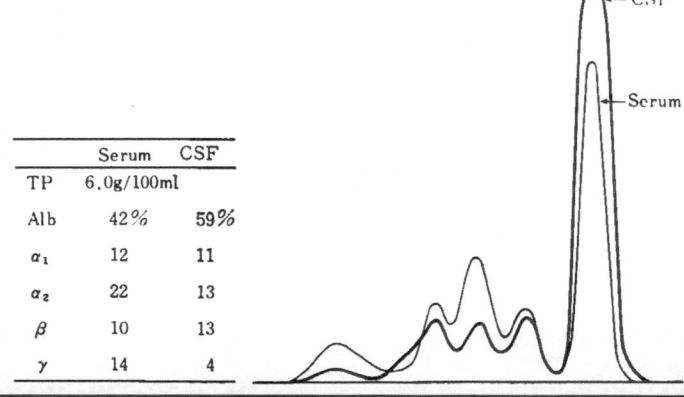

	Serum	CSF
TP	6.0g/100ml	
Alb	42%	59%
α_1	12	11
α_2	22	13
β	10	13
γ	14	4

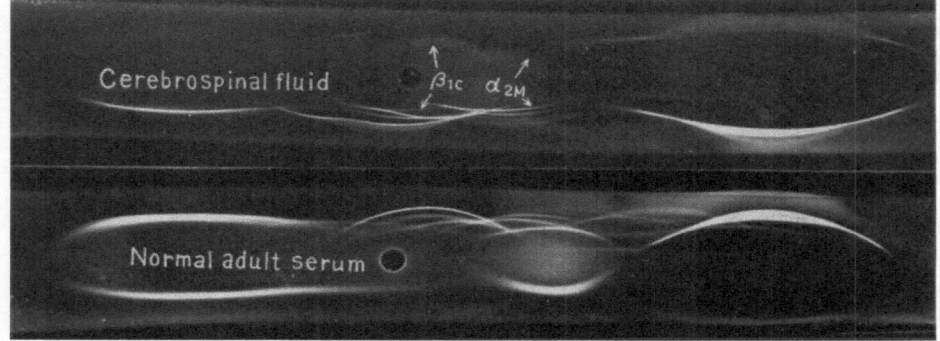

Fig. 75 Protein fractional patterns of the ventricular fluid.
7 m.o., female, hydrocephalus
Because of the secondary infection combined, the serum proteins are of the acute inflammatory pattern.   The electrophoretic pattern of the ventricular fluid shows a lack of the prealbumin fraction, a relative increase of the albumin fraction, and a relative decrease of the α_2 and γ fractions.   The β fraction tends to be increased, clearly demonstrating β_2 peak.
On immunoelectrophoresis, prealbumin is seen only vaguely, and α_1-glycoproteins are prominent.   Several minor lines of precipitation are also seen in the α_2 area.   IgG is distinctly seen, but IgM and IgA are not recognized.   Transferrin (arrow) is biphasic, and β_2-portion is clearly seen.   In addition, hemopexin, β_{1c}-globulin and α_2-macroglobulin (trace) are seen, but lipoproteins are not identified.

13–20%.[142]   On immunoelectrophoresis, the prealbumin fraction is demonstrated to be the same as tryptophan-rich prealbumin in the blood.   It is not known why this prealbumin is contained in the cerebrospinal fluid in such large quantity.   However, this fraction is not necessarily observed in all persons.

On filter-paper electrophoresis of normal spinal fluid, the prealbumin fraction is frequently missing.

β_2 *fraction* (τ *fraction*, BÜCHER, *et al.*) :   Though the β_2 fraction is sometimes seen clearly, it may not be demonstrated as a separate band (Fig. 75).   It has been confirmed by immunoelectrophoretic patterns that the β_2 fraction is made up partly of transferrin. Therefore, transferrin in the cerebrospinal fluid may show a biphasic precipitin arc.[51,111] There are two theories concerning the origin of the slow-migrating transferrin, but still many points remain uncertain.   According to PETTE and STUPP,[111] neuraminidase reacts on β_1-transferrin which originates in the blood and the β_1-transferrin is gradually transformed into β_2-transferrin. On the other hand, some investigators think that, as PARKER and BEARN say,[106] β_1-transferrin comes from the blood while β_2-transferrin is synthesized in the central nervous system.[43]

Relative decrease of the α_2 and γ fractions: At least, it can be said that this is caused by the molecular sieve effect of the choroid plexus. As is shown in Fig. 74, the M fraction, or macromolecular components, is hardly seen on thin-layer gel filtration patterns. By immuno-electrophoretic and immuno-diffusion methods, the following components are demonstrable: prealbumin, albumin, α_1-acid glycoprotein, α_1-antitrypsin, haptoglobin (Hp$_{2-1}$, Hp$_{2-2}$, uncertain), ceruloplasmin, α_2-macroglobulin (which may not be present), hemopexin, transferrin, $\beta_{1C/A}$-globulin, non-clottable fibrinogen (only a trace), IgA-immunoglobulin (in small quantities), IgG-immunoglobulin and β_{2k} globulin (only a trace). Low-density lipoprotein and IgM-immunoglobulin are not detectable.[133]

In addition, minor protein components which are not detectable in plasma are demonstrated in normal cerebrospinal fluid by using the specific antiserum against cerebrospinal fluid.[77]

2. Pathological Protein Patterns of the Cerebrospinal Fluid

Attempts have been made to classify abnormal changes of the proteins of cerebrospinal fluid into serveral types.[5,38,133] Ordinarily, they must be compared with the serum proteinograms of the same patient. Accordingly, it is convenient to classify them into 4 groups on the basis of the mechanism producing abnormal cerebrospinal fluids.

Normal cerebrospinal fluid production: This category assures the maintenance of selective permeability of the brain-blood barrier and usually no significant pathology is recognized in the central nervous system. It shows the similar protein abnormalities to those seen in the plasma. In the patients who have M-components in serum, the M-components are transferred into the cerebrospinal fluid if they are of low molecular weight (BENCE JONES protein, IgG-type M-protein). However, not transferred into cerebrospinal fluid are the M-components with relatively large molecules such as IgA-type, and polymerized IgG-type.[159] In addition, such characteristic serum proteinograms as those of the nephrotic syndrome, liver cirrhosis, defect-dysproteinemia and alloalbuminemia will be reflected in the cerebrospinal fluid.

Mixing of serum protein components: The normal selective permeability is destroyed and the protein patterns of the cerebrospinal fluid become similar to those of the serum. The following results are noted: an increase of total protein concentration, a decrease or disappearance of the prealbumin and β_2 fractions, an increase of the α_2 and γ fractions, and abnormal appearance of β-lipoprotein, α_2-macroglobulin and IgM-immunoglobulin. In acute inflammatory disease, C-reactive protein is also demonstrated in addition to an increase of acute phase reactants.

As examples of this abnormality ("Mischelectrophorese diagramm" of BAUER), the following may be cited: 1) the cases where there are increased capillary permeability of the meninges or the choroid plexus, 2) the cases where mechanical obstruction of subarachnoidal space is present. As the former case, acute and chronic inflammatory changes (e.g. meningitis, etc.), intracranial tumors and cerebral apoplexy may be cited. As the latter case, on the other hand, the same protein pattern is observed also in the cerebrospinal fluid present peripheral to the obstructed lesion. A disease of particular interest is the GUILLAIN-BARRÉ syndrome, which usually shows the obstructive type of electrophoretic protein patterns.[131]

Local production of immunoglobulins: When compared to electrophoretic serum protein patterns, there is a marked increase of the γ fraction. Here it seems that the immunoglobulins synthesized by the plasma cells infiltrating the central nervous tissues are released into cerebrospinal fluid.[133]   The γ fraction may increase as much as 66%.   In addition to these quantitative changes of the immunoglobulins, even IgM-immunoglobulin may be detected in a significant amount in trypanosomiasis, neurosyphilis and benign lymphocytic meningitis.   Also, more than two or three narrow IgG type M-components may be demonstrated in the γ fraction.[78]   It may be thought that this is caused by reactive proliferation of plasma cells of different clones reacting against local stimulation.

Diseases of the central nervous system accompanying a marked increase of the immunoglobulins include various subacute and chronic infections, such as neurosyphilis, tuberculous meningitis, trypanosomiasis, and chronic lymphocytic meningitis, and brain tumors. In addition, an increase of the immunoglobulins is often observed in multiple sclerosis.[78,129,133]   The same phenomenon may also be observed in various degenerative diseases.   M-components in the γ fraction are sometimes seen in polyneuritis, multiple sclerosis, and the GUILLAIN-BARRÉ syndrome.

Atrophic degenerative processes: As was mentioned previously, β_2-transferrin is observed in healthy cerebrospinal fluid.   In abnormal cerebrospinal fluid of this category, the mobility of transferrin, hemopexin, α_1-acid glycoprotein or α_1-antitrypsin becomes slower.

This type of change is noted in degenerative disease of the central nervous system.[78] However, it is not in infections of the central nervous system and is very rare even in multiple sclerosis.

Chapter 13

Catabolism of the Plasma Proteins

CATABOLIC RATES OF THE PLASMA PROTEINS

Each plasma protein component has a special life span of its own. A part of the body pool is continuously being destroyed while the same quantity is being synthesized anew. This keeps plasma proteins in a state of dynamic equilibrium, thus making it possible to determine the degradation rate or the catabolic rate as described previously. Moreover, the degradation rate and the synthetic rate should be equal, if the plasma protein metabolism in the body of a patient is in a steady state during the period of its study. Therefore, the degradation rate of plasma proteins which are in a steady state is also called as the *turnover rate*, and the degraded quantity per day, as the *turnover*. The turnover rate is sometimes expressed as g/day, but because the turnover differs according to bodily weight, the turnover rate is expressed as g/kg/day, using one kilogram of body weight as a basis for calculation. However, under abnormal conditions where plasma protein catabolism varies, it is of course impossible to obtain the synthetic rate from the degradation rate. Accordingly, to obtain the synthetic rate, the method by GROSSMA et al.[54] who use ^{131}I–labeled protein, or a complicated procedure of measuring biosynthesis directly should be employed.[133]

Plasma proteins must be purified without causing any degeneration in order to determine the degradation rate of each plasma protein component. Besides ^{131}I must be labeled properly. Therefore, reliable results have been obtained with regard to only limited kinds of plasma proteins[133] (Table 21).

Table 21 Biological half-life (T1/2) and catabolic rates for various plasma proteins (modified from SCHULTZE and HEREMANS).

Plasma proteins	T 1/2 (days)	Fractional catabolic rate (%/day)	Absolute catabolic rate (%/day)
Prealbumin	1.9		36.3(33.1–39.5)
Albumin	19(17–23)	10(7–11)	4.5
α_1-acid glycoprotein	5.2(5–7)		11.5–14
α_1-lipoprotein	4.2–4.8	14–16	14–16 ?
Haptoglobin	3.5–4.0		
$S_{f6-9}(\beta)$ lipoprotein	3.1–3.4	20–22	20–22 ?
Transferrin	8.5(7–10)	8–17	4–8
$_{Ig}$G-globulin	18(15–26)	5(3.5–7.2)	2.5
$_{Ig}$A-globulin	5–6.5	37.5	15
$_{Ig}$M-globulin	5.1(3.8–6.5)	18.7(14.1–25.1)	
Fibrinogen	4.0–5.5	12–16	12–16

There are two ways of expressing the catabolic rate. One is the *fractional catabolic rate* which indicates what percentage of the intravascular pool is catabolized daily. On the other hand, the *absolute catabolic rate* indicates what percentage of the entire body pool is catabolized daily.

SITES OF PLASMA PROTEIN DEGRADATION

Roughly two sites of plasma protein degradation may be indicated. One is the site where secretions and excretions are discharged out of the body, including the gastrointestinal tract, the kidneys, and other exocrine organs such as the lacrimal glands, the respiratory organs and the genital organs. The other site of degradation in a strict sense or endogenous catabolism includes the liver, the reticuloendothelial tissues and other tissues.

With regard to albumin in a healthy body, it is supposed that 70% is degraded in the gastrointestinal tract and 10–15% in the kidneys.[133] The gastrointestinal tract and the kidneys play a major role, but they will be discussed in the following chapter, while here attention will be focused on endogenous catabolism.

Endogenous catabolism of plasma proteins is generally thought to occur in almost all tissue cells, and it seems to be controlled under the common mechanism to that of other tissue proteins as well.[133,154] However, only in the liver cells has it been confirmed to occur,[27] and no other tissues have been proved to contribute to it. In the liver, about 12.8% of the albumin turnover is catabolized,[27] and about 30% of the turnover of IgG-immunoglobulin and transferrin.[28]

In the reticuloendothelial tissues, only the degenerated plasma proteins seem to be taken up and to be degraded.[133] Thus, the reticuloendothelial tissue may be useful to remove effectively plasma proteins of non-physiological form present in circulating blood. That is, the antigen-antibody complex, intravascular fibrin clot, chylomicron, degerated plasma proteins and haptoglobin-hemoglobin complex are some of the examples to be mentioned.

MODES OF DEGRADATION OF THE PLASMA PROTEINS

In normal states, each plasma protein is continuously being turned over in a fixed rate of its own, and three different modes of degradation have been observed[41] (Fig. 76). Daily degradation of plasma proteins is closely related to its intravascular pool and it is hardly affected by its extravascular pool.

Mode I: Fibrinogen may be cited as an example of the plasma proteins in this group. It exhibits a fixed fractional catabolic rate, even when its intravascular concentration varies. The total amount of proteins degraded daily is in direct proportion to their intravascular concentration. Therefore, exactly the same result is obtained clinically as that of a normal person, with regard to the biological half-life of injected fibrinogen in the patient suffering from afibrinogenemia.

Mode I degradation may be attributed to the fact that a fixed clearance is maintained by simple mechanisms like pinocytosis or gastro-intestinal loss.

Mode II: Proteins known to belong to this group include hatoglobin and transferrin.[41] Their fractional catabolic rate varies in inverse proportion to the protein concentration in plasma and the daily turnover shows a fixed value independent of its plasma concentration.

Mode II degradation may be atrributed to the fact that the plasma proteins are involved in the normal metabolic process and that a certain quantity of plasma proteins is consumed.

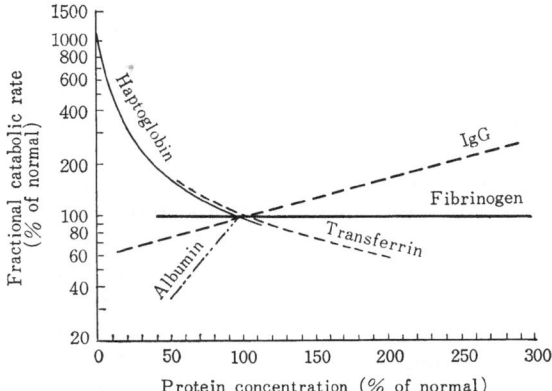

Fig. 76 Different modes of the catabolism of the five major plasma proteins (from FREEMAN).
The plasma proteins seem to be divided into the three groups depending upon the mode of their catabolism. The first group includes fibrinogen, the second group haptoglobin and transferrin, and the third group albumin and IgG.

In the case of haptoglobin, it is conceivable that haptoglobin is involved in hemoglobin metabolism and that it is consumed in reticulo-endothelial tissues.[98,133]

Mode III: Plasma proteins of this group include albumin and IgG-immunoglobulin. The fractional catabolic rate varies in direct proportion to the protein concentration in plasma.   It is conceivable that the catabolic rate is under the control of a feedback mechanism.   When albumin or IgG-immunoglobulin has been injected into the patient with analbuminemia or agammaglobulinemia, it may be noted clinically that its biological half-life is remarkably prolonged.[11,85]   Also, in the cases of hypoalbuminemia, the fractional catabolic rate of albumin decreases notably, while in the case of hyper-gamma globulinemia, that of IgG-immunoglobulin increases.[133]

MECHANISMS OF CATABOLIC ABNORMALITIES OF THE PLASMA PROTEINS

Abnormal degradation of plasma proteins include, 1) a decrease of degradation and, 2) an increase of degradation.   However, it is hardly encountered where a decreased catabolic rate results in an increase of the body pool or the intravascular pool of plasma proteins.   Clinically speaking, as an important example of abnormal changes in plasma protein degradation, hypoproteinemia occurs due to an increased degradation rate.

That is, when the synthetic rate in cells cannot keep up with abnormally accelerated degradation, the body pool and the intravascular pool will become decreased.

1.   Decreased Catabolism of the Plasma Proteins

As was previously mentioned, the fractional catabolic rate of the plasma proteins included in Mode III is decreased through a feedback mechanism when the intravascular pool decreased abnormally exhibiting a certain compensatory activity. When hypoalbuminemia is associated with malnutrition, liver cirrhosis, dextran ingestion, or experimental plasmapheresis, there is a decrease in the catabolic rate of albumin.[133]

2.   Increased Catabolism of the Plasma Proteins

Abnormal acceleration of plasma protein degradation takes place in cases where the following factors are present clinically; (1) increase of external protein loss due to various causes, (2) excessive consumption within the body of some particular plasma proteins, (3) acceleration of endogenous catabolism within tissue cells. The first factor will be discussed in the following section.

Increased consumption within the body.: This is met with fibrinogen, haptoglobin, hemopexin, and β_{1c}-globulin. These plasma protein components have their own biological functions in the living body, and they are consumed excessively in pathological conditions where these biological functions operate excessively.

Fibrinogen is consumed in large quantities as the result of widespread intravascular blood coagulation or the coagulation within various tissues. Also, so-called fibrinogenolysis occurs in the living body as a result of the accentuation of fibrinolytic activity followed by defibrination state.[136] This is observed in various complications in pregnancy, such as premature detachment of the placenta, *in utero* retention of dead fetus, amniotic fluid embolism, infected abortion, and so on.[84,112,115,130] Major thoracic surgery, incompatible blood transfusion, thrombocytopenia associated with hemangioma may also be noted.[15,46]

Haptoglobin is selectively combined with free hemoglobin *in vitro* and *in vivo*, and forms the haptoglobin-hemoglobin complex. It is thought to be taken within the reticuloendothelial system, and removed from circulating blood quickly. Therefore, in various hemolytic anemias where large quantities of hemoglobin are freed from erythrocytes, often a lack of haptoglobin in serum is observed.[62,101,103,163] It may be thought that this occurs because freed hemoglobin is bound to haptoglobin which will be consumed significantly in the body.

Hemopexin binds specifically with heme which is freed from hemoglobin, and, like haptoglobin, it decreases or becomes deficient in various hemolytic anemias.[96,100,135,162] It is thought that this is because hemopexin is consumed in the body by heme which has been freed in large quantities. β_{1c}-globulin is the third component of complements and it is to be consumed at the time of *in vivo* antigen-antibody reaction, such as systemic lupus erythematosus and other auto-immune processes.[66,95]

In addition, the same situation may be applied with respect to antibody globulins. However, their *in vivo* consumption may not be significant enough to cause a decrease of the total immunoglobulin concentration in plasma.

Transferrin is physiologically a carrier protein for iron, but in contrast to the haptoglobin-hemoglobin complex, it binds iron reversibly and may have different metabolic behavior. In fact, transferrin is catabolized in the similar way to albumin.

Increased endogenous catablism in tissue cells: Plasma proteins of this type include albumin and transferrin. Some of the other plasma proteins may behave metabolically in the same way, but albumin is the one studied most extensively, because of its large quantity present in the body. In general, plasma protein catabolism is accelerated in the pathological conditions where tissue protein catabolism is increased.[154]

Increased endogenous catabolism of albumin is encountered in acute inflammatory diseases, stress syndrome (major surgery and trauma) and fever, and hypoalbuminemia is recognized along with an increase of acute phase reactants.[72,147,153] Also, found is an increased albumin turnover when the thyroid hormone and anabolic steroids are administered[69,125,126] JARNUM and LASSEN[72] indicated that transferrin catabolism increased markedly in acute inflammation as albumin catabolism. It is not certain how these two proteins are related in their metabolism. However, it is apparent that transferrin is not only related to iron metabolism.

Chapter **14**

External Loss of the Plasma Proteins

SITES OF EXTERNAL LOSS OF THE PLASMA PROTEINS

As previously noted, plasma proteins are also widely distributed outside the blood vessels and they are metabolized constantly. Not only are they degraded within tissue cells through endogenous catabolism, but they are also lost into almost every bodily secretion or excretion. These secretions and excretions contain plasma proteins in varying amounts and their representative ones are listed in the following:

1) Urine and male genital secretions
2) Digestive fluids (saliva, gastric, intestinal and pancreatic juices, bile)
3) Tracheobronchial secretions, nasal fluid, sputum
4) Female genital secretions and maternal milk
5) Miscellaneous fluids, such as tears, sweat.

Among these, urine and digestive fluids are most important. When the external loss of plasma proteins is abnormally increased and surpasses their synthetic rate, hypoproteinemia results.

Abnormally marked loss of plasma proteins occurs in the following cases: 1) bleeding and hemoptysis, 2) punctures removing large amounts of effusions accumulated in serous cavities, 3) proteinuria, 4) loss into digestive fluids (protein-losing gastroenteropathies), 5) exudation due to strong inflammatory or necrotic changes in the body surface (skin or mucosa).

URINARY PROTEINS

Degradation of the Plasma Proteins in the Kidneys

It is a well-known fact that nearly all plasma constituents except blood cell elements are tranferred into the glomerular filtrate through renal glomeruli. It has been confirmed that some plasma proteins are also transferred into the glomerular filtrate through the capillary wall of the glomerulus. Plasma proteins in the filtrate are, like other plasma constituents, reabsorbed through the renal tubules, but it is thought that certain plasma proteins are partially degraded in the proximal tubular cells after having been reabsorbed. Besides, considering that the glomerular filtrate amounts to approximately 180 l every day, it is conceivable that considerable amounts of plasma proteins are subject of filtration, reabsorption, or degradation in the kidneys. SCHULTZE and HEREMANS[133] have suggested, based on various experiments, that approximately 10–15% of daily albumin degradation is catabolized in kidney tissues.

Protein Pattern of Normal Urine

In healthy individual also, small amounts of plasma proteins are transferred into glomerular filtrate through the glomerular capillaries and some 40–80 mg daily are further excreted out of the body as urinary proteins, although the ordinary qualitative tests for urinary proteins show a negative result.

For the purpose of quantitative or electrophoretic analyses, urine sample needs to be concentrated because normal urine contains only an extremely small amount of proteins. Ordinarily, for clinical purposes, urine is dialyzed in concentrated solutions of high-molecular weight substances like polyvinyl pyrrolidone, polyethylene glycol, carboxymethyl cellulose, and dextran, while for the purpose of researches, gel filtration or ultra-filtration is recommended.   Various methods are being empolyed for measuring total protein content in urine.   However, it should be noticed that the dip sticks using the "protein error of indicators" is not suitable for the demonstration and quantitative analysis of BENCE JONES protein.[137]   On the other hand, such proteins as hemoglobin and myoglobin are not positive in the heat test.

Urinary proteins may come from the plasma, the kidney tissues (the renal tubular epithelium in particular) and the lower urinary tracts.   However, the origins of every urinary protein component have not been confirmed.   For the purpose of convenience, SCHULTZE and HEREMANS classify them into the following three groups:[133]

1)   Plasma proteins (the molecular weight, more than 30,000)
2)   Urinary proteins of low molecular weight
3)   Uromucoids or urinary glycoproteins of TAMM-HORSFALL.

Of these, the greatest in quantity are the urinary proteins of low molecular weight. They constitute approximately 50% of the total urinary protein content in normal urine and are usually not demonstrated in plasma.   Therefore, most of these proteins are thought to come from the urinary tracts.[129,133]   Uromucoids originate in the renal tissues and its characteristics were first elucidated by TAMM and HORSFALL.[148]

The plasma protein components always observable in normal urine by immunological techniques include albumin, a_1-acid glycoprotein, a_1-antitrypsin, a_1-lipoprotein, Zn-a_2-glycoprotein, a_{2HS}-glycoprotein, transferrin, hemopexin, β_{1E}-globulin, IgA-immunoglobulin, and IgG-immunoglobulin; prealbumin, ceruloplasmin, haptoglobin (Hp 1–1), and a trace of a_2-macroglobulin may be detected.[127,133]   But fibrinogen, β-lipoprotein, and IgM-immunoglobulin are not detected in normal urine (Fig. 77).   One may conclude from these facts that plasma proteins with the molecular weight of less than 200,000 can be detected rather frequently, and that macro-molecules are not observed.   Like the capillary walls of other organ tissues, the molecular sieve effect may be exhibited on the process of glomerular filtration (Fig. 77).   Electrophoretic separation of proteins in normal urine is often very poor.   By filter paper electrophoresis, the albumin fraction is a little less than 40% and the globulin fraction is about 60%, A/G ratio being 0.51–0.65.   However, by cellulose acetate electrophoresis, the albumin fraction is 60% or so.   The low percentage of the albumin fraction on filter paper electrophoresis is presumably caused by albumin tailing.   The majority of the globulin fractions is occupied by the a fraction, a_2-fraction in particular, which is mainly made up of protein components coming from the urinary tracts.   There is a report which describes the prealbumin fraction on TISELIUS method,[19] and it seems to be composed of acid mucopolysaccharides.[133]   Ordinarily, the albumin fraction in normal urine has a relative mobility somewhat higher than the albumin in serum.   It is thought that this is because acid mucopolysaccharides present in urine in large amounts have been combined with albumin.   However, according to MERLER et al.,[90] the albumin in urine is the serum albumin which has been partially

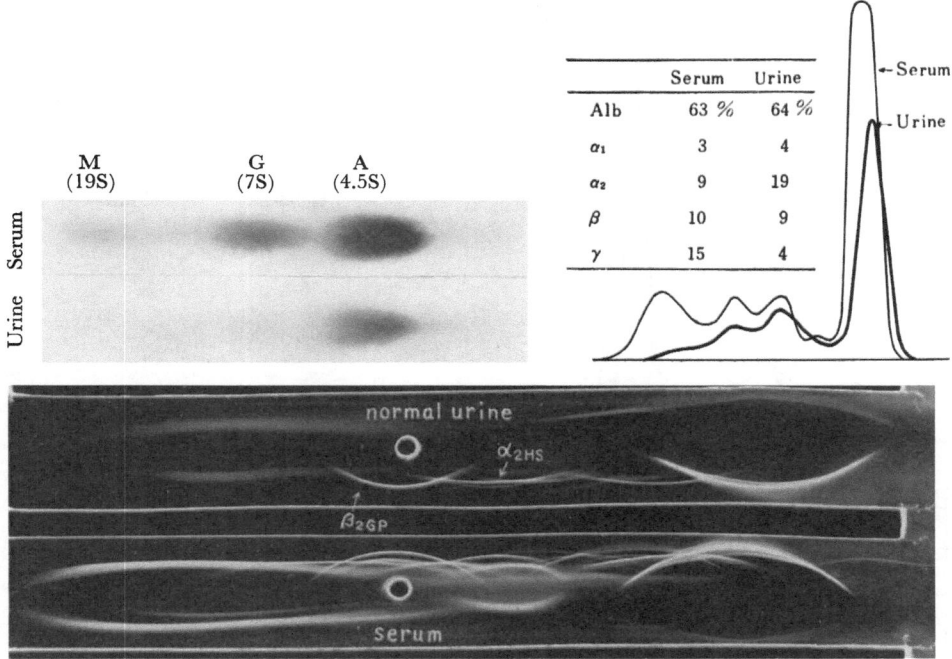

	Serum	Urine
Alb	63 %	64 %
α_1	3	4
α_2	9	19
β	10	9
γ	15	4

Fig. 77 Protein fractional patterns of normal urine.

On cellulose acetate electrophoresis, in comparison to the normal serum protein pattern, the globulin fractions are separated indistinctly, showing a predominant α_2 fraction and a flat γ fraction. The α_2 fraction is mainly composed of various tissue proteins. The albumin fraction is separated clearly, but migrates a little faster than that of the serum, apparently due to the binding of polysaccharides. On immunoelectrophoresis, distinctly identified are albumin, α_1-antitrypsin, α_{2HS}-glycoprotein, and transferrin, and also prealbumin, Gc-globulin, ceruloplasmin, α_2-glycoprotein and IgG. α_2-macro-globulin and IgA are only identified weakly. The line reacting against the anti-IgG is different from those of the serum IgG, migrating in a different zone, and it may indicate the presence of denatured IgG molecules in normal urine. On thin-layer Sephadex G-200 gel filtration, the most of the normal urinary proteins are present in the A fraction. The M fraction (19 S) is entirely lacking, and the G fraction is scarcely recognized (upper left).

degraded and it is of only 2.6S. The γ fraction constitutes approximately 10%, but, in ultracentrifugal analysis, those of more than 4S are not demonstrated. It is thought that there is only a trace of so-called 7S γ-globulins, if any, being present in less than 10% of the total urinary γ-globulins identified immunologically.[127,133] The greater parts of urinary γ-globulins are of considerably low molecular weight, 1.1–2.4S, being demonstrated by immunoelectrophoresis to form several precipitin arcs against anti-human immunoglobulins.[133] At least, part of these low-molecular weight γ-globulins are equivalent to BENCE JONES proteins in physical and chemical characteristics. The author has also confirmed the normal urine concentrated by more than 100 times becomes turbid at 56°C and re-dissolves at above 100°C. It is said that a trace amount of these low-molecular weight γ-globulins is also contained in normal plasma.[12,146] Moreover, TAKATSUKI and OSSERMAN believe that BENCE JONES proteins of K and L types are included in the low-molecular weight γ-globulins.

Protein Patterns in Various Proteinuria

Proteinuria is observed in many pathological conditions. However, considering the

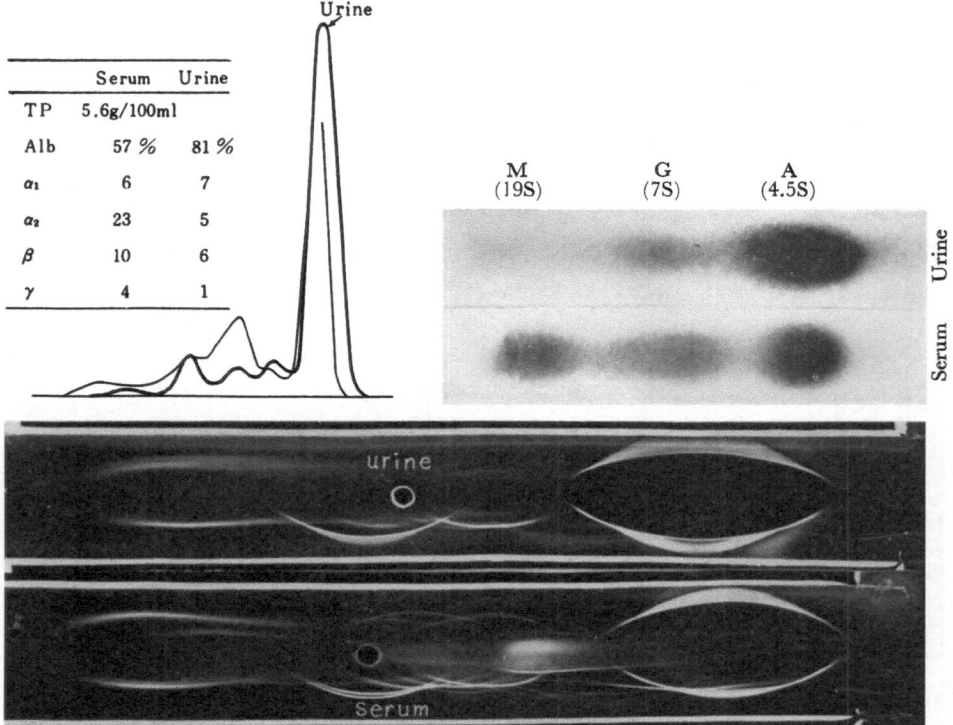

	Serum	Urine
TP	5.6g/100ml	
Alb	57 %	81 %
α_1	6	7
α_2	23	5
β	10	6
γ	4	1

Fig. 78   Protein fractional patterns in glomerular proteinuria.
12 y.o., female, lipoid nephrosis.
The serum protein fractional patterns are typical of the nephrotic type. In comparison to that of the serum, the urinary protein pattern shows a relative increase in the albumin fraction, and a relative decrease in the α_2 and γ fractions. On immunoelectrophoresis, prominently seen are albumin, α_1-antitrypsin, α_{2HS}-glycoprotein, transferrin, and IgG; also vaguely recognized are prealbumin, α_1-acid glycoprotein, Gc-globulin, ceruloplasmin, hemopexin, and IgA. α_{2M} is faintly seen, but LDL is not demonstrated.

mechanism of its occurrence, it may be classified into the following four groups. These are: 1) glomerular proteinuria, 2) prerenal proteinuria, 3) tubular proteinuria, and 4) proteinuria due to vascular rupture, i.e., hematuria and chyluria.

1. Glomerular Proteinuria

A portion of the plasma proteins is filtered through the glomerular basement membrane and is excreted into urine. Therefore, if the selective permeability of the glomerular basement membrane is increased by a certain cause, a considerable amount of plasma proteins may be excreted into urine. This type of proteinuria is called, "glomerular proteinuria."

The proteinuria of this type exhibits a characteristic protein pattern and it has been called albuminuria because the greater part is made up of albumin. On electrophoretic pattern, in comparison to blood serum, it is characterized by a relative increase of the albumin and α_1 fraction, and a relative decrease of the α_2, and γ fractions. It also has a relatively narrow β fraction. On immunoelectrophoresis, albumin, α_1-acid glycoprotein, α_1-antitrypsin and transferrin form major precipitin arcs (Fig. 78). The pattern indicates the presence of the selective permeability of the glomerular basement membrane. In fact, such macromolecules as α_2-lipoprotein, IgM-immunoglobulin, α_2-macroglobulin and

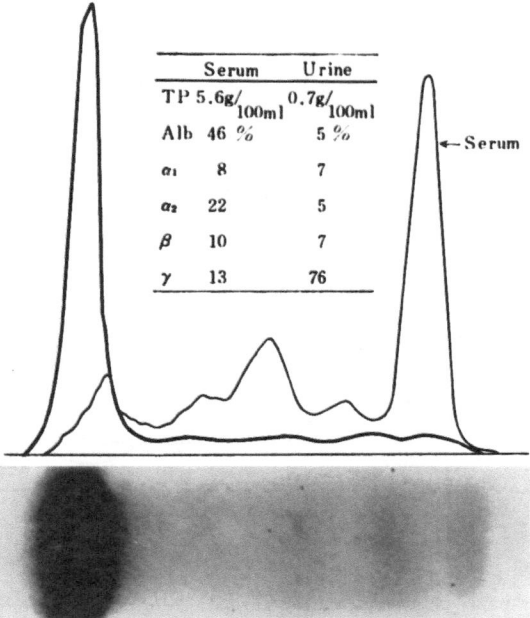

	Serum	Urine
TP	5.6g/100ml	0.7g/100ml
Alb	46 %	5 %
α_1	8	7
α_2	22	5
β	10	7
γ	13	76

←Serum

Fig. 79 Filter paper electrophoretic pattern in prerenal proteinuria
(BENCE JONES proteinuria).
65 y.o., male, multiple myeloma of BENCE JONES type.
The γ fraction of the serum proteins is decreased, showing no M-
protein band. The urinary protein pattern shows a distinct M-protein band (BENCE JONES protein) at the γ zone.

fibrinogen are hardly demonstrated. With respect to IgG-immunoglobulin, its amount
in urine is related to its plasma concentration, but it never exceeds its plasma level.
Besides, IgA-immunoglobulin, Gc-globulin and hemopexin are often detected.

The permeability of the glomerular basement membrane will increase in the following
cases: 1) when there is an inflammatory change of the glomeruli, 2) when there are abnormal deposits in the glomerular basement membrane, 3) when the glomerular capillary
blood pressure is markedly increased.

Severe glomerular proteinuria is encountered in nephrotic syndromes associated with
membranous, lobular or exudative glomerulitis. In nephrotic syndromes, 5–25g or more
of serum proteins are lost in urine every day. Additionally, in glomerulonephritis, also
proteinuria occurs with the same mechanism.[154]

In amyloid kidney or diabetic nephropathy, where there are abnormal deposits in the
glomerular basement membrane, its permeability increases also.

In renal vein thrombosis, it is thought that increased glomerular permeability comes
from markedly increased glomerular capillary blood pressure caused by venous stagnation
in the kidneys. Moreover, the proteinuria which is induced by renin and angiotensin is
also thought to be caused by increased glomerular capillary blood pressure.[34]

Postural (or lordotic, orthostatic) proteinuria may belong to this group. This proteinuria occurs frequently in young, thin persons with visceral ptosis. This holds true also
for patients having glomerular lesions with proteinuria, and proteinuria is aggravated if
the patient is in a standing position. LATHEM et al.[79] think that this is due to increased
glomerular filtration of plasma proteins caused by significant stagnation of glomerular

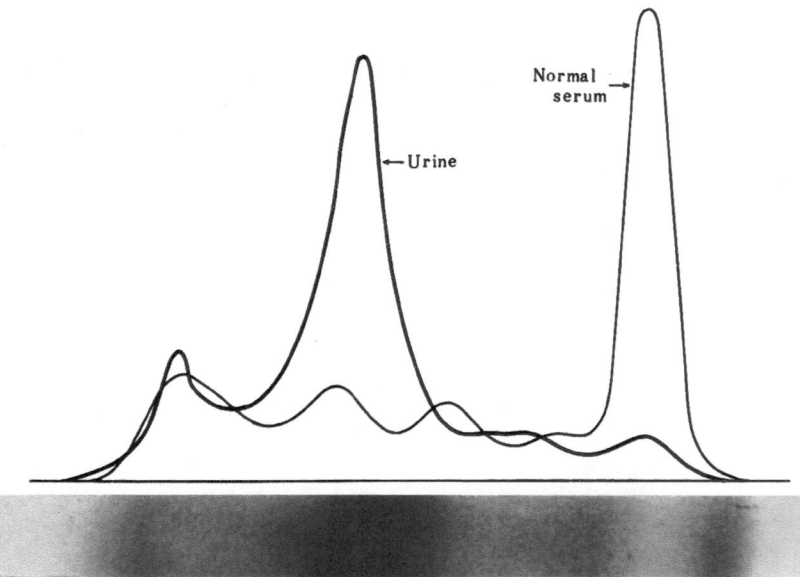

Fig. 80   Filter-paper electrophoretic protein pattern in prerenal proteinuria (hemoglobinuria).
16 y.o., male, tetanus with hemoglobinuria.
Six days before admission, a needle was pierced through the right foot.   Two days
before admission, complained of muscular pain and hemoglobinuria.
The total urinary protein is 200 mg/100 ml, and on the filter-paper electrophoresis,
the albumin 5.8%, the α_1 and α_2 fractions 7.5%, the β 66.5%, and the γ 20.2%.
Before the BPB staining, the β fraction is distinctly red and benzidine-positive. Normal
adult hemoglobin (HbA) migrates at the β fraction on the routine electrophoretic
fractionation.

blood circulation in a standing position.   In this case, the total ordinary protein excretion
is usually small, but its protein pattern resembles that of the nephrotic syndrome showing a
characteristic selective permeability.   Also included in this group are febrile proteinuria
and the proteinuria associated with shock and congestive heart failure.

2.   Prerenal Proteinuria

In this group, the glomerular permeability remains normal.   However, the serum
proteins of relatively low molecular weight are excreted into urine when they are increased
pathologically in blood.   Examples of this type are myeloma, macroglobulinemia WALDEN-
STRÖM and heavy chain disease where immunoglobulin fragments (BENCE JONES protein,
heavy chain fragments) increased in the blood (Fig. 79).   Besides, hemoglobulinuria is
observed when plasma hemoglobin concentration increases significantly as a result of intravascular hemolysis (Fig. 80).   In the same way,   a large amounts of myoglobin is discharged into the blood in widespread muscular necrosis caused by crash syndrome, arterial
embolism, convulsion, myositis, HAFF's disease, muscular dystrophy, familial paroxysmal
myoglobulinuria and fetal myoglobulinuria, and is excreted into urine through the
glomerular basement membrane.

On the other hand, a marked increase of α_1-antitrypsin and α_1-acid glycoprotein in
acute inflammation or acute stress causes these low-molecular weight proteins to be filtered
into urine easily resulting slight proteinuria.

3. Tubular Proteinuria

Although the mechanism is not completely understood, the proteinuria of this type is called "tubular proteinuria" because of the fact that it is accompanied with renal tubular damage.

Protein patterns in tubular proteinuria differ strikingly from those of glomerular proteinuria. On electrophoresis using various supporting media, the α and β fractions are conspicuous, while the albumin fraction is comparatively small. The electrophoretic patterns resemble that of normal urine. On immunoelectrophoresis, the patterns are nearly the same as those of normal urine, as long as the anti-human whole serum antisera are used.[133] The proteinuria of this type is not precipitated by the heat test or the picric acid method of ESBACH.[16,42] But it is precipitated by 4% trichloroacetic acid, sulfosalicylic acid and nitric acid. It seems therefore that the greater part of urinary proteins excreted in tubular proteinuria is of low-molecular weight, being 2–2.5S.[133] Tubular proteinuria is encountered in chronic cadmium poisoning, phenacetin intoxication, vitamin D intoxication, severe chronic hypokalemia, myeloma kidney (induced by BENCE JONES protein) chronic pyelonephritis and various hereditary tubular disorders (WILSON's disease, galactosemia, and FANCONI syndrome).

In addition, the physiological proteinuria which is observed after hard physical exertion, exposure to cold or a bath is a particular case which is thought to fall in this category.

4. Chyluria

Chyluria is observed when the intestinal or retroperitoneal lymphatics are open into the urinary tract. Ordinarily, it is due to (1) inflammatory obstruction of lymphatics usually caused by filariasis, (2) compression of retroperitoneal lymphatics by tumors or the like, and (3) trauma of urinary lymphatics or lymphangioma.

The protein patterns of chyluria are almost the same as the serum patterns, but the albumin fraction is comparatively high. The presence of fibrinogen is rather marked. It contains a large amount of neutral fat, appearing lactescent, but cholesterol is practically absent.[80,139]

PLASMA PROTEINS IN VARIOUS SECRETIONS

Plasma Protein Compositions in Various Types of Secretions

The exocrine organs produce various secretions, each having its characteristic functions. Proteins found in these secretions may be divided roughly into two groups: 1) those specific to each particular secretion and 2) plasma proteins. The former includes the specific proteins with various biological activities, depending upon the physiological functions oɪ each exocrine gland. In contrast, plasma proteins are independent of the type of exocrine cells, presumably being filtered out of the surrounding capillary vessels and mix into secretions through interstitial tissues. Accordingly, they are found in nearly every secretion, a major constituent being serum albumin. Also, observed are prealbumin, α_1-acid glycoprotein, α_1-antitrypsin, transferrin, and IgG-immunoglobulin. In addition, a very small amount of the following may be detected: hemopexin, haptoglobin, ceruloplasmin, α_2-macroglobulin, $\beta_{1C/A}$ globulin, IgA-immunoglobulin, IgM-immunoglobulin and lipoproteins.

In analysis of these secretions, care should be taken for collection of test samples. Because secretions contain a large amount of enzymes and acids, their protein constituents are easily destroyed. Particularly, the acid and the proteolytic enzymes destroy proteins in gastric and intestinal juices. Thus, before gastric juice is collected, it is necessary to wash the gastric mucosa with a phosphate buffer of pH 7.2. On the other hand, when collect-

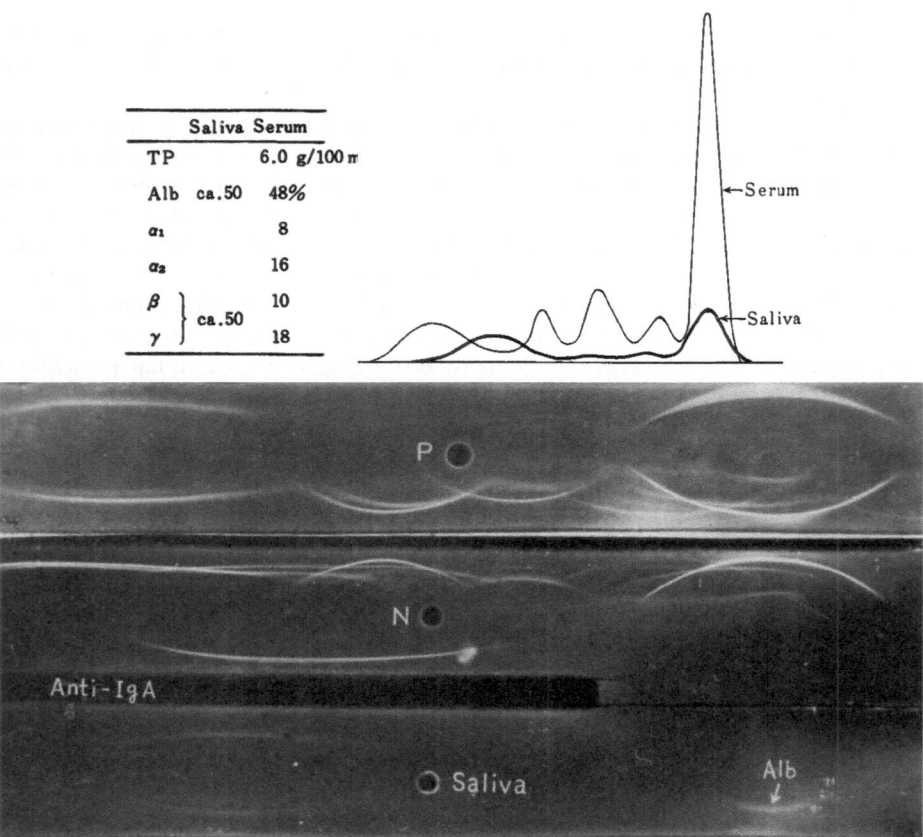

	Saliva	Serum
TP		6.0 g/100 m
Alb	ca.50	48%
α_1		8
α_2		16
β	} ca.50	10
γ		18

Fig. 81 Protein fractional patterns of the saliva.
21 y.o., male, ulcerative colitis.
Complained of frequent diarrhea for 3 months prior to admission, in association with emaciation and anemia (Hb 9.3 g/100 ml). On X-ray examination, the abnormal findings of the mucosal folds are seen in the entire descending colon. Improved rapidly with the administration of steroid hormones. On admission, the serum protein electrophoretic pattern is of the acute inflammatory type, showing a marked increase of α_1-antitrypsin and haptoglobin. IgA is not identified on immunoelectrophoresis, but is only immunochemically in a trace amount. The cellulose acetate electrophoretic pattern of the saliva shows only the albumin and β_2 fractions, and immunoelectrophoretically identified is a significant amount of the secretory IgA.

ing intestinal juices, the intestine should be washed previously with trypsin inhibitor. Thimerosal or sodium azide should be added to the juices collected so that proteins may be preserved as well as possible. Generally speaking, albumin is destroyed, but immuno-globulins are resistant.

Secretions fall into two groups, depending on the content of "secretory IgA-immuno-globulin".

Group I: The secretions of this group contain secretory IgA-immunoglobulin in large amounts and show a distinct IgA precipitin arc on immunoelectrophoresis. In this group, the IgG/IgA ratio is characteristically low in comparison to that in serum.

The following secretions are included in this group: (1) saliva (Fig. 81,[24,44,133,150,151]) (2) gastric juice,[102,156] (3) intestinal juice,[24,31,67] (4) bile,[22,59,117,128] (5) tracheobronchial secretions,[3,33,86] (6) nasal secretions,[120,121] (7) tears[1,24,150] and (8) milk and colostrum.[40,57,58]

The exocrine organs which produce these secretions may be found to have common anatomical characteristics. With the exception of (8), they are directly exposed to the external environment, and are poorly resistant to mechanical and chemical damage. This point is very interesting, in view of the fact that local immunity is related to secretory IgA-immunoglobulin, as will be shown later. On the other hand, maternal milk seems to increase local immunity in infants, being transferred orally into the digestive tract.

Group II: The secretions of this group contains practically no secretory IgA-immunoglobulin or in a small quantity, if any. The IgG/IgA ratio of this troup is equal to or somewhat higher than that of serum. Included in this group are: (1) seminal fluid,[63,81,83] (2) cervical and vaginal secretions,[93] (3) aqueous and vitreous humor[36,39,108] (4) labyrinthine fluid,[23,45] and (5) sweat.[26,120]

It is presumed that, in contrast to those which produce the secretions of Group I, the exocrine organs which produce these do not easily admit the direct entrance of foreign substances or that mechanical resistance of the mucosa is comparatively strong. The plasma proteins demonstrated in the secretions of this group are mainly albumin, transferrin, and IgG-immunoglobulin.

Structure and Functions of the Secretory IgA-Globulin

DAVIES observed the appearance of specific agglutinin in the feces of a patient who had recovered from bacterial dysentery and he proposed the idea of "copro-antibodies"[32]. Since then, various antibodies have been demonstrated in gastrointestinal fluids and various secretions coming from other mucosa, and they have been named "muco-antibodies", which were thought to play the role of a certain local immunity[113]. Knowledges in this field have been progressing rapidly since TOMASI and ZIGELBAUM demonstrated in 1963 the existence of "secretory IgA-immunoglobulin" in parotid saliva and colostrum.[151]

Secretory IgA is a particular immunoglobulin that is contained in comparatively large amounts in various exocrine secretions, as shown in Table 22. They fall into four groups, depending on whether the IgG/IgA ratio is high or low.[24] In Group I, IgA constitutes a major, proportion, while IgG is hardly contained in the secretions including parotid

Table 22 Immunoglobulin concentration in various secretions
(modified from CHODIRKER and TOMASI).

Fluids	Immunoglobulin concentration (mg/100 ml)		
	IgG	IgA	IgG/IgA ratio
Group I			
Saliva	0	28	< 1
Colostrum	0	151	< 1
Tear	0	7	< 1
Group II			
Bile	143	53	2.6
Intestinal fluid	153	74	3.8
Group III			
Prostatic secretion	157	26	10
Vaginal secretion	37	6.3	—
Group IV			
Amniotic fluid	21	1.6	15
Serum	1335	178	8

saliva, colostrum and tears.   In Group II, IgG constitutes a major proportion but the IgG/IgA ratio is lower than that in serum.   Secretions in this group include bile and intestinal fluid.   Group III includes the prostatic and vaginal secretions in which IgA and IgG are contained in the similar proportion to serum.   In Group IV, the IgG/IgA ratio is higher than that in serum, and they include amniotic fluid.   IgM-immunoglobulin is practically absent in these secretions, and only the colostrum contains a trace amount of it.

Up to the present, the secretory IgA in parotid saliva has been examined more thoroughly than others.[149,151]   It differs from the serum IgA in that its sedimentation constant is 11.4S and its molecular weight is 390,000.   It is very strongly resistant to proteolytic enzymes and is not easily digested.   Secretory IgA is composed of two molecules of serum 7S IgA with one molecule of the "secretory (or transport) piece" of approximately 50,000–60,000 molecular weight (Fig. 82).   This secretory piece is antigenically specific for the secretory IgA.[9,68,140,141]

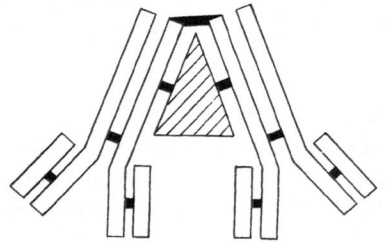

Fig. 82   A schematic model of the secretory IgA molecule (from Tomasi).
The secretory IgA is seen as a dimer to which the secretory or transport piece has been added.

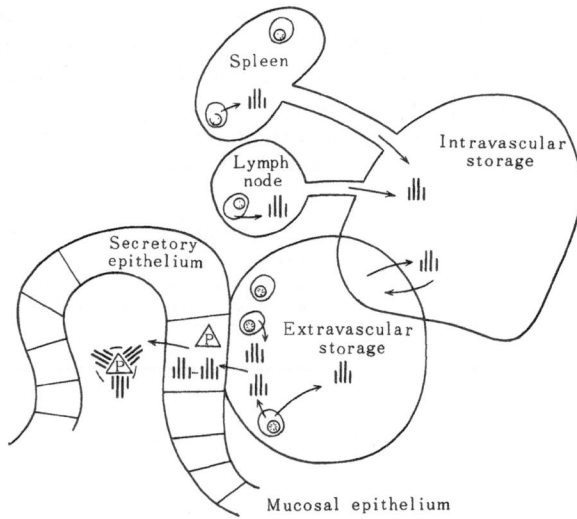

Fig. 83   A synthetic model of the secretory IgG (from South et al.).

The site of synthesis for the secretory IgA seems to be the epithelial cells in secretory glands or mucosa.[9,31]   It is presumed that the 7S IgA molecules produced in plasma cells are combined with the "secretory piece" to form secretory IgA during the secretion process.   Moreover, it is thought that not only does 7S IgA come from serum but that it is synthesized in plasma cells scattered in local tissues (Fig. 83).

Synthesis of secretory IgA is thought to be regulated independently from serum IgA. It has an important function at least for local immunity in the gastrointestinal tracts and the respiratory organs.[9,140,141] In addition, it is said that these local antibodies may regulate normal bacterial flora and that it may be related to the appearance of allergic or autoimmune diseases.[149]

Degradation of the Plasma Proteins in the Gastrointestinal Tract

Of all plasma proteins, albumin has been studied in the greatest depth. It is thought that its daily turnover amounts to approximately 10–15g and that 70% of the turnover is secreted into the gastro-intestinal tracts and is destroyed there.[14,133] In the kidneys and the liver, 10% and 15%, respectively, seems to be degraded. Thus, the amount of albumin degraded in the exocrine organs other than the digestive tracts may not reach 0.5 g daily.

Almost all plasma protein components, with the exception of macromolecules, are more or less found in the digestive fluids secreted and are degraded in the gastrointestinal tract. This has been proved by the experiments using ^{131}I-labelled proteins.[7,17,107] Experiments made on dogs suggest that when ^{131}I-labelled albumin is infused into the digestive tract, about 10% is degraded in the stomach, about 18% in the duodenum, about 25% in the jejunum, about 22% in the ileum and about 5% in the colon.[47,161] As to IgG-immunoglobulin, the experiments similar to those of albumin have been carried out,[2] and it is possibly true for other plasma proteins, also.

Protein Patterns in Pathological Digestive Fluids

It is possible that a great amount of plasma proteins is lost through the gastrointestinal tract under any of the following pathological conditions: 1) gastrointestinal bleeding, 2) cystic fibrosis, 3) villous adenoma, and 4) protein-losing gastroenteropathy.

In gastrointestinal bleeding, whole blood is lost, and therefore, all the plasma proteins are equally lost.

1. The Protein Pattern of the Intestinal Fluid in Cystic Fibrosis

Cystic fibrosis was once referred to as cystic fibrosis of the pancreas, but recently it is known as generalized exocrinopathy, mucoviscidosis, mucoviscoidosis or mucosis. As these names indicate, it is thought of as a congenital disorder involving the exocrine system. The electrolyte concentration of sweat and tears becomes increased, being close to that in serum. In addition, mucus becomes viscous and its secretion is disturbed mechanically. The pancreas, bronchi and biliary tracts are the organs that bring about the most remarkable clinical symptoms, causing lack of pancreatic exocrine secretion, obstructive bronchitis, and biliary cirrhosis. Because of the lack of pancreatic juice and a deficiency of pancreatic proteolytic enzymes, the plasma proteins lost to intestinal fluid remain relatively intact and albumin is found in large amounts.[25] Moreover, it is said that there is a large amount of albumin also in the meconium of a sick infant and that it constitutes 80–90% of the protein found.[53]

2. The Protein Pattern of the Intestinal Fluid in Villous Adenoma

Unlike the polypoid adenoma, villous adenoma is a benign tumor with a characteristic velvety appearance and it shows papillomatous proliferation in a wide area, encountered mainly in the rectum. It is a specific tumor which may secrete watery mucus in large amounts and causes shock as a result of electrolyte loss and dehydration.[56,61] Proteinograms of the intestinal fluid which is secreted in large amounts in this condition resemble those of normal intestinal fluid. The IgA-immunoglobulin having electrophoretic mobility in β_2 (probably secretory IgA-immunoglobulin) is contained in the largest

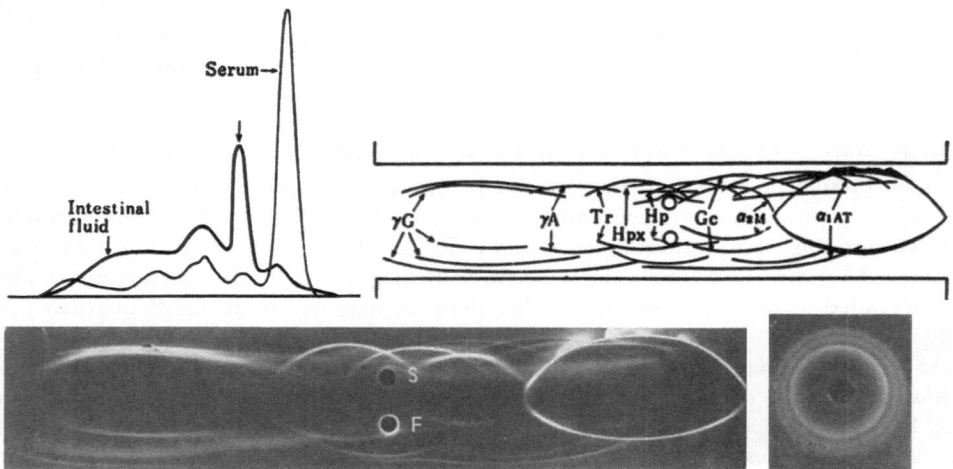

Fig. 84 Protein fractional patterns of the intestinal fluid.
2 m.o., male, congenital megacolon.
A large amount of exudate is excreted from the artificial anus, postoperatively. The exudate is distinctly separated from the feces, and is of watery character.
The serum protein electrophoretic pattern is of the malnutritional type; TP 5.1 g/100 ml, Alb 62%, α_1 7%, α_2 15%, β 8%, and γ 8%.
The intestinal exudate shows prominent α_2 and β_2 fractions on cellulose acetate electrophoresis. A sharp peak (arrow) at the point of application is due to denatured proteins. On immunoelectrophoresis, prominently seen are albumin, α_1-antitrypsin, hemopexin and transferrin; and also, faintly recognized are Gc-globulin, α_2-macroglobulin, haptoglobin and IgA. IgM is also recognized faintly on single radial immunodiffusion. On the single radial immunodiffusion method using the anti-γ, three different rings are seen eccentrically, indicating the presence of IgG-fragments having different molecular weights. Because of poor development of the IgA system, the IgA precipitin line is rather faint.

proportion and in addition, albumin and IgG-immunoglobulin are contained in small amounts.[87]

3. The Protein Patterns of the Gastric and the Intestinal Fuids in Protein-Losing Gastroenteropathy

As described previously, the protein components in gastrointestinal secretions are rapidly digested by proteolytic enzymes. This makes it difficult to be collected without any degeneration of the protein components in secretions even if washing by alkaline buffer solution or trypsin-inhibitor is employed. Therefore, the protein patterns of gastric and intestinal fluids vary significantly. At any rate, the fact that plasma proteins are lost in large amounts has been confirmed by various studies. It is presumed, moreover, that the protein patterns are almost the same as those in plasma[67,102,116,133] (Fig. 84).

4. The Analytical Methods of Proving Gatrointestinal Loss of the Plasma Proteins.

As described previously, the demonstration of plasma protein components by direct analysis of digestive fluids is neither reliable nor stable. To clinically verify plasma protein loss into the gastrointestinal tract, the following procedures using isotopes are employed.

Procedure Employing Labelled-Plasma Proteins: ^{131}I-labelled serum albumin is intravenously injected and radioactivity is measured. In a healthy man, an average 0.14% (0–0.39) of all the radioactivity which has been intravenously injected is demonstrated in feces.[71,144]

In protein-losing gastroenteropathy, a larger amount of activity is detected in feces. Actually, precise evaluation is often difficult because the labelled albumin lost into digestive fluid is digested and then freed ^{131}I is reabsorbed.

JEEJEEBHOY and COGHILL worked out a way to prevent the re-absorption of freed ^{131}I in the digestive tract by orally administering Amberlite IRA-400 (Cl$^-$)[74]. Also, WALDMANN recommends $^{51}CrCl_2$-labelled albumin[158].

Procedure employing artificial colloids: GORDON devised a procedure in which ^{131}I-labelled PVP (polyvinyl-pyrrolidone) is used[49,50]. Here ^{131}I-PVP is intravenously injected and the total fecal radioactivity collected during a period of four days is measured. Even if it is lost into the digestive tract, it remains intact and is excreted into feces. According to this procedure, in a healthy man, an averge 0.8% (0–1.60%) of ^{131}I-labelled PVP injected is excreted into feces in four days, but this amounts to 6–20% when there is an excessive loss of proteins.[49,50] Recently, PVP has been found to deposit in various tissues and the diagnostic value of this procedure is questioned. On the other hand, a similar procedure using ^{59}Fe-labelled dextran is attracting much attention.[73]

REFERENCES

1) ALLERHAND, J., KARELITZ, S., ISENBERG, H. D., PENBHARKKUL, S. and RAMOS, A.: J. Ped., **62**, 234, 1963.
2) ANDERSEN, S. B., GLENERT, J. and WALLEVIK, K.: J. Clin. Invest., **42**, 1873, 1963.
3) ANZAI, T., IBAYASHI, J., CARPENTER, C. M. and HYDE, L.: Am. Rev. Resp. Dis., **88**, 503, 1963.
4) ARMSTRONG, S. H., BRONSKY, D. and HYMABN, S.: Fed. Proc., **16**, 75, 1957.
5) BAUER, H.: Deut. Z. Nervenh., **170**, 381, 1953.
6) BAUER, W., SHORT, C. L. and BENNETT, G. A.: J. Exptl. Med., **57**, 419, 1933.
7) BEEKEN, W. L., and NORMAN, M.: Proc. Soc. Exptl. Biol. Med., **117**, 24, 1964.
8) BEEKEN, W. L., VOLWILER, W., GOLDWORTHY, P., GARBY, L. E., REYNOLDS, W. E., STOGSKILL, R. and STEMLER, R. S.: J. Clin. Invest., **41**, 1312, 1962.
9) BELLANTI, J. A.: Am. J. Dis. Child., **115**, 239, 1968.
10) BENNETT, H. S., LUFT, J. H. and HAMPTON, J. C.: Am. J. Physiol., **196**, 381, 1959.
11) BENNHOLD, H. and KALLEE, E.: J. Clin. Invest., **38**, 863, 1956.
12) BERGGARD, I.: Clin. Chim. Acta, **6**, 545, 1961.
13) BERSON, S. A., YALOW, R. S., SCHREIBER, S. S. and POST, J.: J. Clin. Invest., **32**, 746, 1953.
14) BIRKE, G.: Proc. Xth Congress Intern. Soc. Blood Transf., Stockholm, 1964.
15) BLIX, S. and AAS, K.: Acta. Med. Scand., **169**, 63, 1961.
16) BONNELL, J. A.: Brit. J. Int. Med., **12**, 181, 1951.
17) BORGSTROM, B., DAHLQVIST, A., LUNDH, G. and SJÖVALL, J.: J. Clin. Invest., **36**, 1521, 1957.
18) BORSOOK, H. and KEIGHLEY, L.: Proc. Roy. Soc. London, Ser. B., **118**, 488, 1935.
19) CAMPBELL, R. M., CUTHBERTSON, D. P., MATTHEWS, C. M. and McFARLANE, A. S.: Int. J. Appl. Radiat., **1**, 66, 1956.
20) CAPUTO, A.: Med. Sper., **35**, 1959, Nature, 189; 395, 1961.
21) CARBONARA, A. O., RODAHAIN, J. A. and HEREMANS, J. F.: Nature, **198**, 999, 1963.
22) CHERNICK, W. S., EICHEL, H. J. and BARBERS, G. J.: J. Ped., **65**, 694, 1964.
23) CHEVANCE, L. G., GALLI, A., JEANMAIRE, J. and GÉRARD, G. in "Immunoelectrophoretic Analysis," ed. by GRABAR and BURTIN, Elsevier, 1964.
24) CHODIRKER, W. B. and TOMASI, T. B.: Science, **142**, 1080, 1963.
25) CHODOS, D. D., ELY, R. S. and KELLEY, V. C.: Proc. Soc. Exptl. Biol. Med., **99**, 775, 1958.
26) CIER, J. F., MANUEL, Y. and LACOUR, J. R.: Compt. Rend. Soc. Biol., **157**, 1623, 1963.
27) COHEN, S., and GORDON, A. H.: Biochem. J., **70**, 544, 1958.
28) COHEN, S., GORDON, A. H. and MATTEWS, C.: Biochem. J., **82**, 197, 1962.
29) COONS, A. H., LEDUC, E. H. and CONNOLLY, J. M.: J. Exptl. Med., **102**, 49, 1955.
30) COURTIS, F. C.: J. Physiol. (London), **155**, 456, 1961.
31) CRABBE, P. A., CARBONARA, A. O. and HEREMANS, J. F.: Lab. Invest., **14**, 235, 1965.
32) DAVIES, A.: Lancet, **2**, 1009, 1922.
33) DENNIS, E. G., HORNBROOK, M. M. and ISHIZAKA, K.: J. Allergy, **35**, 464, 1964.
34) DEODHAR, S. D., CUPPAGE, F. E. and GABLEMAN, E.: J. Exptl. Med., **120**, 677, 1964.
35) DRINKER, C. K.: Ann. N. Y. Acad. Sci., **46**, 807, 1946.
36) DUBLER, H. and SCHEIDEGGER, J. J.: Ophthalmologia, **135**, 640, 1968.
37) DUTCHER, T. F. and FAHEY, J. L.: J. Nat. Cancer Inst. **22**, 237, 1959.
38) ESSER, H.: Münch. Med. Wochschr., **94**, 2313, 1952.
39) FAYET, M. T.: Bull. Soc. Chim. Biol., **41**, 1189, 1959.
40) FILIPE de SILVA, J. A. and MONTEIRO, C. C.: Bull. Soc. Chim. Biol., **41**, 1707, 1959.
41) FREEMAN, T.: Proc. Xth Congr. Intern. Soc. Haemat., Stockholm, 1964.
42) FRIBERG, L.: Acta Med. Scand. Suppl., **240**, 138, 1950.
43) FRICK, E.: Klin. Wochschr., **41**, 75, 1963.
44) GABL, F. and WACHTER, H.: in "Protides of the Biological Fluids" ed. by H. PEETERS, Elsevier, 1962.
45) GALLI, M. R., CHEVANCE, L. G., JENMAIRE, J. and GERARD, M. G.: Bull. Soc. Chim. Biol., **41**, 1367, 1959.

46) GANS, H. and KRIVIT, W.: Ann. Surg., **155**, 353, 1962.
47) GLENERT, J., JARNUM, S. and RIEMER, S.: Acta Chir. Scand., **124**, 63, 1962.
48) GOLDSWORTHY, P. D. and VOLWILER, W.: Ann. N. Y. Acad. Sci., **70**, 26, 1957.
49) GORDON, R. S.: J. POLYMER, Sci., **31**, 191, 1958.
50) GORDON, R. S.: Lancet, **1**, 325, 1959.
51) GRABAR, P. and BURTIN, P.: Bull. Soc. Chim. Biol., **37**, 797, 1955.
52) GRABER, P. COURCON, J., ILBERG, P. L., LOUTIT, J. F. and MERRILL, J. P.: Compt. Rend., Soc. Biol. **245**, 950, 1957.
53) GREEN, M. N., CLARKE, J. T. and SCHWACHMAN, H.: Pediatrics, **21**, 635, 1958.
54) GROSSMAN, J., YALOW, A. A. and WESTON, R. E.: Metab. Clin. Exptl., **9**, 528, 1960.
55) HAMASHIMA, Y., KYOGOKU, M.: Immunohistology, 2nd ed., (in Jap.) Igaku Shoin, Tokyo, 1968.
56) HANLEY, P. H., HINES, M. O., RAY, J. E., McPHERSON, F., and HIBBERT, W. A.: South. Med. J., **55**, 233, 1962.
57) HANSON, L. A.: Intern. Arch. Allergy Appl. Immunol., **18**, 241, 1961.
58) HANSON, L. A.: Clin. Chim. Acta. **7**, 828, 1962.
59) HARDWICKE, J., RANKIN, J. G., BAKER, H. J. and PREISIG, R.: Clin. Sci., **26**, 509, 1964.
60) HARPER, H. A.: Review of Physiological Chemistry 11th Ed., Maruzen Asian Ed., 1967.
61) HENSHALL, G. K. Jr.: Am. J. Roent. Rad. Ther. Nucl. Med., **84**, 1105, 1960.
62) HERMAN, E. C. Jr.: J. Lab. Clin. Med., **57**, 825, 1961.
63) HERMANN, G.: in "Immunoelectrophoretic Analysis" ed. by GRABER and BURTIN, Elsevier, 1964.
64) HOCHWALD, G. M., THORBECKE, G. J. and ASOFSKY, R.: J. Exptl. Med. **114**, 459, 1961.
65) HOLLANDER, J. L., McCARTY, D. J., ASTORGA, G. and CASTRO-MURILLO, E.: Ann. Int. Med., **62**, 271, 1965.
66) HOLMAN, H. R.: in "Immunological Diseases" ed. by SAMTER, M., Little Brown & Co., Boston, 1965.
67) HOLMAN, H. R., NICKEL, W. F. and SLEISENGER, M. H.: Am. J. Med., **27**, 963, 1959.
68) HONG, R., POLLARA, B. and GOOD, R. A.: Proc. Nat. Acad. Sc., **56**, 602, 1966.
69) IBER, F. L., NASSAU, K., PLOUGH, I. C., BERGER, F. M., MERONEY, W. H. and FERMONT-SMITH, K.: J. Clin. Invest., **37**, 1442, 1958.
70) JACOB, F. and MONOD. J.: J. Mol. Biol. **3**, 318, 1961, Cold Spring Harbor Symp. Quant. Biol., **26**, 193, 1961.
71) JARNUM, S.: Scandin. J. Clin. Lab. Invest., **13**, 462, 1961.
72) JARNUM, S. and LASSEN, N. A.: Scandin. J. Clin. Lab. Invest., **13**, 357, 1961.
73) JARNUM, S., WESTERGAAARD, H., YSSING, M. and JENSEN, H.: Gastroenterology, **55**, 229, 1968.
74) JEEJEEBHOY, K. N. and COGHILL, N. F.: Gut, **2**, 123, 1961.
75) KABAT, E. A., LANDOW, H. and MOORE, D. H.: Proc. Soc. Exptl. Biol. Med., **49**, 260, 1942.
76) KAWAI, T.: Modern Medecine, (in Jap.) Tokyo, **19**, 2980, 1940.
77) LATERRE, E. C.: Les Protéines du Liquide Céphalorachidien á l'État Normal et Pathologique, Arscia, Brussels, 1964.
78) LATERRE, E. C., HEREMANS, J. F. and DEMANET, G.: Rev. Neurol., **107**, 500, 1962.
79) LATHEM, W., DAVIS, B. B., ZWEIG, P. H. and DEW, R.: J. Clin. Invest., **39**, 840, 1960.
80) LAZARUS, J. A. and MARKS, M. S.: J. Urol., **56**, 246, 1946.
81) LEITHOFF, H. and LEITHOFF, I.: Med. Welt., No. 21, 1137, 1961.
82) LEWALLEN, C. G., BERMAN, M. and RALL, J. E.: J. Clin. Invest., **38**, 66, 1959.
83) LICHT, W. and KEUTEL, H. J.: Z. Urol., **56**, 401, 1963.
84) LONGO, L. D., CAILLOUETTE, J. C. and RUSSELL, K. P.: Obst. Gynec., **14**, 97, 1959.
85) MARTIN, C. M., GORDON, R. S., FELTS, W. R. and McCALLOUGH, N. B.: J. Lab. Clin. Med., **49**, 607, 1957.
86) MASSON, P., HEREMANS, J. F. and PRIGNOT, J.: Biochem. Biophys. Acta. **111**, 466, 1965.
87) MASSON, P. L., et al.: Gastroenterologia, **105**, 270, 1966.
88) McKEE, F. W., WILT, W. G., HYATT, R. E. and WHIPPLE, G. H.: J. Exp. Med., **91**, 115, 1950.
89) MELLORS, R. C. and KORNGOLD, L.: J. Exp. Med., **118**, 387, 1963.
90) MERLER, E., REMINGTON, J. S., FINLAND, M. and GITLIN, D.: Nature, **196**, 1207, 1961.

91) MERRILL, D. A., KIRKPATRIK, C. H., WILSON, W. E. C. and RILEY, C. M.: Proc. Soc. Exptl. Biol. Med., **116**, 748, 1964.
92) MIALE, A.: in "Serum Proteins and the Dysproteinemias" ed. by Sunderman and Sunderman, Lippincott, Philadelphia, 1964.
93) MOGHISSI, K. S. and NEUHAUS, O. W.: Am. J. Obst. Gynec., **83**, 149, 1962.
94) MORGAN, L. M., SCHLESINGER, R. W. and OLITSKY, P. K.: J. Exptl. Med. **76**, 357, 1942.
95) MORSE, J. H., MULLER-EBERHARD, H. J. and KUNKEL, H. G.: Bull. N. Y. Acad. Med., **38**, 642, 1962.
96) MULLER-EBERHARD, U. and Cleve, H.: Nature, **197**, 602, 1963.
97) MULLER, M., FONTAINE, G., MULLER, P. H., GOURGUECHON, A. and OLIVEIRA, F. M.: Pathol. Biol., **10**, 249, 1962.
98) MURRAY, R. K., CONNELL, G. E. and PERT, J. H.: Blood, **17**, 1961.
99) NAIRN, R. C.: Fluorescent Protein Tracing, E. & S. Livingstone, Ltd., Edinburgh, 1964.
100) NEALE, F. C., ABER, G. M. and NORTHAM, B. E.: J. Clin. Path., **11**, 206, 1958.
101) NOSSLIN, B. F. and NYMAN, M.: Lancet, **274**, 1000, 1958.
102) NUSSLE, D., BARANDUN, S., WITSCHI, H. P., KÄSER, H., BETTEX, M. and GIRARDET, P.: Helv. Paediat. Acta., **16** (Suppl. 10), 1, 1961.
103) NYMAN, M.: Scand. J. Clin. and Lab. Invest., **9**, 168, 1957.
104) OEFF, K. and KESSEL, J.: Z. ges. exp. Med., **126**, 278, 1955.
105) OTT, H.: Verhandl. Deut. Ges. Inn. Med., **61**, 270, 1955.
106) PARKER, W. C. and BEARN, A. G.: J. Exp. Med., **115**, 84, 1962.
107) PARKINS, R. A., NIMITRIADOU, R. and BOOTH, F. C.: Clin. Sci., **19**, 595, 1960.
108) PERETZ, W. L. and TOMASI, T. B.: Arch. Ophthalmol., **65**, 20, 1961.
109) PERNIS, B. and CHIAPPINO, G.: Immunology, **7**, 500, 1964.
110) PETERS, T. and ANFINSEN, C. B.: J. Biol. Chem., **182**, 171, 1950.
111) PETTE, D. and STUPP, I.: Klin. Wochschr., **38**, 109, 1960.
112) PHILLIPS, L. L., MONTGOMERY, G. J., and TAYLOR, H. C., Jr.: Am. J. Obst. Gynecol., **73**, 43, 1957.
113) PIERCE, A. E.: Vet. Reus Annot, **5**, 17, 1959.
114) PRENTIC, T. C., SIRI, W. and JOINER, E. E.: Am. J. Med., **13**, 668, 1952.
115) RATNOFF, O. D. and VOSBURGH, G. J.: New Engl. J. Med., **247**, 970, 1952.
116) RASKA, B. and MASOPUST, J.: Ann. Paediat., **198**, 343, 1962.
117) RAWSON, A. J.: Clin. Chem., **8**, 310, 1961.
118) REEVE, E. B. and BAILEY, H. R.: J. Lab. Clin. Med., **60**, 923, 1962.
119) REISS, E., MERTENS, E. and EHRLICH, W. E.: Proc. Soc. Exptl. Biol. Med., **74**, 732, 1950.
120) REMINGTON, J. S. and O'NEAL, PAGE, C.: Clin. Res., **13**, 126, 1965.
121) REMINGTON, J. S., VOSTI, K. L., LIETZE, A. and ZIMMERMAN, A. L.: J. Clin. Invest., **43**, 1613, 1964.
122) RODBELL, M., FREDRICKSON, D. S. and ONO, K.: J. Biol. Chem., **234**, 567, 1959.
123) RODNAN, G. P. and MACLACHLAN, M. J.: Arthritis Rheumat., **3**, 152, 1960.
124) ROTHSCHILD, M. A., BAUMAN, A., YALOW, R. S. and BERSON, S. A.: J. Clin. Invest., **34**, 1354, 1955.
125) ROTHSCHILD, M. A., BAUMAN, A., YALOW, R. S. and BERSON, S. A.: J. Clin. Invest., **36**, 422, 1957.
126) ROTHCHILD, M. A., SCHREIBER, S. S., ORATZ, M. and McGEE, H. L.: J. Clin. Invest., **37**, 1229, 1958.
127) ROWE, D. S. and SOOTHILL, J. F.: Clin. Sci., **21**, 75, 1961.
128) RUSSELL, T. S. and BENNETT, W.: Gastroenterology, **45**, 730, 1963.
129) SANDOR, G.: Serum Proteins in Health and Disease, Chapman & Hall, London, 1966.
130) SCHNEIDER, C. L.: Prog. Hemat., **1**, 202, 1956.
131) SCHNEIDER, G. and WALLENIUS, G.: Scand. J. Clin. Lab. Invest., **3**, 145, 1951.
132) SCHOENHEIMER, R., RATNER, S., RITTENBERG, D. and HEIDELBERG, M.: J. Biol. Chem., **144**, 545, 1942.
133) SCHULTZE, H. E. and HEREMANS, J. F.: Molecular Biology of Human Proteins, Vol. I. Elsevier Publ. Co., Amsterdam, 1966.
134) SCHUR, P. H. and SANDSON, J.: Arthritis Rheumat., **6**, 115, 1963.
135) SEARS, D. A.: J. Lab. & Clin. Med., **71**, 484, 1968.

136) SHERRY, S., FLETCHER, A. P. and ALKJAERSIG, N.: Progr. Hemat., **3**, 244, 1962.

137) SMITH, J. K.: Acta Hematol., **30**, 144, 1963.

138) SOLOMON, A., FAHEY, J. L. and MALMGREN, R. A.: Blood, **21**, 403, 1963.

139) SONNET, J. and BRISBOIS, P.: Acta Clin. Belg., **11**, 42, 1956.

140) SOUTH, M. A., COOPER, M. D., WOLLHEIM, F. A., HONG, R. and GOOD, R. A.: J. Exp. Med., **123**, 615, 1966.

141) SOUTH, M. A., COOPER, M. D., WOLLHEIM, F. A. and GOOD, R. A.: Am. J. Med., **44**, 168, 1968.

142) STEGER, J.: Nünch. Med. Wschr., **96**, 747, 1954.

143) STEIN, W. H. and MOORE, S.: J. Biol. Chem., **211**, 915, 1954.

144) STEINFELD, J. L., DAVIDSON, J. D., GORDON, R. S. and GREENE, F. E.: Am. J. Med., **29**, 405, 1960.

145) STERLING, K.: J. Clin. Invest., **30**, 1228, 1951.

146) STEVENSON, G. T.: J. Clin. Invest. **41**, 1190, 1962.

147) SUKAROCHANA, K., MOTAI, Y., SHIN, M. SHEPARD, R. and KIESEWETTER, W. B.: Surg. Gynec. Obst., **121**, 79, 1965.

148) TAMM, I. and HORSFALL, F. L.: J. Exp. Med., **95**, 71, 1952.

149) TOMASI, T. B., Jr.: Hosp. Prac., **7**, 26, 1967.

150) TOMASI, T. B., TAN, E. M., SOLOMON, A. and PRENDERGAST, R. A.: J. Exp. Med., **121**, 101, 1965.

151) TOMASI, T. B. and ZIGELBAUM, S.: J. Clin. Invest., **42**, 1552, 1963.

152) TSCHIRGI, R. D.: Am. J. Physiol., **163**, 756, 1950.

153) VAUGHAN, O. W., FILER, L. J., Jr., and CHURELLA, H.: Pediatrics, **29**, 90, 1962.

154) VOLWILER, W.: South. Med. J., **54**, 943, 1961.

155) WACHSMUTH, L.: Virchow's Arch., **7**, 330, 1854.

156) WADA, T., SATO, K., TAKAMURA, T., NISHIHARA, T., KONDO, M., IBAYASHI, J., OHARA, H. and ANZAI, T.: Gann, **52**, 27, 1961.

157) WALDMANN, T., TRIER, J. and FALLON, H.: J. Clin. Invest., **42**, 171, 1963.

158) WALDMANN, T. A.: Lancet, **1**, 121, 1961.

159) WEISS, A. H., SMITH, E., CHRISTOFF, N. and KOCHWA, S.: J. Lab. Clin. Med., **66**, 280, 1965.

160) WELTON, J.: Am. J. Med., **44**, 280, 1968.

161) WETTERFORS, J., GULLBERG, R., LILJEDAHL, S. O., PLANTIN, L. O., BIRKE, G. and OLHAGEN, B.: Acta Med. Scand., **168**, 347, 1961.

162) WHEBY, M. S., BARRETT, O. N. and CROSBY, W. H.: Blood, **16**, 1579, 1960.

163) WITSCHI, H. P., BARANDUN, S., and NUSSLÉ, D.: Gastroenterologia, **98**, 65, 1962.

164) ZIMMER, J. and WORINGER, F.: Bull. Soc. Franc. Dermatol. Syphil., **64**, 743, 1957.

DIAGNOSIS AND PATHOGENESIS OF PLASMA PROTEIN ABNORMALITIES

Chapter 15

Diagnostic Approaches in Plasma Protein Abnormalities

CLINICAL SYMPTOMS IN PLASMA PROTEIN ABNORMALITIES

As in other diseases, the diagnosis of plasma protein abnormalities begins with the history-taking and careful physical examination of the patient. It is very important to pick up various clinical symptoms and signs adequately. There are a great variety of these because plasma proteins have a number of physiological functions and their abnormalitites take various forms. MIYOSHI et al.[253] summarize clinical symptoms as shown in Table 23. Nearly all of clinical symptoms are related in some way or other to plasma protein abnormalities, but are not exclusive to them. Still, not only are they useful in their diagnosis, but they are of enormous importance in gaining a precise selection of various laboratory tests to be followed.

Table 23   Clinical symptoms in plasma protein abnormalities
(from MIYOSHI).

1. Generalized edema
2. Accentuated erythrocyte sedimentation rate
3. Proteinuria and renal disorders
4. Hematological disorders (anemia, bleeding tendency, etc)
5. Circulatory disorders
6. Joint disorders
7. Nutritional disorders of skin, nails, and hairs
8. Ophthalmological disorders
9. Infections (poor resistance, antibody deficiency)

The characteristic clinical symptoms of individual plasma protein abnormalities are explained in other sections so it will suffice here to group them generally according to their mechanism as: 1) Those which come from changes in physiological functions of plasma protein components, and 2) Those which come from the basic pathology resulting in plasma protein abnormalities.

The former includes mainly generalized edema and nutritional disorders due to the decrease of albumin, an infectious tendency due to hypoimmunoglobulinemia, a bleeding tendency due to hypergammaglobulinemia, RAYNAUD's signs due to cryoglobulinemia and anemia due to transferrin deficiency. The latter may include numerous clinical symptoms associated with almost every kind of diseases.

DIAGNOSTIC APPLICATION OF VARIOUS LABORATORY TESTS

As previously noted, various laboratory tests are to be done when the presence of a certain disease or syndrome is suspected after careful clinical observation. Sufficient consideration of the following points is important on selecting adequate laboratory tests.

1) The physical and financial burden of the patient should be as light as possible.
2) Those that are simplest, quickest, and micro-techniques having the least possibility of technical errors should be chosen.

When these points are considered, the order of diagnosis of greatest advantage is probably the one shown in Fig. 85.

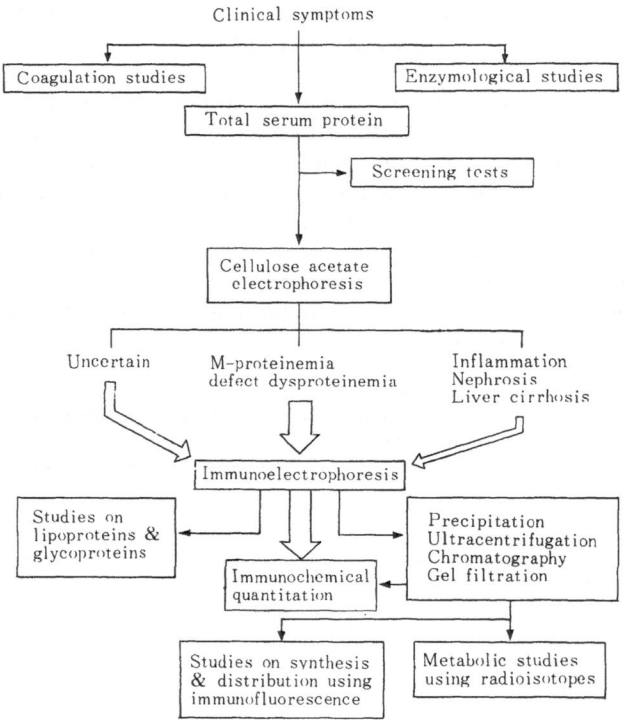

Fig. 85 Laboratory approaches for diagnosis of plasma protein abnormalities.

With regard to those plasma proteins which are contained in extremely small amounts, their characteristic biological activity is used to quantitate their serum levels. For example, in the case of enzymes, enzymological tests are used and the activity of each enzyme is quantitated. With high-molecular weight hormones and antibodies, endocrinological and serological tests are used, respectively. In plasma protein abnormalities involved in blood coagulation and fibrinolysis, blood coagulation studies are first performed. This is because, with the exception of factor XII deficiency, the bleeding tendency is frequently observed in diseases of this nature.

In the cases of a marked decrease of fibrinogen, the bleeding tendency is observed. Accordingly, blood coagulation studies are first done in order to grasp the changes of fibrinogen itself. In electrophoresis using various supporting media, fibrinogen frequently disturbs the migration and separation of other protein components. Therefore, serum samples, instead of plasma, are used for routine analysis. Only in special cases, plasma is used for electrophoretic analysis of proteins.

When the presence of serum protein abnormalities is suspected, the total serum protein concentration is first measured. The frequency of a marked increase or decrease of total serum protein concentration is not so high. As shown in Fig. 86, in the Central Railway Hospital, hypoproteinemia occurred in 14% of all patients, while hyperproteinemia in 3%. This is similar to the cases observed at Juntendo University Hospital (Tokyo) where hypoproteinemia and hyperproteinemia occurred in 14% and 8.2%, respectively. Hypoproteinemia is seen in almost all diseases. When the total serum protein concentration is below 4.5 g/100 ml, the nephrotic syndrome and protein-losing gastroenteropathy are first suspected. This condition may also be observed in cachexia, severe liver damage and acute infectious diseases.

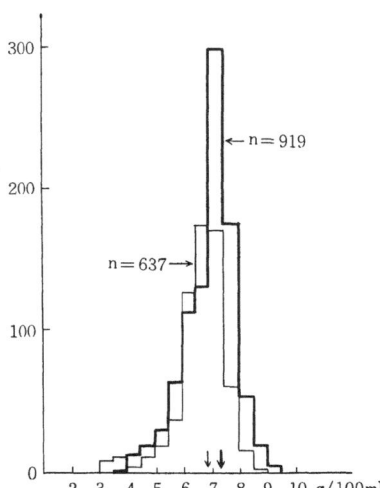

Fig. 86 Frequency distribution of the patients' data on the measurement of the total serum protein.
The thick solid line indicates the values obtained with refractometry, while the thin solid line with the micro-biuret method.

As for hyperproteinemia, it is observable in M-proteinemia, liver cirrhosis, chronic inflammatory diseases and lymphoma. When the total serum protein concentration is more than 10 g/100 ml, M-proteinemia can be suspected. However, diagnostic conclusions are not to be based on total serum protein concentration, alone. Next, an electrophoretic fractionation should be done as the second screening procedure.

When clinical symptoms suggest the presence of a particular plasma protein abnormality, the following simplified laboratory tests may be used for diagnostic purposes:

    1) Hypo- or a-fibrinogenemia——FI test.
    2) Hypo- or a-gammaglobulinemia——GG test.
    3) WILSON's disease——Simplified screening test for ceruloplasmin.
    4) A-betalipoproteinemia, abnormalities of lipid metabolism——β-L test.

Among various electrophoretic procedures, recently the cellulose acetate electrophoresis has rapidly come into wide use. In this procedure, serum proteins are separated into five fractions: the albumin fraction, the α_1 fraction, the α_2 fraction, the β fraction, and the γ fraction. These fractions are made up of more than one serum protein component, especially the α and β fractions consisting of many minor components.   Consequently, only those components which are contained in comparatively large amounts (albumin, α_1-acid glycoprotein, α_2-macroglobulin, haptoglobin, low-density lipoprotein, transferrin and IgG-globulin) reflect changes on cellulose acetate electrophoretic patterns.   In order to recognize the changes of other minor components, other fractional techniques of higher senstivity should be employed.   As a screening test used is the immunoelectrophoretic technique for this purpose.

By means of the immunoelectrophoretic technique, about 30 different kinds of serum protein components may be identified simultaneously.   Of course, the kind of components which may be identified vary with the type of antiserum used.   Nevertheless, thanks to this technique, the identification of individual serum protein components, impossible with the filter-paper or cellulose acetate electrophoretic technique, has become possible. The immunoelectrophoretic technique is used rather for qualitative analysis, and there-fore, another technique is required for quantitative analysis of individual protein compo-nents.

Chapter 16

Variations in the Measurement
of the Plasma Proteins

ARTIFICIAL VARIATIONS IN THE PLASMA
PROTEIN MEASUREMENT

To detect the presence of qualitative and quantitative abnormalities of plasma proteins, the analysis of blood plasma and other similar body fluids is necessary. When analysis is repeated, it will be noted that the results are not always highly reproducible. It is, therefore, necessary to know which factors affects the test results.

Technical Variations

Technical variations are those which originate in analytical techniques and generally fall into the following three groups: ① inherent errors, ② technical errors, and ③ technical failures.

1. The Inherent Errors

These include errors characteristic of a particular analytical procedure. They can be further divided into a) errors due to analytical instruments and b) errors due to analytical methods.

Errors due to analytical instruments: Test results are definitely influenced by the instruments used for the analysis. These include pipettes, refractometers, supporting media, densitometers, and others. For example, the pipettes used for collecting samples should be correctly calibrated. In the case of refractometers, as shown in Fig. 87, the results obtained vary considerably according to the make of the instrument. Even when the same electrophoretic patterns are analyzed, different results are obtained by different densitometers (Table 24). Accordingly, instruments should be carefully examined before purchase. No precise result could be hoped for when instruments of poor quality are used.

As previously stated, on electrophoresis using supporting media, test results vary greatly according to the type of supporting media used. This is also true in the case of the cellulose acetate membrane. That is, as for the albumin fraction, the Millipore membrane

Table 24 Values obtained with different densitometers on the same Oxoid patterns (from SHIMAO).

Densitometer	Wave length (mμ)	Fractions (%)					Total of optical densitites	Maximum optical density
		Alb	α_1	α_1	β	γ		
A	500	50	4	11	11	25	8.0	1.2
B	500	58	4	8	10	21	5.6	0.9

Note: Densitometer A does not show a complete lineality.

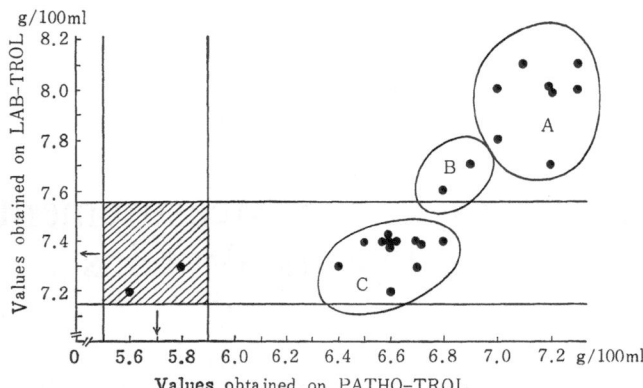

Fig. 87 Comparison of the values obtained in 24 hospitals of the Japanese National
Railways on two different control sera.
The shaded area indicates the acceptable limits for the values obtained with the micro-
biuret method.   The two values in the shaded area were given from two different
laboratories using the biuret method.   The values labelled as "A" were obtained with
refractometry using the refactometer of the A company, "B" using that of the B company and "C" using that of the C company.   Each kind of the refactometers shows
quite different values on the same samples.

shows nearly the same results as in the TISELIUS method.   Somewhat higher results are
obtained with the Oxoid membrane, with the highest results in the Separax membrane.
The α_1-fraction tends to be high with the Oxoid.   Of the α_2-fraction, there is little difference.   The β and γ fractions are measured to be low with the Oxoid and Separax
membranes.   The Sartorius and Sepraphor III membranes show nearly the same tendency as the Millipore membrane.

Errors due to Analytical Methods: In measuring total protein concentration, great variation may be obtained, depending on whether the biuret method or refractometry is used.
This is particularly noticeable in the case of pathological sera (Fig. 87).   A great difference
in electrophoresis results, also, between the TISELIUS method and electrophoresis using
supporting media.   On cellulose acetate electrophoresis, results differ according to the
kind of membranes used.   As long as correctly adjusted densitometers are used, nearly the
same fractional results are obtained by the fractional dye elution method and densitometry (Table 25).

This goes to show that differences in results coming from differences in analytical instruments and methods are quite noticeable.   Evidently, in routine fractionations, care
should be given for selecting instruments and methods.   Or significant errors are inevitable,
no matter how skilled the technologist may be.

2.   The Technical Errors

These originate in technologist's skill in such things as the usage of pipettes, the handling
of the densitometer and the refractometer (particularly in regard to temperature), techniques of electrophoresis.   With experience and effort, great improvement may be hoped
for.

3.   The Technical Failure

As human beings, mistakes due to carelessness or unfavorable mental or physical
conditions are possible.   These could be miscalculations, inaccurate readings of scales,
use of defective instruments and poor preparation of the reagents.

Table 25 Different results obtained with different supporting media (from Committee on Standardization, Society of Electrophoresis, Japan).

Serum sample	Methods	Fractions (%)				
		Alb	α_1	α_2	β	γ
A	Oxoid (elution)	69	3	11	10	8
	Tiselius	58	3	11	14	14
	Filter-paper (elution)	61	4	10	12	13
	Filter-paper (densit.)	63	5	9	14	9
B	Oxoid (elution)	60	3	7	8	23
	Tiselius	47	3	8	11	31
	Filter-paper (elution)	58	2	5	9	26
	Filter-paper (densit.)	60	3	4	5	26
C	Oxoid (elution)	66	3	7	9	16
	Tiselius	54	3	8	11	24
	Filter-paper (elution)	73	3	5	6	13
	Filter-paper (densit.)	72	3	4	7	14
D	Oxoid (elution)	30	2	5	60	2
	Tiselius	19	2	4	69	6
	Filter-paper (elution)	37	3	3	55	2
	Filter-paper (densit.)	32	2	65	—	1

4. Quality Control Program for Technical Errors

As previously stated, technical variations result from a combination of three elements: inherent errors, technical errors and technical failures. These cannot be ignored when analyzing the results of routine laboratory tests. The smaller the technical variation, the more reliable the test. Therefore, pooled serum is usually used in order to keep the technical variation at its lowest, and a quality control program by $\bar{x}$–R control chart is employed (Fig. 88).[219] Measurement of pooled serum of the same lot is made twice a day. Range (R) in a replicate analysis and the average of the measured values are written on the graph and by statistical treatment, the range of m$\pm$2 S.D. is made the allowable limit. As long as the measured results of daily pooled serum remain within the allowable limit, the measurement for the day is considered to be reliable. The frequency with which the same values are obtained by repeated measurements of pooled

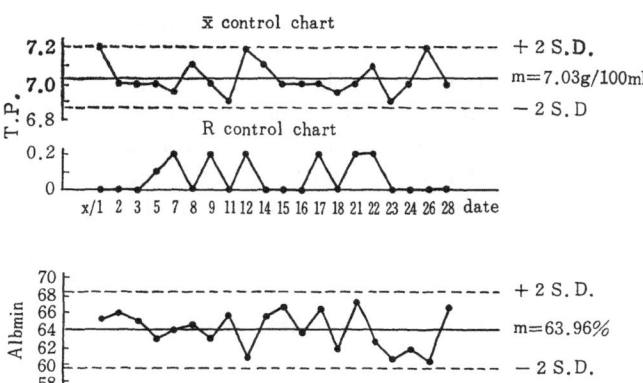

Fig. 88 The control charts obtain on the total serum protein and the albumin fraction values in a laboratory during the October, 1968.

serum of the same lot is usually called "precision" or "reproducibility". The following coefficient of variation (C.V.) is used to express this:

$$C.\ V. = \frac{S.\ D.}{m} \times 100\ (\%)$$

That is, the smaller the C. V., the higher the reliability of test results.

Clinically, the variation within the range of $2 \times$ C.V. is considered to be caused by technical errors. Only the variation which surpasses this range is thought to be biologically significant. For example, when the C.V. of the albumin fraction is 3.4% and the test result of patient's serum is 60%, it is considered that the variation within the range of $2 \times$ C. V. (that is, $60 \pm 4.08\%$, or 55.92%–64.08%) can result from technical variation alone. Therefore, it is possible that the albumin fraction of the patient is considered to be decreased only when the result of the subsequent measurement is below 56%. However, the coefficient of variation does not necessarily remain the same. It varies according to test procedure and analytical conditions. Variation is also due to the following: change of the measurer, lots of reagents, adjustment of instruments, even when testing is carried on in the same laboratory. (Fig. 89, Table 26).

Table 26   Quality control data in an university hospital in Tokyo (October '68).

	Mean	S. D.	C. V. (%)
Total serum protein*	7.03 g/100ml	0.09 g/100ml	1.3
Serum protein fractions**			
Albumin fraction	63.96%	2.16%	3.4
α_1 fraction	3.68	0.72	19.6
α_2 fraction	9.83	0.88	9.0
β fraction	9.56	0.62	9.5
γ fraction	12.90	1.21	9.4

* Measured with the same refractometer
** Analyzed with the standard cellulose acetate (Oxoid) electrophoretic procedure proposed by Society of Electrophoresis, quantitated by densitometry.

In addition, there is the word "accuracy". This word is used to show how close test results are to "true values." In order to control accuracy, standard solutions and standard sera are used. Protein fractional patterns of various kinds of control serum by cellulose acetate electrophoresis are shown in Fig. 90. Lab-Trol and Chemtrol are not to be used. This should be kept in mind when selecting control serum.

Among routine laboratory tests, electrophoretic fractionation of serum proteins is one of those showing the greatest variation. It is laudable that the Japanese Society of Electrophoresis has made a 'film for calibration' by which the lineality of densitometers on the market is calibrated and also the standard procedures proposed. Electrophoresis is a test procedure which is based on complicated physico-chemical phenomena. Thus, routine tests should be done at least by using the standardized procedure in order to keep inter-laboratory variation at the minimum.

Amount of Proteins to be Applied for Analysis

It is a well-known fact that in filter-paper electrophoresis serum albumin is partly adsorbed on filter paper. Moreover, the rate of this adsorption varies according to the absolute quantity of the protein applied to filter paper (Table 27). The adsorption rate is the lowest at 160 μg, while it increases greatly below 80 μg.[276]

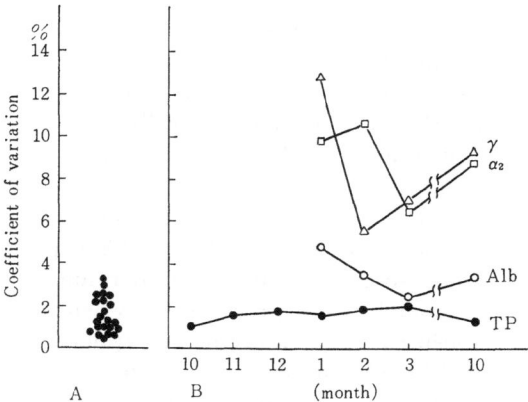

Fig. 89 Variability of the technical C. V. among different laboratories.
The left figure shows different C. V. on serum total protein measurement obtained among 25 clinical laboratories in Tokyo metro-
politan area, ranging from 0 to 3.24%.
The right figure shows different C. V. on serum protein fractionation
obtained in the same laboratory at different months.   The C. V. for
the total protein measurement varies only minimally, but those for
the electrophoretically separated fractions show marked variability.

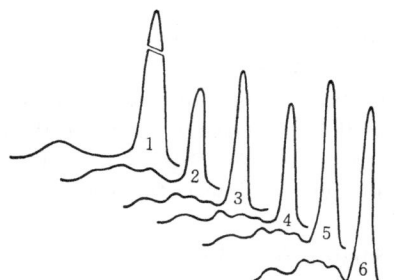

Fig. 90 Cellulose acetate electrophoretic patterns of the serum
proteins on various control sera which are now commercially
available.
1. Lab-Trol 2. Chemtrol 3. Versatol-A 4. Abnormal Clinical
Chemistry Control Serum 5. Normal Clinical Chemistry Control
Serum 6. Moni-Trol 1.

Table 27 Albumin adsorption on filter-paper (SCHLEICHER & SCHUELL 2043 – A mgl) during electrophoresis (from OLIVER and SPORZYNSKI).

Albumin applied (μg)	Albumin adsorbed (μg/cm^2)	Adsorption ratio (%)
239	0.88	6.0
189	0.81	6.5
160	0.41	4.4
80	0.33	13.4
52	0.20	32.0

According to the standard procedures proposed by Society of Electrophoresis, this result is equivalent to the total protein concentration of approximately 4–6 g/100 ml when serum sample is used.   Accordingly, except for the case in which the total protein concentration is either extremely low (below 4 g/100 ml) or high (above 8 g/100 ml), errors are not significantly great.   However, if other body fluids of small protein content is applied, either the fluid volume to be applied should be increased or it should be concentrated before electrophoresis.

Ordinarily, when densitometry is performed, protein contents may be assumed from the reading of an integrator.   The author has found from his experience, using the BPB-methanol staining method and Analytrol (Spinco), that results are best when the total reading of the integrator is around 200.   Error is great when it is below 100 or above 300.[181]   That is, when protein content is low, the adsorption rate of albumin becomes higher and the percentage of the albumin fraction becomes relatively low.   Also, in the a_1-, a_2- and β-fractions which are of relatively low percentage, not only are the densitometer readings inaccurate but the separation of each fraction is indistinct.   On the other hand, when proteins are applied in excessive amount, the separation of each fraction may become indistinct and test results are not entirely reliable.

Cellulose acetate membranes produce less errors than filter paper because protein adsorption onto the membranes is remarkably small.   But it goes without saying that, as stated with respect to the standard procedure, a suitable amount of protein application for the elution method is required.   In accordance with the conditions in each laboratory, a rough estimate of the most suitable amount of protein application obtained from the total number of integrator readings will be helpful.

Protein concentration suitable for analytical capability of the optical system is also selected in TISELIUS electrophoresis.   Moreover, in determining the electrophoretic mobility of albumin, albumin concentration during analysis is particularly important.   In the phosphate buffer of pH 7.8 and the ionic strength of 0.144, its mobility does not change in the albumin concentration range of 1.49–2.379 g/100 ml.   Its mobility decreases with the concentration above 3 g/100 ml, while below 1.20 g/100 ml it increases significantly.[371]

Storage or Heating of Serum Samples

Serum contains many proteins as well as various other inorganic substances and enzymes. As it contains all the nutrients essential to bacterial growth, it is easily contaminated by bacteria with the resultant degeneration of serum proteins.   Even when stored aseptically, serum protein degenerates partially.   Accordingly, it is advisable that serum protein analysis be finished as soon as possible after collection of blood samples.   Actually, there may be cases where serum should be preserved.   The influence of storage on serum protein pattern will be considered next.

The tubes containing test sera must be kept tightly corked.   If they are left uncorked at a room temperature in a period of two to three days, the serum protein concentration rises by 5%; in four days, it rises 10% and in five, 20%.   This is because the evaporation of serum water provokes its protein concentration.   Serum is often thoughtlessly left uncorked in the refrigerator because it is presumed the low temperature prevents evaporation.   However, concentration actually occurs more rapidly there than at ordinary room temperature.   Of course, serum concentration alone does not cause the percentage of serum protein fractions to change, and attempting to determine the concentration of each fraction from the total protein concentration by calculation will lead to erratic results.

If serum is stored in plastic tubes made from acetyl cellulose, a loss of water results because of the porosity of the walls of the tubes. In addition, the pH of the serum decreases markedly, and thus part of the serum proteins degenerate. Plastic tubes made from polycarbonate, polystyrene, and synthetic resin do not show these disadvantages.

Changes of protein fractional values observed during storage differ greatly according to the temperature at which serum is stored. If the serum is separated as soon as possible and is immediately frozen, it will be kept intact longer. If freezing and thawing are not repeated, it may be stored almost semi-permanently without bringing about great changes in serum protein patterns. But, of course, repetition of freezing and thawing will cause immediate degeneration of serum proteins. This tendency is particularly marked with lipoproteins.

If serum is stored in a refrigerator at $5°-10°C$, no remarkable changes in serum protein patterns will occur for a period of abou a week. According to the TISELIUS method, seven days' storage at $4°C$ hardly causes any changes. But detailed observation shows that more than four days' storage at $4°C$ brings about a slight increase of the α fraction, and a slight decrease of the β fraction. By cellulose acetate electrophoresis of fresh serum, β_{1C}-globulin forms a thin, narrow protein band in the β_2 area. But it disappears as a result of storage and it will be included in the β fraction. Of course, the $\beta_{1C}\beta_{1A}$ conversion is clearly observed on immunoelectrophoretic patterns (see page 45).

These degenerative changes are further promoted if serum is stored at $30°C$. On the second day, an increase of the α-fraction and a decrease of the β-fraction is noted. This tendency becomes more pronounced as the days pass.

This is also true for storage at $50-56°C$, with the tendency being more pronounced. In a storage of seven days, a slight decrease of the γ-fraction is an added result.

In any case, changes in serum protein pattern due to long storage of serum include a poor separation of the α_1- and the α_2-fractions, and a tendency to the relative increase of these fractions. They also include a tendency to decrease of the β- and the γ-fractions. When serum degenerates further, the albumin fraction also decreases and the separation of each fractions becomes less distinct (Fig. 91).

As described in the foregoing, long storage of serum is to be avoided for quantitative analysis of each fraction. There is on record the case where serum, after having been frozen for more than 20 years, was found to show still the pathological immunoglobulin. I think that storage by freezing is, of course, of great value. I have stored various pathological sera for six years, still using them for abnormal controls. However, it has been my experience that having undergone repeated thawing and freezing during six years' storage some sera brought about mobility changes in albumin and caused the albumin fraction to become biphasic. Also, the precipitin arc of IgG-immunoglobulin split in such a manner as to be similar to trypsin treatment. Therefore, with the exception of lipoproteins, to keep serum proteins intact for as long a period as possible, it is advisable to separate the serum into small tubes and to freeze them aseptically.

Increased Amount of Chylomicron in Serum Samples

After meals, the serum in the body is in a transient hyperlipemic state and neutral fat increases. This neutral fat exists as chylomicron in the blood stream and is therefore opalescent. In filter-paper electrophoresis, chylomicron remains in the origin and hardly moves. Accordingly, as shown in Fig. 92, chylomicron, when increased markedly, is observed in the origin as a sharp peak on BPB staining. Of course, in the hyperlipemic state which occurs after a meal, the observed increase is not so marked. Still, rather sharp peak may be formed at the origin in the γ-fraction. Thus, the test serum must be taken from a person who is in a fasting state.

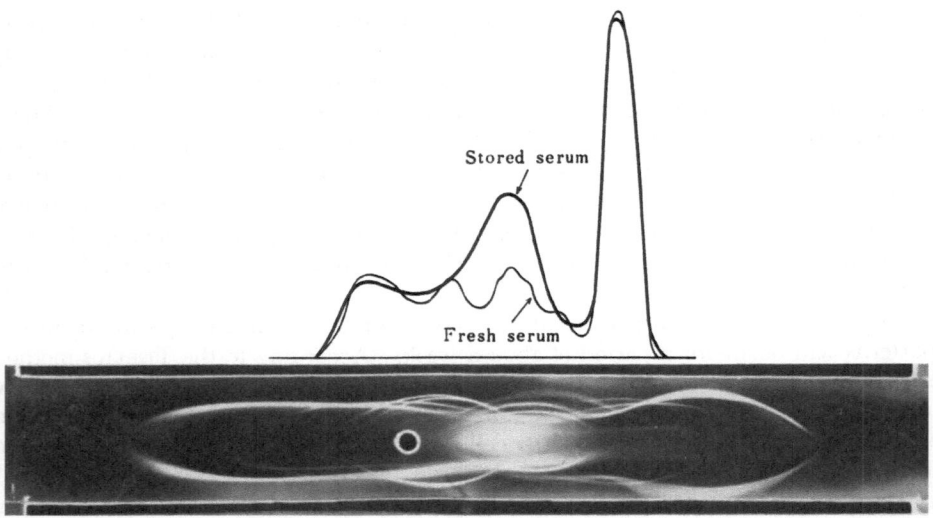

Fig. 91   Electrophoretic protein patterns of a stored serum sample.
The thick solid line and the immunoelectrophoretic pattern are obtained on a serum sample which was stored through repeated freezing and thawing after inactivated at 56°C, showing poor separation of the α fractions.   On immunoelectrophoresis, the denatured proteins are precipitated in the α zone. The albumin line shows a tailing toward the cathodal end, and the lines at the α zone are poorly identified.   The major components at the α and γ zones are relatively well preserved.

With the TISELIUS method, chylomicron has mobility from the α to the β zones; it is not included in the γ fraction.   It may be thought that chylomicron is located in the origin on the cellulose-acetate membrane, but this has not been confirmed.   It is not known how serum protein patterns on cellulose acetate strips are influenced by dietary (or postprandial) hyperlipemic states.

Serum which has been stored for a long period or which has become turbid often shows the sharp peak indicated previously when it is fractionated by means of filter-paper or cellulose acetate electrophoresis.   This results from the fact that, being adsorbed by supporting media, degenerated protein components do not move.   Especially affected are degenerated components of lipoproteins which have been stained out strongly by lipid staining (Fig. 93).   Their electrophoretic patterns resemble, apparently, those of chylomicron. But when the supporting media themselves are looked at, differentiation of degenerated component is easy because it adheres to the surface of the supporting media.

In vitro Hemolysis

Two types of changes may be roughly distinguished.   They are: 1) influence of hemoglobin freed as a result of hemolysis, and 2) influence of non-hemoglobin stromal proteins of erythrocyte.

1.   Effects of Hemoglobin

Changes caused by hemoglobin which has been freed from red blood corpuscles include the poor separation of the α_2 and β fractions.   When hemoglobin is further freed in large quantity, the β fraction increases.

Figure 94 shows that the greater part of the hemoglobin observed under normal conditions is Hb A_1 in adults and HbF in neonates.   Under pH 8.6 and the ionic strength of 0.05–0.1, these two are migrated at β fraction.   Moreover, the haptoglobin contained in normal serum is capable of binding free hemoglobin amounting to 40–160 mg/100 ml[213].

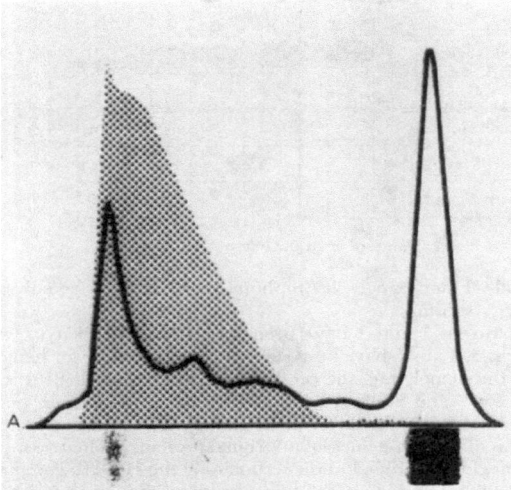

Fig. 92 Filter paper electrophoretic pattern of the serum proteins in hyperlipemia. The serum sample is lactescent, being taken from the patient with diabetic acidosis. The neutral fat in the serum is increased, being 94.0 mEq/l. The solid line indicates the densitometric pattern of the proteins (stained with BPB), and shows a discrete peak at the origin. The Oil Red O staining of the same electrophoretic strip shows a markedly increased lipoprotein, indicated by the shaded peak.

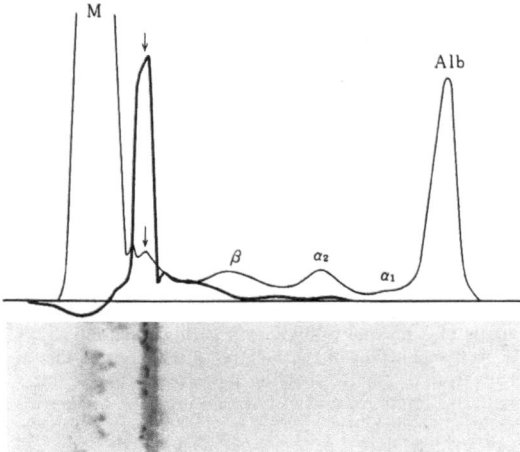

Fig. 93 Filter paper electrophoretic pattern of the stored serum.
The thin line indicates the densitometric pattern of the serum protein fractionation (BPB stain), while the thick line indicates that of the serum lipoprotein fractionation (Sudan Black stain). The thin line pattern shows a minor peak (arrow) at the origin, corresponding to the denatured lipoproteins fixed onto the surface of the filter paper. The thick line pattern shows a curve below the base line, corresponding to the M-protein band. A large amount of M-protein has been jellified on heating, resulting in a decrease of the optical density.

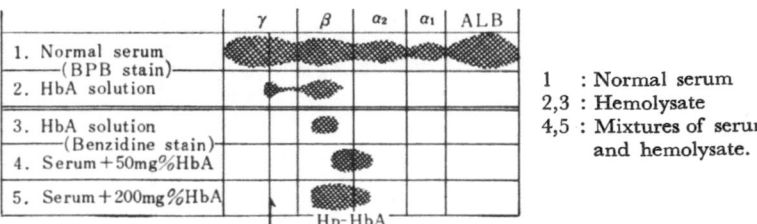

	γ	β	α₂	α₁	ALB
1. Normal serum —(BPB stain)—					
2. HbA solution					
3. HbA solution —(Benzidine stain)—					
4. Serum + 50mg%HbA					
5. Serum + 200mg%HbA					

Hp-HbA

Point of application

1   : Normal serum
2,3 : Hemolysate
4,5 : Mixtures of serum
       and hemolysate.

Fig. 94 Filter paper electrophoretic patterns of hemolysate and serum.
The patterns 1 and 2 have been stained with BPB dye, while the patterns 3, 4 and 5 have been stained with benzidine for hemoglobin. With the hemolysate, the protein band stained with BPB is seen at the β zone, and a minor band at the origin. However, only the β spot is stained positively with benzidine, and a minor band at the origin is of the non-hemoglobin stromal protein. When 50 mg/100 ml of hemoglobin is added to the serum, only the Hp–Hb complex is recognized, migrating at the zone between the α₂ and β fractions. With the addition of 200 mg/100 ml hemoglobin, both the Hp–Hb complex and the free hemoglobin are recognized.

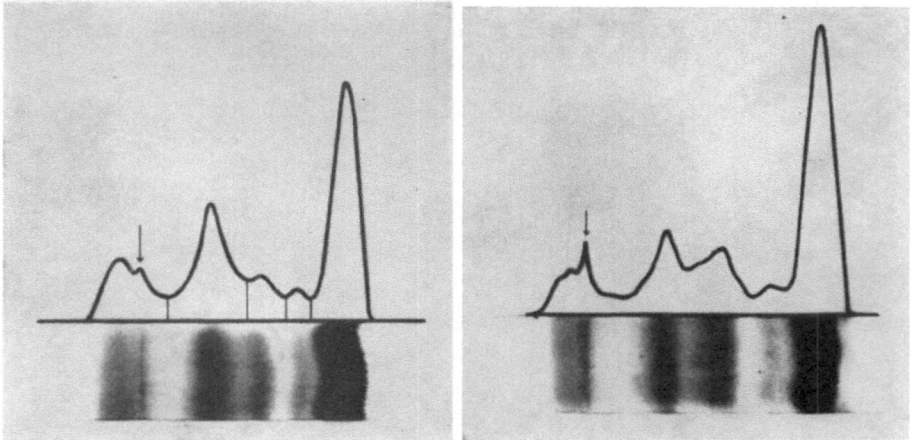

Fig. 95 Filter paper electrophoretic patterns of the hemolyzed sera.
The left pattern is of a normal adult serum strongly hemolyzed, showing a large β fraction and a poor separation of the α₂ and β fractions. The right pattern is of hemolyzed serum from an infant, showing a prominent β fraction, a poor separation between the α₂ and β fractions, and a sharp minor peak at the origin.

Electrophoresis is performed by adding a hemoglobin solution of various concentrations to serum, with the results shown in Fig. 94.[182] That is, haptoglobin migrates in the α₂ fraction, and the Hp-Hb A complex has the mobility between the α₂ and β fractions. However, when hemoglobin of 200 mg/100 ml is added, benzidine-positivity is demonstrated in a wide area ranging from the α₂ fraction to the β fraction. Obviously this implies that the Hp-Hb A complex and free hemoglobin are mixed.

In this way, hemoglobin which is released into serum in *in vitro* hemolysis is first bound to haptoglobin, then moves to the middle zone between the α₂ and β fractions. As a result, the separation between the two fractions becomes indistinct. If hemoglobin is released in such a large amount as to exceed the binding capacity of haptoglobin in serum, free

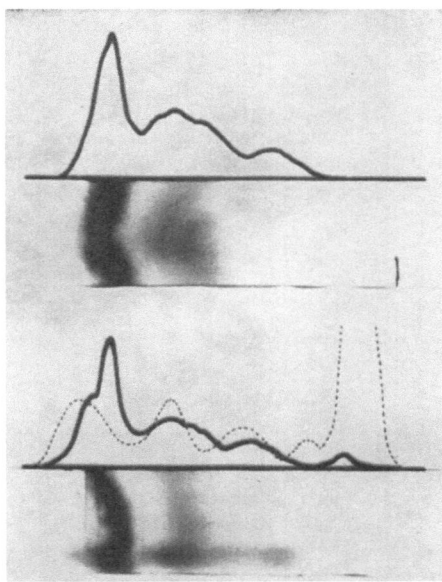

Fig. 96 Filter paper electrophoretic patterns of the non-hemoglobin erythrocyte stromal proteins.
The hemolysate is mixed with CM-Sephadex in the 0.05M phosphate buffer of pH 6.8 to remove hemoglobin. The hemoglobin-free supernatant solution is concentrated, and it is analysed with filter paper electrophoresis. The non-hemoglobin erythrocyte stromal proteins are separated into 3–4 fractions, but the major fraction is migrated at the mid-γ zone.

hemoglobin is seen to move at the β-fraction and a significant increase of the β fraction results. (Fig. 95)

2. Effects of Non-hemoglobin Erythryocyte Stromal Proteins

Along with the liberation of hemoglobin from red blood corpuscles in hemolysis, a small quantity of water-soluble proteins constituting the stroma of erythrocytes is released.[58,140,160,182,185] The author removed hemoglobin by treating hemolysate with CM-Sephadex and obtained a concentrated clear supernatant fluid. By analyzing this sample by means of filter-paper and agar-gel electrophoresis, non-hemoglobin stromal proteins of erythrocyte were separated into 3–4 firactions (Fig. 96, 97). Further, ffve precipitin arcs were confirmed by using the anti-serum obtained from rabbits immunized by the stromal proteins (Fig. 97). HAUT et al.[140] have observed six fractions by using starch-gel electrophoresis with samples obtained through the same procedure as those used by the author. Moreover, HOWE et al[160] demonstrated at least twelve different kinds of non-hemoglobin stromal proteins by using the anti-serum obtained by hemolysate immunization. However, the fraction existing in the greatest quantity moves to the γ zone and is observed clearly when filter-paper, agar gel or cellulose acetate membranes are used (Fig. 96, 97, 98). It is located in the origin on filter-paper electrophoresis.

An amount of water-soluble stromal proteins of erythrocyte released in hemolysis does not seem to be so great, as stated above. But in hemolyzed serum, in which there is a marked physiological or abnormal decrease of the γ fraction, a sharp protein peak may be observed.

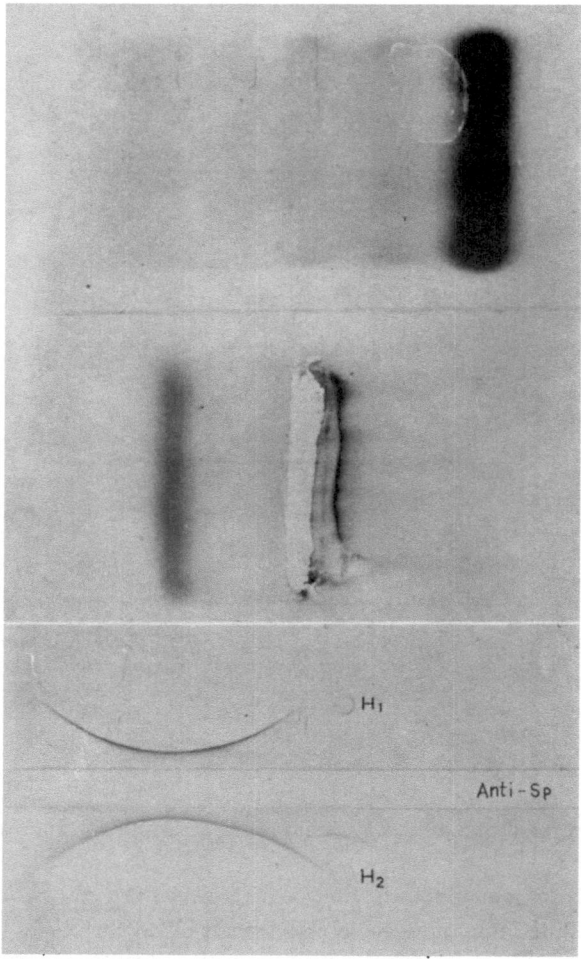

Fig. 97 Agar gel electrophoretic and immunoelectrophoretic patterns of the non-
hemoglobin erythrocyte stromal proteins.
The uppermost pattern is of the normal serum, and the middle pattern is of the non-
hemoglobin stromal proteins (Sp), showing the major protein band at the mid-γ zone,
as seen on filter paper electrophoresis.   The lower pattern is obtained on immunoelectrophoresis, using the anti-Sp rabbit immune serum.   At least, two distinct precipitin lines are identified.

The influence of *in vitro* hemolysis which has already been mentioned is observed in the
same form in filter-paper, agar gel and cellulose acete membranes alike.   Also, on proteinograms by the thin-layer gel filtration technique, an increase of the A- and the M-fractions
is observed in hemolyzed serum (Fig. 99).   That is, the haptoglobin-hemoglobin complex
forms a giant molecule and is separated in the M-fraction.   On the other hand, free
hemoglobin is observed at the post-A zone, slightly behind the A-fraction.

However, by immuno-electrophoresis, confusing precipitin arcs may not be seen
because the antibodies against hemoglobin and stromal proteins are not contained in
ordinary antisera used.

S   A₂   A₁

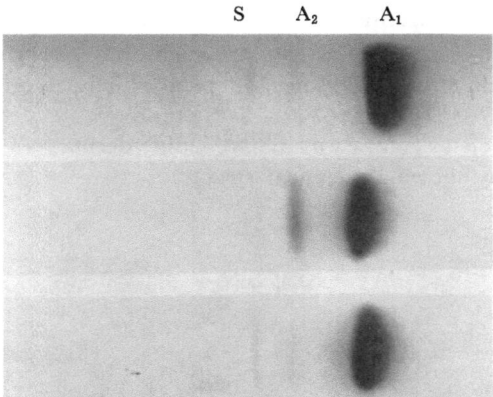

Fig. 98 Cellulose acetate electrophoretic patterns of hemolysates.
The electrophoretic separation was performed on Sepraphor III membrane in the
Tris-borate buffer of pH 9.2 for 75 minutes with 500 V. In this condition, Hb A₁
and Hb A₂ are clearly separated, and also the non-hemoglobin erythrocyte stromal
protein (S) is recognized on protein staining.
The upper pattern is of normal adult, the middle pattern is of the patient with thalassemia minor, and the lower pattern is of the patient's father.

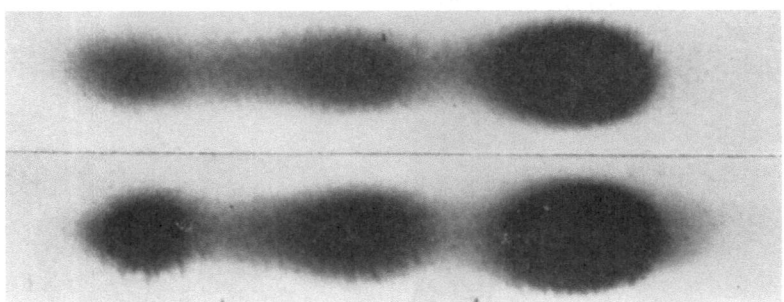

Fig. 99 Thin-layer gel filtration pattern of the hemolyzed serum.
The upper pattern is of the non-hemolyzed serum, and the lower pattern is of the
hemolyzed serum. The latter pattern demonstrates two benzidine-positive spots;
the free hemoglobin is separated just after the A fraction or as a tailing of the A fraction, and the Hp–Hb complex is at the M fraction.

Anticoagulants

Ordinarily, serum samples are used for electrophoresis using supporting media. But
in the TISELIUS method, plasma is frequently used and therefore anticoagulants will be
discussed.

When heparin is added, because of the interaction between albumin and heparin, a
minor spike appears in the albumin fraction. As a result, the albumin fraction becomes
3–4% higher than when double oxalate is used. Also, it is said that oxalated plasma
shows poor separation of the β and the ϕ fractions.

Also, in measuring total plasma protein concentration, the addition of anticoagulants
may influence test results. In specific gravity method such as the copper sulfate method
and refractometry, double oxalate causes a higher value than heparin. Whole blood or
plasma specific gravity is frequently used for screening blood donors, but it is regrettable
that the influence of anticoagulants has not been considered seriously.

Intravenous Contrast Media

In recent years various intravenous contrast media have appeared on the market. It has been made clear that their side effects depend chiefly upon their interaction with serum proteins.[181,188,202,209,291]

All important intravenous contrast media are iodine derivatives of benzoic acid or their dimers. They are the following chemical structures.

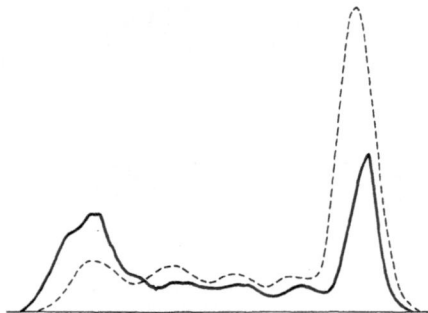

Urokon   Hypaque

After these contrast media are intravenously injected, they are combined chiefly with albumin. The binding activity is strong with Urokon and Cholografin (dimeric form of Urokon) in which the hydrogen atom adhering to the fifth carbon is not replaced by any other chemical group.[202, 209]

Fig. 100   Effect of the intravenous contrast medium (Urokon) on the serum proteins.
37 y.o., female, anemia with possible renal tumor.
Complained of severe vertigo and vomiting during the intravenous injection of Urokon. Six hours after its discontinuation, the subjective symptoms disappeared completely. The solid line indicates the densitogram of the serum protein electrophoretic fractionation for the serum obtained immediately after the side effects appeared. The dotted line is of the serum obtained when the subjective symptoms had completely disappeared. Increased γ fraction and decreased albumin fraction are recognized on the first serum sample.

According to KURT and McDOWELL,[202] who made *in vitro* and *in vivo* experiments on intravenous contrast media, a different electrophoretic pattern is observed, depending upon the absolute quantity of the injected contrast media.[202] But a tendency of albumin to decrease markedly and the pronounced tendency to increase of the γ fraction are seen (Fig. 100). The iodine content is the greatest in the albumin fraction but is hardly demonstrated in the γ fraction. In addition, these changes disappear after *in vitro* dialysis of the serum sample. These artificial influences may be easily avoided by selecting the time and the site of blood collection. However, intravenous injection of a large quantity of contrast media into a patient with pronounced serum protein abnormalities such as myeloma may result in death due to acute renal failure by drastic changes in serum proteins.[188, 291]

PHYSIOLOGICAL VARIATIONS IN THE PLASMA PROTEIN MEASUREMENT

In the previous section, problems concerning the techniques of measurement were considered. Before collecting blood, it must be kept in mind that plasma protein patterns also vary according to the various conditions and phenomena in the living body. This is called physiological variation. It is divided into two groups: 1) Between-individual variations and 2) Within-individual variations.

Variations within the same individual is referred to as within-individual variations. These include such variations as those based upon the kinds of blood drawn (arterial blood and venous blood, plasma and serum), diurnal variation, seasonal variation, variations due to exercise and posture.

Between-individual variations are those which occur among different individuals. They include such variations as those based on sex difference, racial difference and others of unknown origin. Ordinarily, in the presence of within-individual variation, conditions which most minimize this variation are selected. Between-individual variation can be measured only under this condition.

Difference between the Plasma and the Serum

Total protein concentration and protein fractional values differ depending on whether the sample is plasma or serum used. Fibrinogen, prothrombin, factor V and factor VIII which are consumed during the blood coagulation process are not contained in serum. But the only one that matters quantitatively is fibrinogen. As for the other blood coagulation factors, they may ordinarily be neglected in the case of plasma (serum) protein analysis because they are subject to exceedingly minor variations.

As far as total protein concentration is concerned, it is, on the average, higher in plasma than serum by 0.38 g/100 ml (0.3–0.5 g/100 ml). This is almost equal to the normal concentration of fibrinogen.[367]

A considerably different variation is shown among different electrophoretic techniques. In the TISELIUS method, the β fraction of plasma is wider than that of serum. A new ϕ fraction is observed between the β and the γ fractions. The ϕ fraction contains mainly fibrinogen. It should be borne in mind, however, that beside fibrinogen approximately 0.15 (0.10–0.18 g/100 ml) γ_1-globulin is included.[367]

With filter-paper electrophoresis, if the same method is employed as in serum protein fractionation, fibrinogen remains in the origin and hardly moves, as is shown in Fig. 101. This is because not only does fibrin formed adhere to the filter paper but its intrinsic viscosity is great. Moreover, fibrinogen is observed as a relatively sharp peak or band and therefore it is quite indistinguishable from the M-component (Fig. 102). Of course, an experienced observer is able to distinguish it clearly. Accordingly, serum samples are used for protein separation by filter-paper or cellulose acetate electrophoresis. However, the ϕ fraction is clearly separated from the γ fraction by the use of excess blood anti-coagulants added to plasma samples and buffer solution.

Difference between the Arterial and the Venous Bloods

The results obtained by comparing bloods obtained from the brachial artery and that

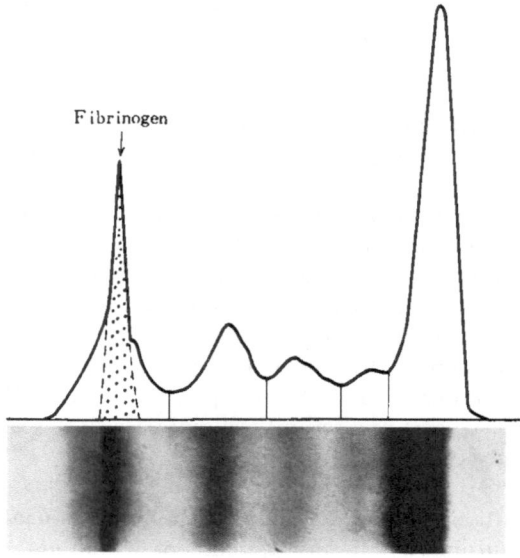

Fibrinogen

Fig. 101 Filter-paper electrophoretic pattern of the normal plasma proteins.   At the point of sample application seen is a somewhat wavy, irregular, narrow protein band of fibrinogen.   On densitometry, a sharp fibrinogen peak may be confused with the M-protein band.

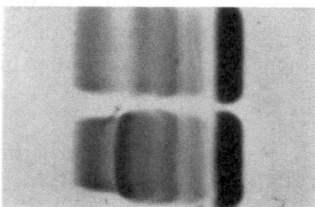

Fig. 102 Cellulose acetate electrophoretic protein patterns of the serum and the plasma.   The upper pattern is of normal serum, while the lower pattern is of the plasma taken from the same individual.   The plasma protein pattern shows a distinct ϕ fraction, appearing a characteristic curved band and a deformed γ fraction band.

from the median vein reveal that total protein concentration is higher in venous blood by 0.2–0.5 g/100 ml, and that there is not much variation to be perceived in the relative percentage of each fraction.   This difference can possibly be explained by relative hemoconcentration occuring in the venous system.[367]

The same relation may also be observed between venous blood and capillary blood from the ear lobe.   Total protein concentration may be significantly higher when capillary blood is used.   In recent years, the refractometer is used for measuring total protein concentration and thus the ear lobe is frequently used.   Puncturing of the ear lobe should be performed after rubbing the ear lobe sufficiently in order to make the blood circulation well.

Diurnal Variation

There are a number of reports on diurnal variation with regard to serum protein concentration, and the results are nearly all the same.   That is, the concentration continues to rise during the daytime with the passage of time.   During sleep at night, it becomes

lower (Fig. 103).   It is said that, on the average, the range of diurnal variation is
1.1 g/100 ml (1.0–1.3 g/100 ml).   Moreover, on serum proteinograms carried out according to the TISELIUS method, the concentration of each fraction shows the same variation
as the total protein level with the passage of time, but the relative percentage of each fraction is hardly changed.   This fact suggests that diurnal variation bears a close relation to
hemodilution and hemoconcentration.   In the case of night laborers, diurnal rhythm
of protein concentration shifts somewhat.   The protein concentration is influenced in a
diversified way by the variation in water flow between tissue fluid and blood, and by the
variation in the metabolism of serum proteins.   The functions of the endocrine system and
the autonomous nervous system play a major role for the variation.[367]

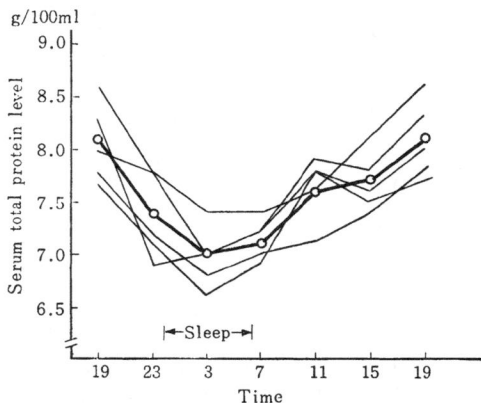

Fig. 103 Diurnal variation of the serum total
protein concentration (from HAGA) The thick
line indicates the average values among 5
different individuals.

Seasonal Variation
Many researches point out that serum protein concentration varies according to the
season, being generally low in summer and high in winter.   In Tokyo, results show that
there is a variation of 0.6 g/100 ml or approximately 10% on the average (Fig. 104).
Serum proteinograms according to the TISELIUS method are shown in Fig. 105.   In A/G
ratio and albumin fraction concentration, their variation is biphasic, being high in spring
and autumn.   In summer, the concentration of the α fraction is rather low, while that of
the β fraction hardly varies throughout the year.   The concentration of the γ fraction
shows the most remarkable change, increasing markedly in summer, 1.5 times higher than
in winter.
The mechanism of these seasonal differences is still unknown but what is certain is that
these are closely related to changes in the environment and temperature.   It is unlikely
that they are caused by hemoconcentration alone.   Rather, the cause is probably in diversified influences of a balance between the function of the endocrine system and the function of
the autonomous nervous system.   On the other hand, it appears that in an environment
such as that of the Antarctic where seasonable differences are not notable, serum protein
concentration and fractional variations are less marked throughout the year than they are
in Tokyo.[265]   Thus, it seems that these variations occur not so much in association with
seasonal variations as those of environment and temperature.

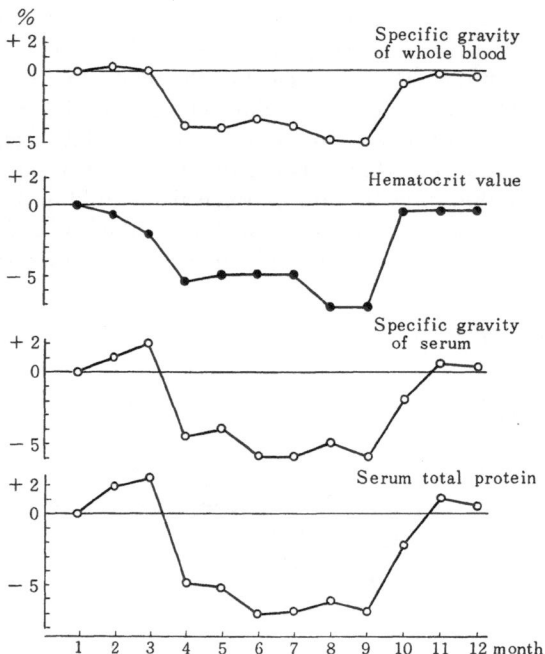

Fig. 104   Seasonal variation of the serum total protein concentration
and hematocrit value (from YAMADA and KAWAI).
The average values are obtained among 12 young girls living in the
same dormitory.

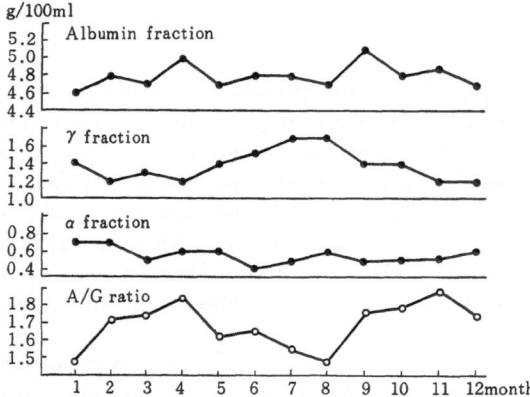

Fig. 105   Seasonal variation of the serum A/G ratio and the major
serum protein fractions (from HAGA).
The average values are obtained from 5 healthy male adults, living
in the Tokyo district.

Postural Variation

It is a well-known fact that various physiological functions may change on different postures. This holds true also for plasma protein concentration. As shown in Fig. 106, five young males were made to stand after lying down for 30 minutes. After remaining in a standing position for 20 minutes, they lay down again, maintaining this position for 20 minutes. This resulted in as much as a 17% increase of total plasma protein concentration after the standing position. Even as long as 20 minutes after the men had reverted to a recumbent position, the previous value of their protein concentration did not return fully. There was little change in the A/G ratio and a relative percentage of each fraction. But the plasma volume measured at the same time decreased in the standing position. Thus, extravasation of plasma water causes relative hemoconcentration in the standing position. In contrast, water flows into the blood vessels when a standing position changes to a recumbent position, resulting in hemo-dilution. This may cause a decrease in protein concentration in a recumbent position[367]

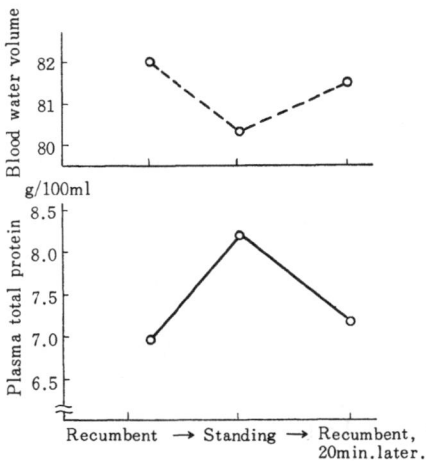

Fig. 106 Postural variation of the plasma total protein concentration (SUGIMOTO et al.)
The first blood sample was drawn after resting for 30 minutes at a recumbent position, the second sample after standing for 20 minutes, and the third sample after resting again for 20 minutes at a recumbent position.

Exertional Variation

It is generally assumed that hemo-concentration occurs with exercise, resulting in a temporary increase of total protein concentration as well as the concentration of various inorganic constituents.[367] Six long-distance relay runners ran 20 km in 80 minutes, and five persons who were not in training skipped rope for 5 minutes. Blood samples were collected from both groups immediately before and after the running and rope-skipping. The results are shown in Fig. 107. Serum protein concentration was generally higher by 0.6 g/100 ml in the trainees, but no significant difference in A/G ratio and relative percentage of each fraction was observed between the two groups. In rope-skipping and bicycle-riding, serum protein concentration increases by 3–9% due to hemo-concentration after exercise, but an hour later, it nearly returns to the value immediately preceding exercise. Besides, no marked change was observed in the A/G ratio and relative percentage of each fraction. However, in marathon racing, which is hard exercise, there was still a 4% increase of serum protein concentration one hour after the exercise. Proteinograms showed that there was an increase of the α and β fractions. It is possible that, unlike rope-skipping, in marathon racing other factors besides hemo-concentration deserve consideration.

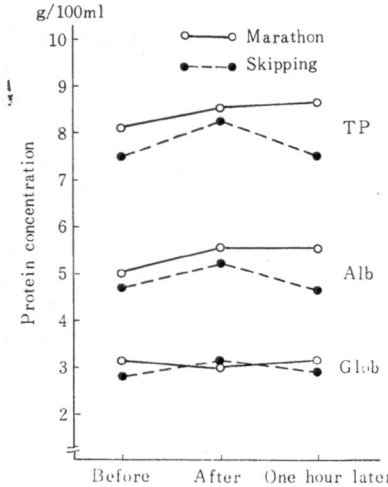

Fig. 107 Exertional variation of the serum total protein concentration (from SUGIMOTO et al.)
TISELIUS electrophoretic fractionation of the serum proteins are done on 6 trained men running marathon and 5 non-trained men skipping a rope. On severe exercise, the γ fraction decreases and the elevation of the total protein concentration continues longer.

Variation in Exhausion

There are various types of exhaustion, but here it will be divided into: 1) pronounced psychological and 2) pronounced physical exhaustion. Of course, both elements are present in any case, so strict separation of the two is difficult.

Here is a report by YAMADA on variations of different blood components observed in the course of a summer excursion of a nursing school in Tokyo.[393] The excursion was conducted in two groups. The students were all of the same age group and lived in the same dormitory but 19 of them made a six-day trip to Kyushu in southern Japan while the other 19 made a 9-day trip to Hokkaido in northern Japan. The former group experienced hot weather and a tight schedule, while the latter group's excursion was made in a cool weather with a comparatively loose schedule. The similar variations of blood components were observed in both groups, although the variation was more marked in the Kyushu group which took a longer time to recover. As shown in Fig. 108, with respect to total serum protein concentration, no variation was observed in the Hokkaido group while there was a 5% increase in the Kyushu group immediately upon their return to Tokyo and a 4% increase even a week after. However, as far as the A/G ratio was concerned, there was a notable variation in both groups, together with a decrease of albumin and an increase of globulins.

As regards exhaustion with pronounced psychological elements, YAMADA made a report on the variation seen in nurse students who were just begining ward training. This study revealed noticeable variation of blood components before and after they started ward training. It is interesting to note that this variation is different in quality from that observed during the trips (Fig. 109, 110). Total serum protein concentration shows a tendency to increase slightly and to parallel changes in the urinary DONADIO reaction which is used for measuring the degree of exhaustion. There is, also, an increase of albumin accompanied by a decrease of globulin. This resembles the case of the marathon runner discussed in the previous section. It may be concluded that approximately five weeks are necessary for a person to adapt to rapid environmental changes.

At any rate, in the case of exhaustion, variations of blood components are conditioned by not only seasonal differences or differences due to exercise but also by changes in the environment which influence a complicated mechanism in the body.[349]

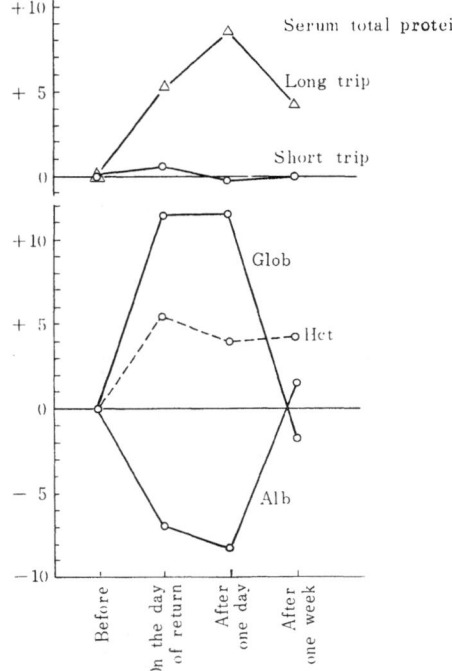

Fig. 108 Variations of the serum proteins after long trips. (from YAMADA)
The group of 19 girl students took a long trip to hot Kyushu area, and the other group took a shorter trip to cool Hokkaido area. In both groups, the similar tendency is noted on the variations of the serum proteins, but the Kyushu group shows more variations than the Hokkaido group.

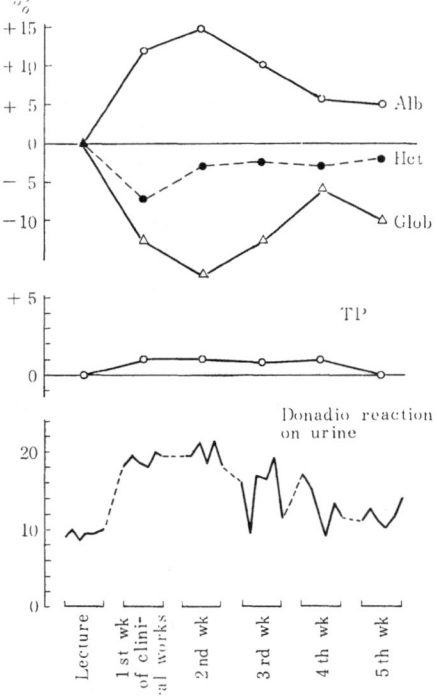

Fig. 109 Variations of the serum proteins on the sudden change in the training environment (from YAMADA).
The studies were made on 14 nurse students during the lecture period and the period where clinical works were just started. The DONADIO reaction on urine was also done, and their mental stress seems to be stabilized within 5 weeks after the clinical works were started.

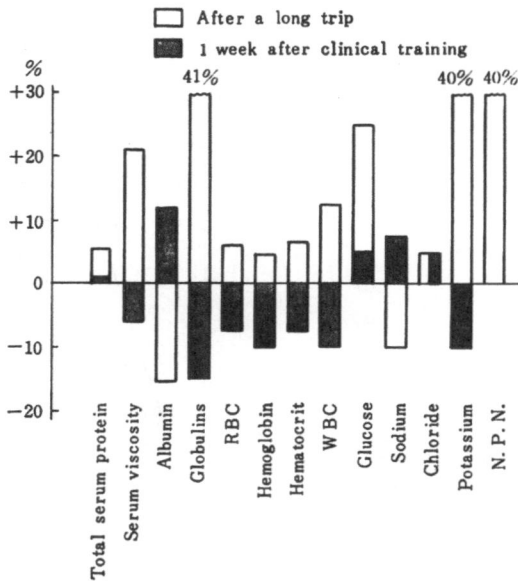

Fig. 110 Variations of the blood constituents during the clinical
training period and after a long trip (from YAMADA).
Various blood constituents were measured among student nurses
immediately after a long trip and also one week after the clinical
training was started. A significantly different tendency is recognized
between the two groups.

Nutritional Variation

It has been pointed out that various blood chemical tests yeild different results when
carried on with healthy people of differeing geographical and age backgrounds. This
difference is marked in plasma protein concentration, especially in the albumin fraction
concentration. This differences appears to be attributable to dietery habits and environmental differences, but the average nutritional condition of a group is also thought to be a
contributing factor. Japan which suffered very poor environmental conditions as a result
of defeat in World War II, has been unique in attaining marvellous economic development within the last twenty years. In Fig. 111, the results obtained by means of the
TISELIUS method (which is least subject to technical variation) are compared during the
last 20 years.[1] An increase of the albumin fraction and a decrease of the γ fraction are
observed. Of course, when interpreting these results, it must be borne in mind that they
are based upon comparison between tests performed on different groups and by different
studies, but it may be presumed, at least, that the improvement of living standards of the
entire nation played a major role.

Racial Variation

It is frequently asserted that normally results of blood chemical tests vary from race
to race. However, the demonstration of this fact is, in reality, rather difficult, for all
physiological variations must be eliminated before an analysis is made, and to do this is
extremely difficult. YOKOTA's research in 1953 is of comparative value in Japan. He
compared the relative percentage of plasma protein fractions of Japanese and American
adults living in Tokyo. The measuring apparatus and the measurer were the same.

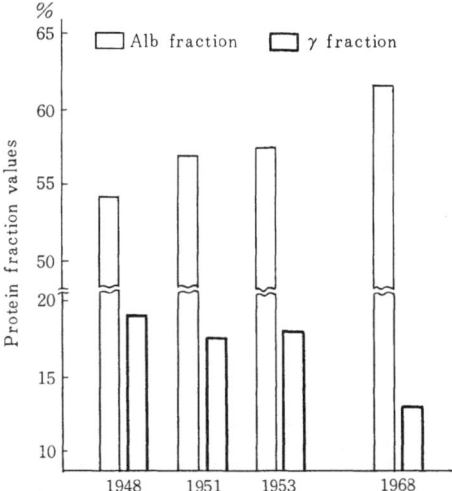

Fig. 111 Changes of the electrophoretic fractions of normal adult
serum proteins (from ABE et al.).
The values were obtained on normal adults in different years, using
the standard TISELIUS electrophoretic method which was proposed
by the Society of Electrophoresis.

However, average values with the Japanese adults were obtained by testing 55 pooled
sera prepared from a total of 4,500 individuals while in the case of American adults, average values were obtained by testing 8 individuals separately (Table 28).   Of course, this
difference may not be solely due to racial difference.

Table 28   Normal plasma protein values on both Japanese and American (from SUGIMOTO, ABE and
YOKOTA).

Fractions (%)	Japanese (n: 4500)	American* (n: 8)
Alb fraction	57.7±3.1	58.2±2.4
α_1 fraction	5.6±2.3	7.6±2.1
α_2 fraction	11.8±1.2	13.2±2.6
β fraction	7.5±1.2	6.4±1.5
γ fraction	17.4±1.2	14.6±1.3

* Healthy American living in Tokyo.
  Electrophoretic fractionation was performed with the TISELIUS method by the same investigator at
  the same time in 1953.

Sexual Variation

The variation of plasma protein values due to sex difference originates mainly in the
qualitative and quantitative difference of serum lipoproteins, and it is generally implied
that there is no significant difference in the percentage of plasma protein fractions.[75,174,181]   In normal adults, the total serum protein concentration is 0.1–0.2 g/100
ml lower in females.[207,275,319,336]   As shown in Fig. 112, during infancy, it is higher
in females with the variation comparatively notable in puberty.   In adulthood, variation due to sex difference is marked in the second and third decades.   But there have

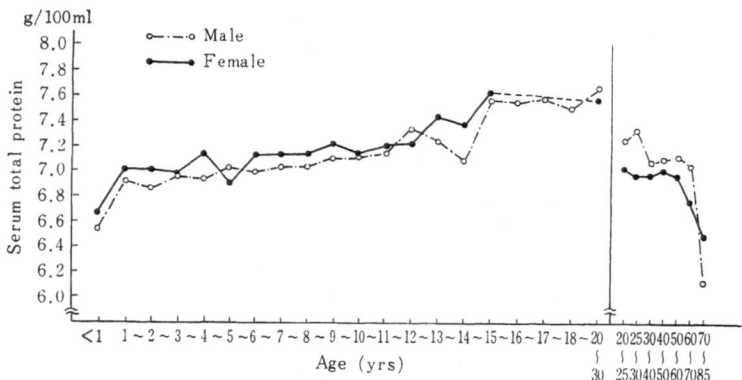

Fig. 112 Variations of the serum total protein concentration in different ages and sexes.
The data on children are from OBA et al. and those on adults are from SANDOR. The difference between the male and famale individuals is most prominent during the reproductive stage.

been some reports stating that it is somewhat higher in females, and it has also been reported that no significant difference is observable.[174,181]   At any rate, it is probable that the difference due to sex is not great.

According to OBA et al. who compared the albumin concentration, that of males is 4.18 g/100 ml while that of females is 4.06 g/100 ml.[275]   The same result was reported by STEINFELD.[361]   Filter-paper electrophoretic analysis also shows that the albumin fraction is lower and the γ fraction is higher in females.

Age Variation

It is a known fact that the body shows various changes due to aging.   A marked variation due to age is also seen in plasma protein concentration.

The total protein concentration shows a characteristic variation as noted in Fig. 112, 113, 114.[257,275,367]   In the neonatal period, it is slightly higher than in infancy, but it is, on the average, lower by 1.4 g/100 ml than in adulthood.   It increases gradually with growth, attaining the peak in the second and third decades.   It decreases again thereafter and becomes the lowest when one is in the sixties.   However, many report the lower values in old age, although there is a different opinion.[131]

With regard to the plasma protein fractions, a large number of reports are available on results obtained through various procedures.   The albumin fraction shows the same variation as in the total protein concentration, naturally because albumin constitutes the predominantly large proportion of the plasma proteins.   Such variation as this in albumin may be presumably explained by insufficient hepatic protein synthesis during the neonatal period, and the active movement of proteins into various tissues during the growing period.   It is possible that the decrease of the albumin fraction in old age is caused by latent decrease of hepatic functions.   However, according to the results obtained recently by YAN and FRANKS,[416] the fractional catabolic rate and body distribution of albumin in the aged (17–80 years of age) are the same as those in the young.   On the other hand, the total intravascular and interstitial albumin is low in the aged and thus the daily turnover is decreased.

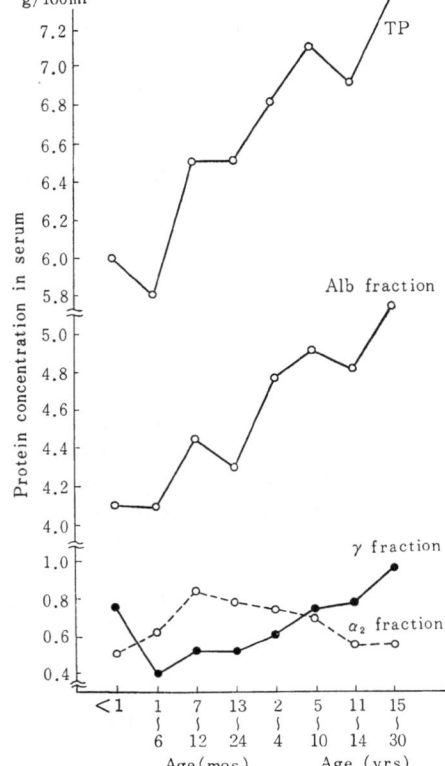

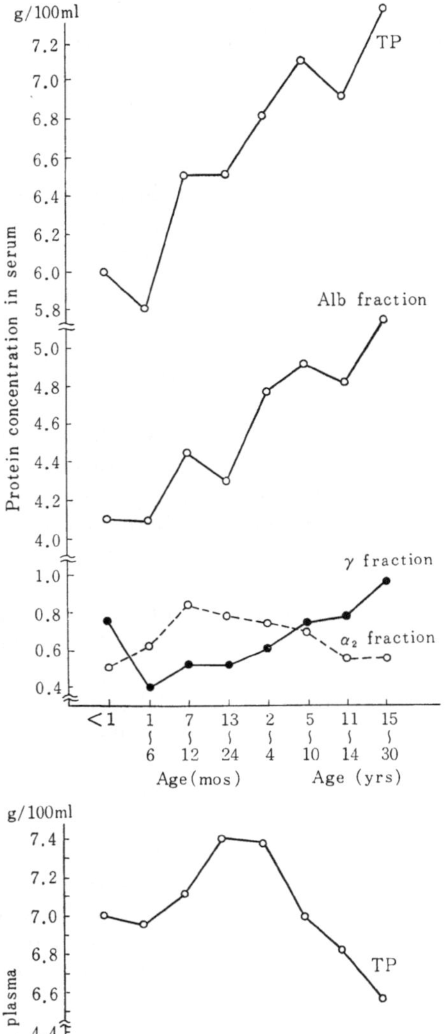

Fig. 113 Age variations on the fractional values of the serum proteins (from Monma et al.)
The fractional values were obtained with the cellulose acetate (Oxoid) electrophoretic analysis. The α_1 and β fractions are not shown in the figure, since they are not significantly different among various age groups.

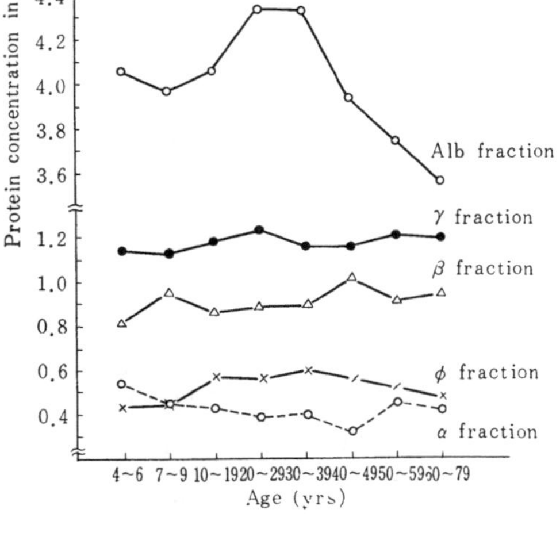

Fig. 114 Age variations of the fractional values of the plasma proteins (Sugimoto et al.)
The fractional values were obtained with the Tiselius electrophoretic method using the phosphate buffre of pH 7.8. Thus, the α fraction is almost equivalent to the α_2 fraction obtained with the standard cellulose acetate electrophoretic method.

The concentration of the a_1 fraction is rather low in infancy but it hardly shows any variation among different age groups.   The a_2 fraction shows notable variation in childhood.   Its concentration increases markedly about six months after birth to one year, becoming one and a half times as high as that in adulthood.   It remains high until four or five years of age, and then it gradually decreases.   Around ten years of age, it reaches the adult's level.   This variation of the a_2 fraction reflects the variation of a_2-macroglobulin which is a main component in this fraction.   The concentration of a_2-macroglobulin in the children of around 2–4 years becomes two to three times as high as that in adulthood.[106]

The concentration of the β fraction is slightly lower in the neonatal and infantile periods, showing a tendency to increase slightly when one is 40 years of age.   It is possible that this is caused by the variation of β-lipoproteins.   In cellulose acetate electrophoresis, the variation of the β fraction tends to be smaller since a major part of lipoproteins is included in the a_2 fraction.   However, it shows a tendency to increase slightly during the period from infancy to adulthood, reflected by the variation of transferrin.

The ϕ fraction increases from the age of ten years and again tends to decrease after 40 years.   This is parallel to the variation of fibrinogen.   However, a major part of IgA- and IgM-immunoglobulins are included in the ϕ fraction.   It is presumed that the increase of IgA-immunoglobulin in childhood is in part reflected here.   But in old age, IgA immunoglobulin shows a tendency to increase while IgM-immunoglobulin hardly varies in old age.[365]

IgG-immunoglobulin constitutes the largest portion of the γ fraction.   In the neonatal period, the concentration of this fraction is almost identical to that of adulthood. It decreases rapidly after birth and is the lowest at the age of 2–4 months.   Then it gradually increases and in adolescence, it becomes as high as in adulthood.   As the years advance, it gradually increases and reaches its highest value in old age.[131,365]

(APPENDIX) PLASMA PROTEINS IN THE CADAVERS

A large quantity of blood may be obtained on post-mortem examination but there are only a few reports on protein analysis of this blood.   According to the results obtained in the author's laboratory, even in the absence of hemolysis, the total protein concentration of post-mortem serum (by refactometry) increases, and its increase is, of course, marked in hemolyzed sera (Fig. 115).   When cellulose acetate electrophoresis is used, the A/G ratio

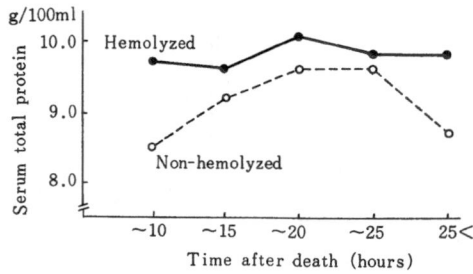

Fig. 115 Serum total protein concentrations in cadaver blood.
    (from Aoki et al.).
The serum total protein concentration was determined with refractometry on 48 cadavers autopsied in the Medical Examiner's Office.   The average values are plotted among 22 hemolyzed sera and 26 non-hemolyzed sera.

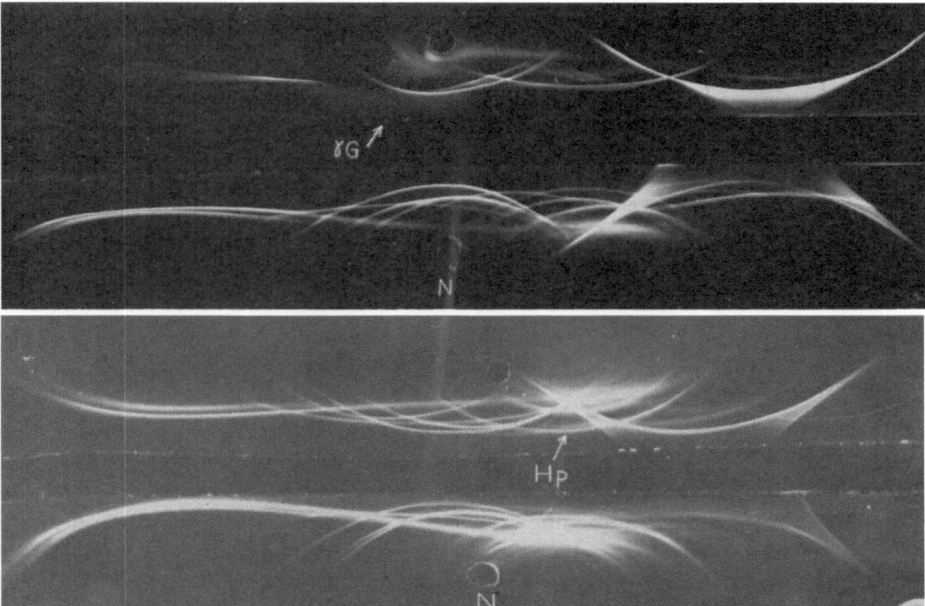

Fig. 116 Immunoelectrophoretic patterns of cadaver sera.
The upper pattern is from the patient with IgG type myeloma, showing a M-bow formation of the IgG line.   The lower pattern is from the patient with septicemia, showing a significant increase in haptoglobin and α_1-antitrypsin.

tends to decrease and a pronounced tendency to increase of the β and γ fractions is observed.   However, the cause is unknown.

As stated previously, a marked change is seen on routine electrophoretic fractionation following death, but no notable change is observed on immunoelectrophoretic patterns 36 hours after death[238].   Even when there has been notable change in electrophoretic patterns using supporting media due to hemolysis, immuno-electrophoretic patterns were not influenced by hemolysis alone.   Accordingly, immunoelectrophoretic patterns of cadaver blood are of diagnostic significance with reference to special serum protein abnormalities after death, when serum protein fractionation was not performed before death (Fig. 116).

Chapter 17

Interpretation of Serum Protein Patterns

NORMAL VALUES

It is said of all laboratory tests that in interpreting laboratory results, they should necessarily be compared to their normal counterparts. This is also true in the case of serum protein fractionation. It is "normal value" that serves as the control. But as is frequently discussed, there are numerous questions with regard to the definition of normal values, the procedure of obtaining them and their use.

Normal values are not treated in great detail here, but major problems in serum protein analysis will be discussed.

Determination of Normal Values

According to SUNDERMAN and BOERNER,[370] "normal values" signify those values or properties characteristic of healthy persons. Then, what is a healthy person? According to SCHNEIDER[329], the term refers to a person who has values of specific and selected attributes not characteristic of those defined states which seem important for the immediate purposes of the physician making the classification. This is a rather ambiguous definition and actually all it says is that a normal person is one whom doctors regard, objectively, as not being ill.

HAYASHI asserts tentatively that a normal person is one who fits to the following description:

1) He has no history of diseases, with the exception of those that everyone is likely to contract such as measles and common colds;
2) His physical examination reveals nothing abnormal;
3) Nothing abnormal is found in chest X-rays, ECG (electrocardiograph), urinalysis (protein, sugar, urobilinogen);
4) His blood pressure is below 159/89 mmHg, and there are no subjective symptoms.

Only 10–25% of healthy persons fulfill these conditions though. In general, "normal values" refer to those values obtained by the measurement of groups of normal individuals selected on the basis of the foregoing.

HOFFMAN et al.[155] propose another method for determining "normal values" from the data obtained on all hospital patients. (Fig. 117). However, these tend to be broader than ordinary "textbook ranges". Usually, in obtaining normal values, a group or groups of many normal individuals are tested collectively in a very short period of time. But as shown in Fig. 118, the daily reproducibility obtained by measuring consecutively day after day is higher than the repeatability obtained in a certain batch of measuring. For this reason, HOFFMAN et al. argue that normal values which are in general use do notreflect true daily technical variation.

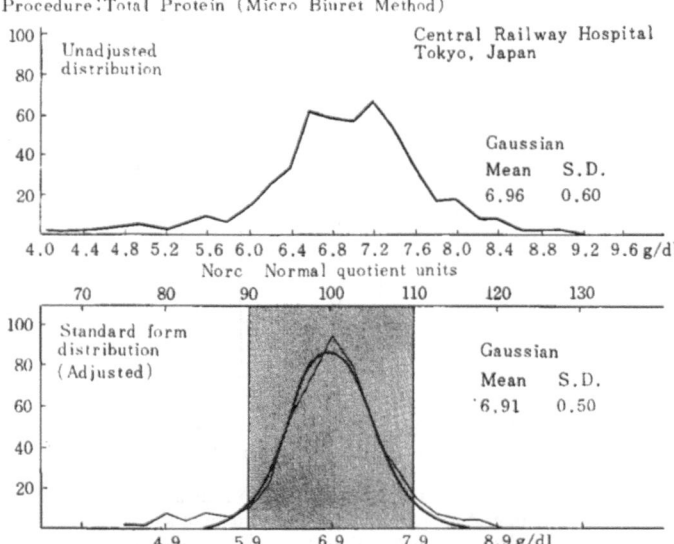

Procedure: Total Protein (Micro Biuret Method)

Fig. 117 Frequency distribution of patients' tests for serum total protein concentration (KAWAI and HOFFMANN).
The 500 consecutive patients' tests performed in the Central Railway Hospital are statistically treated with the method of HOFFMANN using a computer. The normal ranges obtained with HOFFMANN's method are 5.9–7.9 g/100 ml, the mean value being 6.91 g/100 ml.

Characteristics of Normal Values

"Normal values" of a group as obtained in this way have two characteristics: range and variability.

The concentration of total serum protein and serum protein fractions show a normal or Gaussian distribution when measured on a group of normal individuals (Fig. 117). Using M (mean) and S. D. (standard deviation) statistically obtained from the above results, the following range is conventionally referred to as the normal range: M±2 S.D.

Thus, so-called normal value is not a single value, but a range of values comprising 95.4 % of a group of normal individuals. The following three factors influence this range:

1) Factors which make for true differences between individual persons: between-individual or interindividual variation.

2) Factors which make for true differences from time to time in each single person: intraindividudal or within-individual variation.

3) Factors which make for true differences from measurement to measurement of each sample which may be measured: technical variation.
This is mathematically represented as follows:

$$S_0 = \sqrt{S_I^2 + S_W^2 + S_E^2}$$

S_0 is the standard deviation for the range obtained by actual measurement. S_I is the standard deviation for between-indivudal variation and S_W is for within-individual variation. $(S_I + S_W)$ represents physiological or biological variation as stated previously. S_E is the standard deviation for technical variation. When S_E is extremely small as compared to the others,

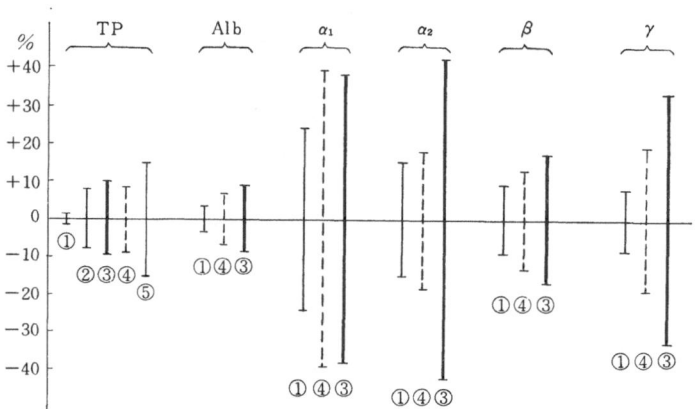

Fig. 118 Technical and biological variations in serum protein fractionation (the total protein with refractometry, the fractionation with the standard cellulose acetate electrophoresis).
The variation ranges are represented by $2 \times$ C.V. ① indicates the technical variation on 20 repeated testing of a single sample analyzed by a single technician, using the same method. ② the seasonal variation on a single individual measured by a single technician, using the same method. ③ the normal range on 40 normal individuals, measured by a single technician in 2 days, using the same method. ④ technical variation of a single pooled serum, measured twice a day for 20 consecutive days (in routine hospital laboratory). ⑤ normal range obtained from patients' tests with HOFFMANN's method.
A significant difference can be noted between two of the technical variation indicated as ① and ④. The "textbook normal ranges" ③ are usually obtained with the technical variation indicated as ①. However, the patients' values obtained in routine hospital laboratories are based on the technical variation indicated as ④. Therefore, strictly speaking, the "textbook ranges" cannot be a true reference, not considering many of the factors which may affect patients. The normal range (⑤) obtained with HOFFMANN's method seems to be adequate including both the technical variation (④) and the within-individual variation (②).
The technical variation on the α_1, α_2 and γ fractions are relatively large, and the physiological variations on the α_2 and γ fractions are also large.

$$S_o \doteqdot \sqrt{S_I{}^2 + S_W{}^2}$$

is formed, and thus physiological variation is accurately evaluated.

Samples of blood are usually collected when the subject is in a fasting state so that dietary, postural, diurnal and exertional variations which are included in S_W may be kept at a minimum. Also, when variations due to sex, age, race and geographic factors are considerable, the normal range of each is given separately. It may be concluded, therefore, that normal range is determined by combining all the influences of artificial and physiological variations, as stated in the previous chapter.

It is also stated therein that S_E arising from the kind of supporting media, electrophoretic conditions and apparatus for measurement are pretty clear. Therefore, normal values should be obtained for each method. On the other hand, S_I or S_W also show a marked variation. Accordingly, it may be necessary at the present state that "normal values on electrophoretic fractionation of serum proteins are determined according to one's technical conditions." For this reason, normal values reported by various researchers are not enumerated here.

Clinical Use of the Normal Values

Using normal values in clinical practice, the following points should be taken into consideration.

1) How were the individuals on whom the normal values were based?
2) Which methods and apparatus were employed for the measurement?
3) What is the range of technical variation obtained by the method used?

Ideally, normal values in the laboratory rather than in a monograph should be obtained and they should be used as a "control".

In addition to the above three points, the normal values obtained from a group have another pitfall. That is, even after various factors causing physiological variation have been eliminated as much as possible, there still remains individuality of an unknown origin. Therefore, as shown in Fig. 118, the range of normal values (③) in the group is always larger than that of physiological variation (②) of one individual. In clinical practice where patients are treated individually the test results of the individual patient as he was quite healthy should be a true "control". A health check-up which meets this demand has recently come into wide use. But since this method is not applied to all patients at the present, the normal values in a group are only used as references.

REPRESENTATION OF SERUM PROTEIN FRACTIONAL VALUES

The increase or decrease of each fraction which is electrophoretically separated may be semi-quantitatively observed by comparing its pattern macroscopically with normal serum protein pattern. But before the increase or decrease is represented quantitatively, each fraction should be expressed by the numerical value. It is ordinarily expressed in the following two ways: 1) relative percentage (%) and 2) concentration (g/100 ml) which is obtained by multiplying the concentration of total serum protein by the relative percentage.

Representation with the Relative Percentage (%)

The proportion of each fraction in total serum protein, which is put at 100%, is represented as %. On desitometry, each fractional value is calculated with integrator or planimeter. On the other hand, in the fractional dye elution method, the amount of

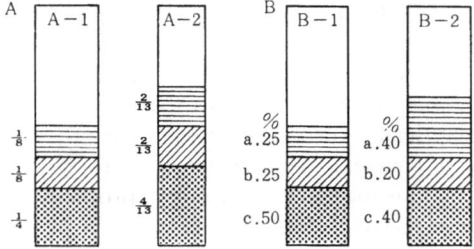

Fig. 119 Comparison of the fractional values expressed in different ways. The columns designated as A show the relation between the total volume and the concentration. There is no difference in the absolute amounts of each fraction, but the concentration of each fraction varies between A–1 and A–2 since the total volume is different. Similarly, the columns designated as B show the relation between the absolute amount and the relative percentage on each fraction. Only the a fraction is larger in B–2 than in B–1, the relative percentages of the b and c fractions are decreased in the column B–2.

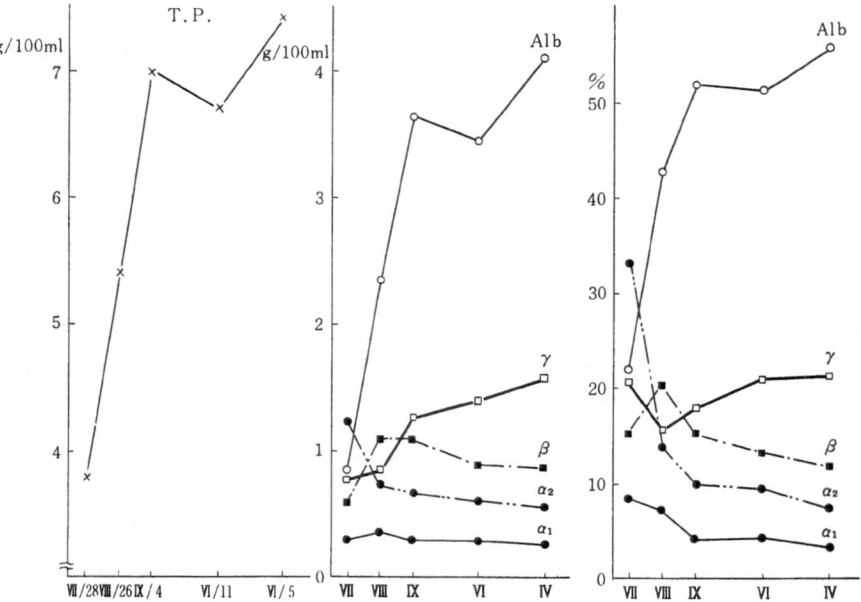

Fig. 120 Changes of the fractional values for serum proteins.
8 y.o., female, nephrotic syndrome.
The serum protein fractional values are plotted. When the total serum protein level
is significantly decreased, the relative percentage of each fraction may not give a true
pathophysiological changes in the serum protein fractions, as seen with the γ fraction
values.

dye eluted from each fraction is obtained colorimetrically and then the relative percentage
is calculated.   In many laboratories, each fraction is expressed as relative percentage since
here the concentration of total serum proteins need not be further multiplied.   However,
as shown in Fig. 119, in the use of relative percentage, the absolute amount of each fraction
is left out of consideration and only the interrelations of five fractions are observed.   Thus,
when a certain fraction markedly increases (or decreases), the other fractions by per cent
tend to decrease (or increase) despite the absence of their change in the absolute amount,
as shown in Fig. 120.   In this case, if the variation of the γ fraction is assessed by relative
percentage, an increase of the γ fraction is mistakenly observed to occur in the worst period
of the disease.   On the other hand, an observation of the γ fraction expressed in concentration shows clearly that it decreases most at the worst period of the disease and that it returns to a normal state along with recovery.   In this way, expression of fractional values
by relative percentage may lead to erroneous assessment of pathophysiological variations
of serum proteins.   This, of course, is the case when the total serum protein concentration
is especially high or low.   This may not be of major problem when the total serum protein concentration is within normal limits (Fig. 121).

Thus, expression in relative percentage also contributes to evaluating roughly the
variation of each fraction when the concentrations of the total serum proteins are within
the range of normal values.   When they are abnormally high or low, relative percentage
should be changed into concentration before the evaluation of each fraction.   Even when
there are no marked increase or decrease in the total serum protein concentration, the
relative percentage of other fractions becomes relatively high in the case where a particular
fraction decreases markedly.   For example, when the γ fraction shows a marked decrease

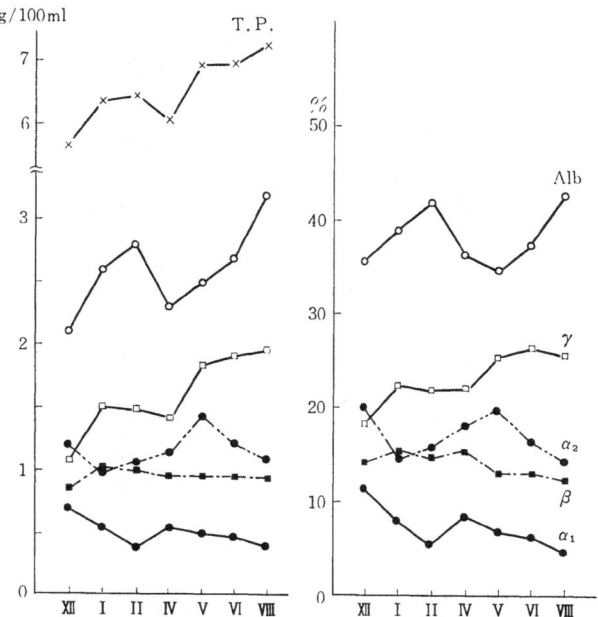

Fig. 121 Changes of the fractional values for serum proteins. 2 y.o., male, pneumococcal septicemia.
In this case, the total serum protein concentrations do not vary significantly, and thus the pattern expressed with the relative percentages is not different from that expressed with the concentrations.

in infancy, the relative percentage of the albumin fraction exceeds 75% in many cases although the concentration of the albumin fraction does not always increase.

Representation with the Concentration (g/100 ml)

As shown in Fig. 119, the concentration is under the influence of the absolute quantity of protein components and of the plasma volume in which they are dissolved. As seen in A–1 and A–2 in the Figure, even if protein quantities are the same, protein concentration increases when the circulating plasma volume decreases. Thus, the circulating plasma volume should be measured in order to obtain the total amount of body or intravascular protein which is of the greatest clinical value. However, it actually is very difficult to measure the circulating plasma volume by following the course of the disease of a patient. Therefore, except in special cases, the circulating plasma volume is accounted normal and protein concentration is substituted to know its absolute quantity. Actually, except in the case of markedly increased or decreased circulating plasma volume, patho-physiological changes can be clearly understood by expressing in g/100 ml as seen in Fig. 120, 121.

As stated above, the concentration (g/100 ml) has certainly greater merit than the relative percentage in representing fractional data. However, attention should be paid to the following points. That is, the relative percentage of each fraction is multiplied by the total protein concentration to obtain the concentration of each fraction. As a result, in addition to the error in protein fractionation, the error in measuring the total protein concentration is added. Therefore, a significant error in the measurement of total protein concentration may cause greater confusion.

OBSERVATION OF SERUM PROTEIN
ELECTROPHORETIC PATTERNS

Important Macroscopic Observation of Electrophoretic Patterns

In the electrophoretic fractionation of proteins in serum and other body fluids, analytical results may be interpreted in two ways.   First, as stated in the previous paragraph, each fraction is represented in numerical value and thus its increase or decrease is observed. Second, after electrophoresis, either stained strips themselves or densitometric patterns are observed.   However, clinicians who are not engaged directly in the analysis of these patterns often neglect the qualitative observation of densitometric patterns and pay attention to the increase or decrease of each fraction.   In addition to the quantitative representation of each fraction, the observation of electrophoretic patterns should be carried out for, as explained later, there is much to be said for this observation.   Therefore, in clinical laboratories this should be done by clinical pathologists or well-experienced senior technologists. Qualitative findings as well as quantitative values should be included in laboratory reports, if necessary.   If possible, densitometric patterns should be added to the reports.   It is desirable that clinicians be also able to interpret densitometric patterns sufficiently.

Informations Obtained through the Macroscopic Observation of Electrophoretic Patterns

Macroscopic observation of electrophoretic patterns alone is not sufficient for quantitative analysis, of course.   However, an approximate increase or decrease of each fraction is recognizable.   Information derived from a macroscopic observation of electrophoretic patterns is indispensable in distinguishing qualitative abnormalities of each fraction.

The following may be assumed also from densitometric patterns, but if possible, the stained strips themselves should be observed.   The information to be interpreted by the observation of stained strips is divided into that of 1) *in vitro* changes and 2) *in vivo* changes.

1.   *In Vitro* Changes

An observation of stained strips themselves should be made to distinguish especially the following changes.   Densitometric patterns alone may not be enough to distinguish from them *in vivo* changes.

   1) If protein samples are applied in an in sufficient amount.
   2) If electrophoretic patterns are not suited for densitometry.
   3) If there is hemolysis. Of course, this is easily differentiated by observing the serum itself (Fig. 94).
   4) Cases indistinguishable from the M-component.
       a.   if chylomicrons are contained in a large amount, the peak appears at the origin in filter paper electrophoretic patterns (Fig. 94).
       b.   if fibrinogen exists, a characteristic protein band appears at the origin on filter paper electrophoretic patterns; at the β_2 or γ_1 area on cellulose acetate electrophoretic patterns (Fig. 101, 102).
       c.   if degenerated proteins are adsorbed by the strip, they are observed at the origin irrespective of the kind of supporting medium (Fig. 95).

2.   *In Vivo* Changes

The changes described below are caused by abnormal changes occurring in patient's serum.   These changes are important ones which cannot be evaluated solely by reading the figures of each fraction.   At the same time, each change is often important or even diagnostic for certain clinical conditions. A detailed description of each change is not given here but references are noted in the parentheses.

1) Abnormal mobility of the albumin fraction (see page 377)
2) Biphasic albumin fraction (see page 11)
3) M-component (see page 305)
4) Wavy Pattern (see page 341)
5) Abnormal mobility of the γ fraction (see page 268, 397)
6) β–γ bridging (or β–γ linking) (see page 238)

Differences between Filter Paper and Cellulose Acetate Electrophoresis

At present, the methods most frequently used for routine serum protein analysis are the filter paper and the cellulose acetate electrophoretic methods.

Both are the same kind of zone electrophoretic method, sharing many common points. However, this is not to say that they are alike in every way, and it is necessary to consider these differences when evaluating electrophoretic patterns.

Technically, the cellulose acetate membrane has the advantage of requiring only a small amount of serum sample to be tested. Also, it takes only a short time for separation, while many samples may be examined simultaneously. However, it may require some skillfulness for handling. Generally speaking, the cellulose acetate membrane gives better separation, so this method is rapidly becoming more prevalent. In Table 29, there is a summary of the main differences between the filter paper method and the cellulose acetate method.

Table 29 Main differences on electrophoretic protein separation between the filter paper and the cellulose acetate methods.

	Filter paper method	Cellulose acetate method
Separability	good	excellent
Albumin fraction value	low	high
γ fraction value	high	low
Protein adsorption	significant (as albumin tailing)	insignificant (very little at β and γ zones)
Prealbumin	not recognizable	recognizable
α_1 fraction	not clearly separated	clearly separated
Low-density lipoportein	at β zone	variable $(\alpha_2$–$\beta_2)$*
β_{1C}-globulin band	not recognizable	often at β_2 zone
Wavy pattern	very rarely seen	often seen**

```
 *  α₂:  Separax, Oxoid
    β :  Cellogel, Selecta, Separaphor III
    β₂:  Millipore
**  Definitely seen on Separax and Oxoid membranes.
```

CLASSIFICATION OF THE SERUM PROTEIN ELECTROPHORETIC PATTERNS IN DISEASES

From about 1940, when TISELIUS' electrophoretic method began to be introduced to clinical medicine, many investigators have attempted to classify serum protein abnormalities into many different groups. With developments in analytical methods, the classification has become more and more complicated. Changes of the serum protein patterns themselves may be diagnostic for certain diseases, but in most cases, serum protein patterns rather reflect the pathophysiology of each patient much more than the disease itself.

Classifications Proposed by Different Investigators

Up to the present, serum protein abnormalities have been classified from various points of view by many clinicians.   In some cases, the same researchers have changed their system of classification with the development of analytical methods.   Of course, systems of classification should be improved in accordance with the advances of the times.   Here, classifications by several authorities are quoted.

1.   Classification of BENNHOLD et al.[29]

Charakteristische Dysproteinaemie: Serum protein patterns which show characteristic changes among those secondary to various diseases. This group is divided into the following 2 subgroups: ① Dysproteinaemie bei Leberkrankheiten (Changes in liver diseases) and ② Dysproteinaemie beim nephrotischen Syndrom (Changes in the nephrotic syndrome)

Paraproteinaemie: Cases in which pathological immunoglobulins appear. Included in this group are plasmocytoma, macroglobulinemia WALDENSTRÖM, and lymphatic leukemia.

Allgemeine uncharakteristische Dysproteinaemie: Differing from the groups described above, this includes many serum protein changes in various diseases in which the characteristic patterns are not observed.

2.   Classification of EMMRICH[83]

This is a simplified form of the concept of "Reaktionskonstellation" proposed by WUHRMANN-WUNDERLY.   In this classification, the nephrotic type is classified as a special variant of the "Entzündungskonstellation".

Mangelkonstellation: Cases in which the deficiency of certain protein fractions is a major change.   "Defektproteinaemien" and "Hypalbuminaemien" are included.

Entzüdungskonstellationen: This is thought to be a change which is caused by the presence of inflammatory lesions, characterized by the decrease of albumin and the increase of all the globulin fractions. This is further divided into an acute type, a subacute-subchronic type, and a chronic type.

Paraplastische Konstellationen: This includes plasmacytoma and macroglobulinemia.

3.   Classification by RIVA[307]

Plasma protein abnormalities are classified according to changes in each fraction of the electrophoretic patterns.   Plasma protein abnormalities are generally called "Pathoproteinaemie" (Dysproteinemia in a broad sense).

Dysproteinaemie im engeren Sinne

1) Dysproteinaemie Typ I.   Cases in which γ globulin increases.   Liver damage, chronic and subchronic inflammatory diseases are included in this category.

2) Dysproteinaemie Typ II.   Cases in which the increase of α globulin is the main change.   The nephrotic syndrome, pregnancy, so-called α-myeloma, malignancies, andacute inflammatory diseases are included.

3) Dysproteinaemie Typ III.   Cases which are a mixture of Typ I and Typ II, including cases in which α globulin and γ globulin fractions increase or all globulinfractions increase.   This is observed in cases of inflammatory diseases, malignancies, etc.

4) Dysproteinaemie Typ IV.   Cases in which β globulin fraction alone increases. It is thought to have no clinical significance.

Defekt-dyspoteinaemie: Cases in which deficiency of each fraction occurs. Agammaglobulinemia, analbuminemia, abeta-globulinemia (?) are included in this group.
Paraproteinaemie

4. Classification of WADA and YACHI[398]

WADA et al. attempted from an early date to fractionate proteins of serum and gastric juice by polarography. They combined results of this effort with data obtained by electrophoretic methods and made the following classification.

  1) Inflammatory type . . . acute, subacute, chronic
  2) Malignancy type . . . early type, metastatic type, cachectic pattern
  3) Liver injury type . . . acute and chronic hepatitis, liver cirrhosis
  4) Nephrotic type

5. Classification of WUHRMANN and WUNDERLY[414]

In the early period, this classification had six groups, but later, considering also the data regarding erythrocyte sedimentation rate, WELTMANN reaction, nephelogram, A/G ratio and various serum colloidal reactions, WUHRMANN-WUNDERLY re-classified into nine groups.[413,414]

  1) Typus der akuten Entzündungen (Acute inflammatory type)
  2) Typus der subakut-chronischen entzündlichen Prozesse (Subacute chronic inflammatory type)
  3) Typus der Hepatitis-diffuse Leberparenchymaschädigung (hepatitis type due to diffuse injury of liver parenchyma)
  4) Typus der Leberzirrhose (Liver cirrhosis type)
  5) Typus des Okkulusionsikterus (Obstructive jaundice type): This is characterizedby the increase of α_2 fraction and marked increase of β fraction on electrophoreogram.
  6) Typus des nephrotischen Symptomkomplexes (Nephrotic type)
  7) Typus der malignen Tumoren (Malignancy type): This is characterized by marked increase of α_2 fraction and the increase of β and γ fractions.
  8) Typus des γ-Globulinplasmozytoms (γ-plasmocytoma type)
  9) Typus des β_1-Globulinplasmozytoms (β-plasmocytoma type)

7. Classification of BRACKERNRIDGE and CSILLAG[48]

Basically, this classification seems to follow WUHRMANN-WUNDERLY' "Konstellationstheorie." In addition to changes of electrophoretic patterns, the reactivity of living body was taken into consideration.

  1) Normal response pattern—Normality, benign tumors, neuropsychiatric diseases, etc.
  2) Immediate response pattern—Acute bacterial infections, metastatic carcinomas, myocardial infarction, etc.
  3) Delayed response pattern—Liver cirrhosis, acute hepatitis, systemic lupus erythematosus, etc.
  4) Augmented response pattern—Angina pectoris, cholelithiasis, etc.
  5) Depleted response pattern—Chronic renal diseases, cachexia, etc.
  6) Irregular response pattern—Normal pregnancy, essential hypertension, etc.
  7) Paraproteinemic response pattern—Myeloma, macroglobulinemia, etc.
  8) Mixed response pattern
       a. Immediate and delayed response pattern: Chronic inflammation, bronchial asthma, etc.

    b. Immediate and augmented response pattern: Degenerative cardio-vascular diseases with cardiac failure, etc.

    c. Immediate and depleted response pattern: Chronic myelogenous leukemia, acute hepatic necrosis, etc.

8.   Classification of MIYOSHI[253]

After long years of study in this field, MIYOSHI had already attempted, in 1953, to classify serum protein abnormalities and in 1967 he gave a special lecture to the General Assembly of the Japanese Medical Association. He called pathological changes of body proteins "protein diseases", and proposed to classify them into the following six categories: (1) blood protein disease, (2) hemophilia and related diseases, (3) hemoglobinopathy, (4) myoglobinopathy, (5) enzymopathy, (6) endocrinopathy. However, there are some which cannot be classified so clearly as described above. There also is some possibility that new categories of body protein abnormalities will be found in the future. At any rate, MIYOSHI's attempt at classification has merited attention.

Among the categories presented by him, blood protein diseases are classified into the following nine groups, according mainly to their pathogenesis.

    1) Malnutirition type

    2) Protein depletion type (Nephrotic syndrome type)

    3) Liver injury type (Liver cirrhosis type)

    4) Infectious and sensitized type

    5) Collagen disease type (Lupus erythematosus type)

    6) Hyperglobulinemia type of unknown etiology

    7) Macroglobulinemia type

    8) Myeloma type

    9) Protein defective and anomaly type—Hypoproteinemia, analbuminemia, double albuminemia, fetal albumin, afibrinogenemia, immunoglobulin deficiency, etc.

Among these nine fundamental types, the following are considered secondary changes when taking into consideration their etiologies: 1) malnutrition type, 2) protein depletion type, 3) liver injury type, 4) infectious and sensitized type. The etiologies of Type 5 (collagen disease type) and Type 6 (hyperglobulinemia) are unknown. At present, Type 8 (myeloma) is thought to be neoplastic. Type 9 (protein defective and anomaly type) is often of familial occurence and refers to hereditary consitutional diseases. As MIYOSHI noted, among the clinical cases observed, there are many cases of several combined forms of these 9 fundamental types. There have also been cases or diseases which cannot be clearly categorized by the foregoing classification. But in spite of inadequacies, this classification is a unique attempt.

CLASSIFICATION OF THE SERUM PROTEIN ELECTROPHORETIC PATTERNS

As described in the previous section, various classifications have been reported up to now. All of them are based on the results of serum (or plasma) protein fractionation by the electrophoretic method.

Serum proteins are divided into five fractions: albumin, α_1, α_2, β, and γ fractions. Clinically speaking, the albumin fraction is either normal or decreased. The other globulin fractions may be normal, decreased or increased, respectively. Therefore, theoretically there are $2 \times 3 \times 3 \times 3 \times 3 = 162$ different patterns to occur. Moreover, as described before, increase in the γ fraction can be subdivided according to whether it shows a broad band or a compact band.   Thus, serum protein fractionation by the electrophoretic method can be said to be one of the most informative laboratory tests of this kind.   As stated above, theoretically, more than 200 different abnormal patterns may be differentiated through a single electrophoretic analysis.   Practically, however, more than twenty patterns can be encountered in association with known clinical conditions, because of a limited technical reproducibility and other complex factors involved.   However, it is easily understandable that more pathophysiological conditions can be distinguished if clinical findings and other laboratory findings are added to these electrophoretic patterns.

If plasma protein abnormalities can be classified according to their pathogenesis, it is

Table 30   Classification of serum protein abnormalities based on their filter-paper or cellulose acetate electrophoretic patterns (by KAWAI).

I.	Hypoproteinemic pattern
	I–1. Malnutritional pattern
	I–2. Protein-losing pattern of non-selective type
II.	Nephrotic pattern (Protein-losing pattern of selective type)
III.	Diffuse acute hepato-degenerative pattern (severe hepatitic pattern)
IV.	Cirrhotic pattern
V.	Acute inflammatory and stress pattern
VI.	Chronic inflammatory pattern
VII.	Broad hypergammaglobulinemic pattern (Polyclonal immunoglobulinopathy)
VIII.	M-proteinemic pattern (Monoclonal immunoglobulinopathy)
IX.	Hyperbetaglobulinemic or hyperalpha-2-globulinemic pattern (hyperlipidemic pattern)
X.	Pregnancy pattern
XI.	Defect dysproteinemias
	XI–1. An-albuminemic pattern
	XI–2. An-alpha-1-globulinemic pattern
	XI–3. A-betaglobulinemic pattern
	XI–4. Agammaglobulinemic pattern

Table 31   Characteristics of fundamental serum protein electrophoretic patterns.

Types		Total protein	Albumin	α_1	α_2	β	γ
I.	Hypoproteinemic	↓↓	↓↓	N,↑	N,↑	↓	↓(N)
II.	Nephrotic	↓↓	↓↓		↑↑		↓(N,↑)
III.	Severe hepatitic	↓,N	↓↓	↓	↓	↓	↑
IV.	Cirrhotic	↓,N,↑	↓↓			↓	β–γ bridging
V.	Acute inflammatory		↓	↑	↑		N
VI.	Chronic inflammatory		↓	↑	↑		↑
VII.	Broad hypergammaglobulinemic	↑	↓				↑
VIII.	M-proteinemic					←— M-peak —→	
IX.	Hyperlipidemic*				(↑)	↑	
X.	Pregnancy	↓	↓			↑	
XI.	Defect dysproteinemia						
	XI–1. An-albuminemic		↓↓↓				
	XI–2. An-alpha-1-globulinemic			↓↓			
	XI–3. A-betaglobulinemic					↓↓	
	XI–4. Agammaglobulinemic						↓↓↓

* $\alpha_2 \sim \beta$ increase on cellulose acetate electrophoresis.
**   Blank indicates insignificant or variable change.

surely helpful to understand their pathophysiological backgrounds.   However, at the present stage, not all cases have clear causes, while in some diseases, many complicated factors are often involved.   Serum protein analysis is only one of the laboratory tests and it forms just part of the basis to make a general clinical diagnosis.   For that reason, the author purposely avoids the etiological classification, but evaluates carefully the informations obtained from serum protein patterns.   As the result of such evaluation, the author attempts the classification shown in Table 30.   This classification is a revision of the one proposed in 1964.

These types are, in spite of all, fundamental and so in some cases more than two types are combined.   Also, at certain stages of an illness, there no doubt exist types which are transitional when the patient is on recovery.   It should not be forgotten that pathological conditions in the living body are always changing and that, as explained in Fig. 52, blood proteins show only a part of pathological changes in the body.   This fact should be noted in order to evaluate serum protein patterns.

A detailed explanation of each fundamental type is made in the section devoted to each one, and Table 31 simply summarizes the changes characteristic for each type.

Chapter 18

Plasma Protein Changes in Malnutritional Conditions

CHARACTERISTICS OF THE MALNUTRITIONAL SERUM PROTEIN ELECTROPHORETIC PATTERN

Serum protein electrophoretic pattern of the malnutritional type is encountered pathophysiologically in deficiency of amino acids which are necessary to plasma protein synthesis.

This pattern always accompanies hypoproteinemia (Fig. 122), and is characterized by a decrease of the β and albumin fractions, especially of albumin. The percentage of the α_1 and α_2 fractions usually increases; however, their concentration increases slightly or stays within the normal range. The percentage of the γ fraction always increases and its concentration remains within the normal range or shows a slight tendency to decrease. However, complications may sometimes cause the γ fraction to increase, and also it frequently decreases remarkably in infancy.

	(g/100ml)	(%)
TP	5.1 ($\downarrow\downarrow$)	100
Alb	2.8 ($\downarrow\downarrow$)	55($\downarrow\downarrow$)
α_1	0.21(N)	4($\uparrow$)
α_2	0.61(N)	12($\uparrow$)
β	0.46($\downarrow$)	9(N)
γ	1.02(N)	20($\uparrow$)

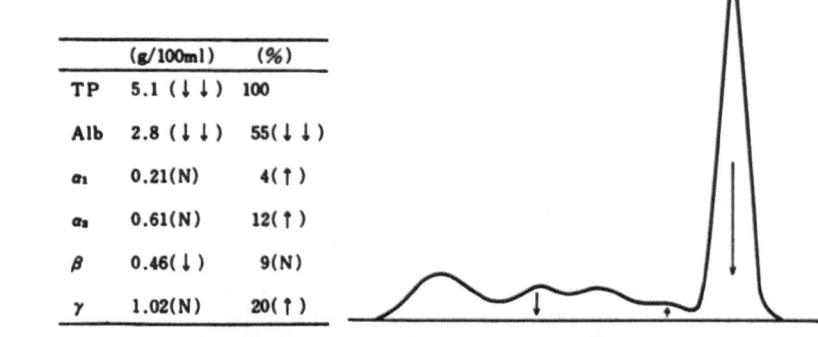

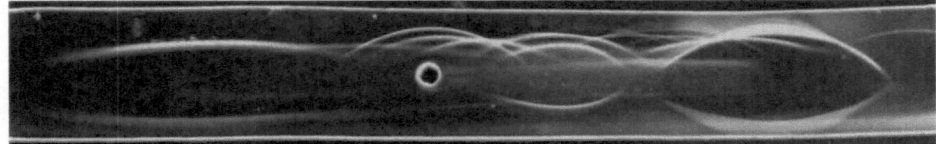

Fig. 122 Serum protein electrophoretic pattern of the malnutritional type.
S. K., 66 y.o., male, cerebral arteriosclerosis and hypertensive cardiac disease.
Admitted with poor appetite, emaciation, renal glycosuria and cataracta. Blood pressure 190/70.
PSP 32% at 15 min. NPN 33.2 mg/100 ml, uric acid 5.4 mg/100 ml, glucose 68 mg/100 ml, total cholesterol 91 mg/100 ml, GOT 25 u., GPT, 16 u., Hb. 12.9 g/100 ml, RBC 3.44, WBC 4300, CRP (−) TTT 0.7 u., ZnTT 6.5 u.

There is no remarkable change on immunoelectrohoretic patterns. When the total protein concentration is very low, the precipitation lines of the minor components sometimes become obscure. Plasma fibrinogen concentration shows the tendency to decrease and other plasma protein components except α-glycoproteins decreases also. The decrease of albumin is especially notable, and transferrin decreases also in parallel to albumin.

DIFFERENTIATION OF THE MALNUTIRITIONAL SERUM PROTEIN ELECTROPHORETIC PATTERN

This pattern should be differentiated from the following two.

Differentiation from the Non-Selective Protein-Losing Pattern.

It is impossible to distinguish this pattern from the non-selective protein-losing pattern described in the next chapter. Therefore, the distinction is made clinically by observing 1) the malnutiritional state (anamnesis, emaciation, dry skin, anemia, low blood pressure, bradypnea, bradycardia etc.), 2) generalized edema, 3) the absence of protein loss into urine or digestive fluids.

Differentiation from the Acute Inflammatory Pattern Associated with Liver Damage.

In acute inflammation or stress the α_1 and α_2 fractions tend to increase; however, if liver damage is accompanied with it, their increase may not become significant, and the serum protein pattern will certainly be similar to that of the malnutritional type. Of course, in this case, the C-reactive protein becomes positive, and liver function tests, anamnesis, physical findings are helpful for the differentiation.

Besides this, malnutritional serum protein patterns themselves have transitional types and also the existence of complications cause atypical protein patterns.

Case   65 y.o., female, liver abscess

	Concentration (g/100 ml)	Percentage (%)
Total protein	5.3 ($\downarrow\downarrow$)	100
Alb fraction	2.88($\downarrow\downarrow$)	54 ($\downarrow$)
α_1 fraction	0.34($\uparrow$)	6 ($\uparrow$)
α_2 fraction	0.73(N)	14 ($\uparrow$)
β fraction	0.51($\downarrow$)	10 (N)
γ fraction	0.84(N)	16 (N)

Suffering from fever with leukocytosis (ca. 15,000) and CRP (3+) for the past one month. Liver palpable 2 f.b., with moderate liver injury: alkaline phosphatase 20 K.A.U. and no jaundice.

REPRESENTATIVE DISEASES ASSOCIATED WITH THE MALNUTRITIONAL SERUM PROTEIN ELECTROPHORETIC PATTERN

This type of protein changes is caused by deficiency of amino acids available for syntheis of plasma proteins in the protein-synthesizing organs. SCRIMSHAW and BEHAR[000] indicate various factors determining protein nutritional conditions. From the clinical standpoints, these changes can be divided into the following two groups, 1) low intake of proteins as the nitrogen source, 2) malabsorption of the nitrogen source into the body.

Poor Intake of Protein Foods

Low protein intake is clinically encountered in starvation, anorexia of various etiology, organic discorders of the gastrointestinal tract (cf. malignancy, stenosis of the gastrointestinal tract, etc.), neuro-psychiatric conditions, and endocrinopathies. Among them, cachexia and Marasmus–Kwashiorkor show the most serious malnutritional state.

Kwashiorkor is the protein malnutritional state often observed in children of tropical and subtropical areas. Marasmus is the serious malnutrition often observed in infants under one year of age, and it is caused by protein malnutrition and low intake of other nutrients and calories[334].

Table 32 Classification of the malabsorption syndrome.

I. Inadequate mixing of foods with bile acids
    1. Postoperative pyrolic insufficiency
    2. Partial and total gastrectomy

II. Inadequate lipolysis (lipase deficiency)
    1. Pancreatic insufficiency
        a. Cystic fibrosis
        b. Chronic pancreatitis
        c. Carcinoma of pancreas and papilla of Vater
        d. Pancreatic fistula
        e. Protein deficiency

III. Inadequate emulsification (bile acid deficiency)
    1. Obstructive jaundice
    2. Severe liver damage

IV. Inadequate absorption of the intestines
    1. Shortening of the intestines
        a. Surgical resection
        b. Intestinal fistulas
    2. Stagnation of the mesenteric lymphatics
        a. Malignant lymphomas
        b. Carcinomas
        c. Whipple's disease
    3. Inadequate absorptive surface
        a. Inflammation (tuberculosis, regional ileitis, etc.)
        b. Malignancy
        c. Amyloidosis
        d. Scleroderma
    4. Biochemical abnormality
        a. Celiac disease
        b. Tropical sprue
        c. A-beta-lipoproteinemia
        d. Deficiency of sugar splitting enzymes
        e. Severe starvation
    5. Altered bacterial flora
        a. Blind loops
        b. Multiple intestinal diverticula
        c. Multiple intestinal strictures
        d. Small fitulas
        e. Neomycin therapy

V. Miscellaneous causes
        a. Carcinoid syndrome
        b. Diabetes mellitus
        c. Hypoparathyroidism
        d. Hypothyroidism
        e. Islet cell tumor of the pancreas
        f. Hypogammaglobulinemia
        g. Mesenteric artery insufficiency

Cachexia is characterized clinically by emaciation, anemia, dry grey-yellowish skin, bradypnea, bradycardia, low body temperature and low blood pressure. This disease occurs most frequently in cases of malignant tumor, especially of the gastro-intestinal tract. It is also observed in cases of endocrine abnormalities like ADDISON's disease and SIMMOND' disease, anorexia nervosa, and chronic inflammation (especially tuberculosis).

Malabsorption Syndrome

The malabsorption syndrome is not a single disease, but a group of the diseases as listed in Table 32. They are often combined with protein-losing gastroenteropathy which will be described later, and they tend to show rather complex pathophysiology.

PATHOPHYSIOLOGICAL BACKGROUNDS OF THE MALNUTRITIONAL PLASMA PROTEIN CHANGES

Hypo-proteinemia is mainly due to a decrease of serum albumin. This decrease is particularly prominent in patients who have clinically severe edema (Table 33.[342]) Generalized edema occurring in cases of malnutrition is caused chiefly by the decreased colloid osmotic pressure due to hypoproteinemia.

Table 33 Filter paper electrophoretic fractional values of serum proteins in the patients with Kwashiorkor (from SENECAL et al.).

Fractions	Alb	α_1	α_2	β	γ
Before treatment					
Edema $(+++)$	1.43	0.41	0.65	0.51	1.40
Edema $(\pm)$	1.83	0.51	0.82	0.68	1.88
Recovery after treatment					
Excellent	3.2	0.56	0.83	1.0	1.66
Poor	2.12	0.50	0.82	0.68	1.60

Unit: g/100 ml

There is a report by JAMES and HAY[171] dealing with albumin metabolism during the malnutritional state. In man, the catabolic rate of albumin starts to decrease three to five days after a low-protein diet begins. After about two weeks on a low-protein diet, the catabolic rate reaches its lowest point. It is said that when a high protein diet is given to malnutritional infants, the catabolic rate becomes normal in three weeks. The albumin synthetic rate in malnutritional infants is definitely decreased. That is, when the amino acids for protein synthesis are insufficient, 1) the albumin synthetic rate is lowered, and 2) the intravascular albumin pool is decreased and subsequently is compensated for immediately by the movement of extravascular albumin into the blood vessels, and later 3) the decreased pool is also compensated for by the decreased albumin catabolic rate. As the lowering of the albumin catabolic rate proceeds, the movement of extravascular albumin into the blood vessels becomes less prominent and will reach to a pathologic equilibrium. Therefore, in the state of malnutrition, the albumin synthtic rate and the albumin catabolic rate decrease, and the intravascular and extravascular albumin pools decrease. However, the extravascular albumin pool decreases more markedly and thus the maintenance of the intravascular pool seems to be of primary significance.

The A/G ratio usually decreases becuase of the significant albumin decrease and a slight increase of the globulin fractions. The α_1 and α_2 fractions remain within the normal range or even increase slightly. The mechanism for the increased α fractions is not well known, but it seems to compensate for a marked decrease in the colloid osmotic pressure caused by the low albumin concentration. This will be discussed in more detail in the section on the acute phase response pattern.

Also, because the malnutritional protein pattern resembles that of the patients receiving adrenalin injection, they may have the pathophysiological backgrounds similar to those of the so-called stress syndrome[319]. In rare instances such as the most serious cases of malnutrition (Kwashiorkor, etc.) the α_1 and α_2 fractions show a slight decrease.

The β fraction usually decreases, especially in severe cases[319]. This is caused mainly by the decrease of transferrin. As described often in other sections, transferrin shows changes parallel to those of albumin, and during convalescence, its increase parallels that of albumin[264]. Further, the decrease of hemopexin seems to parallel that of transferrin.

In many cases, the γ fraction shows hardly any change. In man, it seems that after the nitrogen reserve has been consumed a homeostatic mechanism seems to be effective to maintain the normal serum level of the γ globulin fraction[319]. This phenomenon plays a great role in the maintenance of immunity. However, it is known that in the case of Kwashiorkor there is often an increase of the γ fraction, while in Europe the increase of the γ fraction is rather rare[319]. Therefore, many scientists believe that the increase of the γ fraction is not caused by malnutrition itself but by fatty liver or secondary infection which often combines with Kwashiorkor and other malnutritional states.

Fibrinogen often decreases during malnutrition but in cases of slight malnutrition, it tends to increase.

Serum lipids decrease in simple nitrogen deficiency states, but when the caloric deficiency is added they rather increase[60]. This is probably due to preserve the energy source by movement of stored fat[319]. Also, it is known that various protein hormone syntheses are lowered on low protein diet, and this may be one of the causes for metabolic abnormality encountered in malnutrition[357].

Chapter   19

Plasma Protein Changes in Protein-Losing Conditions

CHARACTERISTICS OF THE PROTEIN-LOSING SERUM PROTEIN ELECTROPHORETIC PATTERNS

Serum protein electrophoretic pattern of the protein-losing type is typically encountered when hypoproteinemia appears due to marked external loss of plasma proteins.   These conditions are further subdivided into two groups, depending on the degree of "molecular sieve effect" influencing to the protein loss: 1) selective protein-losing type and 2) non-selective protein-losing type.

Serum Protein Electrophoretic Pattern of the Selective Protein-Losing Type (the Nephrotic Type)

In this case, the molecular sieve effect is clearly observable. Low molecular weight serum proteins are lost in large amounts and macroglobulin components remain within the blood.   Therefore, the total serum protein concentration decreases markedly. Especially notable is the decrease of albumin which is of relatively low molecular weight. The α_2 fraction which contains macroglobulins in comparatively large amounts increases characteristically (Fig. 123).   Since this pattern is characteristic of the nephrotic syndrome, it is also called the nephrotic pattern.   A detailed explanation of this is given in the section dealing with the nephrotic syndrome.

Serum Protein Electrophoretic Pattern of Non-Selective Protein-Losing Type

This appears when the molecular sieve effect is not clearly influencing for protein loss. In this case, each component of plasma proteins is lost almost at the same rate.   The loss of serum albumin is the largest and the other components also tend to decrease.   Therefore, the decrease of serum total protein and albumin fraction is remarkable and the other fractions tend to decrese or remain within the normal range (Fig. 123).

DIFFERENTIATION OF THE PROTEIN-LOSING SERUM PROTEIN ELECTROPHORETIC PATTERNS

Differentiation of the Non-Selective Protein-Losing Pattern

This type has to be differentiated from the malnutritional pattern.   As described in the preceding chapter, serum protein analyses are not sufficient for differentiating the two. It is necessary to evaluate sufficiently the medical history of the patient, clinical symptoms, and other laboratory findings.

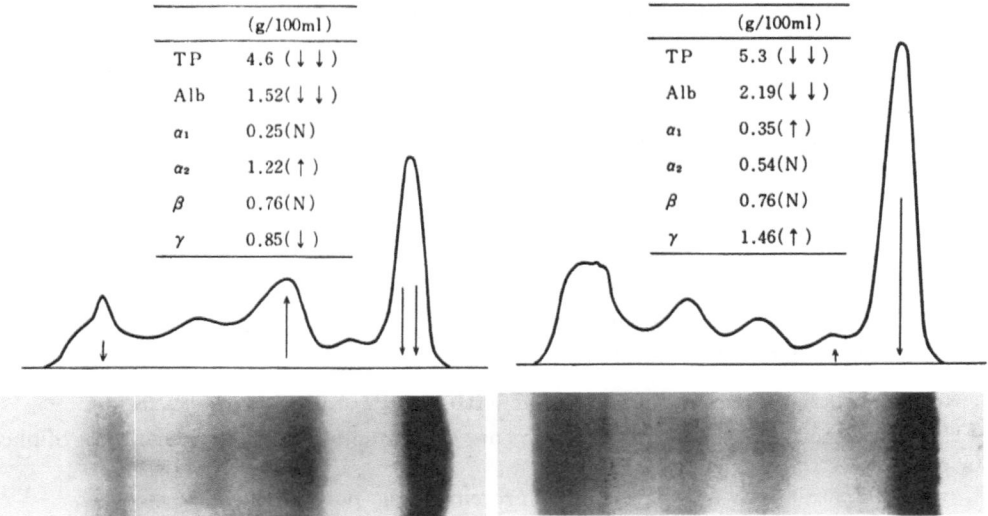

	(g/100ml)
TP	4.6 (↓ ↓)
Alb	1.52(↓ ↓)
α_1	0.25(N)
α_2	1.22(↑)
β	0.76(N)
γ	0.85(↓)

	(g/100ml)
TP	5.3 (↓ ↓)
Alb	2.19(↓ ↓)
α_1	0.35(↑)
α_2	0.54(N)
β	0.76(N)
γ	1.46(↑)

Fig. 123 The serum protein electrophoretic pattern of the protein-
losing type.
The left pattern is of the selective protein-losing type, while the right
pattern is of the non-selective protein-losing type.

Differentiation of the Selective Protein-Losing Pattern

1. Differentation from the α_2 Type M-proteinemic Pattern

When M-proteins, especially BENCE JONES proteins migrating to the α_2 zone exist in serum, it is necessary to differentiate it from the selective protein-losing pattern. In this case, differentiation is achieved by other laboratory and clinical findings. With regard to serum protein analysis by thin-layer gel filtration, the M-protein spot is observed in the G, A and Post-A fractions. This fact contrasts with a remarkable increase of the M fraction in cases of the nephrotic type. By immunoelectrophoretic methods, in contrast to the increase of α_2 macroglobulin in the case of the nephrotic type, M-protein is usually identified immunochemically.

2. Differentiation from the Serum Protein Pattern Seen on Renal Transplantation.

After renal transplantation, the serum protein electrophoresis shows a transitory increase of the α_2 fraction, and it may have to be differentiated from the nephrotic pattern. In this case, however, the albumin fraction usually does not show a marked decrease. Differentiation of the two conditions is easily done from other clinical findings.

3. Differentiation from the Non-Selective Protein-Losing Pattern Associated with Acute Inflammatory Conditions.

When the non-selective protein-losing pattern combines with acute inflammatory pattern, a remarkable decrease of the albumin and an increase of the α fractions are observed. Thus, the electrophoretic pattern becomes similar to that of the selective protein-losing pattern. However, on immunoelectrophoresis, α_2 macroglobulin increases remarkably in cases of the nephrotic type, while haptoglobin increases significantly in cases of the pattern combined with acute inflammatory conditions. Also, by thin-layer gel filtration, the difference in the molecular sieve effect helps in making the differentiation.

SITES AND SELECTIVITY OF THE PLASMA PROTEIN LOSS

Sites of the Plasma Protein Loss

The subject is discussed in detail elsewhere (page 135).   Clinically important sites of plasma protein loss are listed as follows:
1) Urinary loss
2) Gastrointestinal loss
3) Percutaneous loss
4) Pulmonary loss
5) Pleural and ascitic puncture
6) Loss into the extravascular pool
7) Bleeding and lymphorrhea

Representative Diseases Associated with the Protein-Losing Patterns

Hypoproteinemia and generalized edema are encountered due to significant loss of plasma proteins in the following diseases:
1) Diseases showing the selective protein-losing pattern
     a. Nephrotic syndrome
2) Diseases showing the non-selective protein-losing pattern
     a. Protein-losing gastroenteropathy
     b. Exudative dermatopathy
     c. Exudative pulmonary diseases
     d. Bleeding and plasmapheresis
     e. Lymphorrhea and chyluria
     f. "Essential hypoproteinemia" (?)
     g. Diseases with abundant ascites and pleural fluid

Selectivity of the Plasma Protein Loss

When the plasma proteins are lost into the extravascular pool or outside the body, not all plasma protein components are removed in an equal quantity.   As shown in Fig. 124, there are cases in which only the small molecular weight components are lost, while in some other cases almost all protein components are lost similarly.   When plasma protein loss shows significant molecular sieve effect, it is said that selectivity is high, as seen in urinary loss.   In removing the pleural fluid, selectivity is apparently low.

It is the renal glomerulus that shows high selectivity.   On the contrary, the gastrointestinal tract, skin and lungs show very low selectivity.   In the case of blood loss, they show no selectivity at all.   Even in the case of glomerular filtration, although in lipoid nephrosis it shows very high selectivity, its selectivity will be lost as the destruction of glomerular basement membrane advances.

Quantitative Estimation of Selectivity of the Plasma Protein Loss

1.   Electrophoretic Measurement

The protein fractionation is done on both blood serum and excretions which are taken from a patient at the same time, and the permeability ratio of each fraction is calculated according to the following formula.[39]

$$\text{Permeability Ratio of Each Fraction} = \frac{\text{Relative percentage of the fraction in excretion}}{\text{Relative percentage of the fraction in serum}}$$

The permeability ratios of the albumin fraction, the α_1 fraction, the β fraction, the α_2

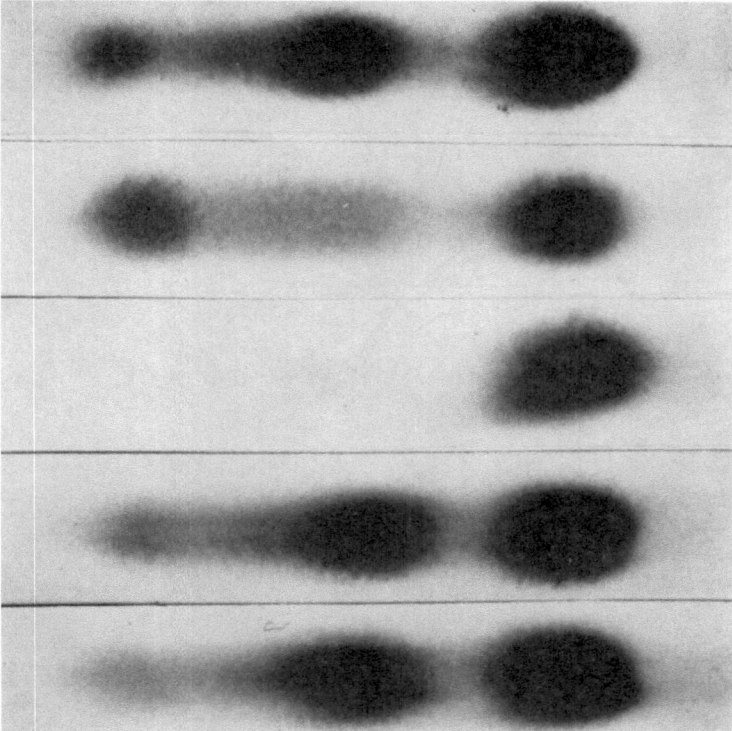

Fig. 124 Thin-layer gel filtration patterns of the proteins in urine and pleural fluid, showing the selectivity of their protein loss.
Sephadex G–200 superfine is used for analysis. 1: normal serum, 2, 3: serum and urine from the patient T. K. with nephrotic syndrome, 4, 5: serum and pleural fluid from the patient with carcinomatous pleuritis.
In nephrotic syndrome, urinary protein loss shows a high selectivity, being composed mostly of the small molecular weight A fraction. Speaking of the pleural fluid, however, the pattern is almost similar to that of the serum, indicating a low selectivity.

fraction, and the γ fraction which are obtained by the above formula are plotted from left to right, as in Fig. 125. In this case, the degree of selectivity for plasma protein loss is evaluated by the slope of the line plotted from the albumin fraction to the α_2 fraction. Permeability ratio for the albumin has been used clinically from early days; that is, the albumin permeability ratio is >2 in the nephrotic syndrome, and it is <2 in "nephritis".[39,332]

The author makes a graph such as the one shown in Fig. 125, and its clinical evaluation is sought by the permeability ratio of the albumin fraction and the slope of the straight line which is drawn between the points showing the albumin permeability ratio and the α_2 permeability ratio. The steeper the line, the higher the selectivity. When the permeability ratio of the γ fraction is below 1, the ratio is in inverse proportion to the degree of selectivity. When it is more than 1, it is thought that this reflects the presence of local immunological defense mechanism or tissue reaction at the sites of protein loss.

2. Measurement Using Gel Filtration Techniques

Gel filtration technique is an analytic method using the principle of molecular sieve effect. Therefore, it is easily understood that this method is very useful for analyzing in

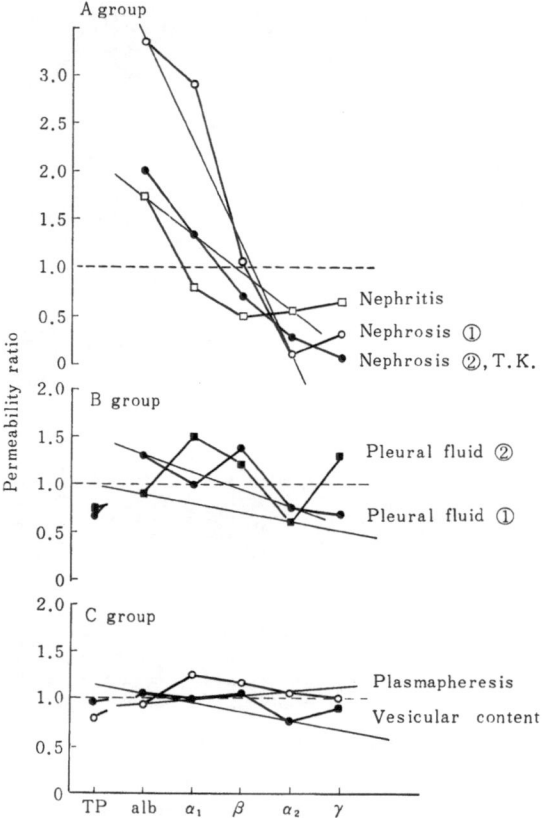

Fig. 125   Selectivity of protein loss, analyzed with zone electrophoresis.
The white dots were obtained with filter paper method, while the
black dots with the cellulose acetate method.

$$\text{Permeability ratio} = \frac{\text{Fractional value (\%) in excretion}}{\text{Fractional value (\%) in serum}}$$

In the A group, urinary protein loss has a relatively high selectivity,
being less selective in nephritis than in nephrosis.   In the B group,
its selectivity is low, and almost no selectivity is seen in the C group.

this field.   The author applies thin layer gel filtration technique using Sephadex G–200
(Fig. 126).   The permeability ratio of each fraction is calculated according to the formula
described in the above section.   The slope of the straight line between the A and M fractions reflects the degree of selectivity very well.   If the M fraction is observed in urine
taken from the patients with the nephrotic syndrome, prognosis of the patients is very
unfavorable.   In the same way, if α_2 macroglobulin is found on immunoelectrophoresis
the selectivity can be considered to be very low.

   3.   Measurement Using Multiple Clearances on Various Plasma Protein Components.
   The renal clearances of urea, creatinine and other substances have been used from early
days as renal function tests.   BLAINEY et al.[43] applied the clearance measurement for the
analysis of serum protein components.   Recently, an improved method by JOACHIM et al
(Fig. 127) has been widely accepted for the analysis of selectivity of urinary protein

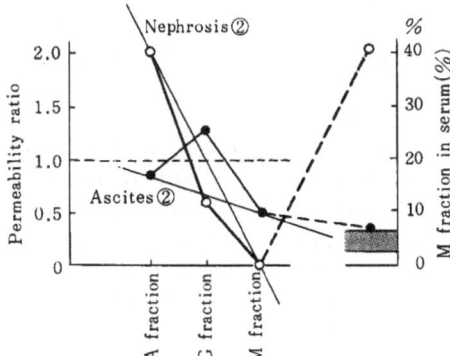

Fig. 126 Selectivity of protein loss, analyzed with thin-layer Sephadex gel filtration.

$$\text{Permeability ratio} = \frac{\text{Fractional value (\%) in urine or ascites}}{\text{Fractional value (\%) in serum}}$$

In the nephrotic patient showing a high selectivity, the permeability ratio of the M fraction between the serum and urine is almost 0, but the M fraction value in the serum is markedly increased. In contrast, the pemeability ratio of the M fraction between the serum and ascites is relatively high, indicating a low selectivity of protein loss. The permeability ratio of the M fraction analyzed with thin-layer gel filtration techniques is closely related to the selectivity of protein loss into various excretions.

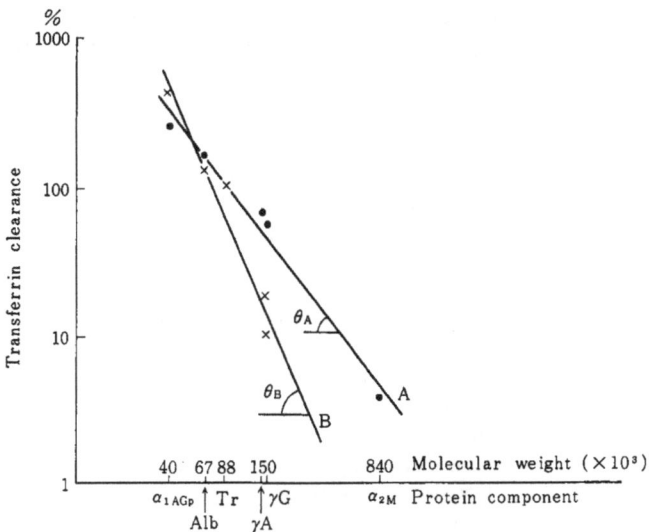

Fig. 127 Quantitative representation of the selectivity of protein loss (from JOACHIM et al.).

The renal clearances of serum protein components having different molecular weights were plotted on logarithmic paper against their molecular weights in two nephrotic patients. The slope (θ_A) for the patient A is 53°, indicating low selectivity of protein loss, and the slope (θ_B) for the patient B is 68°, indicating high selectivity of protein loss.

loss.[332] This method measures immunochemically the concentration of serum protein components of various molecular weight contained in serum and urine which have been collected at the same time.   In this way, the clearance of each component is determined.   Among serum protein compnents, there are a_1-acid glycoprotein (a_{1AGp}; mol. wt. 40,000), albumin (Alb. 67,000), transferrin (Tr; 88,000), IgG-immunoglobulin (IgG; 150,000), IgA-immunoglobulin (IgA; 160,000), and a_2-macroglobulin (a_{2M}; 840,000). The clearance of each component is expressed as percentage of the clearance of transferrin (100%), and this is plotted on doubly logarithmic graph against the corresponding molecular weights.   In this graph, the slope of the straight line (θ) indicates the selectivity of protein loss.   The degree of renal selectivity is classified into four groups by JOACHIM et al. as follows: 1) high selectivity, $\theta = >67°$, 2) intermediate selectivity, $\theta = 63.1°-67°$, 3) average selectivity, $\theta = 54°-63°$, 4) low selectivity, $\theta = <54°$.

NEPHROTIC SYNDROME

Definition and Classification

The nephrotic syndrome refers to the pathological condition showing the markedly increased renal permeability for plasma proteins due to injury of the renal glomeruli.   Clinically, a typical nephrotic syndrome is manifested by: 1) marked proteinuria (containing mainly serum albumin), 2) hypoalbuminemia, 3) hyperlipidemia, and 4) generalized edema.[36]

Formerly, the nephrotic syndrome was distinguished pathologically as nephrosis from glomerulonephritis.   At present, it is widely held that the main abnormality exists not in the renal tubules but in the glomeruli.   Therefore, the nephrotic syndrome is the general term for the renal abnormalities characterized functionally by the increase of renal permeability.   The nephrotic syndrome is classified eitologically as in Table 34.

Table 34 Classification of the nephrotic syndrome.

I.   Idiopathic nephrotic syndrome*
A.   Chronic nephrotic glomerulonephritis (Type II ELLıs)**
1.   Chronic nephrotic glomerulonephritis
2.   Membranous nephropathy
B.   Glomerulonephrosis (Lipoid nephrosis)
II.   Lupus nephritis (Nephritis due to collagen diseases)
III.   Diabetic glomerulosclerosis (KIMMELSTIEL-WILSON disease)
IV.   Renal amyloidosis
V.   Syphilitic nephrosis
VI.   Nephropathy due to toxemia of pregnancy
VII.   Radiation nephritis
VIII.   Toxic nephritis (trimethadione, paramethadione, penicillamine, heavy metals, poison oak, bee stings, etc.)
IX.   Malignant nephrosclerosis
X.   Venous stasis
A.   Bilateral renal vein thrombosis
B.   Constrictive pericarditis
C.   Congestive heart failure

   * Classification by OSHIMA (1968)
  ** According to ALLEN (1962), it is divided into membranous and lobular glomerulonephritis, and membranous nephropathy by OSHIMA is included in membranous glomerulonephritis.

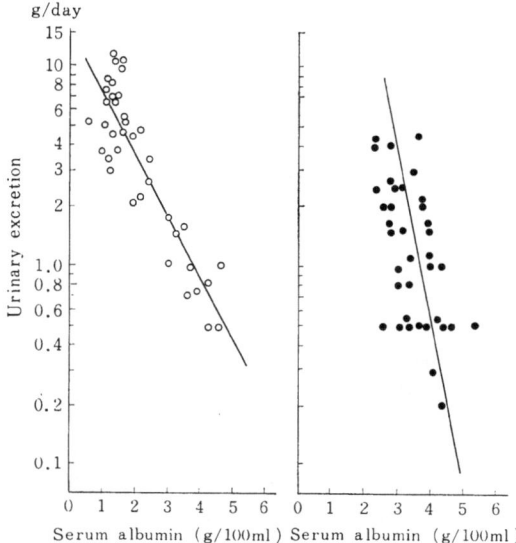

Fig. 128 The relation between the urinary protein excretion and
the serum albumin concentration.
The left is the figure obtained among the nephrotic patients, and
the right is that obtained among the patients with non-nephrotic renal
diseases. The slopes of the regression lines in both groups are distinctly different. In the non-nephrotic patients, the serum albumin concentration of less than 2 g/100 ml is not encountered.

Proteinuria in the Nephrotic Syndrome

In nephrotic syndrome, large amounts of plasma proteins are excreted into urine.
Usually, they amount to more than 5 to 10 grams daily. Generally, in children the
excreted amount of proteins is smaller than that in adults. However, considering the
size of the kidneys and the volume of glomerular filtration, it is clear that the smallness
of the amount of urinary protein excretion in children cannot be neglected. There is
a certain relation observed between the amount of urinary protein and the albumin
concentration in serum, as shown in Fig. 128.

When this relation is compared with renal diseases other than the nephrotic syndrome,
the serum albumin concentration of less than 2.0 g/100 ml is seen only in the nephrotic
syndrome. This coincides with the report by MOORE and VAN SLYKE that generalized
edema is likely to occur when serum albumin concentration becomes less than 2.3%.[260]
Therefore, the amount of urinary protein excretion which corresponds to serum albumin
concentration of less than 2.3 g/100 ml is around 3.0 g/day (Fig. 128). That is, when urinary proteins of over 3.0 g/day are excreted, the possibility of the nephrotic syndrome
becomes very high. The similar data were also reported by BERMAN.[34] However, as
shown in Fig. 128, even when the urinary protein excretion exceeds 3.0 g/day, other
diseases can not be completely excluded, but they are certainly not likely to be present.

The greater part of urinary proteins excreted is occupied by albumin (Fig. 124–128).
However, as described before, the protein composition changes if renal permeability
changes. With the permeability of average selectivity, the albumin fraction occupies
the largest portion, and the β fraction follows. The α_2 fraction is of the least. On

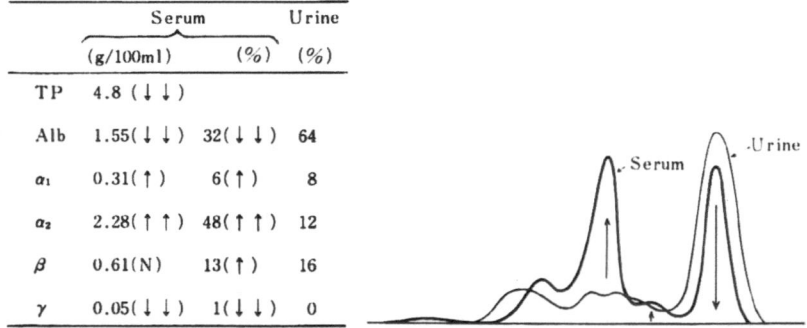

	Serum		Urine
	(g/100ml)	(%)	(%)
TP	4.8 (↓↓)		
Alb	1.55(↓↓)	32(↓↓)	64
α_1	0.31(↑)	6(↑)	8
α_2	2.28(↑↑)	48(↑↑)	12
β	0.61(N)	13(↑)	16
γ	0.05(↓↓)	1(↓↓)	0

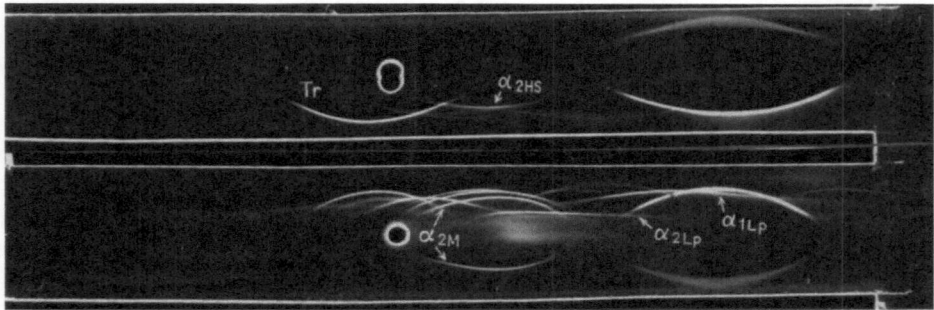

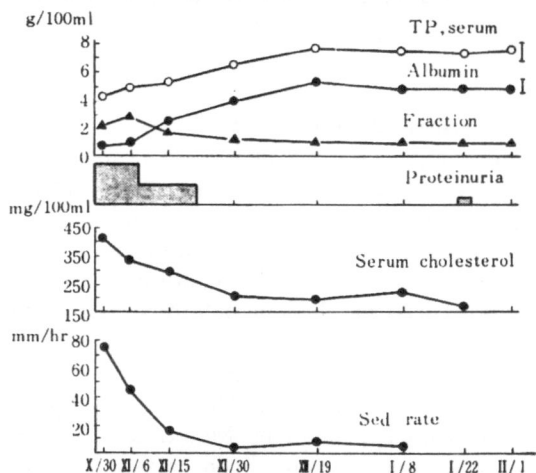

Fig. 129   Serum protein patterns of the nephrotic (selective protein-losing) type.
T. K., 12 y.o., male, lipoid nephrosis.
At the age of 9 years, diagnosed as nephrosis, and treated thereafter with corticosteroids.   Admitted repeatedly in the past.
No anemia (Hb 14 g/100 ml); serum urea N, normal.   Proteinuria (3.1 g/day).   Urinary sediment, containing several hyaline casts.   Responded quickly to steroids, and discharged 3 months later.
The immunoelectrophoretic pattern of the serum proteins is typical of the nephrotic type., showing marked increase in α_2-macroglobulin and lipoproteins, and marked decrease in albumin, transferrin, IgG and IgA.   The urine contains a significant amount of albumin, α_1-antitrypsin, α_{2HS}-glycoprotein, and tranferrin, and also a small amount of prealbumin, ceruloplasmin, and IgG.   However, α_2-macroglobulin and lipoproteins are not demonstrated in urine.

immunoelectrophoresis, the lines of albumin and transferrin are pronounced, and α_1-glycoproteins, α_{2HS}-glycoprotein, IgG-immunoglobulin are also observed. However, lipoproteins, IgM-immunoglobulin and α_2-macroglobulin are not found. On the other hand, with low selectivity, even α_2-macroglobulin comes to be identified in urine.

Besides marked proteinuria, many hyaline and granular casts are observed, and in many cases, fat droplets and epithelial cells showing fatty degeneration are buried in these casts. Sometimes, even free fat droplets mainly composed of cholesterol ester are found in urinary sediments.

Plasma Protein Changes in the Nephrotic Syndrome

In typical cases of the nephrotic syndrome, the serum protein pattern of the selective protein-losing type is seen. The characteristics of the pattern are marked hypoproteinemia associated with a marked decrease of the albumin fraction and a significant increase of the α_2 fraction (Fig. 130).

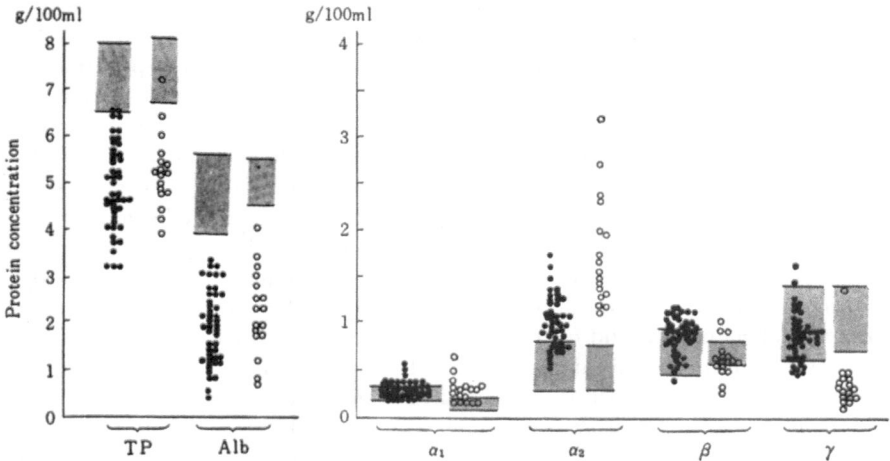

Fig. 130 Fractional values of the serum proteins obtained with both the cellulose acetate and filter-paper electrophoretic techniques in the nephrotic patients. The black dots are the values obtained with the cellulose acetate method, and the white dots are obtained with the filter paper method. The shaded areas indicate the normal ranges for each fraction. A significant difference is noted (mean ± 2 S.D.) with the α_2, β and γ fractions.

The decrease of serum total protein concentration is caused mainly by the decrease of serum albumin and in some cases, it decreases down to about 3 g/100 ml. This tendency is pronounced in lipoid nephrosis of children.

1. The Albumin Fraction

A decrease of the albumin fraction is inevitable and when this goes below 2 g/100 ml generalized edema always occurs. On various immunochemical analyses, a remarkable decrease of serum albumin is also observed, but prealbumin tends to increase.[289,319]

The decrease of the albumin concentration in serum is in logarithmic inverse proportion to the amount of proteins excreted into urine, and this fact shows that the decrease of serum albumin is mainly caused by urinary protein loss. However, as shown in Fig. 128, in the case of renal diseases other than the nephrotic syndrome, even when urinary

protein excretion exceeds 3 g/day, serious hypoalbuminemia seldom occurs.   Also, since
the replenishment of the proteins lost through urinary excretion does not affect the degree
of hypoproteinemia, it seems that in the nephrotic syndrome the albumin may be destroyed significantly in the liver or the kidneys.   Metabolic studies on albumin turnover
in the nephrotic syndrome have been made by many scientists.   In brief, it can be said
that because of the large amount of albumin excreted into urine, the half-life is remarkably shortened, and the turnover rate and the synthetic rate increase to 2.5 times the normal.   On the other hand, in renal insufficiency there is rather a decrease in the albumin
turnover rate.   This is a fundamental difference in the albumin turnover between renal
insufficiency and the nephrotic syndrome.   Also, it has been reported that in normal
rats, albumin metabolism did not change significantly after nephrectomy, and yet in nephrotic rats, the albumin degradation rate decreased about 50% after nephrectomy.[169)]
Therefore, also, it may be assumed that albumin degradation in kidneys is remarkably
increased in the nephrotic syndrome.

2.   The α_1 Fraction

With the filter paper electrophoresis, the α_1 fraction remains within the normal range
in most of the nephrotic patients, and it may decrease in only a few cases (Fig. 130).
However, with the cellulose acetate method by which the clearer separation of the α_1
fraction is usually obtained, it is found to increase in many cases.   As a matter of fact, the
relative percentage of the α_1 fraction is increased in almost all cases.   With the immunoelectrophoretic method, the increase of α_1-lipoprotein is remarkable, and α_1-anti-
trypsin and α_1-acid glycoprotein remain within the normal range or tend to increase.
This seems to be odd at a glance, considering that a comparatively large amount of low-
molecular weight α_1-glycoproteins are excreted into urine.   It is very interesting to note
that in all conditions including those of the malnutritional type which shows hypoproteinemia, a similar change in the α_1-globulins is observed.   When the relation between serum
total protein concentration and α_1-fraction concentration in serum is sought, it is difficult
in many cases to find a certain interrelation.   However, when each case is followed up
continuously, it can be assumed that a certain relation exists as in Fig. 131.   That is, the
α_1 fraction increases as if attempting to compensate for decreased serum colloid osmotic
pressure due to severe hypoproteinemia or hypoalbuminemia, particularly when it becomes
below 10 mmHg or when the albumin osmotic pressure becomes less than about 5 mmHg.
Of course, most of the α_1-globulins are of relatively low molecular weight and exerts a
great influence on colloid osmotic pressure.   Therefore, it is thought that in spite of their
urinary loss the colloid osmotic pressure level tends to be compensated by the increase
of their synthetic rate.

3.   The α_2 Fraction

As shown in Fig. 132, the relative percentage of the α_2 fraction in serum increases in all
nephrotic cases.   With the filter paper method, however, about one third of the patients
show a normal concentration of the α_2 fraction in their serum (Fig. 130).   There is a dif-
fernce between the serum concentrations obtained by the filter paper method and the
cellulose acetate method, and the α_2 level tends to be higher with the latter method.   This
is caused by the difference of mobility of low-density lipoproteins; that is, LDL is migrated
in the α_2 fraction with some of the cellulose acetate methods and in the β fraction with the
filter paper method.

Beside the low-density lipoproteins, an increase of the α_2 fraction in the nephotic patients is mainly due to the increase of α_2-macroglobulin.[319)]   The increase of α_2-macro-
globulin is thought to occur secondary to its stagnation in serum, resulted from molecular

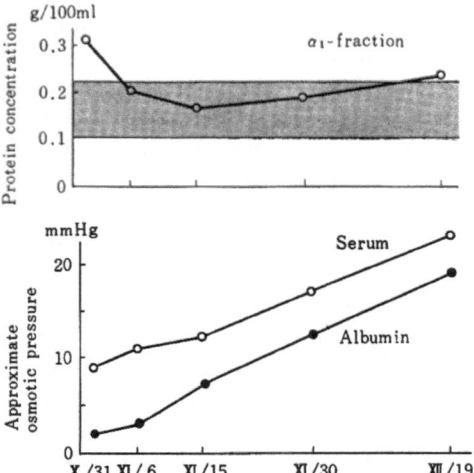

Fig. 131 The relation between the colloid osmotic pressure and the α_1-globulin fraction level in the serum taken from a nephrotic patient (T. K.).

The colloid osmotic pressures for the serum and the albumin are calculated depending on the data reported by OTT (Fig. 66). At the time when the patient is in the most marked hypoproteinemic state, the serum concentration of the α_1 fraction is increased.

Fig. 132 The α_2 and β fraction values obtained with both the cellulose acetate and filter paper electrophoretic techniques in the nephrotic patients.

The relative percentage of the α_2 fraction is increased in all instances, but the values tend to be higher with the cellulose acetate method (white circles).

In contrast, the β fraction values tend to be lower with the cellulose acetate method than those with the filter paper method (black dots).

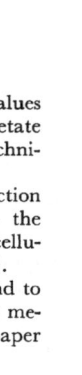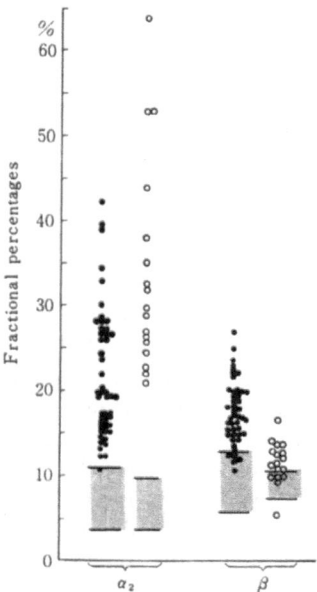

sieve effect of the renal permeability.   However, it is impossible to explain the marked increase of α_2-macroglobulin only by this mechanism and it is also thought that its synthetic rate is increased.[146]   The serum concentration of haptoglobin remains normal in many patients.   However, WHITTAKER stated that haptoglobin shows significant increase and WADA et al. said that it increased in the case of lupus nephritis.[319,407]   With regard to ceruloplasmin, its serum concentration varies among different patients.

Thyroxine-binding globulin (TBG) and transcortin are usually decreased and this seems to be caused mainly by their urinary loss.   Due chiefly to a TBG decrease, in the nephrotic syndrome, the Triosorb test shows a high value, thus contrasting with the decreased protein-bound iodine.

4.   The β Fraction

The change of the β fraction differs according to the electrophoretic methods which are used for serum analysis (Fig. 130, 132).   This is because the low-density lipoproteins which increase markedly in the nephrotic syndrome change their mobility with different electrophoretic methods used.   The β fraction increases markedly when analyzed with the TISELIUS method.[111]   Since the β fraction value is notably lowered when the serum is treated beforehand by ether, the remarkable increase of the β fraction in the TISELIUS method is due principally to lipids contained in lipoproteins.[225]   That is, in the case of the TISELIUS method, the difference of each fraction concentration is measured by the Schlieren optical system using light refraction, and so the increase of lipoproteins which contain great amounts of fat is accentuated.   In the filter-paper method, the peptide moiety of lipoprotein molecules is stained by Brom Phenol Blue or by Amido Black 10B.   Since about 20% of low-density lipoproteins is occupied by the peptide moiety, the increase of the β fraction is slight (Fig. 130, 132).[319]

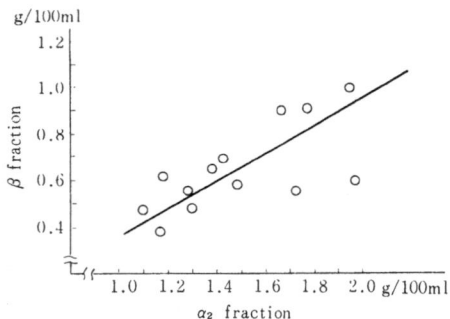

Fig. 133   The correlation between the α_2 and β fraction levels in the nephrotic sera, obtained by the cellulose acetate electrophoretic fractionation.
On cellulose acetate (Separax) electrophoresis, the low density lipoproteins are mostly included in the α_2 fraction, but a portion of the LDL seems to be included also in the β fraction.

On the other hand, in the cellulose acetate method and in the agar gel method, low density lipoproteins are mainly contained in the α_2 fraction and so the β fraction hardly increases, remaining within the normal range or tending to decrease in most cases.   However, as shown in Fig. 133, the relation between the α_2 fraction concentration and the β fraction concentration in the cellulose acetate method is almost in direct proportion. It is therefore thought that most low density lipoproteins are contained in the α_2 fraction, but they exert considerable influence on the β fraction.

Transferrin which is the major component of the β fraction decreases and not rarely atransferrinemia is encountered[143] (Fig. 134).   This is, the metabolic pool of transferrin decreases and its half-life is shortened.   However, it is not clear whether this fact is related to the increased catabolism of transferrin in the body in addition to the excess urinary transferrin loss.[195]   However, bodily catabolism of transferrin may be thought to increase, because transferrin shows very similar turnover to albumin.   The author experienced two cases of atransferrinemia which followed the nephrotic syndrome and both of them reacted to steroid treatment dramatically and recovered quite well in two to three

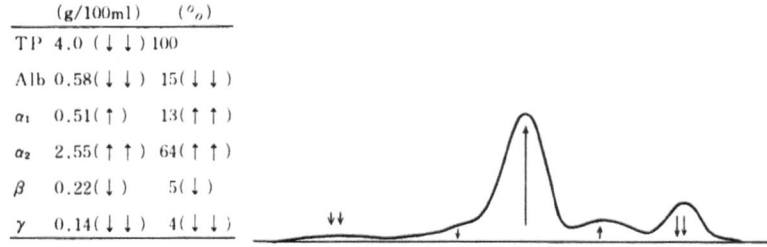

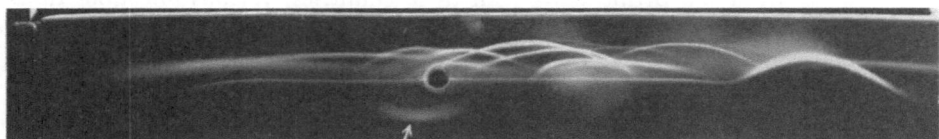

(g/100ml)	(%)
TP	4.0 (↓↓) 100	
Alb	0.58(↓↓)	15(↓↓)
α₁	0.51(↑)	13(↑↑)
α₂	2.55(↑↑)	64(↑↑)
β	0.22(↓)	5(↓)
γ	0.14(↓↓)	4(↓↓)

Fig. 134 Severe hypotransferrinemia in lipoid nephrosis.
R. I., 14 y. o., male, lipoid nephrosis.
The cellulose acetate electrophoretic pattern of the serum proteins (upper) is typical
for the nephrotic condition.   On immunoelectrophoresis, only a weak line for trans-
ferrin is recognized against the anti-transferrin rabbit immune serum (lower trough).
IgG is markedly decreased, but IgA is rather prominent.

days. SANDOR considers the decrease of transferrin to be a cause of anemia which accompanies the nephrotic syndrome[319].

It is said that β_{1C}-globulin, or the third component of the complement system (C3)
generally does not show any noticeable change, but in the case of lupus nephritis and some
cases of lipoid nephrosis, it is often decreased.[125,406]   Moreover, in the case of membranous nephropathy (Table 34, I-A-2), the immune deposit is a crucial factor.   This
deposit is found on the basement membrane stained in a chain-like pattern, against the
anti-human IgG, IgM, β_{1C}-globulin, by the use of the fluorescent antibody technique.
Also, in the case of lupus nephritis, many immune deposits are observed in the subepithe-
lial and the subendothelial regions.[278]   It can be supposed that there exists a close relationship between the decrease of β_{1C}-globulin in the serum and the renal findings obtained with fluorescent antibody technique.

5.   The γ Fraction

The serum concentration of the γ fraction varies greatly according to the types of the
basic pathology causing nephrotic syndrome.   Most of the cases with lipoid nephrosis
in children show a significant decrease in the γ fraction; IgG and IgA immunoglobulins
are decreased while IgM immunoglobulin is normal or slightly increased[388]. ANDERSEN
et al.[11]   reported on the turnover of IgG immunoglobulin, stating that IgG decrease
was due to its urinary loss and its increased catabolism.   They observed that IgG synthetic rate was increased in 11 out of 16 cases with nephrotic syndrome.   However, as shown
in Fig. 129, the γ fraction may become almost zero in occasional cases.   In these cases,
no significant urinary loss of IgG is recognized, and thus the IgG decrease may be thought
to occur either from its decreased synthesis or from its increased catabolism.   However,
in two nephrotic cases of our own having agammaglobulinemia, the γ fraction becomes
decreased gradually after many recurrences during steroid therapy.   Thus, the marked
decrease of the γ fraction may be assumed to occur due to the immuno-suppression after
steroid therapy in addition to its increased catabolism (see Fig. 142).   In the cases with

chronic nephritis, the γ fraction is usually decreased, but it tends to increases when a sclerotic glomerular change increases.

In lupus nephritis, the γ fraction is increased significantly, and all major immunoglobulins, IgG, IgA and IgM, are increased. In the cases of diabetic glomerulosclerosis, the similar changes are noted, but they are less prominent than in lupus nephritis.[388] The γ fraction is increased also in the nephrotic cases with amyloidosis or LIBMAN-SACKS disease.[296,319] VIGUELLOUX and SANKALE[110] reported that the γ fraction was frequently increased in the nephrotic cases in Africa. SOOTHILL and HENDRIKSE[356] also reported that in Nigeria, many of the nephrotic cases were resistant to steroid therapy and showed low selectivity of the renal permeability. In these countries, various infectious diseases are frequently encountered, and in facts, GILLES and HENDRIKSE[110] indicated a close correlation between nephrotic syndrome and malarial infection in Nigeria.

6. Fibringogen

Fibrinogen shows remarkable increase, especially in the chronic cases of the nephrotic syndrome. It increases to about 1.5–2.0 times as much that of the normal value.[22]

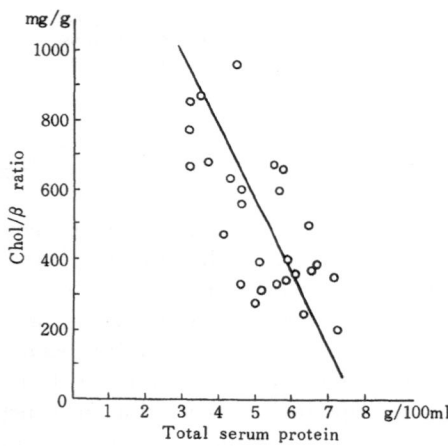

Fig. 135   The relation between the total serum protein levels and the total cholesterol levels in nephrotic patients. Chol/β (mg/g) = the ratio between the serum total cholesterol concentration and the β fraction level obtained by the filterpaper electrophoretic method. The ratio may indicate the approximate amount of cholesterol carried by a unit concentration of the β-globulins.

Serum Lipid and Lipoprotein Changes in the Nephrotic Syndrome

In the nephrotic syndrome, serum total lipids increase, and lactescence of the serum often appears even in a fasting state. It is cholesterol which increases most remarkably among serum lipids, reaching over 600 mg/100 ml. However, in many cases, it is accompanied by hyperlipemia or hyperglyceridemia in which neutral fat increases. This is socalled mixted type hyperlipidemia.[99,319] SANDOR says that the serum hardly differs from that of normal persons, and only the cholesterol/phospholipid ratio tends to increase, when serum cholesterol level rises. Hyperlipidemia seen in the cases of nephrotic syndrome is usually less prominent than that in diabetes mellitus, primary hyperlipidemia, obstructive jaundice, etc.

When the same case is observed continuously, it will be seen that the increase of serum lipids, especially that of cholesterol, is in inverse proportion to the serum albumin concentration and to the total protein concentration. When the relation between the two is investigated in many cases, almost similar inverse proportions are found. Serum total cholesterol concentration change runs parallel to that of the α_2 and β fractions. However, the best correlation is found, as shown in Fig. 135, between the serum total protein concentration and cholesterol/β fraction ratio (chol/β). The chol/β ratio is obtained by dividing the serum total cholesterol concentration by the β fraction concentration which is obtained by the filter paper method. The significance of the ratio is unknown, but it may be thought, at least for the present, to be equivalent to that of the serum lipid or cholesterol amount carried by the unit concentration of the β globulin (much of the β globulin occupied by low-density lipoproteins). This may be a finding coincident with the increase of what SANDOR calls "giant" molecule lipoproteins.

β lipoproteins increase remarkably with the increase of serum total lipids, and the β/a ratio increases notably.[108,304] The high density lipoprotein (a_1-lipoprotein) does not show any remarkable change; rather, it tends to decrease.[108,304] However, the author has observed a significant increase in a_1 lipoprotein on immunoelectrophoresis. At any rate, in the nephrotic syndrome, the change of a_1 lipoprotein is slight and in the cases in which serum cholesterol concentration is high, it tends rather to decrease. Further, there is a report stating that the catabolic rate of a_1 lipoprotein increases in nephrotic syndrome.[109]

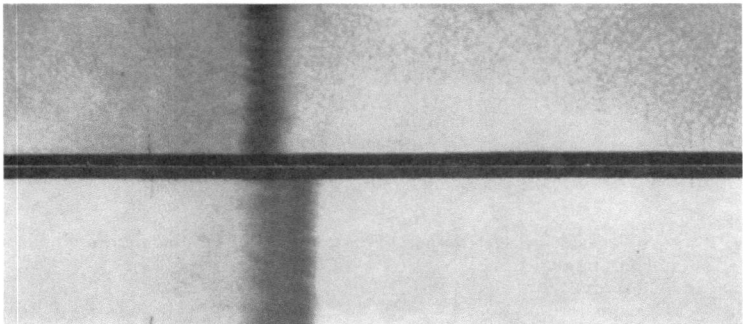

Fig. 136 The filter-paper electrophoretic patterns of the serum lipoproteins in the nephrotic patients.
The upper pattern indicates an increase of the β lipropotein fraction(TypeII), and the lower pattern indicates an increase of the β lipoprotein and the pre-β lipoprotein fractions (Type IV).

It is well recognized by many authorities that low-density lipoproteins increase notably in nephrotic syndrome. As shown in Fig. 136 they do not show any constant tendency in the lipoprotein pattern obtained through the imporved filter paper electrophoretic technique. However, frequently encountered are Type II hyperlipoproteinemia or hyperbetalipoproteinemia in which only β-lipoprotein increases and Type IV hyperlipoproteinemia or hyper-prebetalipoproteinemia in which pre-β lipoprotein increases remarkably, and sometimes Type III hyperlipoproteinemia which may be thought of as an intermediate type.[99] Low density lipoproteins change their mobility on electrophoretic analysis, but the general tendency is that the higher the serum concentration the higher the electrophoretic mobility. This tendency is remarkably noted, especially when agar or agarose gels are used[274] (Fig. 137).

a_1 lipoprotein also shows the similar tendency. Such changes in their mobility may be thought to be caused by the binding of free fatty acids to the lipoproteins. This binding may occur due to the marked decrease of serum albumin which serves as a carrier of free fatty acids in a normal state.

Low-density lipoproteins differ from other serum protein components that the catabolic rate of their peptide and lipid moieties remains normal while cholesterol is said to rather decrease.[111,372]

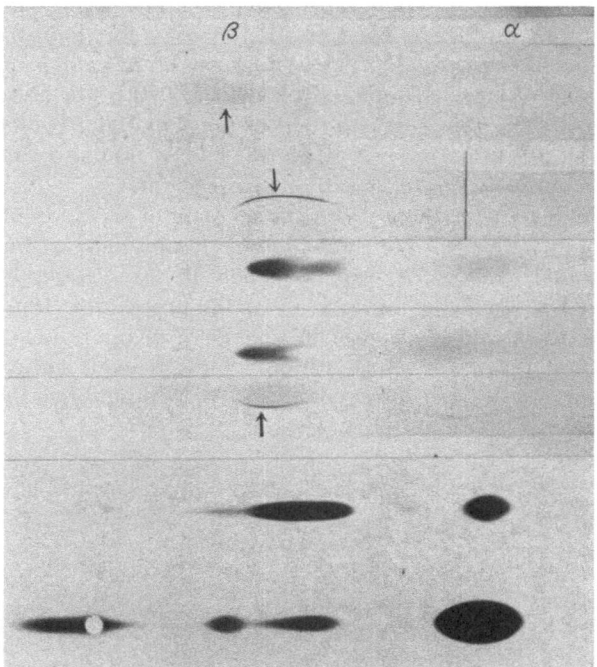

Fig. 137 The agarose-gel electrophoretic patterns of the serum lipoproteins in the
       nephrotic patients.
1,7 = normal serum.
2,3, 6 = nephrotic serum (No. 1)
4,5 = nephrotic serum (No. 2)
1,2,5 = Immunoelectrophoretic patterns stained with Oil Red O.
3,4 = Electrophoretic patterns stained with Oil Red O.
6,7 = Electrophoretic patterns stained with Amido Black 10 B.
The arrows indicate the precipitation lines of the low-density lipoproteins. The
LDL spot may split occasionally due to a significant increase of the pre-β lipoprotein.

Thus, the remarkable change of lipid metabolism is observed in the nephrotic syndrome,
but its mechanism is unknown although there are many theories concerning it.   It is
thought that this change is caused when the transformation from pre- β lipoprotein to β-
lipoprotein is blocked and in addition, the uptake of neutral fats in fat tissues is lowered.[98,321]   Another theory is that the increase of neutral fat is caused by the increased
release of free fatty acids from fat tissues.

Mechanisms of the Nephrotic Edema

The mechanisms of edema may be divided into local and systemic ones (Table 35).
In the nephrotic syndrome, generalized edema appears, but its mechanism is not clear
entirely.   In spite of this, there is the elaborate research of METCOFF et al. and the
following theory is propounded.[249]

In the case of nephrotic edema, systemic factors are thought to be of principal one, but
there is no factual foundation for stating that capillary blood pressure is aggravated by
increased capillary permeability and by generalized circulatory disturbance.   It is
therefore thought to be due to hypoproteinemia and retention of sodium and water.

First, a large amount of serum proteins is excreted into urine because the glomerular

Table 35 Mechanisms of edema formation.

I. Local Factors
    1. Increased water from capillary to tissues
        1) Increased capillary pressure
        2) Decreased tissue pressure
        3) Decreased colloid osmotic pressure in plasma
        4) Increased colloid osmotic pressure and increased hydration in tissues
        5) Increased capillary permeability
    2. Decreased lymphatic return
II. Systemic Factors
    1. Generalized circulatory disturbances
    2. Endocrinological disturbances
    3. Decreased colloid osmotic pressure in blood
    4. Increased capillary permeability

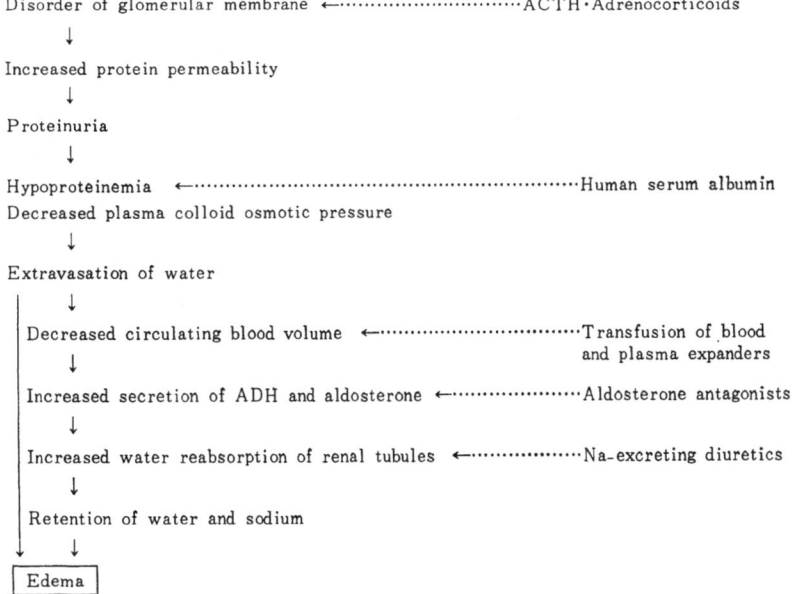

Fig. 138 Schematic representation on mechanisms of nephrotic edema and its related therapeutics.

permeability increases. In addition, albumin is further decreased because of the increased catabolism of albumin. This results in the lowering of blood colloid osmotic pressure, and intravascular water moves out easily into the interstitium, thereby causing tissue hydration. The decrease of circulating plasma volume caused in this way may be compensated by the increased secretion of antidiuretic hormone, and this is thought to aggravate renal water retention.[64] On the other hand, there is sodium retention caused by the increased secretion of corticoids. This aggravates also water retention.

The above mentioned process is summarized in Fig. 138.

Plasma Protein Changes on Treatment in the Nephrotic Syndrome

As shown in Table 34, there are many primary diseases which cause the nephrotic syndrome. Therefore, treatment of the primary diseases is essential. In any case, adequate treatment is to be chosen from, considering its true working mechanisms as shown in Fig. 138. However, the true working mechanisms of ACTH and adrenocorticoid treatment are still unknown in many points, and moreover, their effectiveness is expected only in the idiopathic case and lupus nephritis among those listed in Table 34.

ACTH and adrenocorticoids introduced a great epoch in the treatment of the nephrotic syndrome. When this treatment induces dramatic response the proteinuria decreases in a few days and the serum protein abnormality improves rapidly. The decrease of urinary protein loss is, of course, related directly to the increase of serum albumin concentration, and in addition, it is said that the improvement of albumin metabolism also has an important role.[55] The improvement of serum lipoprotein pattern is recognized similarly to that of serum protein patterns, and the former tends to occur earlier than the latter. Among the cases of idiopathic nephrotic syndrome, only about 70% responds to ACTH-steroid treatment. Moreover, as the use of steroids becomes widespread and is carried in greater doses and for longer periods, it has become very important to predict the effect of steroid treatment for adequate care of the nephrotic patients.

Prediction of the effects of ACTH-steroid treatment can be met with the following two ways; one which involves histological findings through renal biopsy, and the other, serum urinary protein findings. Although both methods are not 100% reliable, the degree of reliability is high reasonably.

Prediction through renal biopsy: Generally speaking, the cases in which normal renal functions are maintained and the change of renal glomeruli is slight, respond very well to steroids. Steroids are also effective for proliferative glomerulonephritis showing only proliferative inflammation. It is especially effective during the acute phase. However, when renal function is impaired and azotemia is present, and as the glomerular change is compounded by sclerotic process the effect of steroids becomes poor.[5,51]

According to OSHIMA, steroids are remarkably effective in glomerulonephrosis (lipoid nephrosis) but membranous nephropathy shows poor response.[278]

Prediction through serum and urinary protein patterns: There are many reports which relate this prediction to the selectivity of urinary protein loss. Generally, the cases in which the selectivity of renal permeability is high, respond well to steroids, and as the selectivity becomes lower, resistance to steroid therapy appears.[43,146,176]

UEDA et al. reported that when the alb/γ ratio in urinary protein patterns is above 6, good response to steroids may be expected.[389]

PROTEIN-LOSING GASTROENTEROPATHIES

Difinition and Classification of the Protein-Losing Gastroenteropathies

Protein-losing gastroenteropathy is the general term for diseases or abnormal states associated with hypoproteinemia, caused by the excessive loss of plasma proteins into the gastrointestinal tract. Clinically speaking, it is characterized by edema or peritoneal effusion. Thus, rather than a single disease, protein-losing gastroenteropathy is a so-called syndrome.

Included in this syndrome are many diseases, classified in various ways, enumerated in Table 36. Besides these, gastrointestinal bleeding may be cited as an example of protein-losing gastroenteropathy considered in a broad sense, but this particular one is described elsewhere.

Considering the mechanism of its occurrence the protein-losing gastroenteropathy may be classified into the following three groups.

1) Cases wherein exudation of blood or plasma components from the ulcerated lesion of the mucosa.   Typical examples are ulcerative colitis and cancers.

2) Cases wherein lymph is lost into the gastrointestinal tract because of dilated lymphatics and lymphatic stagnation resulting from various local lesions.

3) Cases wherein increased leakage of proteins is caused by increased capillary permeability in the mucosa, resulting from various local and systemic factors.

Mechanisms of the Intestinal Lymphangiectasia

Essential intestinal lymphectasia or lymphangiectasia is the general term for dilatation of the intestinal lymphatics of unknown etiology, but lymphangiectasia occurring in the intestinal mucosa is considered very rare.   Rather, in many cases, it is caused by lymphatic obstruction occuring at the thoracic duct and other lymphatic channels.   However, it is quite difficult to examine accurately the existence of lymphatic circulatory disturbance and sometimes an exhaustive study may be necessary to diagnose the condition.   Clinically, such measures as the following are used: lymphangiography, peritoneoscopy, determination of lymphatic output from the thoracic duct, and biopsy.[270]

Primary intestinal lymphectasia may be encountered rarely, but most of the cases with intestinal lymphetasia are secondary to obstruction of the thoracic duct or of the mesenteric and retroperitoneal lymphatic plexus.   The following may be considered as the four causes of lymphatic circulatory disturbance.

Embryological malformation: POMERANZ and WALDMANN[297] reported four cases of lymphangiectasia in which the cause may be attributed to congenital malformation of the lymphatic tissue.   Also, in cases reported by IRINO, NISHIKAWA, and others, stenosis of the thoracic duct at the venous angle has been confirmed.   The cause of the disease in these cases is regarded as embryological mal-development.

Mechanical compression: There are cases in which obstruction is caused by intrinsic and extrinsic compression of the lymphatic trunk due to such disease as leukemia, lymphosarcoma and carcinoma.   This condition is infrequently associated with lymphatic edema and chylous ascites, probably because of sufficient development of lymphatic collateral circulation.

Inflammation and scar formation: A case of chronic inflammatory obstruction has been reported by McDONAGH and others.[243]   The author also has observed a case which is supposed to have been caused by filaria infection (Fig. 139).

Autoimmune lesions: MATSUMOTO[239] has demonstrated severe lymphatic obstruction by repeating experimental sensitization with homologous lymphatic tissues in rats.   This is of great interest but as yet no one has reported that the similar lesion is caused by immunological mechanisms in man.   However, there is a report stating that cellular immunity is lowered in patients suffering from intestinal lymphangiectasia[366].   At the present stage, direct loss of peripheral lymphocytes into the intestine is regarded as a cause of lymphopenia, and this seems to cause decreased hypersensitivity and transplantation reactions.   However, further immunological investigation is certainly required.

Plasma Protein Changes in the Protein-Losing Gastroenteropathy

As mentioned previously at page 141 the protein composition of digestive fluids is almost the same as that of blood serum, with the exception of secretory IgA.   Therefore, in protein-losing gastroenteropathy also, the serum protein composition does not change significantly and the selectivity for protein loss seen in the nephrotic syndrome is not observed.

This is the reason why protein-losing patterns of the non-selective type are found.   In

Matabolic date using RISA

Plasma volume	3400ml	(Normal)
Halflife of albumin	8.2days	(Decreased)
Albumin turnover rate	8.4%day	(Increased)
Albumin turnover	16.2g/day	(Normal)
Total exchangeable albumin	192g(3g/kg)	(Decreased)
Total circulating albumin	59g(0.9g/kg)	(Decreased)
Total albumin	251g	(Decreased)

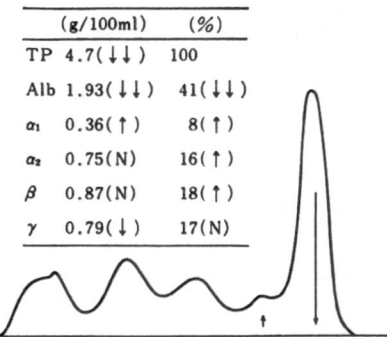

	(g/100ml)	(%)
TP	4.7(↓↓)	100
Alb	1.93(↓↓)	41(↓↓)
α_1	0.36(↑)	8(↑)
α_2	0.75(N)	16(↑)
β	0.87(N)	18(↑)
γ	0.79(↓)	17(N)

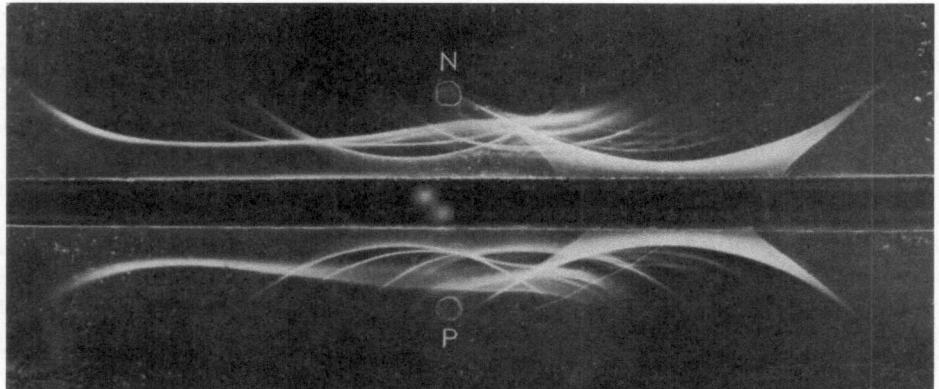

Fig. 139   Protein-losing gastroenteropathy.

Y. W., 50 y.o., male, elephantiasis of the right leg.

Military service in China from 1934 to 1937, and complained of eczematous rashes after the service. Since several years before admission, suffered from elephantiasis of the right leg.

The total serum protein concentration ranged from 3.9 to 4.7 g/100 ml, but there was no other chemical abnormality. As shown in the figure, the metabolic studies using RISA indicate a decreased half-life and an increased catabolic rate. The GORDON test using ^{131}I-PVP was 14% fecal excretion in 3 days, indicating an increased protein loss. On X-ray examination of the intestinal tract, nothing abnormal was noted. Lymphangiography was unsuccessful in two repeated occasions. None of the lymph vessel was identified in the right foot, and the skin biopsy showed marked atrophy of the lymph vessels. The tissues obtained from the right inguinal region failed to show any lymphoid structure both macroscopically and microscopically. (see the next page)

Upon surgical exposure of the retroperitoneum, poorly demarcated loose connective tissues were surrounding the inferior vena cava and the adjacent major veins. The loose connective tissue fragment contained microscopically a few almost completely obstructed lymph vessels, associated with scanty lymphocytic infiltration. The lumina of the lymph vessels were apparently occluded by a PAS-weakly positive material, without calcification or parasites. After surgically removed some of the loose connective tissues in the retroperitoneum, the clinical symptoms improved slightly, but swelling of the right leg and hypoproteinemia have became worse again.

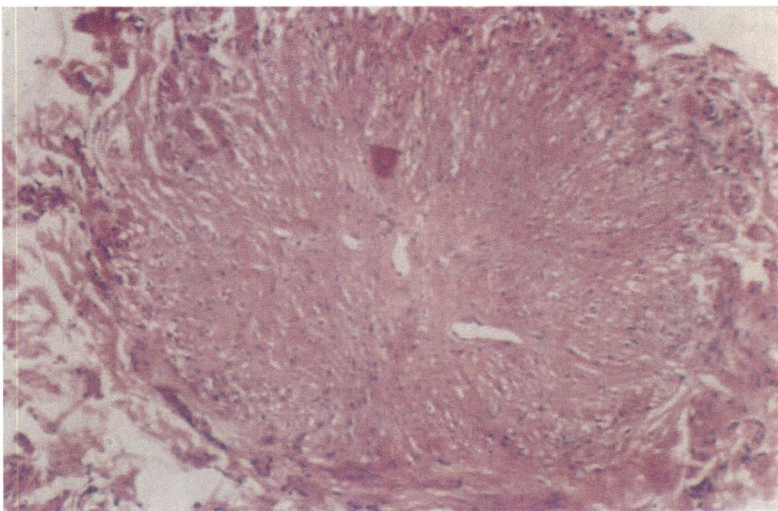

Photomicrograph Inflammatory cell infiltration in the connective tissue pad around the inferior vena cava, and degenerated lymph vessels with luminal obstruction.

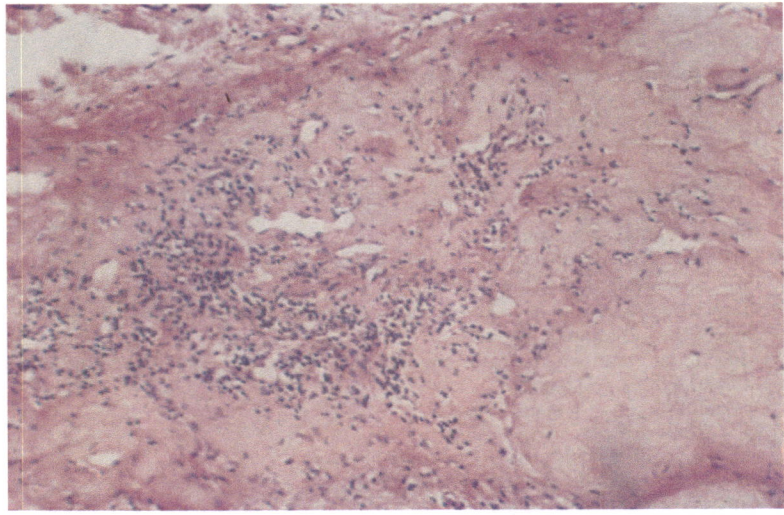

Photomicrograph Higher magnification of the same area shown in the other, showing marked luminal stenosis due to abnormal deposits.

Table 36 Disorders accompanied by excessive losses of plasma proteins through the gastrointestinal tract (modified from SCHULTZE and HEREMANS).

A. Diseases of the stomach
    (1) Gastric carcinoma and lymphosarcoma
    (2) Ventricular polyp and disseminated polyposis
    (3) MÉNÉTRIER's disease (giant-fold gastritis)
    (4) Occasional cases of simple hypertrophic gastritis and diffuse ulcerative gastritis
    (5) Gastric retention after partial gastrectomy
    (6) Atrophic gastritis (cystica?)

B. Diseases of the intestine
    (1) Sprue syndrome
        (a) With recognizable allergic basis (juvenile and adult celiac disease = gluten enteropathy, other digestive allergies, aminopterin enteropathy, allergic gastroenteropathy)
        (b) Without recognizable allergic basis (idiopathic steatorrhea).
    (2) Infectious and auto-immune (?) inflammatory conditions
        (a) Acute infantile enterocolitis
        (b) Acute transient exudative gastroenteropathy
        (c) Enteritis associated with the immuno-deficiency syndrome
        (d) Regional enteritis (CROHN's disease)
        (e) Ulcerative colitis
        (f) Intestinal tuberculosis
        (g) Colitis cystica profunda
    (3) Parasitic diseases: *strongyloides stercoralis*, trichinosis (?)
    (4) Mechanical disturbances of the intestinal tract
        (a) Intestinal stenosis
        (b) Jejunal diverticulitis
        (c) Megacolon (HIRSCHSPRUNG's disease)
    (5) Intestinal amyloidosis
    (6) WHIPPLE's disease (?)
    (7) Carcinoma of the biliary tract and the intestine
    (8) Cystic fibrosis
    (9) Gastro-colic fistula

C. Disorders of the intestinal lymphatic circulation
    (1) Lymphomas
        (a) HODGKIN's disease
        (b) Lymphosarcoma
        (c) Chronic lymphatic leukemia
        (d) Miscellaneous granulomas of the abdominal lymph nodes (tuberculosis, sarcoid, WHIPPLE's disease)
    (2) Retroperitoneal lymphangioma
    (3) Chylous fistula communicating with the duodenum, as a sequel of pancreatitis
    (4) Obstruction (ligation) of the thoracic duct
    (5) Obstruction of the thoracic duct and the retroperitoneal lymphatics of unknown etiology (filariasis, auto-immune disease?)
    (6) Mesenterial lymphatic obstruction with pigmentation of the intestinal wall (adhesion?): idiopathic hypoproteinemia
    (7) Essential intestinal lymphectasia
        (a) Adult form
        (b) Congenital chylous ascites with lymphedema

D. Disorders of the intestinal venous return
    (1) Congestive right-heart failure
    (2) Constrictive pericarditis (also associated with lymphatic obstruction)
    (3) Familial cardiomegaly and primary myocardial disease
    (4) Interatrial septal defect
    (5) Pulmonary stenosis
    (6) Obstacles to normal flow in the areas of the large venous trunks

E. Disorders of intestinal capillary permeability
    (1) Congenital form associated with tuberous sclerosis
    (2) Circulatory shock
    (3) Post-irradiative syndrome
    (4) Nephrotic syndrome
    (5) Kwashiorkor (?)
    (6) Extensive burns

order to differentiate from other types of protein-losing conditions clinically, it is necessary to demonstrate protein loss into the digestive tract by using radioisotopes.

The total serum protein concentration decreases markedly (1.92–4.6 g/100 ml), particularly associated with marked decrease of the albumin concentration (1.0—3.3 g/100 ml) often accompanied by generalized edema. No remarkable change is found in other serum protein fractions, but the α fractions tend to increase slightly. The γ fraction shows a slight tendency to decrease as long as there are no complications.

A marked increase in daily albumin degradation (or synthesis) is recognized. On the basis of his experiments, IWASAKI concluded that its increase could go up to as much as 2.5 times of normal (Fig. 140).[167] This is very similar to what happens in the nephrotic syndrome, but as Fig. 141 shows, there is a distinct difference between the two with regard to the relation of fractional catabolic rate and serum albumin concentration. Therefore, it may be said that protein-losing gastroenteropathy differs in some ways from the nephrotic syndrome, although both exhibit the same type of serum protein loss.

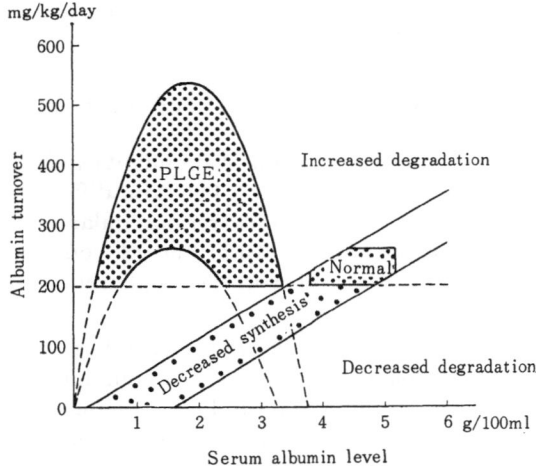

Fig. 140 The relation between the serum albumin level and albumin turnover in protein-losing gastroenteropathy (NISHIKAWA et al.).
The peak of the parabola indicates the theoretical maximal turnover of the albumin in man, being approximately 2.5 times of the normal.

As for the metabolism of transferrin, the same tendency is present as in the case of albumin; that is, there is an increase of the synthetic rate. But there are cases in which transferrin maintains its normal level, though examination of the same patients shows albumin decrease. Consequently, it is recognized that the increase of the synthetic rate of transferrin tends to be greater than that of albumin.[175]

According to studies by STROBER and others,[366] immunoglobulins also show a decrease of 40–70%. The mean values in 18 cases of lymphangiectasia are: IgG-globulin 4.5± 2.1 mg/100 ml (normal value, 12.1±2.7), IgA-globulin 1.15±0.32 mg/100 ml (normal value 2.61±1.07), IgM-globulin 0.76±0.26 mg/100 ml (normal value 1.45±0.68). These results show that the half-life becomes shorter as in the case of albumin and transferrin, and the synthetic rate increases. It is also recognized that their intravascular and extravascular pools show remarkable decrease.

In the case of lymphangiectasia, lymphopenia exists.   This probably causes a decrease of skin hypersensitvity reaction and transplantation immunity.   At the same time, however, humoral antibody response is maintained at almost normal level.[366]

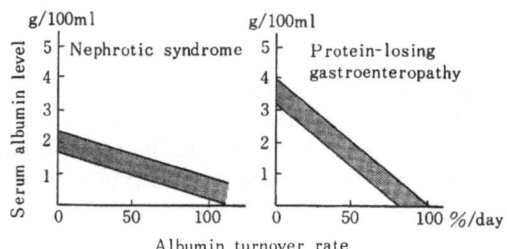

Fig. 141 The relation between the serum albumin level and the albumin turnover rate in protein-losing gastroenteropathy and nephrotic syndrome (from NISHIKA-WA et al.).
In both pathological conditions, the albumin turnover is increased significantly, but the albumin turnover rate is different in both the conditions.

EXUDATIVE DERMATOPATHIES

In cases of exudative dermatopathy, because of abundant exudation due to inflammatory destruction of skin tissues, protein loss occurs.   The skin, differing from the gastrointestinal mucosa, is solid tissue, and therefore only serious inflammatory descruction causes protein loss.   Inflammatory exudate, as shown in Fig. 72, comes to have almost the same total protein concentration as serum, and with electrophoretic analysis, it is found that the same components are present in both.

This protein loss occurs in a non-selective way.   Therefore, the serum protein pattern in exudative dermatopathy is found in the form of a combined protein-losing and acute inflammatory pattern.   That is, in addition to hypoproteinemia and hypoalbuminemia, remarkable increases of the α_1 and α_2 fractions are noted, as shown in Fig. 142.   With the decrease of the total protein concentration of serum (Table 37), the β fraction tends to decrease.   The change of the γ fraction is influenced by various factors.   In infants, it decreases easily, and as in the cases shown in Fig. 142, the decrease of serum immunoglobulins occurs by the administration of a large dose of adrenocoticoids.

Cases of exudative dermatopathy showing hypoproteinemia can be divided into the following three groups:
   1)   Cases of physical injury, such as burns;
   2)   Cases of chemical injury, such as cantharidin;
   3)   Cases of wide-spread vesicular dermatopathies, such as pemphigus vulgaris, dermatitis herpetiformis (DUHRING), erythema multiforme bullosum, epidermolysis bullosa, and bullous lichen planus.[332]   Besides these, as shown in Table 37 conditions such as diffuse eczema and drug eruptions are also included in this group.

Serum Protein Patterns in Burn
Widespread burn causes to exudate abundant plasma constituents through the burned skin lesion.   Ordinarily, in about eight hours after the occurence of burn, hemoconcentration appears temporarily.   This is caused by the formation of vesicles containing abundant plasma water leaked from the burned skin lesion.   But as the tissue fluid moves from

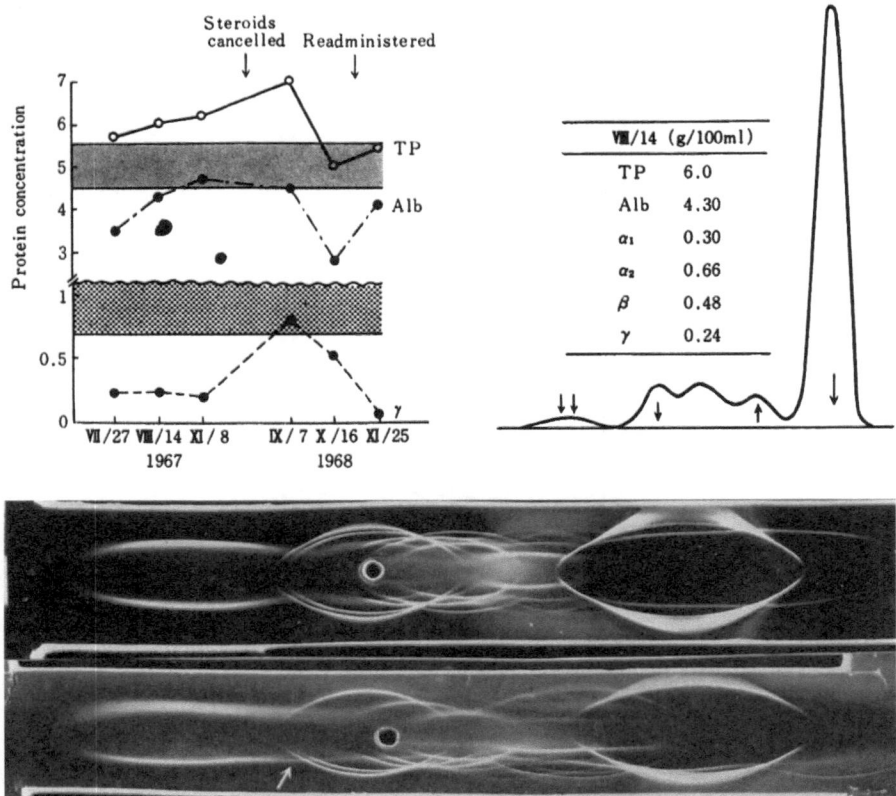

VIII/14 (g/100ml)	
TP	6.0
Alb	4.30
α_1	0.30
α_2	0.66
β	0.48
γ	0.24

Fig. 142 Serum protein patterns in exudative dermatopathy.
G. T., 64 y.o., male, pemphigus vulgaris.
On the administration of adrenocorticoids, a marked improvement was obtained in the exanthema, the serum protein and albumin concentrations, but hypogammaglobulinemia existed. In 1967, the immunoelectrophoretic pattern of the serum proteins (upper) shows a decreased IgG, and IgA and IgM are not recognized. On immunochemical quantitation, IgG was 36%, IgA 19% and IgM 25% of the average normal value. In November, 1967, the exanthema disappeared, and the serum protein concentration returned to a normal level. The steroid administration was discontinued and the patient was uneventful thereafter. The serum level of the γ fraction returned to a normal level. However, in October, 1968, the clinical symptoms recurred, showing a slight decrease of the γ fraction and a marked decrease of the albumin fraction. Upon the readministration of corticosteroids, the albumin fraction increased rapidly, but the γ fraction decreased markedly again. The immunoelectrophoretic pattern of the serum proteins (lower) at that time shows a marked decrease of the IgG, but only a slight decrease of the IgA. The serum level of β_{1c}-globulin is always within a normal limit.

the extravascular pool into the blood vessels, the plasma protein constituents are diluted. Moreover, rupturing of the vesicles results in siginificant protein loss, and this aggravates hypoproteinemia (Fig. 143). Ordinarily, the critical point is reached one week after the occurrence of burn.[14] This is true both experimentally and clinically.

In cases of burn, the albumin decreases most remarkably. Among the causes of hypoalbuminemia, exudation is the principal one. But there is another report that cites the accentuation of albumin catabolism and points out the possible influence of the decrease of the synthetic rate caused by liver damage.[14,40] Experiments show that when no medical treatment is given, at least three weeks are required for the complete

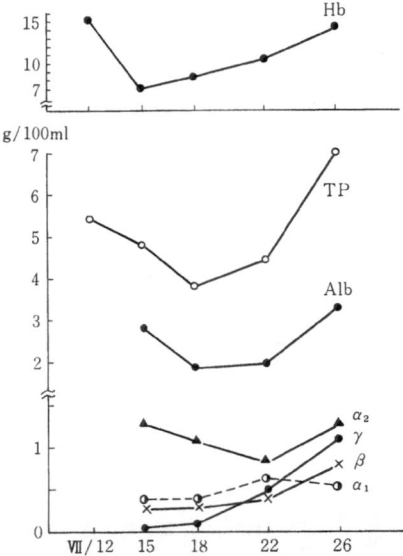

Fig. 143 Exudative dermatopathy.
S. T., 4 y.o., male, burn (Grade II).
Burned with boiling water in approximately 57% of the body surface.
From the fifth hospital day on, the cutaneous exudation decreased
significantly, and the serum protein changes were also improved.
However, the patient suffered continuously from high fever. The
total serum protein, the albumin fraction and the γ fraction decreased
markedly. The β fraction was slightly decreased, and the α_1 and α_2
fractions were increased significantly. Thus, the protein pattern is
of the mixed acute inflammatory and protein-losing type.

Table 37 Serum protein patterns in the cases with exudative dermatopathy.

Cellulose acetate patterns (Nihon Univ. Hospital)　　　　　　　　　　　unit: g/100 ml

Name	Age	Sex	TP	Alb	α_1	α_2	β	γ	Diagnosis
G. T.	65	M	5.0	2.8	0.32	0.89	0.47	0.52	P. V.
K. S.	69	M	5.0	2.6	0.27	0.57	0.71	0.85	P. V.
M. W.	49	M	6.4	4.07	0.33	0.77	0.65	0.58	P. V.
S. I.	54	F	5.6	2.85	0.29	0.79	0.63	1.04	D. H. D.
S. T.	4	M	3.8	1.93	0.38	1.1	0.28	0.11	Burn (2°)
H. T.	22	M	6.3	4.22	0.35	0.58	0.4	0.75	Burn (2°)
Control			6.7–8.1	4.5–5.5	0.1–0.22	0.3–0.75	0.56–0.8	0.7–1.4	

Filter paper patterns (Central Railway Hospital)

Name	Age	Sex	TP	Alb	α_1	α_2	β	γ	Diagnosis
Y. S.	42	M	7.4	2.89	0.51	0.96	1.15	1.89	Burn (2°)
A. H.	37	M	7.0	3.2	0.30	1.02	1.09	1.39	P. V.
T. S.	49	F	6.7	2.83	0.32	0.70	0.84	2.02	P. V.
I. O.	39	M	4.5	1.67	0.40	0.80	0.59	1.05	D. H. D.
T. M.	55	M	5.0	2.3	0.38	0.52	0.69	1.13	P. E.
M. K.	45	M	6.0	2.99	0.38	0.59	0.86	1.18	A. E.
Control			6.5–8.0	3.9–5.6	0.2–0.36	0.3–0.81	0.45–0.93	0.58–1.38	

P. V.:　Pemphigus vulgaris, D. H. D.:　Dermatitis herpetiformis Duhring, P. E.:　Erythrodermia
　　　(Penicillin exanthema), A. E.:　Autosensitized eczema

recovery of the serum total protein concentration. The complete recovery of albumin takes much more time. The α_1 and α_2 fractions show marked increase, and the β fraction shows almost the same tendency as albumin, but in the period of recovery, it tends rather to increase.[14] The γ fraction tends to decrease, especially in infants. This decrease of the γ fraction may accelerate susceptibility to mixed infection in the case of burn. The recovery of blood hemoglobin concentration after burn precedes that of serum protein (Fig. 143).

According to the report by LYTLE and others,[228] in the serum taken from infants suffering from burns, the prealbumin component is abnormal. The use of the barbital buffer which is used in daily routine analysis seems to indicate that this prealbumin component belongs to the α fraction, but its origin is still unknown.

EXUDATIVE PULMONARY DISEASES

The sputa of normal men contain a very small amount of protein components. When diffuse exudative lesion occurs in the pulmonary parenchyma, protein loss becomes a problem. In this case, protein-losing is of non-selective type. As in the case of exudative dermatopathy, it is caused by pulmonary inflammation, and so basically marked hypoproteinemia and hypoalbuminemia accompany the combined serum protein pattern of both acute inflammatory and protein-losing types.

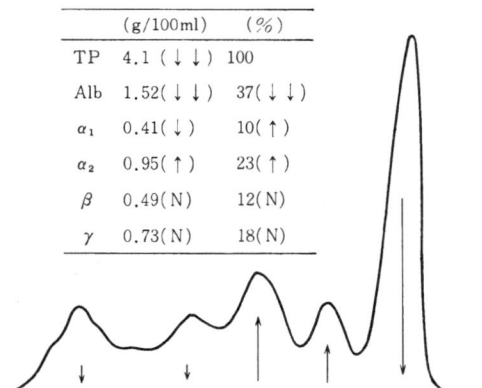

	(g/100ml)	(%)
TP	4.1 ($\downarrow\downarrow$)	100
Alb	1.52($\downarrow\downarrow$)	37($\downarrow\downarrow$)
α_1	0.41($\downarrow$)	10($\uparrow$)
α_2	0.95($\uparrow$)	23($\uparrow$)
β	0.49(N)	12(N)
γ	0.73(N)	18(N)

Fig. 144 Exudative pulmonary disease.
N. W., 71 y.o., male, lobar pneumonia.
Discharged 3 weeks before, but readmitted with high fever, and leukocytosis (12,000). The pneumococcus was cultured from his sputum, and the chest X-ray showed diffuse shadow throughout the right lung.

Among exudative pulmonary diseases which cause hypoproteinemia, the following may be mentioned: lobar pneumonia (Fig. 144), pulmonary gangrene, bronchiectasis, pulmonary abscess, pulmonary cancer and tuberculosis with mixed infections, pulmonary edema, etc. In recent years, cases of hypoproteinemia caused by pulmonary diseases are very rare because the use of antibiotics has become popular. Some of the cases experienced by the author are shown in Table 38. It should be noted that all the patients are of advanced age, so in addition to the protein loss caused by exudation, there may be at the same time combined with decreased synthetic rate.

Table 38 Serum protein patterns in lobar pneumonia.

Filter paper patterns

Name	Age	Sex	TP	Alb	α_1	α_2	β	γ
Y. M.	78	M	5.3	1.34	0.52	0.82	0.66	1.97
T. T.	70	M	5.3	1.88	0.34	0.60	0.60	1.88
T. M.	68	F	5.5	1.86	0.43	0.66	0.89	1.66
Control			6.5–8.0	3.9–5.6	0.2–0.36	0.3–0.81	0.45–0.93	0.58–1.38

BLOOD LOSS AND EXPERIMENTAL PLASMAPHERESIS

There are detailed studies regarding anemia induced by acute bleeding, but the turnover of the serum protein component is still unclear. However, it is thought that anemia is associated with hypoproteinemia and the recovery of the serum protein abnormality as in the case of burn is slow in comparison to the recovery of hemoglobin concentration. As to the serum protein pattern resulted from experimental bleeding or plasmapheresis, the results may be summarized as follows.[60,319,345,419] In both bleeding and plasmapheresis, the change of serum proteins is basically similar, but usually is more prominent in the former. Fig. 145 is made in accordance with the report by SHIMIZU[345]. He reports that in experiments using rabbits, hemoglobin and total serum protein concentration decrease, and they become normal in one month.

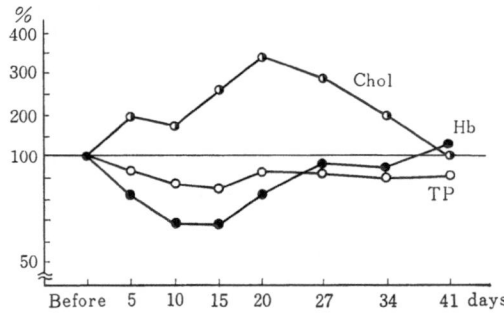

Fig. 145 Changes in blood constituents in experimental rabbits with
         plasmapheresis (from SHIMIZU).
Using 5 rabbits, 20 ml plasmapheresis was repeated for 20 days. The
values obtained before the plasmapheresis are set as 100%.

At the same time, a marked increase of serum total cholesterol and β-lipoproteins is noted. Hemoglobin concentration recovers most quickly, and it is followed by the recovery of serum protein abnormality. The change of serum lipids remains for the longest period. "Bleeding hyperlipemia," observed in experiments with rabbits, is said to be not so clearly observed in other experimental animals[319]. As to the serum protein pattern, the albumin fraction decreases markedly while the α_1, α_2, and β fractions increase. The increase of the β fraction is regarded to be mainly the reflection of the increase of β-lipoproteins. As to the γ fraction, there is no remarkable change, but in cases of the combined bleeding and nitrogen deficiency, a slight increase is seen[319]. In fact, the injection of trithium-labelled albumin into rabbits which have been submitted to plasmapheresis results in a 99% disappearance of the injected albumin from the serum in a short period of time.   It is taken into the body tissues and part of it is degraded immediately and is utilized for synthesis of new globulin molecules.

LYMPHORRHEA AND CHYLURIA

Lymphorrhea usually occurs as the retention of chyle in pleural, peritoneal and pericardial cavities, caused by the rupture of the thoracic duct or of the main lymphatics.   It may also be noted through skin lesions.

Among the causes of lymphorrhea, there are, 1) trauma, 2) congenital malformation, 3) compression due to granuloma or neoplasm.

Chyluria occurs when lymph is lost from lymphatics into urinary tracts.   It is caused by 1) filaria infection and 2) obstruction of the thoracic duct or abdominal lymphatics. As described in page 141 chyle contains a lot of plasma protein components as well as lipids, and when it is leaked outside the body continuously, hypoproteinemia results.   The protein pattern of chyle is almost the same as that of serum and it also contains fibrinogen. Therefore, it shows a non-selective protein-losing pattern (Fig. 146).

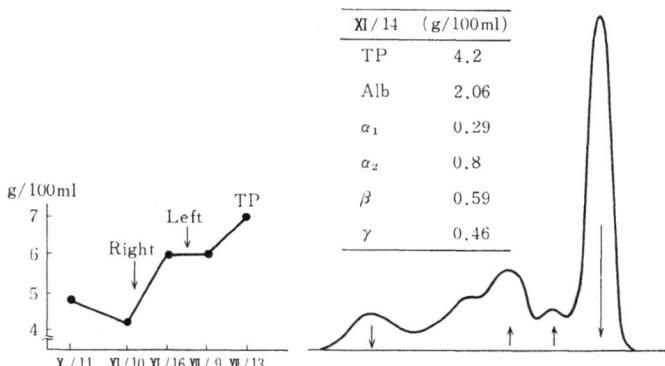

Fig. 146 Serum protein electrophoretic pattern in the patient with chyluria. K. H. 63 y.o., male, chyluria.
Noted milky urine for the past 4 years. On lymphangiography, marked dilatation was demonstrated bilaterally with the lymphatics around the renal hilus.   Hypoproteinemia was improved dramatically after the removal of the dilated lymphatics.

ESSENTIAL HYPOPROTEINEMIA

Essential or idiopathic hypoproteinemia is the general term given to hypoproteinemia of unknown etiology.

Clinically, until about 1957, cases of hypoproteinemia without any apparent liver damage nor abundant urinary protein loss were called essential or idiopathic hypoproteinemia. In 1957, SCHWARTZ and THOMSEN made metabolic studies using RISA and found that hypoproteinemia was caused by marked acceleration of albumin catabolism, calling "hypercatabolic hypoproteinemia." Almost at the same time, in the article "The mechanism of hypoproteinemia associated with giant hypertrophy of the gastric mucosa", CITRIN et al.[62] reported that abundant loss of serum albumin into the gatrointestinal tract through the gastric mucosa resulting in hypoproteinemia. Since then, protein-losing gastroenteropathy has been separated from essential or idiopathic hypoproteinemia.[62] Therefore, among the patients of essential hypoproteinemia reported before 1959–60, there might be some cases of protein-losing gastroenteropathy. But, although very few, some cases are reported in which protein loss or hypercatabolism cannot be proven even by metabolic studies of albumin using radioactive isotopes.

The real nature of this disease is quite unknown.   However, we can classify hypoproteinemia into six groups as is shown schematically in Figure 147.

    1) Mal-nutritional hypoproteinemia, resulting from marked deficiency of protein and amino acids.

    2) Hepatic hypoproteinemia, resulting from decreased hepatic synthesis of serum proteins.

    3) Protein-losing hypoproteinemia, resulting from excessive protein loss.

    4) Hypercatabolic or endocrine hypoproteinemia, resulting from increased endogenous catabolism of serum proteins.

    5) Hemodilutional or pregnancy type hypoproteinemia, resulting from increased circulating blood volume.

    6) Mal-distributional hypoproteinemia, resulting from abnormal distribution of serum proteins in the body.

Among the causes of hypoproteinemia described above, 1) – 5) have been known up to the present.   Though the author has suggested the possibility of 6), the only report of mal-distributional hypoproteinemia until now is a case reported by WEINBREN et al.[403]

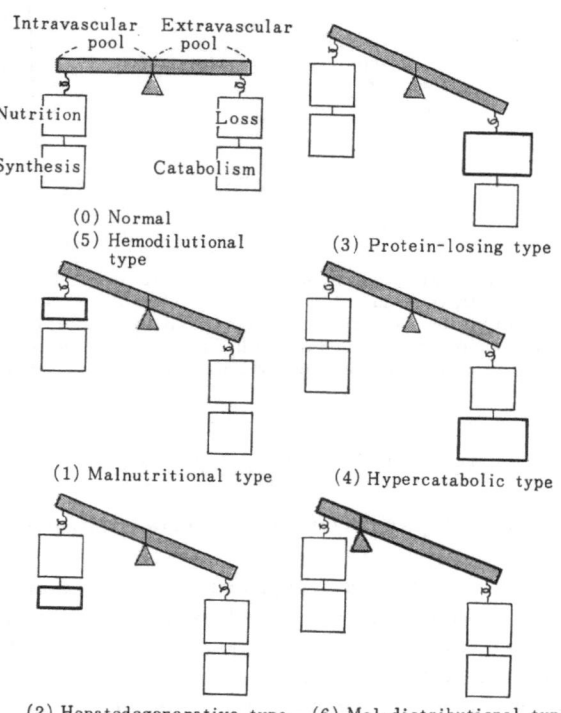

Fig. 147 A schematic diagram showing the mechanisms of hypoproteinemia.

Table 39 Albumin metabolic data in "mal-distributional" hypoproteinemia. (from WEINBREN et al.)

	before treatment	after treatment
Total exchangeable albumin	188 mg/kg	243 mg/kg
Albumin turnover	203 mg/kg/day	315 mg/kg/day
Intravascular pool	918 mg/kg	1286 mg/kg
Extravascular pool	1170 mg/kg	1275 mg/kg
Extra./intravascular pool ratio	1.275	0.99
Capillary permeability	152.5 %/day	290.0 %/day
Weight	48 kg	46 kg
Fecal excretion	0.843 %	0.516 %
Urinary excretion	3.780 %	3.700 %

A 65 y.o. female with SLE, having edema and hypoproteinemia, with normal renal function tests. Edema and hypoproteinemia improved significantly after prednisone treatment.

The results of their albumin metabolic studies are quoted in Table 39. It is clear here that the ratio of the intravascular/extravascular albumin is so small that the albumin transfer from the extravascular space into the intravascular space must be interfered significantly. More precise metabolic studies may contribute to find many cases of mal-distributional hypoproteinemia.

Chapter 20

Plasma Protein Changes in Hepatic Disorders

CHARACTERISTICS OF THE HEPATO-DEGENERATIVE SERUM PROTEIN ELECTROPHORETIC PATTERN.

The liver injury type is resulted from the following two changes: 1) that caused by the decreased hepatic synthesis of serum proteins with the exception of immunoglobulins, and 2) the change caused by the increased immunoglobulin synthesis. Therefore, the characteristic electrophoretic pattern of serum proteins show the decrease in the albumin and a fractions and the increase in the γ fraction (Fig. 148).

The total serum protein concentration usually remains within normal range, but it decreases when liver cirrhosis and fulminant hepatitis result serious liver damage.

On filter paper electrophoresis of serum proteins, the albumin fraction is markedly decreased in cases of serious liver damage, the a_1 and β fractions tend to be decreased and the γ fraction is increased significantly. Generally, the a_1 fraction increases slightly, but sometimes it decreases (Fig. 148). The γ fraction rarely shows M-component, and generally it shows a broad increase. The author classifies the serum protein electrophoretic pattern of the liver injury type into the following three groups according to their electrophoretic patterns (Fig. 149).

Group I: It is impossible to separate the β fraction from the γ fraction, and the densitogram characteristically shows a gentle "ski-jump slope" between the β and γ fractions. In 1953, WIEME found the pattern of this kind and called it "β–γ linking", and SUNDERMAN later called it "β–γ bridging".[369)]

Group II: This group includes those cases with the increased γ fraction accompanying the β–γ valley to be very shallow, showing the tendency of forming β–γ linking.

Group III: In this group are those cases in which the β fraction is clearly separated from the γ fraction.

On cellulose acetate electrophoresis, the decrease of the a_2 fraction is less prominent while the decrease of the β fraction is more prominent, when compared with the data obtained with filter paper electrophoresis. As described later, it is because the low-density lipoproteins migrate in different zones with different electrophoretic methods. That is, on cellulose acetate electrophoresis, the low-density lipoproteins mostly migrate in the a_2 zone and they compensate the decrease of other a_2 globulins. β–γ bridging is observed most clearly on filter paper electrophoresis, especially by using the horizontal electrophoretic cells. On cellulose acetate electrophoresis, β–γ bridging is infrequently seen, apparently because the clearing of the cellulose acetate membrane after staining can be done sufficiently and thus the β and γ fractions can be separated more clearly on the densitograms. In cases of markedly increased γ fraction, the cellulose acetate strip fre-

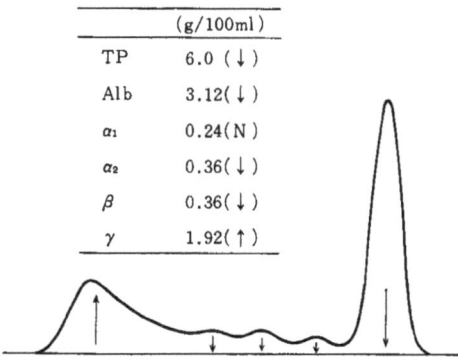

	(g/100ml)
TP	6.0 (↓)
Alb	3.12(↓)
α_1	0.24(N)
α_2	0.36(↓)
β	0.36(↓)
γ	1.92(↑)

Fig. 148 Protein changes in liver damage.

G.G., 58 y.o., male, acute serum hepatitis.

Seven months prior to admission, received multiple blood transfusion and irradiation for carcinoma of the esophagus. Since 10 days before admission, complained of general malaise, poor appetite, nausea, vomiting and jaundice. The liver was palpable at 2 f.b., and was soft, smooth and somewhat tender. Upon admission, slight anemia (Hb 12.6 g/100 ml), jaundice (total bilirubin 15.5 mg/100 ml; direct 7.5 mg/100 ml). SGOT 1780 u., SGPT 1100 u., LDH 930 u., Al-P 10.9 K.A.u., the plasma prothrombin time 21.8 sec. (control, 11.8 sec.). Died of hepatic coma in 3 days after admitted. The cellulose acetate electrophoretic pattern of th serum protein shows a decrease in the albumin, α_2 and β fractions. The γ fraction is increased broadly.

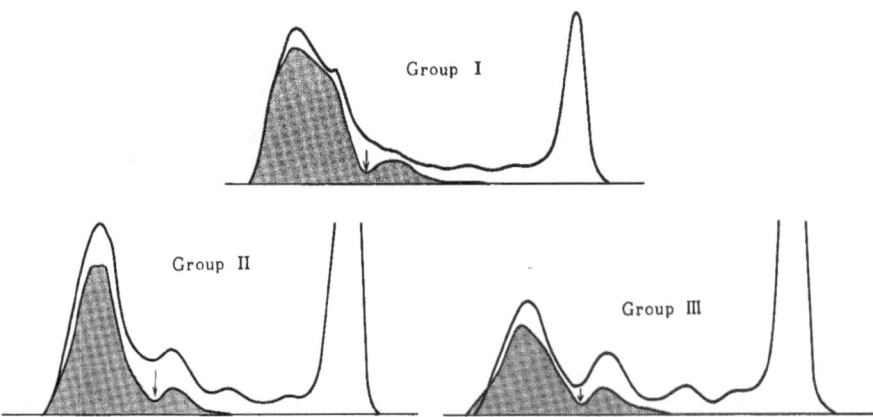

Fig. 149 Various types of the serum protein electrophoretic patterns in liver disorders.

The solid lines indicate the densitometric patterns of the serum protein in various types. The shaded areas are the electrophoretic patterns of the supernatant serum after Rivanol precipitation, showing mainly the IgG fraction and the transferrin fraction. IgM and IgA are removed almost completely with Rivanol precipitation. Group I shows a complete β–γ linking, and Group II shows a definite tendency of the β–γ linking. In Group III, both the β and the γ fractions are separated clearly.

quently shows the distorted shape of the γ fraction, as shown in Fig. 150. In cases with severe jaundice, the albumin fraction tends to show a tailing toward the anodal end. This is because significant amounts of bilirubin and bile acids combine to serum albumin, causing an increase in the electrophoretic mobility of albumin. This phenomenon is especially notable on cellulose acetate strips. (Fig. 150)

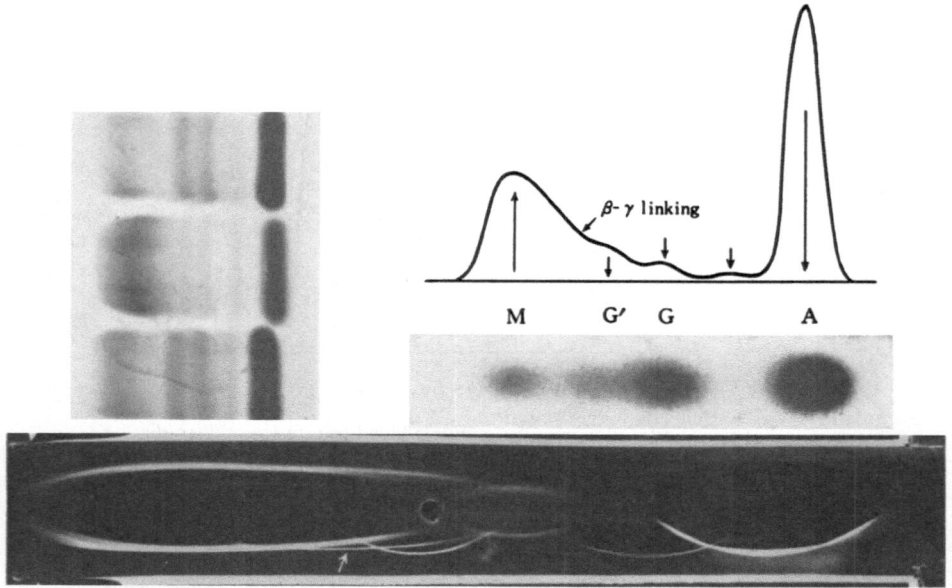

Fig. 150 β-γ linking in the serum protein electrophoretic pattern of the liver injury type.
K. Y., 58 y.o., female, liver cirrhosis and arteriosclerosis.
On the cellulose acetate electrophoretic separation, a characteristic β-γ linking is noted. Because of
strong jaundice, the albumin fraction is migrated faster (left upper).
TP 5.8 g/100 ml, alb 2.74 g/100 ml ($\downarrow$), α_1 0.07 g/100 ml ($\downarrow$), α_2 0.20 g/100 ml ($\downarrow$), β 0.35 g/100 ml
($\downarrow$), and γ 2.44 g/100 ml ($\uparrow$), on cellulose acetate electrophoresis.
On the thin-layer gel filtration, the A fraction is decreased, the G fraction increased, and the M fraction slightly decreased. In addition, because of a marked increase in the serum IgA, the G' fraction
is prominently noted between the G and M fractions.
On immunoelectrophoresis, the patterns in the α and β zones look definitely empty. α_1-antitrypsin is
relatively clear, and hemopexin is not recognized. All the major classes of the immunoglobulins, especially IgA (arrow), are increased prominently. β_{1C}-globulin is not clearly seen.

DIFFERENTIATION OF THE HEPATO-DEGENERATIVE
SERUM PROTEIN ELECTROPHORETIC PATTERN

Differentiation from the Hyperimmunoglobulinemic Pattern Due to Non-hepatic Pathologies.

Unless the serum protein pattern shows the decrease of the α_1, α_2 and β fractions, it is
impossible to differentiate it from the other conditions showing broad increase of the γ
fraction. But the β-γ linking is helpful to distinguish it from other diseases, if present.

Differentiation of the Hepato-Degenerative Pattern with β-γ Linking

1. False β-γ linking on the densitograms may be falsely recognized due to some
technical failures. For examples, 1) incomplete separation of the β and γ fractions due to
short electrophoretic migration; 2) the distorted electrophoretic separation of each fraction; 3) excessive diffusion after electrophoretic separation due to delayed fixation. These
changes caused by technical failures can be apprehended by directly examining the electrophoretic strips, and they, of course, should not be screened with densitometers.

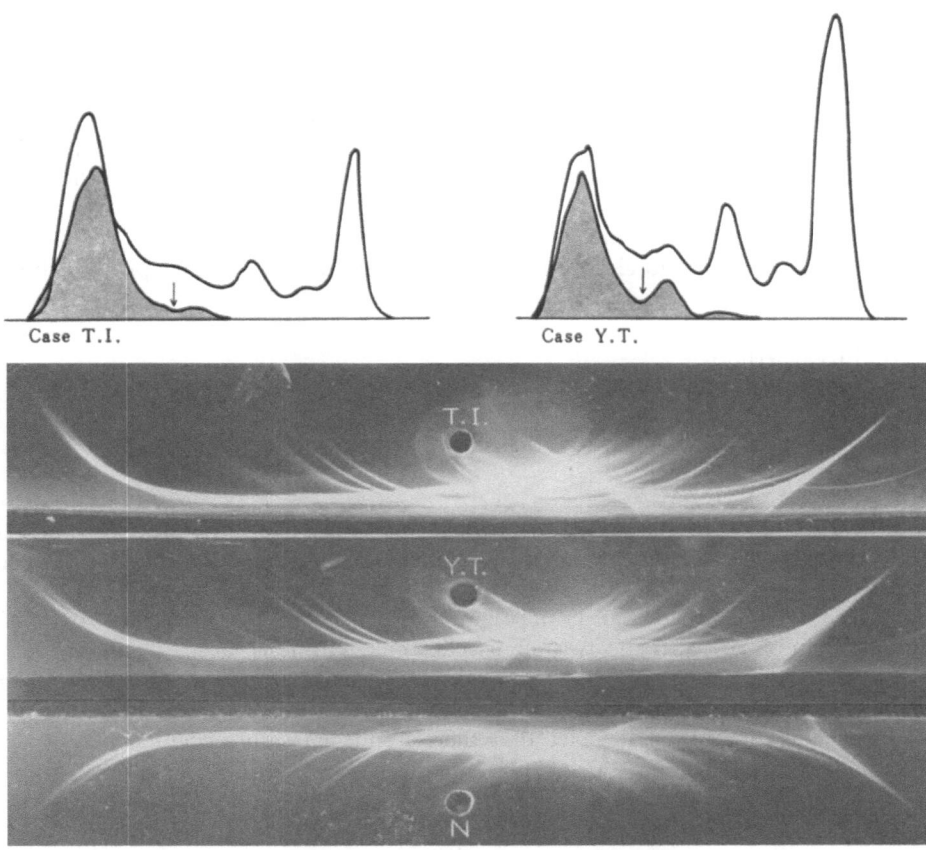

Fig. 151 β–γ Linking in non-hepatic diseases.

Case T. I., 51 y.o., female, squamous cell carcinoma of the urinary bladder with wide-spread metastasis. The carcinoma invades into the vagina, but not to the liver.   TP 7.8 g/100 ml, alb 1.51 g/100 ml ($\downarrow\downarrow$), α_1 0.40 g/100 ml, α_2 0.84 g/100 ml ($\uparrow$), $\beta+\gamma$ 3.90 g/100 ml ($\uparrow\uparrow$).   On immunoelectrophoresis, IgG, IgM, IgA are all increased significantly, IgA being 4 times that of the normal.

Case Y. T., 28 y.o., male, tuberculous epidydimitis.

T. P 8.5 g/100 ml ($\uparrow$), alb 2.93 g/100 ml ($\downarrow$), α_1 0.50 g/100 ($\uparrow$), α_2 1.22 g/100 ml ($\uparrow\uparrow$), β 1.0 g/100 ml, γ 2,85 g/100 ml ($\uparrow$).   On immunoelectrophoresis, IgG and IgA are markedly increased.

On both cases, the β–γ zone (arrows) becomes deep after treated with Rivanol precipitation.

2. "β–γ Linking" Seen in Chronic Inflammatory Diseases and in Malignant Tumors

As shown in Fig. 151, in some cases of chronic inflammatory diseases and malignant tumors entailing the marked increase of the γ fraction, the tendency of β–γ linking is occasionally seen.   However, the complete β–γ linking is very rarely encountered in diseases other than liver cirrhosis.   But in these cases, being different from that of liver cirrhosis, the α_1 and α_2 fractions are increased, showing the characteristic chronic inflammatory pattern.

REPRESENTATIVE DISEASES MANIFESTED BY SERUM PROTEIN PATTERNS OF LIVER INJURY TYPE

The cases which show the characteristic serum protein patterns are, of course, the primary and secondary liver diseases associated with hepatocellular damage.   But the liver has a great functional reserve, and thus all liver diseases do not necessarily show the characteristic serum protein abnormalities.   Liver diseases can be classified on the basis of the serum protein patterns as follows.

1.   Those without Significant Serum Protein Abnormality.

Among these are cases in which there is only slight damage or none at all in liver functions, such as mild cases of acute hepatitis, primary hepatoma without liver cirrhosis, and metastatic tumors of the liver.

2.   Those with Definite Serum Protein Abnormality.

It might be appropriate to classify these into the following three groups according to their pathophysiology.

     a.   Hepatitis
     b.   Liver cirrhosis
     c.   Obstructive liver disease

HEPATITIS

Classification of Hepatitis

Acute liver damage can occur due to various causes, among which are viral and bacterial infections as well as physical and chemical injuries.   Of these, viral hepatitis occurs most frequently.

Viral hepatitis is classified into 1) infectious hepatitis, and 2) serum hepatitis.   The differential diagnosis between the two often relies on anamnesis, and not on the serum protein electrophoretic patterns.   Therefore, these two will be referred to simply as viral hepatitis.   There is no common opinion as to classifying viral hepatitis according to its stages.   Especially notable is the fact that there is no generally accepted definition of "chronic hepatitis".   Therefore, to avoid using the name "chronic hepatitis", and in

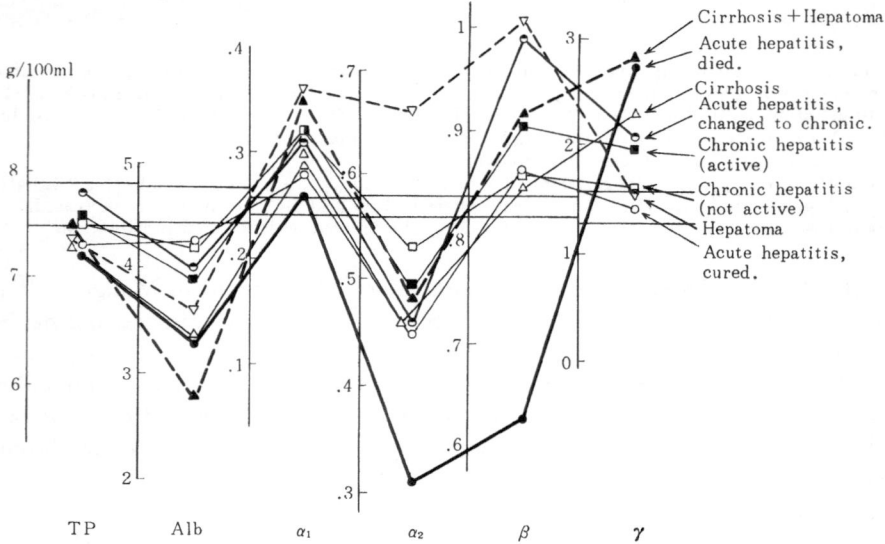

Fig. 152   Filter-paper electrophoretic patterns of the serum proteins in liver diseases.
The average values on each group are plotted.   Excluding the
cases having hepatoma without cirrhosis and the cases suffering from
fulminant hepatitis, the other groups show the similar changes.

consideration of the pathophysiological or functional aspects of the plasma protein abnormality, viral hepatitis is classified as follows: 1) acute (viral) hepatitis, 2) protracted hepatitis, 3) relapsing hepatitis, and 4) progressive hepatitis.

Toxic hepatitis or toxic liver damage is histologically and clinically classified into the following groups: 1) the cytotoxic type, 2) the hepatitic type, and 3) the cholestatic type. This shows a similar plasma protein abnormality as in viral hepatitis.

Lupoid hepatitis is a special type of hepatitis that entails severe chronic hepatic damage. This was observed by MACKAY.[244]   This condition is often found among young women, shows a positive LE cell phenomenon and clinical findings quite similar to those of systemic lupus erythematosus.

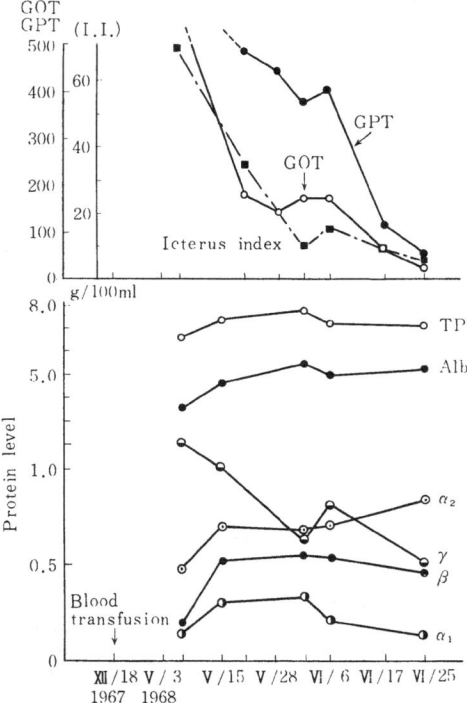

Fig. 153 Follow-up studies of the serum protein electrophoretic fractions in acute hepatitis.
Y. Y., 32 y.o., male, acute hepatitis.
Received 2500 ml of stored whole blood during and after the surgery for osteosarcoma of the left femur.   About 6 months later, developed generalized icterus and hepatomegaly.   The changes in the serum GOT and GPT are parallel to those of the serum γ fraction.

Plasma Protein Changes in Hepatitis

A great deal of investigations have been done on plasma protein abnormality in different stages of hepatitis.   However, because of a lack of commonly accepted classification on the stages of hepatitis, their results vary among different investigators.   In spite of that, generally speaking, the albumin and α_2 fractions are decreased, and the γ fraction is increased (Fig. 152).   In order to understand the changes of plasma proteins in hepatitis, each patient must be followed up carefully (Fig. 153).

1. The Total Serum Protein Concentration

In most cases of hepatitis, the total serum protein concentration remains within normal range, and occasionally shows slight decrease. Therefore, it is not diagnostically important.

2. The Albumin Fraction

The albumin fraction in serum tends to decrease, but it decreases slightly only when the liver damage is severe.   In the cases with acute hepatitis showing uneventful course, metabolic studies using RISA are not remarkable.[313]   In fulminant hepatitis, however, hepatocellular necrosis occurs widely, and the albumin fraction eventually becomes decreased due to significant deficiency of albumin synthesis.   But generally speaking, once the patient is on the way to recovery in bed rest, the normal albumin synthetic rate is rapidly regained.   Since the half-life of serum albumin is rather long, a significant decrease of the albumin fraction will become manifest only when the disease proceeds chronically. Prealbumin is usually decreased in hepatitis.

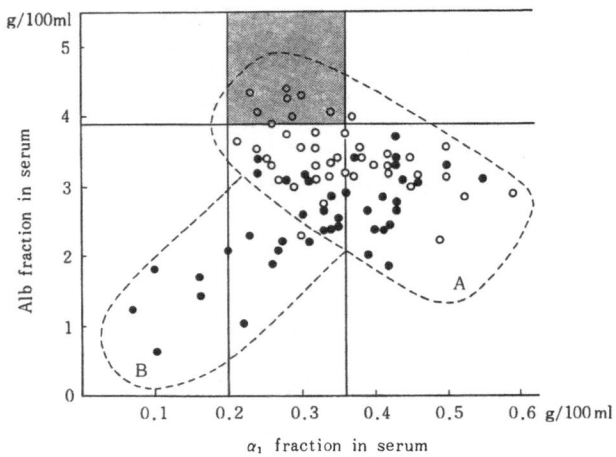

Fig. 154 Relationship between the albumin and α_1 fractions in liver diseases. The white dots are for the patients with hepatitis and the black dots for the patients with liver cirrhosis. The serum concentrations of the albumin and α_1 fractions were obtained with the filter paper electrophoresis. In the A group, both the fractions are related inversely, being similar to that seen in the other types of inflammatory diseases. In the B group, however, they are in the proportional relation, and they include the data obtained from the patients with severe liver cirrhosis.

3. The α_1 Fraction

As shown in Fig. 154, there seems to be two groups of the patients with liver injury, based on the changes of the α_1 fraction: 1) Group A which shows an inverse proportional relation between the albumin and α_1 fraction levels, 2) Group B which shows a directly proportional relation.   Most cases of hepatitis belong to Group A, and have almost the same relation as the inflammatory conditions other than hepatitis.   On the other hand, Group B includes cases of liver cirrhosis with severe liver damage and cases of severe acute hepatitis, both of which have characteristics essentially different from other inflammatory diseases.   Of course, there exists some overlapping between Groups A and B, indicating various transitional forms.

About 90% of the α_1 fraction consists of the globulins of hepatic origin.   It is not known why they show the tendency to increase in liver disease, but some of the α_1 globulins may originate from destruction of tissue protein components. The author has observed that there is a significant increase, at least of a_1-antitrypsin and a_1-acid glycoprotein in cases of acute hepatitis on immunoelectrophoretic analyses. The same result has been reported by BETSUYAKU.[37] Therefore, the increase of a_1-antitrypsin and a_1-acid glycoprotein, both of which are acute phase reactants, is thought to play a major contribution to the increased fraction.

In many cases of hepatitis, the increased a_1 fraction seems to occur as follows: as in the other inflammatory conditions, the mechanism which causes the increase of the a_1 globulin components as acute phase reactants (probably, synthetic increase in the liver cells) is in action. However, the hepatic synthetic reserve of a_1-antitrypsin and a_1-acid glycoprotein may be greater than those of a_2 and β globulin components. So even in cases of liver damage, the liver synthetic system responding to the inflammatory stimulation is thought to fulfill its function above the normal level. As indicated by NAGEL et al,[266] it may be said that if the a_1 fraction shows normal value even under apparent inflammatory stimulation, significant liver damage must be present and it may indicate poor prognosis. Generally, as might be assumed from Figure 154, cases of hepatitis which belong to Group B having less than 3.5 g/100 ml of the albumin fraction and the decreased a_1 fraction are considered to have poor prognosis.

a_1-lipoprotein tends to decrease, being marked in cases of severe liver damage, and it cannot be recognized on immunoelectrophoresis.

4. The a_2 Fraction

The a_2 fraction tends to decrease in cases of hepatitis and is affected mainly by the decrease of haptoglobin. However, a_2-macroglobulin tends rather to increase. Ceruloplasmin does not exhibit any particular change but maintains a normal level in most cases of hepatitis.

The a_2 fraction is usually within normal range during the first week of the disease, and it decreases only in severe hepatitis. In the second week of the disease, when the patient is on the way to uneventful recovery, an increase of a_2 fraction, especially that of haptoglobin, is observed. The cases in which haptoglobin does not increase or rather decreases, are regarded as being of poor prognosis.[145]

5. The β Fraction

In cases of acute hepatitis, the γ fraction tends to increase slightly, but when liver damage is severe, it rather decreases. Transferrin, which exists in the greatest amount in the β fraction, shows little change in cases of acute hepatitis. It is said that in some cases it tends rather to increase.[148] But when the disease becomes chronic, transferrin decreases and hemopexin also shows the same tendency. The serum complement titer becomes greater in the early stage of the disease, but it usually returns to normal after 8 to 10 weeks. However, in protracted cases, it shows a continuously high level.[144] In fatal cases of acute hepatitis, every component of the β fraction shows marked decrease.

On filter paper electrophoresis, the increase of low-density lipoproteins causes the increase of the β fraction. At least in cases of chronic persistent hepatitis, increase of the β fraction is considered to be caused by the increase of low-density lipoproteins. On cellulose acetate electrophoresis, the low-density lipoproteins are included in the a_2 fraction and then the a_2 fraction tends to increase (Fig. 153).

6. The γ Fraction

Generally speaking, in cases of acute hepatitis, the increase of the γ fraction slightly delays the decrease of the albumin fraction. It is proportionate to the severity of hepatitis and returns to normal level during recovery. Its increase becomes more noticeable

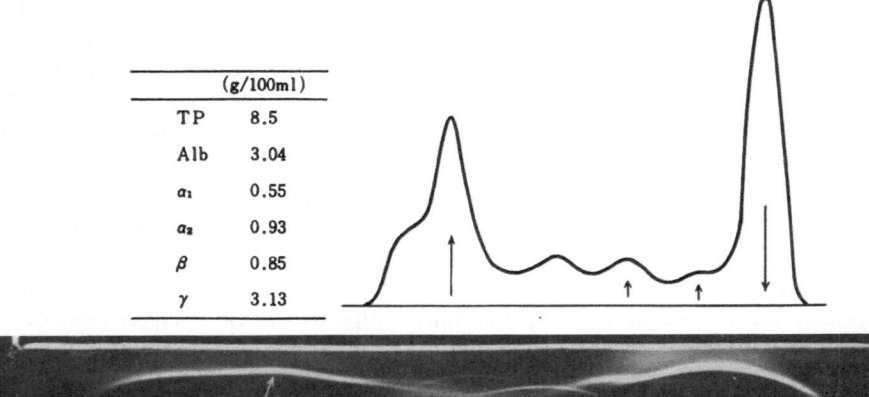

	(g/100ml)
TP	8.5
Alb	3.04
a_1	0.55
a_2	0.93
β	0.85
γ	3.13

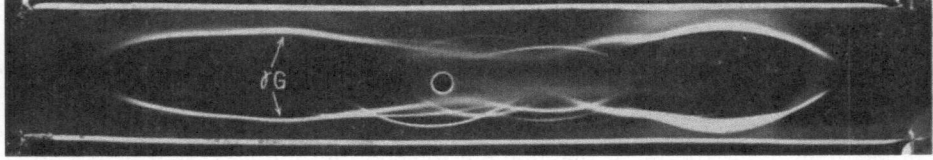

Fig. 155 M-proteinemia in "chronic hepatitis".

M. N., 41 y.o., male, chronic hepatitis.

In 1956, gastrectomy for peptic ulcer, and received 3,000 ml blood transfusion. Uneventful since then, but hepatomegaly found on routine health-checkup in 1962. Two years later, complained of general malaise and edema, and diagnosed as "chronic hepatitis". On admission, slight jaundice (icteric index 11), GOT 73 u., GPT 23 u., serum total cholesterol 146 mg/100 ml, Al-P 2.0 B.u. Bone marrow and bone X-ray examinations showed nothing remarkable.

On filter-paper electrophoresis, slight decrease in the albumin fraction and moderate increase in the γ fraction, demonstrating a M-protein band at the γ zone.

On immunoelectrophoresis, the IgG line shows a typical M-bow (arrow); IgG 3840 mg/100 ml, IgA 340 mg/100 ml and IgM 310 mg/100 ml.

when the disease becomes chronic. In cases of hepatitis, the increase of the γ fraction coincides with the clinical course, and it is helpful to assume the chronicity of the hepatitic condition (Fig. 153).[374] However, β–γ linking can not be observed in cases of hepatitis, with occasional exception of subacute hepatitis.

There have been many reports about the changes of serum immunoglobulins in hepatitis.[148, 215, 411] In the acute stage of hepatitis, IgM-globulin increases, and especially in cases of infectious hepatitis, it increases to around two to five times its normal level. However, after the first week of the disease, it reverts rapidly to the normal level. On the other hand, the increase of IgG-globulin stays in a moderate degree, and continues to show a high level for a comparatively long period. IgA-globulin changes very little, or if it does, it shows only the slightest increase. IgD-globulin also shows little change. But in chronic cases the increase of IgG-globulin becomes more marked and IgM-globulin maintains a high level. The author has observed M-protein of IgG type in a case of protracted hepatitis (Fig. 155). In cases of infantile hepatitis, the γ fraction shows a marked decrease (Fig. 156). The reason for this seems to be that immunoglobulins do not increase because their synthetic capacity has not been developed sufficiently during infancy.

In acute hepatitis the anti-liver antibodies may be demonstrated in about 10–30%, and more frequently in chronic cases.[127,148] Some reports correlate the positivity of the anti-liver antibody with the serum concentration of the γ fraction. However, PASNICK et al.[286] have doubted presence of the significant correlation between the two. As it is known that autoantibody causes cellular destruction, it may be possible that anti-liver antibody injures liver cells further.[148]

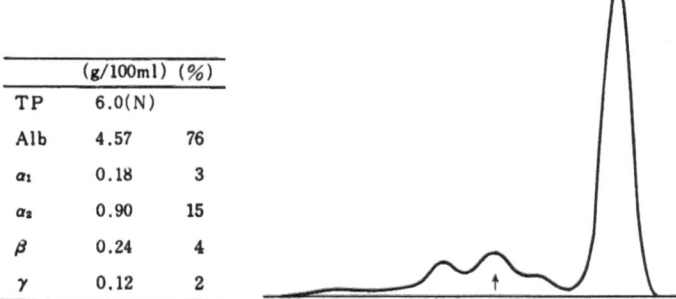

	(g/100ml)	(%)
TP	6.0(N)	
Alb	4.57	76
α_1	0.18	3
α_2	0.90	15
β	0.24	4
γ	0.12	2

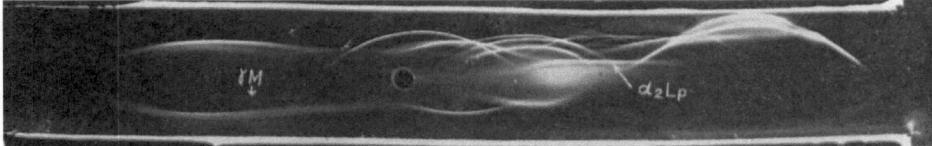

Fig. 156 Serum protein patterns in neonatal hepatitis.
S. S., 3 m.o., male, neonatal hepatitis.
Physiological jaundice disappeared almost completely in 8 days after birth, but jaundice re-appeared in two weeks later. Total serum bilirubin 8.5 mg/100 ml (direct 5.0; indirect 3.5). GOT 105 u., GPT 84 u.; urinary urobilinogen (2+).
On cellulose acetate electrophoresis, slight increase in the α_2 fraction and moderate decrease in the γ fraction. On immunoelectrophoresis, marked increase in the low-density lipoproteins and marked decrease in the IgG; IgG 215 mg/100 ml, IgA 74 mg/100 ml (moderately increased for the age) and IgM 59 mg/100 ml (normal for the age).

It is generally accepted that the immunoglobulins increase in hepatitis because of their increased synthesis in both hepatic and extrahepatic tissues.[148, 285] With the fluorescent antibody technique, the immunoglobulins, as well as fibrinogen, albumin and other plasma protein components, are demonstrated in liver tissues. But the immunoglobulins are the most abundant to be demonstrated, and are thought to be synthesized mostly in the liver. Fibrinogen, albumin and other protein components are present probably on the result of cellular phagocytosis occurring during the acute phase.[285] In cases of acute hepatitis, the increase of the immunoglobulins is observed from the early stage of the disease, but its reason remains unknown.

7. Fibrinogen

In cases of hepatitis, fibrinogen shows little change, and it decreases only when severe liver damage occurs.[15, 374] The decrease in fibrinogen may be caused by its decreased synthetic rate in the liver, and it may also be caused in some cases by accentuated fibrinolytic activity.[91, 203]

8. The Basic Serum Protein Electrophoretic Patterns in Hepatitis

Serum protein electrophoretic patterns are quite variable in cases of hepatitis, depending on the stages of the disease and the difference of the individual bodily responsiveness. However, the patterns must be influenced dynamically by the pathophysiological changes occurring in the body. Therefore, it is worthy of classifying the basic serum protein patterns in order to understand the pathophysiology of hepatitis.[367]

The type without any notable change : This type is usually encountered in the very early stage or the mild cases. The patients with this type are recovered usually.

The type which is manifested by the decrease of the albumin fraction only : In these cases, the liver cells are certainly damaged, but the interstitial tissues must show very little change.

The type which is manifested by the decrease of the albumin fraction and the increase of other frac-

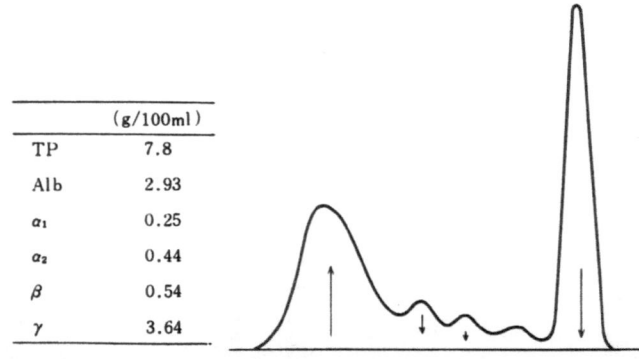

	(g/100ml)
TP	7.8
Alb	2.93
α_1	0.25
α_2	0.44
β	0.54
γ	3.64

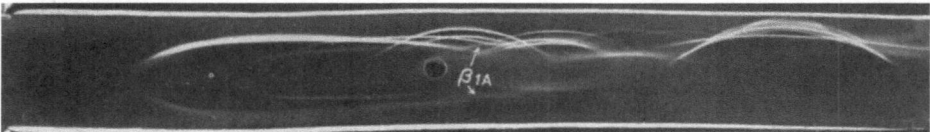

Fig. 157 Serum protein patterns in lupoid hepatitis.
Y. K., 56 y.o., female, lupoid hepatitis and rheumatoid arthritis.
Rheumatoid arthritis since the age of 21 years. Suffering from chronic hepatitis for the past 3 months, and from buccal exanthema. Slight anemia (Hb 10.7 g/100 ml), WBC 5900 (stab. 38%, lymph. 17%), GOT 179 u., GPT 90 u., CCF 3+, LE cell phenomenon (+), ANF (fluorescent antibody technique) (+), sed rate 79 mm/l hr.
On cellulose acetate electrophoresis, slight decrease in the albumin and β fractions, and marked broad increase in the γ fraction. On immunoelectrophoresis, a marked decrease in β_{1C}-globulin, a marked increase in IgG, and moderate increase in IgA and IgM.

tions: This type is encountered most commonly. The cases showing the increase of the α fraction usually have good prognosis, and are not associated with the increase of the γ fraction. However, those in which the β fraction increases often show an accompanying increase of the γ fraction. When liver damage becomes severe, the β fraction either becomes normal or decreases, and the γ fraction increases more prominently, indicating unfavorable prognosis.

The type which is manifested by the increase of the γ fraction, only: In these cases, liver cell damage is slight, while the interstitial lesion is quite prominent.

    9.   The Serum Concentration of the Immunoglobulins in Carriers of the Hepatitis Virus.

BEVAN et al. have reported that among 63 donors who were involved in the dissemination of hepatitis by blood transfusion, 21 showed an increase of one of the immunoglobulins.[38] Moreover, all except one were found to be normal in liver function tests. Therefore, they suggested that the measurement of serum immunoglobulin concentration may help to screen the donors.

Plasma Protein Changes in Lupoid Hepatitis

Plasma protein abnormality in cases of lupoid hepatitis is characterized by the decrease of the albumin fraction and the marked increase of the γ fraction. Sometimes, the decrease of the α_2 and β fractions is also observed (Fig. 157).

ZLOTNICK et al.[422] reported that the following changes are characteristic: the decrease of haptoglobin, the increase of α_2-macroglobulin, the decrease of α_2-lipoproteins, the decrease of β-lipoproteins, and the marked decrease of $\beta_{1C/A}$ globulin. Transferrin remains normal. Besides these, all major classes of the immunoglobulins are increased, most prominent in IgG-globulin. The author has observed the same changes. Interestingly β_{1C}-globulin is markedly decreased, and it is often not demonstrated on immunoelectrophoresis. Since the decrease of β_{1C}-globulin is observed regularly in cases of systemic lupus erythematosus, the close relationship between the two conditions is surmised, and in cases of lupoid hepatitis, β_{1C}-globulin may possibly be consumed in *in vivo* antigen-antibody reaction.

Another characteristic is the marked increase of the γ fraction. In the two cases that the author observed, the γ fraction showed 3.64 g/100 ml and 2.44 g/100 ml respectively. In the six cases studied by ZLOTNICK,[422] it ranged from 3.99 g/100 ml to 8.56 g/100 ml.

In the case of a 46 year-old patient the γ fraction was 5.78 g/100 ml and thymic enlargement was observed.[320] One year and six months after thymectomy, IgM and IgA-globulins decreased remarkably in comparison to that measured before thymectomy. The removed thymus showed hyperplasia of the epithelial cells and increased numbers of HASSAL's corpuscles. These findings, as well as the decreased β_{1C}-globulin, may suggest that some immunological mechanisms are involved in the pathogenesis of lupoid hepatitis.

CIRRHOSIS OF THE LIVER

Definition and Classification of the Liver Cirrhosis

The generally accepted definition of liver cirrhosis is: "a disease characterized by the distorted reconstruction of lobular architecture throughout the entire liver or at least a considerable part thereof, accompanied by nodular hyperplasia or regeneration of liver parenchymal cells as well as irregular fibrous septal formation." Therefore, cirrhosis of the liver itself is diagnosed by morphological changes. Types of liver cirrhosis vary, as will be later described, but in all cases the following characteristics are observed hisopathologically: 1) piecemeal necrosis, 2) infiltration of the lymphocytes, plasma cells and histiocytes, 3) proliferation of bile canaliculi.[322]

There is no generally accepted classification of cirrhosis of the liver though various kinds of classification have been suggested from the etiological and morphological points of view.

Primary cirrhosis
   1) Postnecrotic type
   2) Posthepatitic type
   3) Nutritional, alcoholic, or fatty type.
Metabolic cirrhosis
   1) Hemochromatosis
   2) WILSON's disease
Biliary cirrhosis
   1) Primary
   2) Secondary
Parasitic cirrhosis
Congestive or cardiac cirrhosis
Syphilitic cirrhosis
Biliary cirrhosis will be discussed in the section of obstructive jaundice.

Plasma Protein Changes in the Liver Cirrhosis

Plasma protein changes in liver cirrhosis are fundamentally the same as those seen in

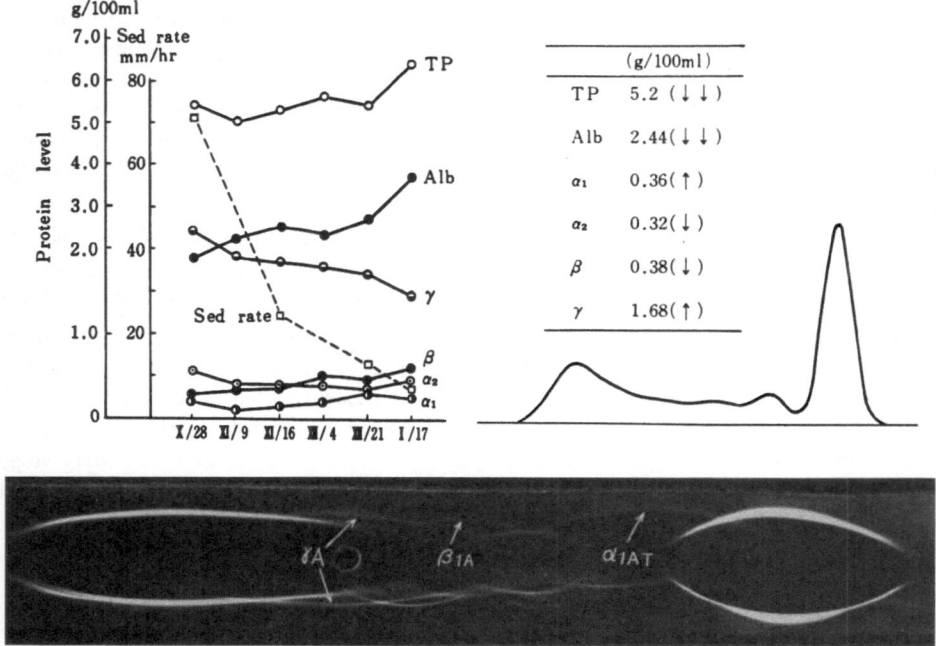

Fig. 158 Serum protein changes in liver cirrhosis.
M. M., 37 y.o., female, liver cirrhosis and pulmonary tuberculosis.
Treated for iron deficiency anemia, and developed later serum hepatitis 6 months before the admission. On the admission, suffered from ascites and hepatomegaly. Moderate anemia (Hb 8.6 g/100 ml), icterus index 8, s-GOT 60 units, s-GPT 34 units, serum Al-P 58.6 K. A. U., total cholesterol 91 mg/100 ml, blood NH_3-N 96 μg/100 ml, plasma prothrombin time 16.0 seconds (control 11.0 seconds).
On the cellulose acetate electrophoresis, a characteristic pattern for the liver cirrhosis was recognized, showing β–γ linking.
On immunoelectrophoresis, IgA is markedly increased. All the serum protein components other than the immunoglobulins are significantly decreased. Hemopexin is not seen, but α_1-antitrypsin is relatively clear.

chronic hepatitis, characterized by the decrease of the albumin fraction and the increase of the γ fraction. Also, the decrease of the α_1, α_2, and β fractions are observed in accordance with the stages of severity of liver damage. However, there exist some remarkably characteristic differences between hepatitis and liver cirrhosis.

1. The Total Serum Protein Concentration

Generally, the total serum protein concentration is within the normal range. However, in cases of severe liver damage, especially the ones with abundant ascites, it may decrease significantly. About half of the cases observed by the author have shown it to be below 6.5 g/100 ml with only about 7% showing below 5 g/100 ml, and one case belonging to Group III showed the lowest concentration of 3.2 g/100 ml. About 7% of the cases observed have shown the increased total serum protein above 8.0 g/100 ml, apparently due to a marked increase of the γ fraction. One case belonging to Group I has shown the highest concentration of 9.5 g/100 ml, and the γ fraction in this case was 6.4 g/100 ml. No apparent difference is recognized among Groups I, II and III.

When the total serum protein is decreased, it usually returns to normal on clinical improvement (Fig. 158).

2. The Albumin Fraction

In most cases, the albumin fraction decreases markedly. Its decrease is the most notable in Group I, followed by Group II, and then Group III (Fig. 159). However, this is the general tendency, and the albumin fraction varies among different cases. There is a close paralelism between the decrease of the albumin fraction and the increase of the γ fraction, and their degree correlates well with the severity of liver damage and clinical findings.[374] In general, when the albumin fraction goes below 2.5 g/100 ml, ascites and edema appear clinically. The albumin fraction sometimes decreases to below 1 g/100 ml.

The decrease of the albumin fraction in liver cirrhosis is caused by the decreased synthesis in the liver. Medical treatments sometimes cause changes in albumin metabolism, with different changes appearing as the result of different treatment. For instance, a high-protein diet has therapeutic significance but it does not always result in clinical improvement. Blood and plasma transfusions cause both the total albumin and its turnover to increase, and they are especially effective for improvement of the cases with ascites. Also, glucocorticoids accelerate albumin metabolism and increase its synthesis and degradation to almost the same degree.

Qualitative abnormalities of serum albumin are thought to exist in cases of liver cirrhosis but up to the present no definitive evidence has been attained.[148]

3. The a_1 Fraction

As shown in Fig. 154, in cases of hepatitis, an inverse proportion is observed between the concentrations of the albumin and a_1 fraction, but in cases of cirrhosis of the liver, on the contrary, a direct proportion is found. This fact implies that there exists some difference in the reactivity of the liver cells between hepatitis and cirrhosis of the liver. That is, in cirrhosis of the liver, the synthetic reserve for a-globulins is thought to be relatively small. Therefore, it seems that the a_1 fraction decreases almost parallel to that of the albumin fraction. The serum concentration of the a_1 fraction tends to be decreased slightly in Group I. a_1-lipoprotein, in most cases, can not be demonstrated on immuno electrophoresis, showing marked decrease.

4. The a_2 Fraction

The a_2 fraction tends to decrease in many cases. As shown in Fig. 159, this tendency is most remarkable in Group I, followed by Groups II and III, respectively. This resembles the albumin and a_1 fractions. In most cases, the a_2 fraction remains within normal range, but decreases significantly in cases of severe liver damage. Generally, the decrease of this fraction appears at an earlier stage than that of the a_1 fraction. On immunoelectrophoresis, haptoglobin is found to be decreased almost always. a_2-macroglobulin remains to be normal until liver damage becomes extremely severe.

5. The β Fraction

In general, the β fraction shows a tendency to decrease. It appears earlier than the decrease of the a_2 fraction, but later than the decrease of the albumin fraction and the increase of the γ fraction. The decrease of the β fraction may be encountered more frequently by using the cellulose acetate method. No apparent difference is noted among Groups I, II and III, as far as the β fraction goes. Transferrin, the major component of the β fraction, generally decreases. Hemopexin also shows significant decrease, and often is not demonstrated on immunoelectrophoresis. No notable decrease of the β_{1C}-globulin is found except at the last stage of the disease.

Low-density lipoproteins do not show any remarkable change, except in cases of biliary cirrhosis. However, it is said that they increase slightly.[148]

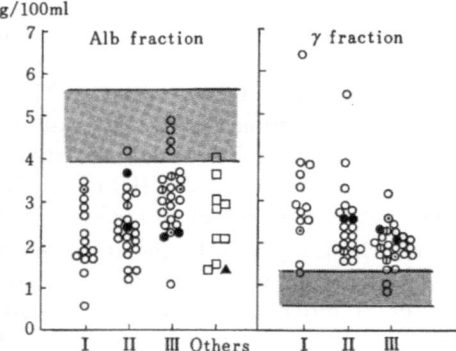

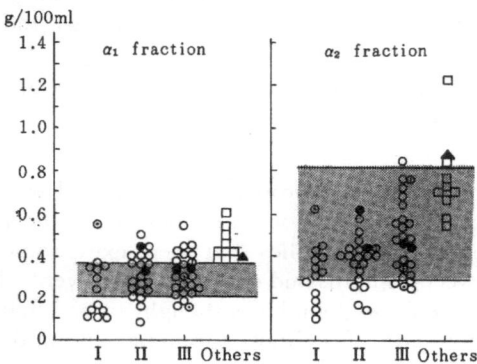

Fig. 159 Serum protein electrophoretic fractional values in liver cirrhosis.
The patients with liver cirrhosis were classified into 3 groups, depending on the serum protein abnormalities.
○ Liver cirrhosis
● Congestive cirrhosis
□ Non-hepatic disorder showing the pattern similar to Group II
⊙ Liver cirrhosis+hepatoma
⊘ Biliary cirrhosis
▲ Non-hepatic disorder showing the pattern similar to Group I
In the non-hepatic disorders showing β-γ linking, the α_1 and α_2 fractions are significantly increased, being fundamentally different from the cirrrhotic pattern.

6. The γ Fraction

The γ fraction shows a marked increase in most cases of cirrhosis of the liver, and the relative percentage of this fraction increases incessantly (Fig. 159).   The tendency of its increase is particularly remarkable in Groups I and II, and some cases may show the γ fraction to increase up to 6.4 g/100 ml.

In cases of liver cirrhosis, all major classes of immunoglobulins increase markedly while the absolute concentration is the most prominent with IgG-globulin.   However, with regard to the rate of their increase against the normal value, IgA-and IgM-globulins are more prominent than IgG-globulin, especially so with IgA-globulin.[148,163] IgDglobulin is also reported to increase significantly.[299]

It is said that the increase of the γ fraction correlates well with the histological findings represented by piecemeal necrosis, interstitial cell infiltration and proliferation of bile canaliculi.[280]   There are reports that the increase of IgG-globulin is quite parallel to fibrosis.[148]   Some investigators think that the increase of IgG-globulin has a much closer relation to the activity of the lesion rather than to the degree of liver damage.   Moreover, in the Western countries, the increase of IgA-globulin is reported to be especially marked in cases of alcoholic cirrhosis.[115, 382]   But no such tendency can be recognized in Japan.[148]

As to the metabolism of immunoglobulins, many researchers have submitted nearly the same data, indicating the increased turnover and synthetic rate.   In cases of liver cirrhosis, the increase of the synthetic rate is most prominent with IgA-globulin, followed by

IgM-globulin and IgG-globulin, respectively.[148] The reason for the synthesis of IgA-globulin and IgM-globulin being more prominent than that of IgG-globulin is unknown.

Increased synthesis of immunoglobulins in cirrhosis of the liver is thought to occur chiefly in the extrahepatic reticuloendothelial tissue, but it is reported to occur also in the liver.[280]

Qualitative abnormality of immunoglobulins has been studied in cirrhosis of the liver. Sometimes, abnormality of the IgG-precipitation line is observed on immunoelectrophoresis. There are cases in which the IgG-line extends more toward the cathode or the anode. In the latter case, since the abnormality almost disappears in the buffer with the ionic strength of 0.075 or 0.1, it is thought to be a change caused by the interaction of slow-migrating components of IgG-immunoglobulin and agar gel. At any rate, the clonal distribution of IgG-synthesizing cells seems to become wider in many cases of liver cirrhosis.

7. β–γ Linking

The most characteristic change in cases of cirrhosis of the liver is β–γ linking on their electrophoretic patterns. The author has classified this disease into the following three groups, according to the degree of β–γ linking observed (Fig. 149).

Group I (diagnostic pattern): β–γ linking is most pronounced in cases of this group, and the β fraction can not be separated from the γ fraction. This pattern is almost diagnostic of liver cirrhosis, and it may be encountered very rarely in cases of other diseases.

Group II (suggestive pattern): In cases of this group, β–γ linking is observed, but the separation of the β and γ fractions is still possible. This pattern, too, is not likely to be observed in cases of liver diseases other than liver cirrhosis.[163] However, DEMEULE-NAERE et al.[72] stated that they found β–γ linking in only five out of 1145 cases of liver diseases. That was in one case of subacute yellow atrophy of the liver, two cases of metastatic carcinoma of the liver which showed innumerable nodular metastatic foci, and two cases of cured subacute yellow atrophy of the liver. Incomplete β–γ linking of this kind is observed, but rarely, in other cases of hypergammaglobulinemia (Fig. 151, 159). So, the β–γ linking of this kind in serum protein patterns does not necessarily lead to the diagnosis of liver cirrhosis, although the pattern certainly suggests its possibility. When β–γ linking coexists with the decrease of the a_1 and a_2 fractions, it is an almost diagnostic finding of liver cirrhosis.

Group III (Non-specific pattern): In the cases of this group, the tendency of forming β–γ linking is not observed. The separation of the β and γ fractions is quite apparent, and cases of liver cirrhosis in this group cannot be distinguished from other hypergammaglobulinemia.

As mentioned above, the frequency of showing the patterns of each group seems to differ according to the type of electrophoretic cells and supporting media used. By filter paper electrophoresis, about 70 to 80% of the cases of primary cirrhosis and observed as belonging to Groups I and II, which show apparent β–γ linking (Table 40)[72, 163] By the cellulose acetate method, the frequency of the cases which show the β–γ linking is lower than that found by the filter paper method. It is because the separation on cellulose acetate strips is achieved comparatively well, and also because complete clearing of the strips is possible after staining. However, the valley between the β and γ fractions is often found to be shallower than normal.

The main cause of forming β–γ linking is the marked increase of IgA- and IgM-globulins, with IgA-globulin playing the more effective role. That is, IgA- and IgM-globulins migrate in the β_2 or γ_1 zone. In all cases in which IgA-globulin increases broadly about three times as much as its normal concentration, β–γ linking definitely will appear. As shown

Table 40 Frequency of different serum protein electrophoretic patterns in cirrhosis of the liver.

Methods of electrophoresis	Group I	Group II	Group III	Cases* examined
Filter paper, horizontal	20 50 %	9 22 %	11 28 %	40
Filter paper, vertical	13 26 %	18 37 %	18 37 %	49
Cellulose acetate Oxoid	3 14 %	4 19 %	14 67 %	21
Filter paper,** vertical	8 40 %	6 30 %	6 30 %	20

* Histologically confirmed through biposy or autopsy.
** Reported by Ichiba in Japan.

in Fig. 149, the author has confirmed by Rivanol precipitation that $\beta-\gamma$ linking is caused by the increase of IgA- and IgM-globulins. No apparent correlation was observed between $\beta-\gamma$ linking and histological findings.

8.   Fibrinogen

In many cases of liver cirrhosis, the decrease of fibrinogen is observed.   This is thought to be due to the decreased synthesis of fibrinogen in the liver.

Plasma Protein Changes in Special Forms of the Liver Cirrhosis

When liver cirrhosis is combined with hepatoma, it has been observed that the α_1 and α_2 fractions tend to increase, while the albumin fraction decreases markedly, and the γ fraction increases (Fig. 159).   However, when liver damage is severe, the increase of the α_1 and α_2 fractions may not be observed.   In cases of hepatoma without cirrhosis, the serum protein pattern is similar to that of other malignant tumors (Fig. 152).

There is no particular difference in serum protein pattern among the cases of liver cirrhosis other than biliary cirrhosis (Fig. 159).   There are some cases of congestive cirrhosis which have shown the $\beta-\gamma$ linking.   In the cirrhotic cases with hemochromatosis and *Schistosoma japonicum* infection, the serum protein patterns are similar to those of primary cirrhosis.

OBSTRUCTIVE LIVER DISEASES

Classification of the Obstructive Liver Diseases

Obstructive liver diseases are classified into two major groups as shown in Table 41. Obstructive jaundice is further divided into intrahepatic and extrahepatic types.   Genuine intrahepatic obstruction is rarely observed, and in many cases it is combined with other liver and biliary diseases.   Extrahepatic obstruction has various causes such as neoplasms, stones, inflammation, parasites, congenital obstruction and so on.

Hepatocellular injury becomes manifest eventually in obstructive jaundice.   It was formerly thought that liver damage did not occur unless there was long-standing obstruction.   However, as studied with modern liver function tests and liver biopsy, it has become known that biliary obstruction will be associated with liver damage from a comparatively early stage.   In particular, extrahepatic obstruction causes liver damage in the various stages, and proceeds to biliary cirrhosis in the final stage.[298]

Biliary cirrhosis occurs when there is long-standing biliary obstruction, as described previously.   When it is caused by extrahepatic obstruction, it is called the secondary one.

	(g/100ml)	(%)
TP	7.0	
Alb	3.97	56.7
α_1	0.22	3.1
α_2	0.22($\downarrow$)	3.1($\downarrow$)
β	0.94	13.4($\uparrow$)
γ	1.65($\uparrow$)	23.6($\uparrow$)

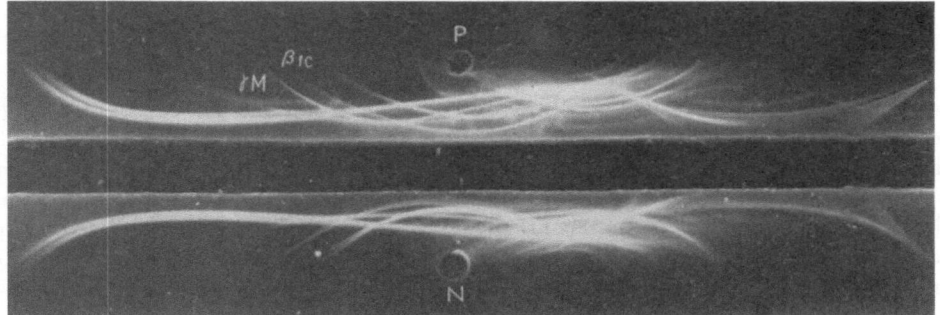

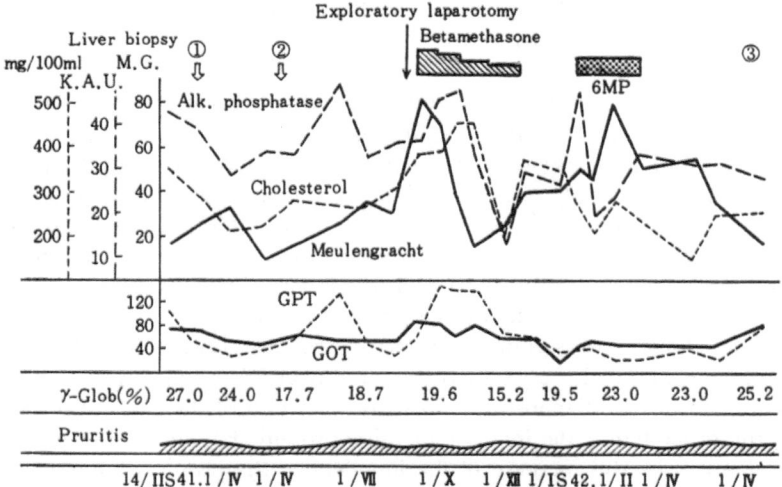

Fig. 160 Serum protein patterns in obstructive liver disease (1).

T. K., 51 y.o., female, primary biliary cirrhosis.

At the age of 18 years, jaudice for one week. At the age of 20 years, urticaria. Since several years ago, suffering from generalized pruritis, worse in the summer time. Treated as chronic hepatitis for the past 3 months. Hepatomegaly, but no splenomegaly nor ascites. Blood chemistry: icterus index 15, BSP 30% in 45 minutes, Al-P 44.5 K.A.u., GOT 67 u., GPT 103 u., LDH 430 u., total cholesterol 343 mg/100 ml. No cholelithiasis on cholangio-cystography, and slightly swollen, dark red liver on exploratory laparotomy. Biliary cirrhosis confirmed histopathologically. On filter-paper electrophoresis, a slight decrease in the α_2 fraction and a slight increase in the γ fraction. The serum concentration of the β fraction is within normal limits, but its relative percentage is slightly increased. On immunoelectrophoresis, a significant increase in IgM and β_{1c}-globulin. IgG 1300 mg/100 ml, IgA 310 mg/100 ml, and IgM 300 mg/100 ml.

Table 41   Classification of obstructive liver diseases.

I.   Obstructive jaundice
    A.   Intrahepatic biliary obstruction
    B.   Extrahepatic biliary obstruction
II.   Biliary cirrhosis
    A.   Primary biliary cirrhosis
    B.   Secondary biliary cirrhosis or cholangitic cirrhosis

When no apparent obstruction in the extra-hepatic ducts is notable, it is diagnosed as primary biliary cirrhosis.   However, the primary one is very rare (Fig. 160).

Plasma Protein Changes in the Obstructive Liver Diseases

In general, plasma protein abnormality in cases of obstructive liver diseases are of a great variety and it is difficult to distinguish any particular tendency.   Considering the pathophysiology of the disease, this is not unexpected, but it is basically of chronic inflammatory patterns.   The changes of liver injury type are added to these protein patterns if liver damage becomes severe.   That is, as liver damage advances, the decrease of the albumin and a_2-fractions becomes more marked and the increase of the γ fraction becomes quite marked.   In cases of biliary cirrhosis, these tendencies are more pronounced and sometimes, they are associated with β–γ linking (Group II).[103]

The β fraction tends to increase as β-lipoproteins increase.   On filter-paper electrophoresis, its concentration is not always high, but its relative percentage is definitely increased in many cases (Fig. 161, 162).

Serum Lipid and Lipoprotein Changes in the Obstructive Liver Diseases

In cases of obstructive liver diseases, the characteristic changes in serum lipids and lipoproteins are found.[79,97,200,315,319]   Total cholesterol and phospholipids in serum increase markedly (Table 42).   Moreover, as the increase of phospholipids is especially prominent, the phospholipids/cholesterol becomes very high.   The increase of serum total cholesterol is caused mainly by the increase of free cholesterol, and cholesterol ester stays normal or decreases.   Neutral fat and free fatty acids do not show any remarkable change and they stay within normal range in most cases.

Remarkable changes of serum lipoproteins are also observed.   High-density lipoproteins decrease markedly or sometimes disappear.[99]   On immunoelectrophoresis, a_1-lipoprotein can not be demonstrated in more than half of the cases of obstructive jaundice.   On the other hand, the increase of β-lipoproteins has been observed.[148]   By ultracentrifugal analysis, the increase of the low-density lipoproteins having the density of 1.019–1.063 has been found, but by COHN's ethanol fractionation, most of serum lipids exist in Fraction IV.   Since a_1-lipoprotein is normally included in COHN Fraction IV,

Table 42   Serum lipids in obstructive liver diseases (from EDER).

Disease	Cases	Phospholipids (mg/100 ml)	Total cholesterol (mg/100 ml)	PL/TC ratio
Primary biliary cirrhosis	7	1510	749	2.01
Obstructive jaundice	11	732	377	1.01
Secondary biliary cirrhosis	5	415	287	1.43

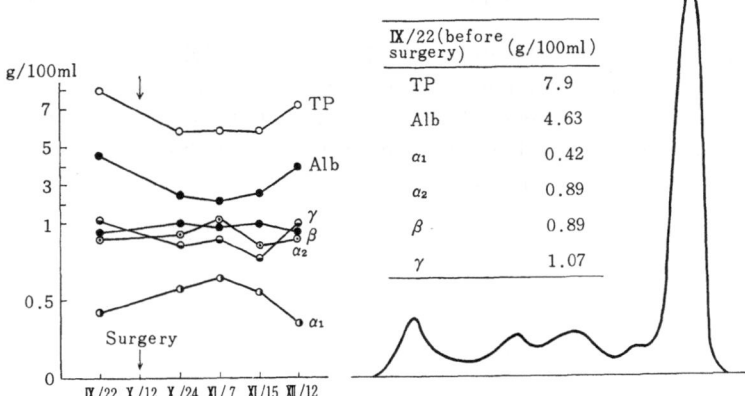

IX/22 (before surgery)	(g/100ml)
TP	7.9
Alb	4.63
α_1	0.42
α_2	0.89
β	0.89
γ	1.07

Fig. 161 Serum protein electrophoretic patterns in obstructive liver diseases.
F. S., 13 y.o., female, cyst of the common bile duct.
Complained of abdominal pain repeatedly since the age of 5 years. About 2 months
before the admission, had vomiting and abdominal pain. Developed icterus for the
past one month. Serum total cholesterol 250 mg/100 ml, icterus index 12, s-GOT
110 u., s-GPT, 204 u. Serum protein electrophoretic pattern was normal, as indicated above. The postoperative course was uneventful.

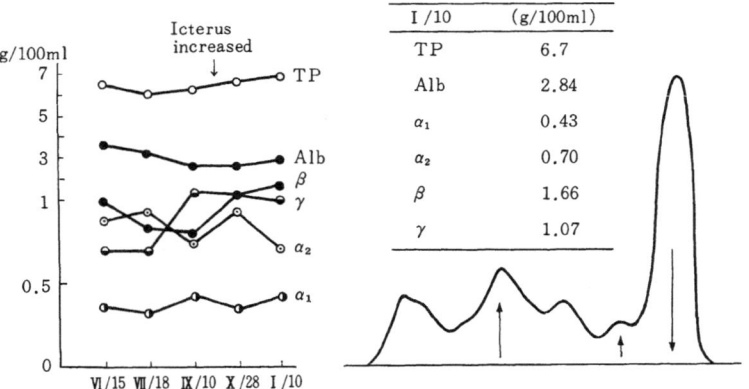

I /10	(g/100ml)
TP	6.7
Alb	2.84
α_1	0.43
α_2	0.70
β	1.66
γ	1.07

Fig. 162 Serum protein electrophoretic patterns in obstructive liver disease.
M. I., 7 y.o., female, cholangiolitic hepatitis.
In March, complained of cough, fever and erythematous rash, and also icterus soon
later. Even with various medical treatments, icterus became worse, and developed
hepatomegaly. On filter paper electrophoresis of the serum protein, only slight
increase in the β fraction and slight decrease in the albumin fraction were noted at the
beginning. However, later developed a significant increase in the β and γ fractions,
as shown above. On immunoelectrophoreisis, IgG was increased, but IgA and IgM
were not demonstrated. IgG 1100 mg/100 ml, IgA 50 mg/100 ml, and IgM 140 mg/
100 ml.

the low-density lipoproteins found in cases of obstructive liver diseases are different from the normal.

Moreover, it is said that the obstructive lipoprotein which is found in serum taken from the cases with obstructive jaundice contains much more phosphlipids and that it does not react against anti-β-lipoproteins.[315,373] The author has experienced in occasional cases with obstructive jaundice to have slow-migrating lipoprotein on cellulose acetate electro-phoresis. The reason for appearance of the obstructive or slow-migrating lipoprotein has not been clear yet.

Chapter 21

Plasma Protein Changes in Acute Phase Responses

CHARACTERISTICS OF THE SERUM PROTEIN ELECTROPHORETIC PATTERN OF THE ACUTE PHASE RESPONSE TYPE

The term acute phase response pattern refers to the serum protein pattern recognized in cases where there is an increase of some of the plasma protein components which are known as acute phase reactants.

In electrophoretic analysis of serum proteins, it is characterized by the decrease of the albumin fraction and the increase of the α_1 and α_2 fractions, as shown in Fig. 163 and Fig. 164.

The total serum protein is usually within normal range. As long as the γ fraction does not increase markedly, the total serum protein concentration does not change notably. However, when these cases are combined with protein loss or malnutrition, hypoproteinemia is observed not infrequently.

The albumin fraction always decreases. Although the albumin fraction decreases in serum, synthesis of albumin in the liver increases. The decrease of serum albumin is apparently due mainly to the increased degradation of albumin in the body.[173]

Of course, in cases of burn and lobar pneumonia, protein loss is combined with its hypercatabolism.

The increase of the α fractions is met with the increase of α-glycoproteins.

The β and γ fractions do not show any constant tendency, and vary in different stages of the disease.

The β fraction is mainly affected by changes of transferrin. Usually, transferrin decreases in proportion to the decrease of albumin, but its decrease occurs prominently less than that of albumin.[173] On cellulose acetate electrophoresis, the β fraction tends to decrease more frequently.

In subacute infections, malignancies, and active collagen diseases, a slight increase of the β fraction is seen apparently due to the increase of β-lipoproteins. Also, the increase of β1c/A-globulin as one of the acute phase reactants may contribute partly to the increase of the β fraction.

The increase of the γ fraction will be discussed collectively under the sections on polyclonal and monoclonal hyperimmunoglobulinemias. The acute phase response pattern is classified into the following two major groups according to whether or not an increase of the γ fraction is observed.

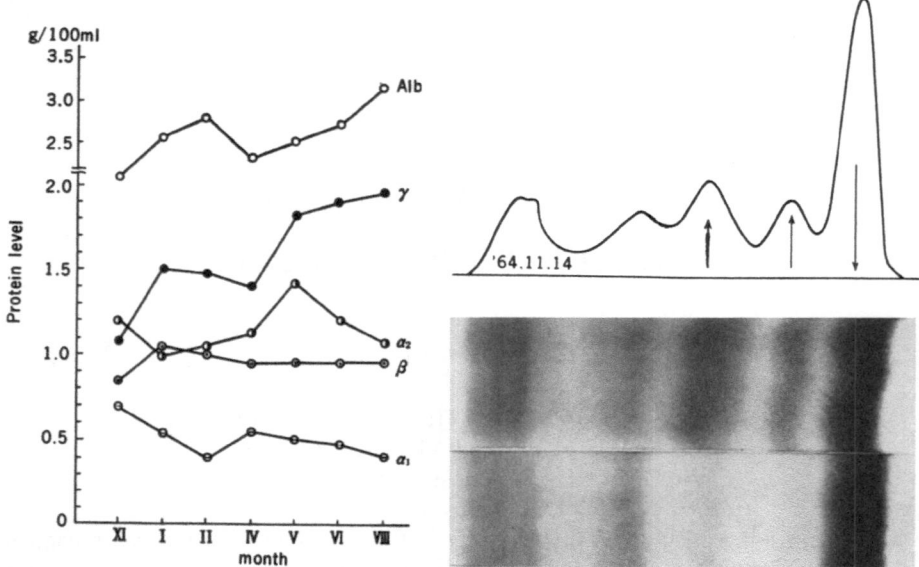

Fig. 163 The serum protein electrophoretic pattern of the acute inflammatory type.
T.I. 2 y.o., male, pneumococcal pneumonia and septicemia.
Upon admission, the serum total protein was slightly decreased, being 5.9g/100ml, and the filter paper electrophoretic pattern was of the acute inflammatory type, as shown at the right. Later the α_2 and γ fractions became markedly increased, showing the chronic inflammatory pattern. Although his clinical symptoms were improved and the α_2 fraction returned almost to a normal level, the γ fraction still remains elevated.

	(g/100ml)	(%)
TP	6.9	100
Alb	2.41	35
α_1	0.41	6
α_2	1.38	20
β	0.41	6
γ	2.29	33

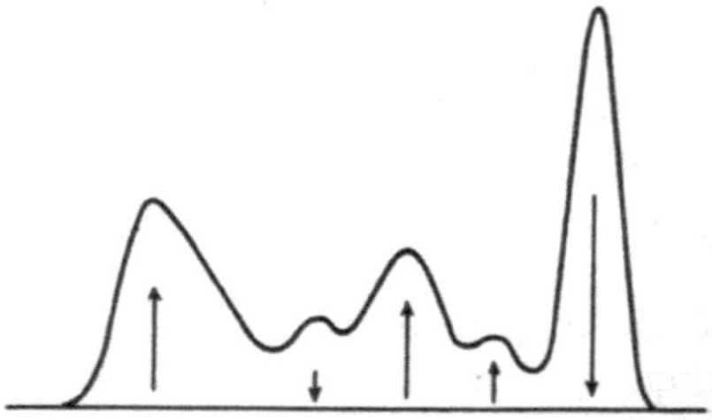

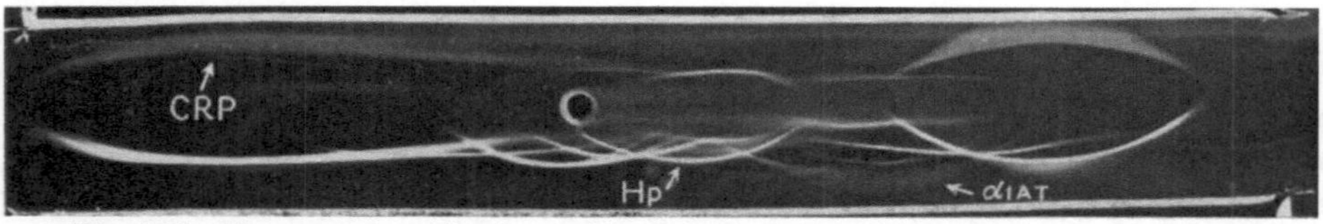

Fig. 164 The serum protein electrophoretic pattern of the chronic inflammatory type.
K.E., 63 y.o., male, carcinoma of the rectum.
The lesion was inoperable, since the tumor had spread widely in the peritoneal cavity.
Although the total protein is normal, the electrophoretic fractionation shows a characteristic chronic inflammatory pattern. On immunoelectrophoresis, noted are a marked decrease in albumin and transferrin and a significant increase in α_1-antitrypsin, haptoglobin, β_{1A}-globulin and immunoglobulins.

1) Acute phase response pattern without increased γ fraction. As representative examples, the acute inflammatory and acute stress patterns may be included in this category.

2) Acute phase response pattern with increased γ fraction. The representative example of this category is the chronic inflammatory pattern.

DIFFERENTIATION OF THE ACUTE PHASE RESPONSE PATTERN

It is not so difficult to distinguish the typical acute phase response pattern from other types of electrophoretic patterns. It shows a moderate or marked increase of both the α_1 and α_2 fractions. However, the patterns with a slight increase of the α_1 and α_2 fractions may have to be differentiated clinically from the following patterns.

Differentiation from the Malnutritional Pattern

In many cases, the acute phase response pattern with slight changes are difficult to be differentiated and clinical findings need to be carefully studied. On immunoelectrophoresis, the increase of α_2-glycoproteins is not usually observed in the malnutritional pattern.

Differentiation from the Non-Selective Prtoein-Losing Pattern

As in cases of the malnutritional pattern, it is impossible to be differentiated from atypical acute phase response patterns. But in the non-selective protein-losing pattern, a marked decrease in the total serum protein is often observed. At any rate, its differentiation must be based on other clinical information.

Differentiation from the Nephrotic Pattern

In the typical nephrotic pattern, hypoproteinemia is prominent and the α_2 fraction shows marked increase. Therefore, the typical nephrotic pattern can easily be differentiated from the acute phase response pattern, in which both the α_1 and α_2 fractions increase definitely. Moreover, on immunoelectrophoresis, α_2-macroglobulin does not increase significantly in the acute phase response pattern.

Differentiation from the Pregnancy Pattern

It is especially important to distinguish the pregnancy pattern on filter paper electrophoresis. However, the differentiation can be made with comparative ease by observing the marked increase of the β fraction, especially of β-lipoproteins.

HYPER-α-GLYCOPROTEINEMIAS AND THE ACUTE PHASE REACTANTS

Pathogenesis and Classification of Hyperalphaglycoproteinemia

Hyperalphaglobulinemia has long been known to be caused chiefly by the increase of various glycoproteins migrating at the α zones.

Up to the present, three possible mechanisms have been proposed as to origin of the increased serum glycoproteins. First, glycoproteins are released into the blood stream from injured, inflamed, or otherwise altered tissues[58]; second, glycoproteins arise as a response of the organism to tissue injury or proliferation[337,344]; third, glycoproteins in serum becomes elevated because certain tissues (for example, cancer tissue) utilize proteins low in bound carbohydrates.[402]   The third theory is difficult to sustain because the increase of α-glycoprotein components is observed at least in the acute phase response patterns mentioned in this chapter.   As yet, there is no definitive proof from any of these three theories.   Perhaps each glycoprotein component may be metabolized with different mechanisms.   Therefore, the author classifies hyperalphaglycoproteinemia, as shown in Table 43.

Table 43   Classification of Hyperalphaglobulinemia.

1. **Hyperalphaglobulinemia due to osmotic response**: the increase of low molecular weight α_1-glycoproteins, probably compensating the decrease of the plasma colloid osmotic pressure resulted from hypoalbuminemia.
    a. Malnurtitional serum protein pattern
    b. Non-selective protein-losing serum protein pattern
    c. Analbuminemia
    d. Nephrotic syndrome
    e. Foetus (?)

2. **Hyperalphaglobulinemia due to hormonal response**: mainly due to the hormonal action, such as ACTH, adrenocorticoids, adrenalin, etc.
    a. Anxiety
    b. Administration of various hormones
    c. Endocrinopathies (chiefly of the adrenal and pituitary glands)

3. **Hyperalphaglobulinemia due to inflammatory response**: primary or secondary inflammatory lesions, causing marked increase in acute phase reactants.
    A. Acute inflammatory serum protein pattern
        a. Acute infectious diseases
        b. Trauma (mechanical, physical, chemical, etc.)
        c. Myocardial infarction, thrombosis, cardiac failure, etc.
        d. Auto-toxicosis (uremia, shock, etc.)
        e. Delivery and pregnancy (?)
    B. Chronic inflammatory serum protein pattern
        a. Chronic infectious diseases
        b. Connective tissue diseases, autoimmune diseases, etc.
        c. Allergic diseases
        d. Malignancies

4. **Hyperalphaglobulinemia due to specific functional response**
    a. Nephrotic syndrome: marked increase of α_2-macroglobulin through distinct molecular-sieve effect of the renal glomeruli.
    b. Pregnancy: increase in a α_2-macroglobulin, thyroxine-binding globulin and transcortin due to female sex hormones.
    c. Infancy: increase in α_2-macroglobulin during growth (?).
    d. Foetus: increase in α-fetoprotein.

In hyperalphaglobulinemia due to osmotic and hormonal responses, the α fractions increase only slightly and C-reactive protein is usually not present.   Hyperalphaglobulinemia due to osmotic response may be pathophysiologically different from the acute phase response pattern, and it is discussed elsewhere.   Therefore, the serum protein pattern of acute phase response type is defined as the pattern which is characterized mainly by the inflammatory hyperalphaglobulinemia.

Acute Phase Reactants

The main cause of inflammatory hyperalphaglobulinemia is the increase of α-glycoproteins.   Besides, there are plasma protein components which increase non-specifically in plasma when active inflammatory lesions exist.   BOLLET referred to these components

Table 44   Representative plasma protein components
included as acute phase reactants.

I.   Components included in the α_1 fraction.
   1. α_1-Antitrypsin
   2. α_1-Acid glycoprotein
   3.   Other minor components

II.   Components included in the α_2 fraction.
   1.   Haptoglobin
   2. α_{2HS}-Glycoprotein
   3. α_2-Macroglobulin
   4.   Ceruloplasmin

III.   Other components
   1. $\beta_{1C/A}$-Globulin
   2.   Fibrinogen
   3.   C-reactive protein

generally as "acute phase reactants".[46]   In Table 44, there is a list of the plasma protein components which are included as acute phase reactants.

1. α_1 Glycoproteins

α_1-antitrypsin and α_1-acid glycoprotein are the most influential factor in the increase of the α_1 fraction.   These two increase side by side in almost all cases showing the acute phase response pattern.   However, as has been mentioned previously, in cases of pregnancy, α_1-acid glycoprotein tends to decrease.   Some of the minor components in the α_1 zone are also increased.

2. α_2 Glycoproteins

Haptoglobin and α_{2HS}-glycoprotein are the major components to increase in the α_2 fraction.   In newborn babies, the synthetic ability of haptoglobin is insufficiently developed. Thus, even when severe inflammatory lesions are present, the increase of the α_2 fraction may not be clearly demonstrated.   However, even in newborn babies, the increase of α_1-glycoproteins is observed and the increase of α_{2HS}-glycoprotein is also remarkable. In many cases, α_{2HS}-glycoprotein is found in urine in comparatively large amounts. Besides, although not so notable as that of haptoglobin, there is a slight increase of both ceruloplasmin and α_2-macroglobulin.   However, in acute leukemia, the increase of ceruloplasmin is quite remarkable (Fig. 165).[198,205]   This may be because ceruloplasmin carries copper which is contained in large amounts in leukocytes and may be necessary for proliferation of neoplastic leukocytes.

3. $\beta_{1C/A}$-Globulin

$\beta_{1C/A}$-globulin often increases in the acute phase response pattern.   In cellulose acetate or agar gel electrophoreses, fresh serum sometimes shows a minor protein band of β_{1C}-globulin at the β_2 zone.   But when stored serum is analyzed, it is converted to β_{1A}-globulin and is buried in the β_1 fraction.   In cases of systemic lupus erythematosus and auto-immune diseases, β_{1C}-globulin frequently decreases markedly in spite of the increase of other acute phase reactants.

4.   Fibrinogen

Fibrinogen usually increases significantly, being parallel to α-glycoproteins.

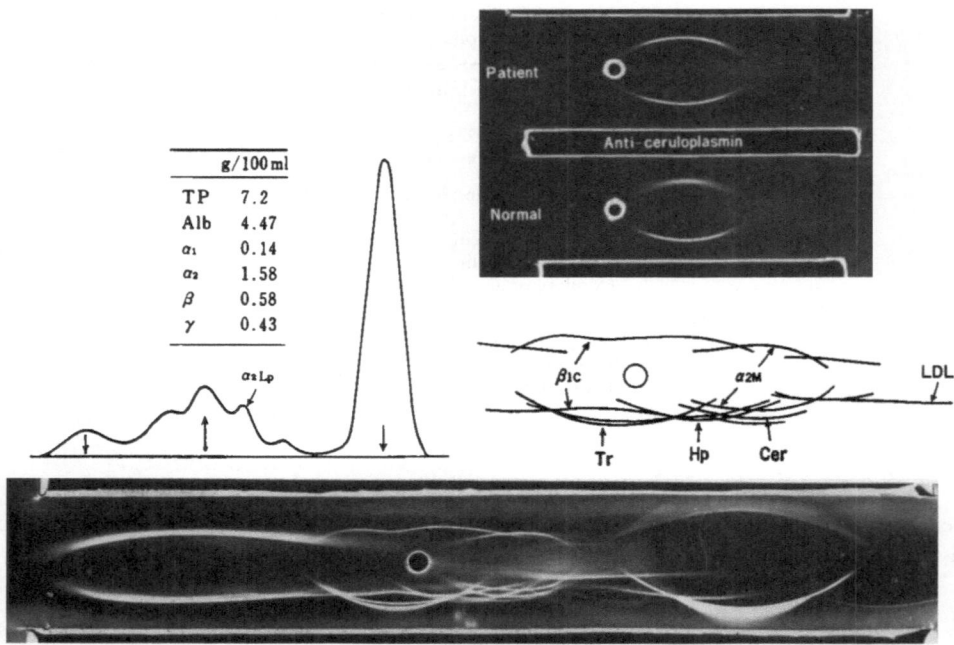

	g/100 ml
TP	7.2
Alb	4.47
α_1	0.14
α_2	1.58
β	0.58
γ	0.43

Fig. 165   The serum protein electrophoretic pattern of the acute phase response type.
H. K., 4 y.o., female, acute lymphatic leukemia.
Diagnosed as acute leukemia 5 months ago, and readmitted with bleeding tendency.   WBC 6,200
with 28% lymphoblasts.   Bone marrow containing 88% lymphoblasts.   Serum LDH 540 u.; sed
rate 59 mm/hr.   On cellulose acetate electrophoresis, noted are a marked increase in the α_2 fraction
and a moderate decrease in the γ fraction.   On immunoelectrophoresis, noted is a significant increase
in LDL, ceruloplasmin and haptoglobin.   The precipitation line of ceruloplasmin is usually difficult to
be identified clearly on a routine immunoelectrophoretic pattern, but its increase is clearly demonstrated by using the specific antiserum against ceruloplasmin (upper right).

C-Reactive Protein

C-reactive protein (CRP) is usually not observed in normal serum.   It is observed
very frequently in the acute phase response pattern.   In 1930, TILET and FRANCIS mentioned its existence for the first time in the serum of a patient with pneumococcal pneumonia.[379]   Since it shows a specific precipitin reaction with pneumococcal C-polysaccharides, it has been named the C-reactive protein.   It has very strong antigenicity and
when injected into animals, it produces anti-CRP.   By the use of this anti-CRP, C-reactive protein in serum is now measured immunologically.[230]

1.   Physico-Chemical Characteristics of C-Reactive Protein

CRP was crystallized by WOOD et al.[412]   Its physicochemical characteristics are still
not well known.   Its molecular weight is said to be about 129,000, and its sedimentation
constant is 7.5S.   It is said to consist of six subunits.   Each subunit is formed by one
polypeptide chain containing one disulfide bond, and has a molecular weight of about
21,500.[126]

C-reactive protein is destroyed by heating at 65°C for over 30 minutes, being relatively
heat-labile.

Electrophoretically, it is classified into three types as shown in Fig. 166,[13]   being the

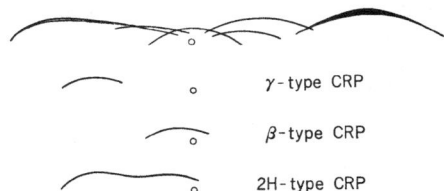

γ- type CRP

β-type CRP

2H-type CRP

Fig. 166 Agar gel immunoelectrophoretic patterns of the C-reactive protein (from Anzai et al.).
The uppermost pattern shows 5 major serum protein components for comparison. The three different patterns of the CRP are shown, γ-type CRP, β-type CRP and 2H-(2 hump) CRP.

γ-type, the β-type and the 2H-type. Later, the γ-type CRP has been found to have the same mobility as that of crystallized CRP (g-CRP). The β-type CRP seems to correspond to m-CRP which is combined with mucopolysaccharides in serum. The CRP proper migrates to the γ zone and is called g-CRP or γ-type CRP. Since CRP has a strong affinity to mucopolysaccharides, part or most of g-CRP is thought to produce m-CRP, the complex of g-CRP and muco-polysaccharides, being migrated as the 2H-type or β-type.[156]

2. Biological Characteristics of C-Reactive Proten

There are no definitive data with regard to the synthetic site of CRP. It is thought that CRP is synthesized mainly in the liver but the possibility that it is synthesized in other organs or in reticuloendothelial tissues cannot be entirely neglected.

CRP is not observed in normal serum, usually. It is found non-specifically in cases showing the serum protein pattern of acute response type. It increases to about 2 % of the total serum protein. CRP is not transmitted from maternal blood to the fetal circulation through the placenta, but is found in comparatively large amounts in body fluids such as peritoneal, pleural, pericardial and synovial fluids.

The biological significance of CRP is still unknown. However, at least, CRP certainly does have a close affinity to exogenous polysaccharides such as pneumococcal C-polysaccharides, and to endogenous polysacchardies in the patient's serum. Considering the fact that mucopolysacchardies are very difficult to be dissolved in water, CRP may be considered to be a carrier of polysacchardides. That is, inflammatory tissue damage releases water-insoluble mucopolysaccharides into blood circulation and CRP may be the one which carry them out of the body.

3. Clinical Significance of C-reactive Protein

CRP is found non-specifically in the serum of patients who have any inflammatory lesion or tissue necrosis in the body. Generally speaking, CRP appears in plasma fourteen to twenty-four hours after inflammation begins, and it disappears during the convalescent stage.

Its appearance is rarely helpful for the diagnosis of particular diseases, but it generally is quite helpful in evaluating the activity, severity, and course of many diseases.

CRP is frequently found in serum of the patients with bacterial infections, rheumatic fever, active rheumatoid arthritis, myocardial infarction, malignancies with widespread metastasis, and postoperative state.

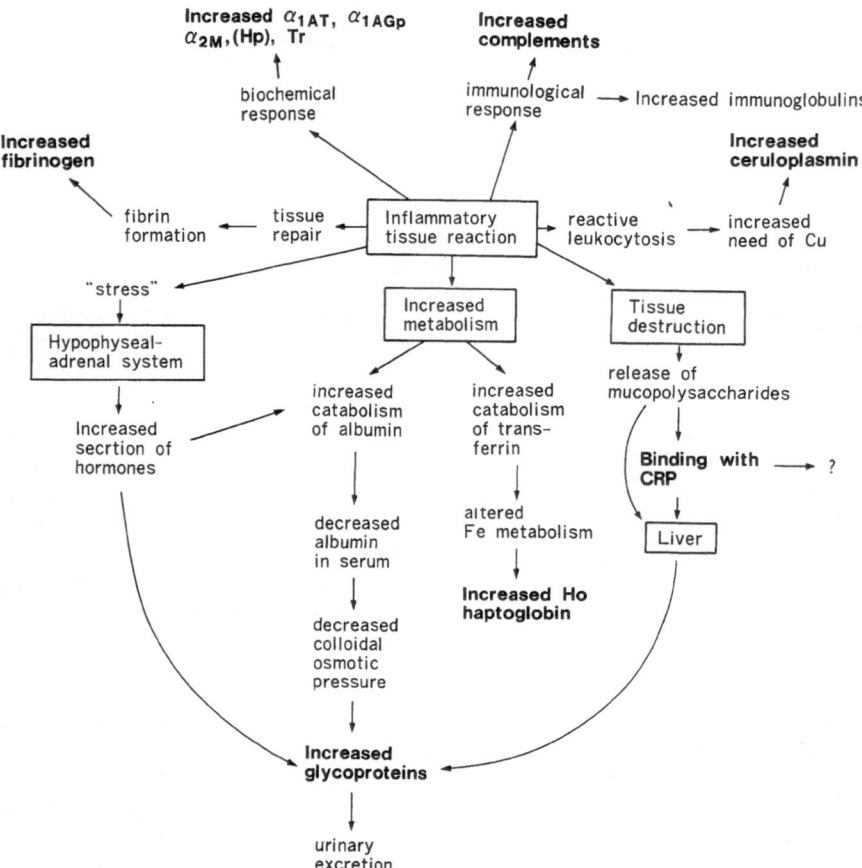

Fig. 167 Hypothetical diagram on the mechanism of increasing acute phase reactants in serum.

Hypothesis on the Mechanism of Increased Acute Phase Reactants

The increase of various acute phase reactants is noted in the acute phase response pattern. However, it is still not known how each reactant increases. In recent years, the biological characteristics of each plasma protein component has become increasingly clear, and the author proposes the following hypothesis on the mechanism of the increased acute phase reactants (Fig. 167).

When, from a variety of causes, inflammatory tissue reaction occurs, "stress" causes alteration in the hypophysio-adrenal system, resulting in the decrease of the albumin fraction and the increase in the α fractions. Part of the tissue proteins released by tissue necrosis is thought to be carried into the blood. In the same way, acid mucopolysacchardies which are the main constituents of interstitial tissues are thought to be released. However, since mucopolysaccharides are water-insoluble, it is thought that they combine with CRP and are carried through the blood circulation. As a result, CRP could appear in large amounts in inflammatory conditions. m-CRP combined with mucopolysaccharides may be carried to the reticuloendothelial system to be destroyed, or they may be transported to the liver and separated into g-CRP and mucopolysaccharides. Mucopolysacchardies thus released could be degraded and used further for the synthesis of new glycoprotein molecules (especially those with low molecular weight, such as α_1-acid glycoproteins, α_1-antitrypsin, α2HS-glycoprotein, and so on). HOKOMA et al.[156] have reported that the mucopolysacchardies which combine with CRP are those which contain sialic acid. In acute inflammatory diseases, sialic acid increases in the blood. Furthermore, interestingly α_1-acid glycoprotein was reported to contain little sialic acid in the cases with chronic inflammatory lesion.[327] On the other hand, inflammatory stimulation accelerates protein catabolism in tissues.[173] Therefore, though the synthetic rate of serum proteins in the liver decreases, the serum concentration of albumin decreases and then the colloid osmotic pressure is lowered. To compensate for this, low-molecular weight glycoproteins seem to increase. At any rate, these low-molecular weight glycoproteins are excreted in the urine in comparatively large amounts (prerenal proteinuria). Ultimately, they are thought to be helpful in removing mucopolysaccharides from degenerated tissues. Many leukocytes in inflammatory lesion are rapidly destroyed, causing reactive leukocytic proliferation in bone marrow. To supply enough copper for leukocytic proliferation, the increase of ceruloplasmin may become necessary. Besides, it is not unrealistic to think that, as an immunological defense mechanism against inflammations, the complements and immunoglobulins may be necessary to increase. Moreover, abundant fibrin or fibrinogen deposits are found in inflammatory lesions being confirmed by the fluorescent antibody technique. This phenomenon is prominent especially in serious fibrinoid degeneration. It has been well known for a long time that these changes play an important role in the regeneration and organization of destroyed tissue. Fibrinogen is thought to increase in order to compensate for fibrinogen consumption in local lesions. As the fractional catabolic rate of fibrinogen is constant,[332] fibrinogen concentration in plasma must in creasein order to maintain the increased consumption of fibrinogen. It is known that haptoglobin, α_1-antitrypsin, α_1-acid glycoprotein, α_2-macroglobulin, and ceruloplasmin have specific biological activities and these biological activities are naturally supposed to have some significant roles in the body's defense mechanism.

(APPENDIX) EEFFCTS OF HORMONES AND VITAMINES ON SERUM PROTEINS

Among various hormones studied, there are three which show some effects on serum proteins: ACTH, adrenocorticoids and adrenalin.

ACTH seems to increase the serum concentration of α-globulins. A single injection of a physiological dosis of ACTH does not alter the serum protein pattern in man, but the seromucoids always rise after two injections in association with the decrease of serum albumin.[201,319] The similar results are also obtained experimentally, and the same pattern is frequently encountered clinically in CUSHING's syndrome.

The results on the effects of adrenocorticoids are contradictory. In normal individuals, the administration of adrenocorticoids may not alter the serum level of α-globulins significantly and causes slight decrease of serum albumin.[319] However, after adrenalectomy and hypophysectomy, seromucoids are found to be decreased.[186] In connective tissue diseases, autoimmune diseases and nephrotic syndrome, the administration of adrenocorticoids usually causes the increase of serum albumin and the decrease of the α fractions, in association with notable clinical improvement.

Adrenalin injections seem to decrease serum albumin and to increase α-globulins in both man and experimental animals.[78,319] Also found is hypertriglyceridemia. The administration of adrenalin results in the serum protein patterns similar to those found in cachexia, and these changes are possibly induced by activation of cellular metabolism.[304]

Other hormones, like growth hormone and insulin, are thought to cause any notable change in the serum proteins.   As to the thyroid hormone, both clinical and experimental findings are contradictory.   Clinically, α-globulins and seromucoids increase in hyperthyroidism, and tend to decrease in hypothyrodidism.[263,304,285)   Some reports state that seromucoids increase in myxedema.[233)

The parathyroid hormone, vitamin C and hyaluronidase are said to participate on the metabolism of connective tissues, usually resulting in the increase of seromucoids.[319)

Chapter 22

Plasma Protein Changes in Polyclonal Hyperimmunoglobulinemia

CHARACTERISTICS OF THE BROAD HYPERGAMMAGLOBULINEMIC SERUM PROTEIN ELECTROPHORETIC PATTERN

Hyperimmunoglobulinemia is observed in various diseases. It is classified as in Table 45 mainly from the electrophoretic characteristics.

A-(hypo-) gammaglobulinemia which shows a remarkable decrease of the γ fraction is discussed elsewhere in the section on the immunologic deficiency syndrome. Other abnormalities of the γ fraction may be divided into those with M-protein and those without it. A detailed description of the former appears in the section on the M-proteinemic pattern.

Broad hypergammaglobulinemic pattern is characterized by a slight decrease of the albumin fraction and the broad increase of the γ fraction.

Broad Increase of the γ Fraction

Even on electrophoretic fractionation of normal serum, the γ fraction is observed as a broad protein band (Fig. 34), and the IgG-line extends toward the α_2 region. It is thought that each immunoglobulin with a different mobility is synthesized by the corresponding clone of immunoglobulin-producing cells. Moreover, since clonal composition may be almost the same among different individuals, the shape of the γ fraction peak keeps to be alike in all normal sera (Fig. 168).

In the broad hypergammaglobulinemic pattern, each clone may proliferate equally, and, as shown in Fig. 169, the characteristic of this pattern is the broad increase of the γ fraction. This is called polyclonal hypergammaglobulinemia or hyperimmunoglobulinemia.[399] The broad increase of the γ fraction is clearly demonstrated when a sufficiently long electrophoretic separation is performed on agar gel medium (Fig. 170).

On immunoelectrophoresis, the immunoglobulin precipitin lines become prominent but are quite similar in their shape to those of normal serum. Usually, all three major

Table 45 Electrophoretic classification of immunoglobulin abnormalities.

I. An-(hypo-)immunoglobulinemia or agammaglobulinemic pattern

II. Polyclonal gammopathy (or immunoglobulinopathy)
    A. Polyclonal hyperimmunoglobulinemia or broad hypergammaglobulinemic pattern
    B. Abnormal distribution of clonal population without an increased γ fraction

III. Monoclonal gammopathy (or immunoglobulinopathy) or M-proteinemic pattern
    A. Monoclonal hyperimmunoglobulinemia
    B. Monoclonal (or discontinuous) immunoglobulinopathy without increased immunoglobulins

---

classes of IgG, IgA and IgM increase euqally, but sometimes, only one or two classes of immunoglobulins is increased. Furthermore, what differentiates polyclonal immunoglobulinopathy from monoclonal immunoglobulinopathy is the increase of both L-type and K-type immunoglobulins (Fig. 169).

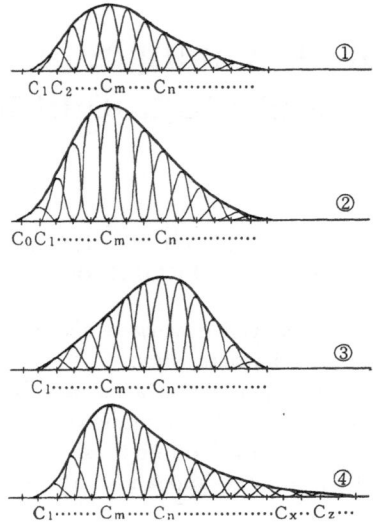

Fig. 168 Clonal distribution in various types of polyclonal hyperimmunoglobulinemia.

Each clone of the IgG-producing cells is expressed by C_0, C_1, ... C_m .... C_n. The IgG molecules produced in each clone may be thought to vary in their electrophoretic mobility. Then, the height of the peak representing each clone may be comparable to the population of the plasma cells belonging to each clone.
① represents a normal clonal distribution, ② hyperimmunoglobulinemia showing almost normal clonal distribution, ③ a transition of the mode from C_m to C_n, (shown in Fig. 169), and ④ broadening of the clonal distribution (shown in Fig. 171).

Changes in the Modal Mobility of the γ Fraction

Innumerable clones of the immunoglobulin-forming cells participate in polyclonal hyperimmunoglobulinemia, but there are sometimes the cases in which the shape of the γ fraction is distorted, probably representing abnormal distribution of clonal population in the immunoglobulin-synthesizing tissues. As shown schematically in Fig. 168, C_m is supposed to be present in the greatest frequency and the modal mobility of the γ fraction corresponds to the position of C_m in normal and many hypergammaglobulinemic sera. Suppose that the population of the plasma cells belonging to the clone C_n is the highest, the modal mobility of the γ fraction will move to the position of C_n.

The modal mobility of the γ fraction is calculated on cellulose acetate strips, as described in detail on page 398. The similar studies were reported, using the immunoelectrophoretic technique.

Clinical significance of abnormal modal mobility of the γ fraction is unknown. In our studies analyzing many clinical cases, no apparent relationship could be established between abnormal modal mobility and certain diseases.

In addition to abnormalities of the modal mobility of the γ fraction, the IgG-precipitin line occasionally extends towards the anode (Fig. 171). This type of immunoelectrophoretic finding is infrequently recognized when electrophoretic separation is performed on agar gel in the buffer with low ionic strength. Therefore, further electrophoretic separation should be done on agarose gel or in the buffer with higher ionic strength. In the particular case shown in Fig. 171 the IgG line does not change even when electrophoresis is done in the buffer with higher ionic strength, and thus, abnormal distribution of clonal population may be one of the possibilities.

Changes in the Serum Protein Fractions Other than the γ Fraction

If the γ fraction shows a marked increase, the total serum protein concentration becomes increased, at times reaching up to 10 g/100 ml.

The albumin fraction usually shows a tendency to decrease, and depending on the condition of the patient, it very often shows a marked decrease. Both the α_1 and α_2 fractions tend to increase, as a result of acute phase response, and this tendency is particularly evident in active stages of connective tissue diseases or malignant tumors. However, in cases of chronic infections, the increase of the α_1 and α_2 fractions are not as evident as in cases of acute infections. The β fraction shows very little change, but tends to decrease on cellulose acetate electrophoretic patterns.

REPRESENTATIVE DISEASES ASSOCIATED WITH POLYCLONAL HYPERIMMUNOGLOBULINEMIA

As can be seen on Table 46, there are many disorders which accompany polyclonal hyperimmunoglobulinemia. However, it is still completely unknown just how polyclonal hyperimmunoglobulinemia arises in these various disorders. In Fig. 172, the author himself has attempted to put into a model form his own thoughts on the subject.

In reference to immunoglobulins and IgG-globulin in particular, there are proportional relations between their serum concentration and the fractional catabolic rates. In other words, if there is increased serum concentration of IgG-globulin, there is likewise an increase in the fractional catabolic rate, and its daily turnover increases thereby. Consequently, unless its synthetic rate is more than its turnover rate, its serum concentration will not rise. Whatever the mechanism may be, the increase of IgG synthesis has been confirmed through studies with the use of isotopes.

Table 46 Various diseases accompanying polyclonal
hyperimmunoglobulinemia.

1. Acute and chronic liver diseases
2. Chronic infectious diseases
3. Protracted hypersensitization • Adjuvant disease • Sarcoidosis
4. Malignancies
   a. Carcinomas or non-reticular malignancies
   b. Reticular malignancies
      (Addendum) Lymphoreticulosis hyperglobulinaemica
                  Splenomegalia hyperglobulinaemica
                  Lymphadenosis hyperglobulinaemica
5. Autoimmune diseases
   a. Organ-specific autoimmune diseases
   b. Connective tissue diseases ("Collagen Diseases")
6. Essential hyperimmunoglobulinemia
      Purpura hyperglobulinaemica (Waldenström)
      Calcinosis hyperglobulinaemica (Miyoshi)
      Neuroencephalopathia hyperglobulinaemica

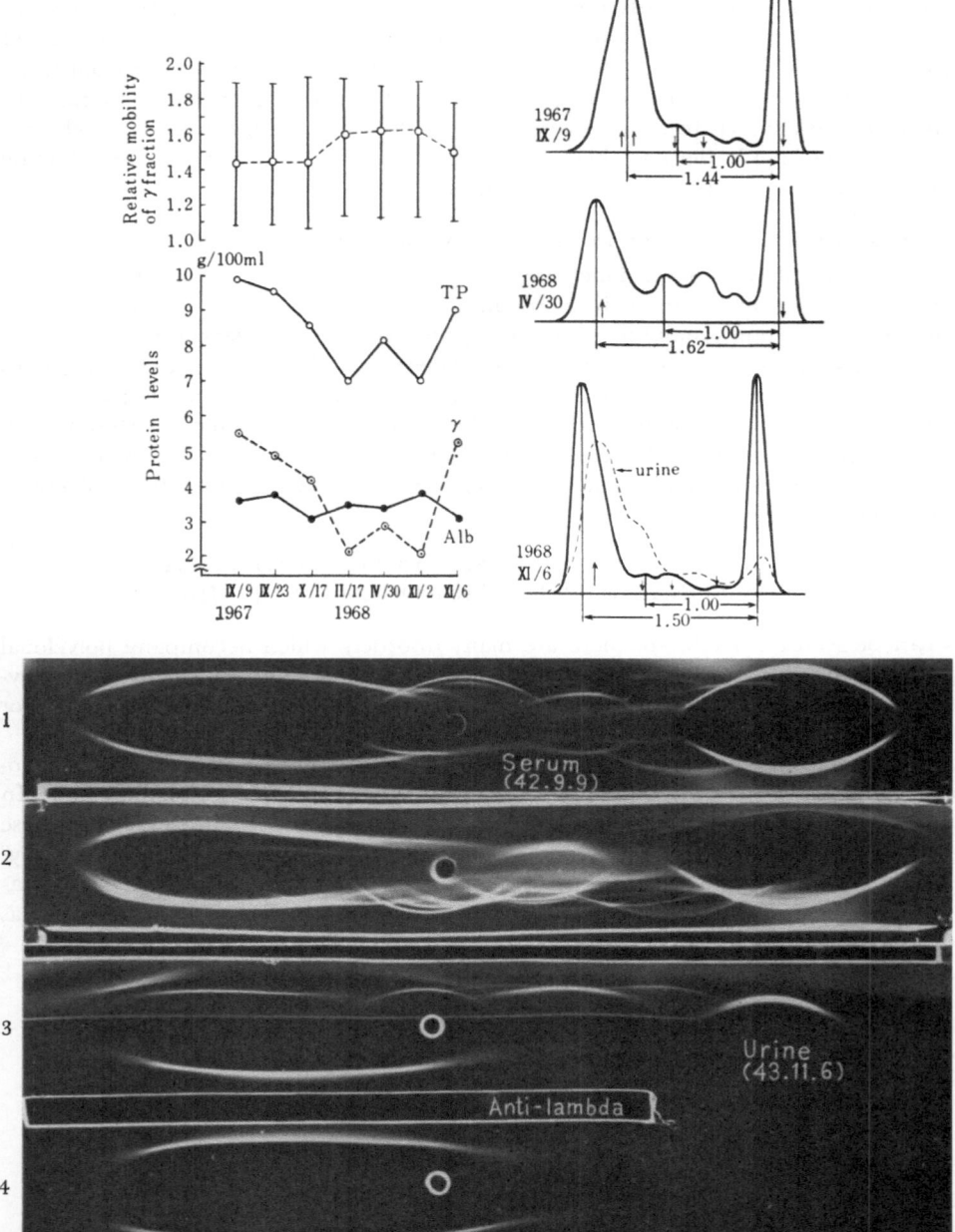

Fig. 169   (See also the next page)

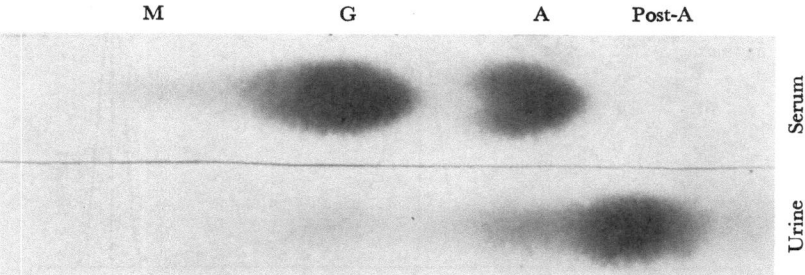

Fig. 169   Serum and urine protein patterns in polyclonal hyperimmunoglobulinemia.
K. S., 20 y.o., male, highly suspicious for SLE.
Developed cervical lymphadenopathy and occasional high fever since April, 1967, and soon developed hepatosplenomegaly.   Lymph node biopsy showed reactive hyperplasia.   Improved clinically with corticosteroid therapy, but became worse again since November, 1968, having proteinuria.   Moderate anemia (Hb 8–10 g/100 ml), no leukocytosis, bone marrow showing no plasmocytosis.   Sed rate 140 mm/hr.   STS, strongly positive (to be BFP).   RA test postivive.   LE cell phenomenon positive, and ANF positive.
On cellulose acetate electrophoresis, the serum protein pattern was that of the broad hypergammaglobulinemic type, the modal mobility of the γ fraction being small, 1.44.   The immunoelectrophoretic pattern of the serum proteins (1) showed markedly increased IgG, but no significant increase in IgA and IgM.
During the clinical remission (Apr. 30, 1968), the γ fraction was lower but its modal mobility became normal, being 1.62.   The immunoelectrophoretic pattern of the serum at that time (2) showed an almost normal precipitation pattern of IgG.
The γ fraction became increased tremendously again on November 6th, 1968, especially a marked increase in IgG. β_{1C}-globulin was decreased markedly, when the γ fraction became increased.
Interesting to mention on urine protein patterns.   On cellulose acetate electrophoresis of the urine proteins, a broad peak was noted at the β and γ zones.   On immunoelectrophoresis (3, 4), the immunoglobulin fragments were noted, reacting against both the anti-kappa and anti-lambda.   On thin-layer gel filtration, the serum protein pattern showed a notable increase in the G fraction, but the urine protein pattern showed a main fraction to be at the position equivalent to around 3.5 S.   Detail immunological analysis of the urine proteins has been in progress, but we feel that the condition is to be called as "immunoglobulin fragmenturia."

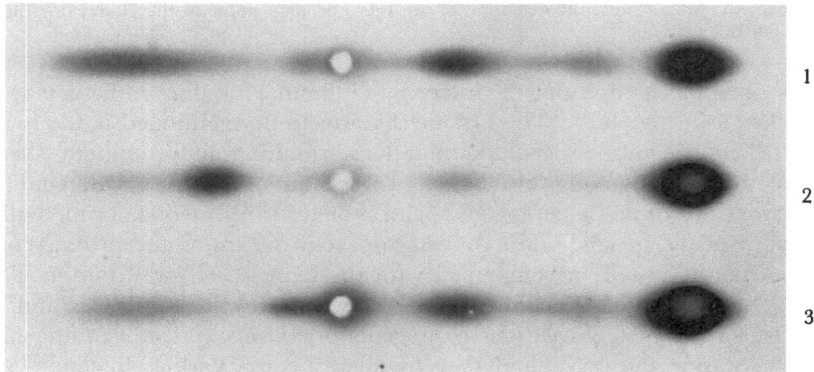

Fig. 170   Agar gel electrophoretic patterns of the serum proteins
in monoclonal and polyclonal hypergammaglobulinemia.
1: polyclonally increased γ fraction.
2: monoclonally increased γ fraction.
3: normal γ fraction.
On cellulose acetate electrophoretic patterns, the differentiation between the monoclonal and polyclonal increases of the γ fraction may be difficult in some instances.   On agar gel electrophoresis where the γ fraction is distributed widely, their differentiation is usually easy, as shown above.

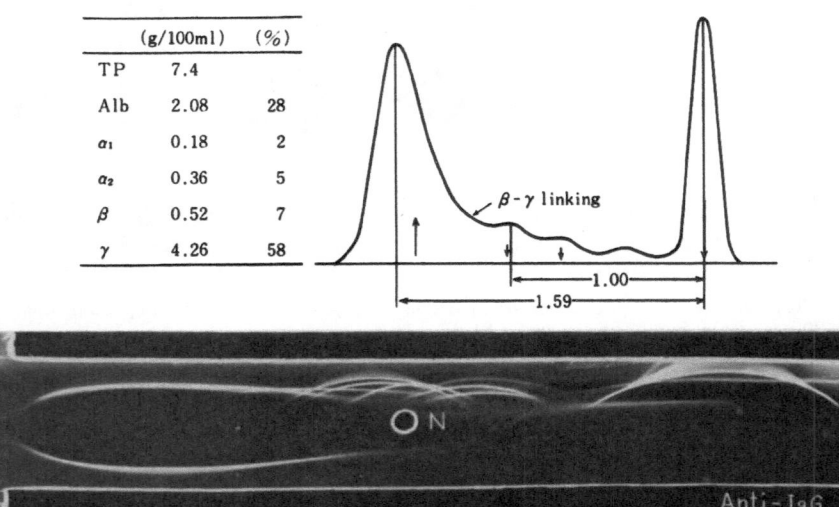

	(g/100ml)	(%)
TP	7.4	
Alb	2.08	28
α_1	0.18	2
α_2	0.36	5
β	0.52	7
γ	4.26	58

Fig. 171 Serum protein electrophoretic patterns in polyclonal hyperimmunoglobulinemia. K. H., 59 y.o., female, suspicious for connective tissue disease.

Developed edema in the face and lower extremities for the past 2 months and suffered from fever intermittently.   No icterus.   Urinalysis, normal.   Slight anemia (Hb 10.1 g/100 ml), sed rate 125 mm/hr, TTT 12.2 units, ZTT 40 units, LE cells positive.

On cellulose acetate electrophoresis, the serum protein pattern showed $\beta-\gamma$ linking.   On immunoelectrophoresis, IgG, IgA and IgM increased markedly.   The IgG line extends towards the anode significantly longer than the normal, and this pattern did not change even with the buffer having a high ionic strength.   Thus, the extension of the IgG line may indicate the clonal distribution being wider than the normal.

The possible mechanisms for the increase of immunoglobulin synthesis may be summarized as shown in Fig. 172.   The first mechanism to be mentioned is the increase of exogenous or endogenous antigenic stimulations, resulting in prominent acceleration of producing the corresponding antibodies.   For example, when an antigenic foreign substance such as one of the heterologous serum proteins is injected experimentally, a significant amount of the specific antibody will appear in serum as the primary or secondary immune response, usually accompanied with the increase of total immunoglobulin level in serum.[139]   Antigenic stimulations may be divided into the exogenous and endogenous ones.   The former includes various pathogenic organisms and other antigenic materials originating outside the body, and they frequently cause various infections and allergic diseases.   The latter includes various antigenic materials originating degenerated or degraded components of its own body, and they may induce some of autoimmune diseases. It is easily understandable that specific antibodies are produced through those antigenic stimulations, but the mechanism of accompanying the increased serum level of nonspecific immunoglobulin components is entirely unknown.   However, the following two possibilities are suggested: Firstly, many clones of the immunoglobulin-producing cells are activated more or less simultaneously upon antigenic stimulation of a certain clone; Secondly, since a certain clone proliferates to produce large amounts of the corresponding antibody molecule, the synthesis of other immunoglobulins may be sacrificed temprarily and then it may be proliferated later through a feed-back mechanism.

The second possible mechanism to produce abnormally increasing immunoglobulin synthesis is that immunoglobulin-producing tissues become hypersensitive against antigenic stimulations.   Of course, it is still unknown whether antibody-producing cells or immunologically competent cells are involved for the abnormality.   At any rate, in this type of the abnormality, prominent antibody formation will be in action even when a certain antigenic stimulation of normally uneffective degree is introduced.   This mechanism may be involved in some of so-called autoimmune diseases, which is sometimes called as "autosensitization".[394]   Also in some of allergic diseases, the similar mechanism is thought to be in action, thus frequently called as "hypersensitivity".

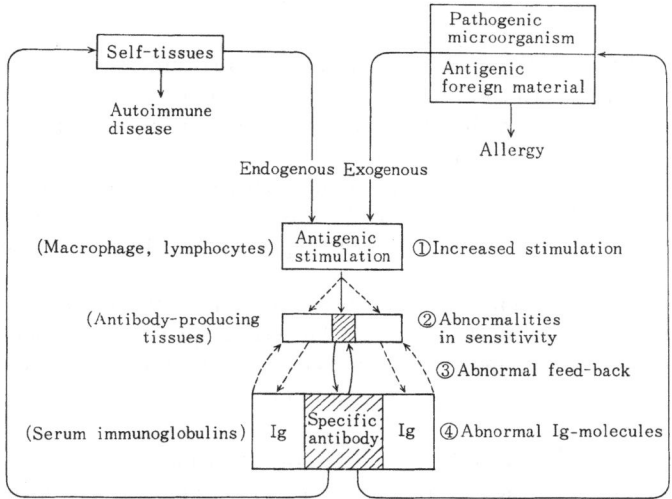

Fig. 172 Hypothetical diagram on the pathogenesis of polyclonal hyperimmunoglobulinemia.

In normal states, the serum concentration of immunoglobulins is kept in a certain constant level, and it is thought that some immunological "feedback" is present to control the synthesis of immunoglobulins.   Therefore, abnormality of this immunological feedback could result in the increase of serum immunoglobulin levels.   The fourth possibility is the abnormality of the immunoglobulin molecules produced in pathological states. That is, because of molecular abnormalities of the immunoglobulins produced pathologically, they may be unable to enter the normal immunological feedback system mentioned above.

The third and fourth mechanisms mentioned above are only of theoretical interest at the present time, and in fact the two possibilities cannot be distinguished experimentally.

Polyclonal hyperimmnoglobulinemia of unknown etiology, therefore, must be discussed pathophysiologically on the basis of the four possible mechanisms listed above.

NON-SPECIFIC PROTEIN REACTIONS AND SEROLOGICAL FALSE POSITIVE REACTIONS IN POLYCLONAL HYPERIMMUNOGLOBULINEMIA

Since immunoglubulins possess various characteristic physico-chemical properties, hyperimmunoglubulinemic sera frequently show many characteristic phenomena.   As a result, some of the non-specific protein reactions or serological reactions are useful for clinical screening of hyperimmunoglobulinemia.

Abnormalities in Non-Specific Protein Reactions

1. Increased Sedimentation Rate of the Erythrocytes (ESR)

The erythrocyte sedimentation reaction is quite a simple phenomenon which has been used widely for the past many years, but its real mechanism is not well understood. Among various factors suggested so far, the two most important ones are the numbers and the morphology of the erythrocytes and the composition of the plasma proteins. Fibrinogen and α_2-globulins are known to influence in the greatest degree for accentuation of the ESR. Immunoglobulins are also fairly effective for its accentuation. However, in hypergammaglobulinemia, since the γ fraction usually increases in much greater quantity than do fibrinogen and α_2-globulins, it plays an important role in the accentuation of ESR.

With various disorders, the ESR is accentuated non-specifically. If the increased ESR of unknown etiology is verified clinically, then abnormalities of serum immunoglobulins must be always suspected and the electrophoresis of serum proteins has to be performed. The accentuated ESR is caused by the following three types of plasma protein abnormalities: (1) acute phase response type, (2) broad hypergammaglobulinemic type; and (3) M-proteinemic type. In the first type, ESR is accentuated mainly due to the increase of both fibrinogen and α-globulins, but the other types involve the increase of immunoglobulins.

In hyperimmunoglobulinemia, a rouleaux formation of the erythocytes easily occurs, thus causing the accentuation of the ESR. Generally speaking, the accentuation of the ESR is more prominent in monoclonal hyperimmunoglobulinemia than in polyclonal type, but it may reach to more than 100 mm/hour.

2. Serum Gel Reactions

In general, serum globulins tend to jellify in different physicochemical conditions, and some of the phenomena include acid gel reaction, alkali gel reaction, formol gel reaction, urea or urethane gel reaction, heat gel reactions, etc. In polyclonal hyperimmunoglobulinemia, frequently seen is the abnormal positivity of such serum gel reactions, and the formol gel reaction has been used clinically in particular. Similarly, many different kinds of serum colloid reactions are used clinically for screening the abnormalities of immunoglobulins.

3. Water Dilution Test

In polyclonal hyperimmunolgobulinemia, all classes of immunoglobulins are increased, including IgM-globulin which is one of the major components of serum euglobulin. Therefore, Sia water dilution test becomes positive not infrequently.

4. Increased Serum Viscosity

The intrinsic viscosity of immunoglobulins is high, particularly of IgM-globulin. Therefore, in severe polyclonal hyperimmunoglobulinemia, the relative viscosity of serum is frequently increased. However, in contrast to M-proteinemic serum, the serum relative viscosity in polyclonal hyperimmunoglobulinemia rarely exceeds over 3.0. In addition, the reduced viscosity of the γ fraction remains within a normal range, and thus the molecular aggregation of the γ-globulin in polyclonal hyperimmunoglobulinemia must be almost the same as that of the normal γ-globulin (Fig. 214).

Serological False Positive Reactions

In polyclonal hyperimmunoglobulinemia, not infrequently the positive tests are encountered non-specifically with serological tests for syphilis, rheumatoid factor and anti-nuclear factors.

1. Reagin

The serological tests for diagnosing syphilis may be classfied by the type of antigen used. The first group includes non-treponemal or lipid antigen test, using cardiolipin as the test antigen, and it is referred to "serologic tests for synphilis" or STS.   The second group includes treponemal antigen tests, using treponemes or treponemal extracts to detect antibody, and some of the most commonly used tests are FTA–ABS (fluorescent-treponemal antibody-absorption) test, TPHA (*treponema pallidum* hemagglutination) test and TPI (*treponema pallidum* immobilization) test.   These treponemal antigen tests are specific for syphilis.   However, non-treponemal antigen tests become occasionally positive in non-syphilitic individuals, and they are called "biologic false positive" reactors (BFP).

Reagin is the one which reacts in STS, and is thought to be the antibody against tissue lipids.   That is, once lipid components are released from destroyed tissues into the blood circulation, they probably bind to some high-molecular weight materials and they may become antigenic as haptenes, resulting the formation of reagin.   *Treponema pallidum* may have strong affinity with tissue lipids, thus causing distinct formation of reagin in syphilitic patients.   Therefore, reagin may be produced in various conditions other than syphilis.   In acute BFP, the reaction lasts a few weeks to a few months, being found in persons suffering from many viral and bacterial infections or who have had certain vaccinations and immunizations or who are pregnant.

Chronic BFP usually lasts for more than 6 months, and is often accompanied with polyclonal hyperimmunoglobulinemia.   It is less frequent than acute BFP.   Among chronic false-positive reactors, about 50% were found initially in persons under 30 years of age, 70 percent are women, and about 10 percent have been associated with SLE.   In other instances, it is encountered with polyarteritis nodosa, rheumatoid arthritis, rheumatic fever and other connective tissue diseases.   Still others are associated with polyclonal hyperimmunoglobulinemia such as trypanosomiasis, subacute bacterial endocarditis and lymphogranuloma venereum.   In some instances, familial occurrence of BFP was reported, suggesting its genetic involvement.[56,399]

One of the abnormal reactions related to STS is anticomplementary activity which is sometimes seen in polyclonal hyperimmunoglobulinemia.   The mechanism of its occurrence is not known, but the following two possibilities have been suggested.   First, the increased immunoglobulins consume various complement components; second, they inhibit the complement fixation reaction in a certain antigen-antibody system.   At least in the serum from the patient with rheumatoid arthritis, the second mechanism has been confirmed experimentally.[144]

2. Rheumatoid Factor

Rheumatoid factor (RF) has been most frequently found in the sera of the patients suffering from rheumatoid arthritis.

This rheumatoid factor is an anti-IgG globulin-like substance which binds specifically to IgG-globulins.   Looking at its immunological nature, it can be divided into two groups. One group reacts with both human and rabbit γ-globulins, and the other reacts only with human γ-globulin. Usually the rheumatoid factor shows a molecular weight of roughly a million and the sedimentation constant of 19S, belonging to the IgM class.   In serum, it combines with several molecules of IgG-globulin, and is thought to form either 22S or even larger 27S soluble complex.   The sites where the rheumatoid factor is produced are not certain.   However, according to experiments using fluorescent antibody techniques, it has been located in the plasma cells infiltrating in the synovial membrane, the cells in the germinal centers, and the subcutaneous nodules.[118]

RA test and WAALER–ROSE test are the two most commonly used methods to detect the rheumatoid factor in serum.   The former is the latex fixation test, using human denatured IgG-globulin, while the latter is the passive hemagglutination test, using rabbit γ-globulin.   These tests are found to be positive most frequently in the patients suffering from rheumatoid arthritis, being positive in 70–90 percent.   However, they come positive in other conditions as well.

RA test becomes positive in 24% of connective tissue diseases other than rheumatoid arthritis, 40% of liver disorders (especially with cirrhosis of the liver, 54%), and in other such diseases as rheumatoid disorders, malignant tumors, etc., from 10 to 20 percent. However, if serum samples are inactivated by heating at 56 °C for 30 minutes, many of the positive sera which are obtained from non-rheumatoid patients becomes negative. This is thought to be due to the existence of heat-labile anti-γ-globulins.   In these situations, very often the WAALER-ROSE test proves to be negative, and it is thought to be a type of the rheumatoid factors that does not react to rabbit γ-globulins.

With liver disorders and in particular, chronic liver disorders, WALDENSTRÖM and others, focusing on the fact that the rheumatoid factor can be detected in high percentages, point out that there is a relation between the rheumatoid factor and hypergammaglobulinemia.[118]   As a result, they indicated that the rheumatoid factor was detected more frequently in hypergammaglobulinemic group than in non-hypergammaglobulinemic group.   The similar results have been confirmed by the author in other disorders showing polyclonal hyperimmunoglobulinemia.

3.   Antinuclear Antibodies

Antinuclear factors are actually autoantibody-like globulins which bind specifically with cell nucleus or nuclear components.   The antinuclear factors can be detected by the following 3 ways, namely; (1) LE cell phenomenon, (2) fluorescent antibody techniques detecting antibodies against the whole cell nucleus, and (3) methods for detecting antibodies against nuclear components.   LE cell phenomenon is quite specific for systemic lupus erythematosus, but some investigators claim that it appears positive rarely with other disorders as well.   On the other hand, however, in situations where fluoresent antibody techniques are employed to detect autobodies either against the cell nucleus or its components, the positive test is found not only in cases of SLE, but occasionally rheumatoid arthritis, diffuse scleroderma, dermatomyositis, and other connective tissue diseases as well.[216]   Thus, a positive test for the antinuclear factors tends to appear also in cases of polyclonal hyperimmunoglobulinemia.   According to a report by LEONHARDT,[216] in the families of patients suffering from SLE, the antinuclear factors were frequently detected, suggesting some hereditary elements.

Table 47   Representative diseases accompanying circulating anticoagulants.

I.   Appearance of anti-coagulants as immune antibodies.
  1.   Hemophilias
II.   Appearance of anti-coagulants as so-called auto-antibodies
  1.   During and after pregnancy
  2.   Systemic lupus erythematosus and related connective tissue diseases
  3.   Other conditions: leukemia and other malignancies, liver cirrhosis, pemphigus, after virus infections, in old age.

4. Circulating Anticoagulants

Normal hemostais is currently thought to be achieved through the physiological coagulation mechanism which seems to be a balanced system of accelerators, inhibitors and fibrinolytic factors. Among various inhibitor components or circulating anticoagulants, only the antithrombin I is thought to be important physiologically, and other anticoagulants are encountered mostly in various pathological conditions, some of them being listed in Table 47.[70,399] In hemophilias and its related disorders, the antibodies against deficient coagulation factors may appear through repeated blood transfusion. Other circulating anticoagulants may be thought to be autoantibodies against the protein coagulation factors. In fact, they have been encountered mostly in the pathological conditions where immunological abnormalities are found to exist, particularly the conditions accompanying polyclonal hyperimmunoglobulinemia.

The above mentioned four serological false positive reactions are closely related to polyclonal hyperimmunoglobulinemia. First of all, four reactants are of autoantibody nature, which react either against denatured tissue protein components or against normal plasma components. That is;

1) Reagin reacts with the lipid component which distributes widely in the body as a tissue constituent.

2) Rheumatoid factor reacts with denatured IgG molecules which seem to be derived from *in vivo* antigen-antibody reactions.

3) Antinuclear factors react with the nuclear components which distribute widely in almost all body cells.

4) Circulating anticoagulants react with the protein coagulation factors which exist normally in plasma.

Further still, the antigenic components against these reactants are most likely to exist in or to be released into blood circulation even in the normal condition. Interestingly, some herediatry contributions are strongly suggested to exist among the individuals who show these serological false positive reactions. Therefore, it is assumed that the serological false positive reactors may have congenital abnormality of the immunoglobulin or antibody-producing tissue which results in production of autoantibodies. Additionally, acquired pathologic conditions accompanying increased tissue destruction may contribute also for developing the serological false positive reactions. As schematically described in Fig. 172, thus, polyclonal hyperimmunoglobulinemia may be likely to accompany the serological false positive reactions.

In monoclonal immunoglobulinopathy, however, the serological false positive reactions are rarely encountered.

ACUTE AND CHRONIC LIVER DISEASES

Polyclonal hyperimmunoglobulinemia occurring in liver disorders is discussed in detail elsewhere (page 232). However, it is necessary to state here that in cases of acute hepatitis recognized is increased γ fraction from a relatively early stage of the disease. The reason is not clear, but it may be said that in acute hepatitis, a relatively large amount of degenerated tissue components are released and taken up by the adjacent reticuloendothelial tissue. During the prolongation of the hepatitic process, the tendency of the γ fraction to increase becomes quite marked, and it is especially true in the cases with cirrhosis of the liver. Furthermore, in cirrhosis of the liver, along with the increase of IgG-globulin, an increase of the IgA-globulin is striking as well. Thus, as indicated previously, $\beta-\gamma$ linking becomes prominent on serum protein electrophoretic fractionation.

CHRONIC INFECTIONS AND THE
HYPERGAMMAGLOBULINEMIC STATE

Should pathogenic micro-organisms invade living body tissues, the resulting body response arise in the following two ways. As microbial infection causes more or less tissue destruction, there occurs non-specific acute phase response against this. Subsequently, immune antibodies are produced as immunological responses against antigenic pathogens. In whatever body responses considered, there is brought about great alterations in the plasma proteins; the abnormality of acute phase response type and of polyclonal hyperimmunoglobulinemic type.

As described in Fig. 173, MIYOSHI has classified the plasma protein abnormalities in infectious diseases into the following four stages.   That is, the acute stage, the subacute (or subchronic) stage, the chronic stage, and "status hypergammaglobulinaemicus.[253]" It does not necessarily mean that all infected patients pass through the four stages, but recovery can be made and return to normal from any of the different stages.

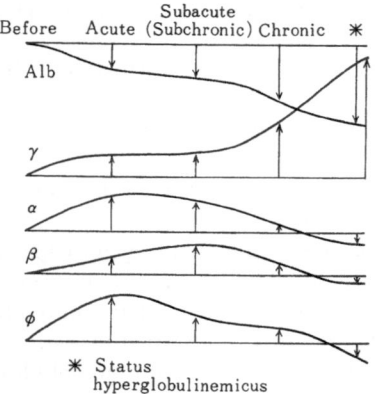

Fig. 173 A diagramatic representation of the serum protein changes in infectious diseases (from MIYOSHI)
The data were obtained from the TISELIUS electrophoretic patterns of many patients with various infectious diseases.

In the acute stage, the plasma protein abnormality of the acute phase response type is recognized, and the serum immunoglobulins is not increased. Two or three weeks after infection has been contracted, gradually, specific antibodies against the infected pathogen become detected in blood circulation. From the subacute to subchronic stages, a rise in the antibody titer becomes apparent along with the increase in serum immunoglobulin levels.   When the disease is prolonged to the advanced stage, there is a marked increase in the γ fraction, representing what MIYOSHI calls "status hypergammaglobulinaemicus", or the hypergammaglobulinemic state.

In the hypergammaglobulinemic state, unlike the chronic stage, there is a significant decrease in fibrinogen, as well as the a and β fractions, thus giving us a deviation essentially from simple infectious disorders, and approaches more the autoimmune disease state.[253] Furthermore, the increase of the γ fraction as seen in the hypergammaglobulinemic state far surpasses the expected γ fraction level due to the rise in the antibody titer.[253]   This seems to indicate that the normal immune mechanism has become distorted to get a new abnormal equilibrium.

Here, in discussing plasma protein abnormalities individually on each infectious disease, it is difficult to avoid unnecessary repetition.   Therefore, the following discussion will be limited to those diseases which give rise to an evident increase of the γ fraction, that is, those infections that tend to proceed to the hypergammaglobulinemic state.

The infectious disorders which most frequently show the hypergammaglobulinemic state to a marked degree are those infected by relatively large parasites, such as leishmaniasis, trypanosomiasis and plasmodial infections.   Among these, Kala-azar has been pointed out for years as one of the diseases often showing a significant increase of the γ fractions.   Furthermore, not only are these pathogenic agents large, but it would be well to note also that in all cases they tend to show intracellular proliferation.   In other words, when certain pathogens proliferate intracellularly, it is usually difficult for them to be attacked by circulating antibodies and thus, the infectious lesion tends to continue for a long period, resulting in prolonged sensitization of the immune system.

Among bacterial infectious diseases, subacute bacterial endocarditis is the one which most often shows severe hypergammaglobulinemic state.   In this disorder, the microorganisms become buried in the endocardial valves where circulating antibodies find hard to penetrate, and thus the endocardial lesions tend to keep a prolonged process.

Again, in lymphogranulomatosis inguinale, severe tissue destruction occurs in association with marked lymphocytic and plasmocytic infiltration.   It is well known that an extremely severe degree of hypergammaglobulinemic state tends to be encountered in this disease.[319]

With chronic infectious diseases, as long as there are no complications having diffuse destruction of the antibody-producing tissues (lymphoid tissues), an increase of the γ fraction in serum occurs invariably.   In the event of a prolonged microbial infection, where the infiltration of both lymphocytes and plasmocytes are quite evident, it is not rare to see a marked increase in the γ fraction.   In instances accompanying the hypergammaglobulinemic state, even after cicatrization of the infected lesions, there is a tendency for the serum protein abnormality to continue for a relatively long period.

HYPERIMMUNIZATION, ADJUVANT DISEASE AND SARCOIDOSIS

If there are antigenic foreign substances brought into the body, as in the situation where pathogenic microorganisms are infected, the same type of body response is likely to occur. Since tissue damage tends to be milder in the former condition, the non-specific acute phase response is kept to exist in the minimum.   However, the production of specific antibodies is exactly the same as in infectious disorders.   Under normal conditions, antibody production plays an important role for development of immune protection against a noxious agent in the body.   In some of the pathologic conditions, however, the repeated introduction of the same antigenic substance may cause abnormal immune response disadvantageous to the host, and it is called allergic state or hypersensitivity.[109]   In man, the similar hypersensitive process may develop spontaneously, and is called the atopic disease.

In normally sensitized and allergic states, unless some secondary infection is associated with, only slight increase of the γ fraction is encountered along with increasing serum titer of the specific antibodies.   In allergic disease, reagin or skin-sensitizing antibody has been known to increase in serum, and recently IgE-immunoglobulin has been drawing much attention.[335]   However, even though reagin and other antibodies increase in serum in a significantly high titer, the serum immunoglobulin level shows hardly any change or very slight, if any.

Allergy or hypersensitivity occurs mainly on the basis of abnormal reactivity of the host. In addition, there is a condition induced by "protracted sensitization", resulting in marked hyperimmunoglobulinemia.[277]

"Protracted sensitization", which is sometimes called also hypersensitization or hyperimmunization, should not be confused with delayed type hypersensitivity. Delayed hypersensitivity reaction is the body immune response closely related to cellular immunity. It has to be noted that the designation "delayed" is applied not to any special chronological facet of induction, but to the body states on the basis of characteristics of elicited reactions.

Experimental Hypersensitization

OKABAYASHI et al. has been studying experimentally on the hypersensitization induced by egg albumin in rabbits.[277] They reported the plasma protein changes throughout all four stages; namely, the initial, the acme, the protracted, and finally the exhaustion stages. The plasma protein abnormalities seen in the initial, acme and protracted stages parallel to those observed as in infectious diseases discussed in the previous section. A detailed histological study was also made, and affirmed that in the acme stage the antibody titer was at its highest, and plasma cell reaction very evident. In the exhaustion stage (after a lapse of 200 days) hyperglobulinemia was still maintained, and eventually some cases developed amyloidosis and nephrosis. Furthermore, the γ fraction in serum became biphasic in some cases, and this may indicate some abnormality related to the clonal distribution of the antibody-producing cells.

In man, the typical hyperimmunization state cannot be affirmed as in the above process; nevertheless, the diseases such as "adjuvant" disease, connective tissue disease and the hypergammaglobulinemic state may be understood in a similar way.

Adjuvant Disease

By the subcutaenous injections of FREUND's complete adjuvant into rats, PEARSON was able to induce arthritis and periostitis similar to those of rheumatoid arthritis in man. He labelled it as "adjuvant disease".[288] Subsequent to that, even with the use of paraffin or the incomplete adjuvant the similar disorder was successfully produced.[208,360] It is still unclear just how the adjuvant participates in immune reaction, but with the use of FREUND's complete adjuvant, at least wax D contained in tubercle bacilli thought to give the adjuvant effect. PEARSON believes that the disorder is a type of hypersensitivity resulted from some substance contained in wax D.[288]

MIYOSHI et al., taking PEARSON's hypothesis on the adjuvant disease, have extended it to the realm of humans, and have reported extremely interesting cases.[255] The following two conditions are included as the adjuvant disease found in man, (1) post-mastoplasty disorder and (2) pulmonary silicosis. In addition, both berylliosis and zirconiosis may be included in this category. Beryllium and zirconium frequently develop the granulomatous lesion quite similar to that of pulmonary silicosis.

1. Adjuvant Disease after Mastoplasty

A 38 y.o. housewife had 100 ml of "Organogen" (mostly composed of paraffin) injected into both breasts for the purpose of mastoplasty 6 years ago. About 2 years ago, she began to complain of a fist-size induration in each breast, and developed fever of unknown etiology and some septicemic symptoms about one year before her hospitalization. Bilateral axillary lymph nodes were enlarged up to 4 cm in diameter, and showed histologically epithelioid cell granuloma containing small numbers of LANGHANS giant cells. The biopsied breast tissue also showed the similar histological changes. The sed rate was markedly accelerated to 142 mm/one hour. The electrophoretic fractionation of the serum proteins showed the total protein concentration to be 7.6 g/100 ml, the albumin fraction 22.9%, the α_1 fraction 6.3%, the α_2 fraction 11.8%, the β fraction 7.4% and the γ fraction 51.7%. The serum γ fraction increased markedly, amounting to 3.9 g/100 ml. The sternal bone marrow contained the plasma cells in 3.5%.   After the foreign material was removed from the breasts on the 49th hospital day, all clinical symptoms improved significantly.   Approximately 6 months later, the sed rate become 9 mm/one hour and the serum γ fraction decreased to 28.9%.

In this case, there seemed to be some deterioration of the tissue proteins surrounding the injected foreign material, and it may become closely bound to paraffin, resulting in the development of a situation thought to be an autoimmune disease as the result of its prolonged sensitization.[255]

2.   Pulmonary Silicosis

Here was a case reported on of a 49 year-old male, who suffered from SLE complicated with pulmonary silicosis.   The γ fraction in serum showed a marked increase to 3.3 g/ 100 ml.[255]

In pulmonary silicosis, the serum γ fraction often increase significantly, and more so in silico-tuberculosis.   In addition, RA test is positive in many of these cases and the LE cell phenomenon is positive in 4 out of 33 cases studied.

Silica has been known to become closely bound to tissue proteins and to show strong adjuvant effects.[292]   Therefore, silicotic nodules in the lungs may sensitize the body for a long period of time, probably causing the development of an autoimmune process.

Sarcoidosis

Sarcoidosis is a vaguely defined condition, because its etiology is unknown and the definitive diagnostic tests are not available at the present time.   An international conference on Sarcoidosis took place in Washington, D.C. in 1960, where a summary statement describing the characteristics of the disease was formulated:

"Sarcoidosis is a systemic granulomatous disease of undetermined etiology and pathogenesis.   Mediastinal and peripheral lymph nodes, lungs, liver, spleen, skin, eyes, phalangeal bones, and parotid glands are most often involved, but other organs or tissues may be affected.   The KVEIM reaction is often positive and the tuberculin test often negative. Other important findings are hypercalciuria and increased serum globulins.   The characteristic histological appearance of epithelioid tubercles with little or no necrosis is not pathognomonic; tuberculosis, fungal infection, berylliosis, and local sarcoid-tissue reactions must be excluded.   The diagnosis should be restricted to patients who have consistent clinical and radiologic features together with biopsy evidence of epithelioid tubercles or a positve KVEIM test."

The granulomatous lesions shown in Fig. 174, are by no means characteristic for sarcoidosis.   The same changes can be produced with KVEIM antigen (extracted from the lymph nodes affected by sarcoidosis), zirconium, beryllium, silica, and some long-chain fatty acids.   Immunologically, it is assumed that some abnormality is occurring in the cell-mediated immune mechanism, evidenced by decreased delayed hypersensitivity, granulomatous formation such as seen with the KVEIM reaction, and normal production of various circulating antibodies.[165]   However its pathogenesis seems to be similar to those of adjuvant disease, silicosis and berylliosis.

The filter-paper electrophoretic fractionations of the serum proteins in sarcoidosis are shown in Fig. 175.   There is practically no change in the total serum protein concentration.   The albumin fraction is decreased in one half of the cases studied, but in a few cases an increase can be seen in the a_1, a_2 and β fractions.   In roughly two thirds of the cases, the γ fraction shows somewhat of an increase.   According to NORBERG,[271] in the patients suffering from bilateral hilar lymphadenopathy the γ fraction is almost normal, but in the patients with some pulmonary lesion it shows an increase.   As a result, the electro-

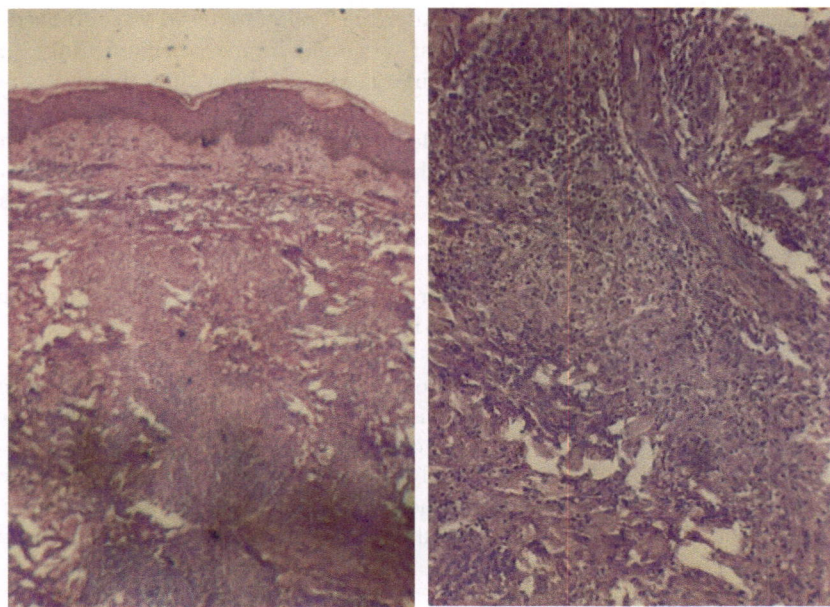

Fig. 174 Granulomatous lesions in sarcoidosis.
Histological examination was done for the skin nodule developed 28 days after the intracutaneous injection of KVEIM antigen. The left photomicrograph shows a characteristic histological pattern of the KVEIM-positive lesion, and the right is a higher magnification of the same lesion showing epithelioid cell proliferation and lymphocytic infiltration.

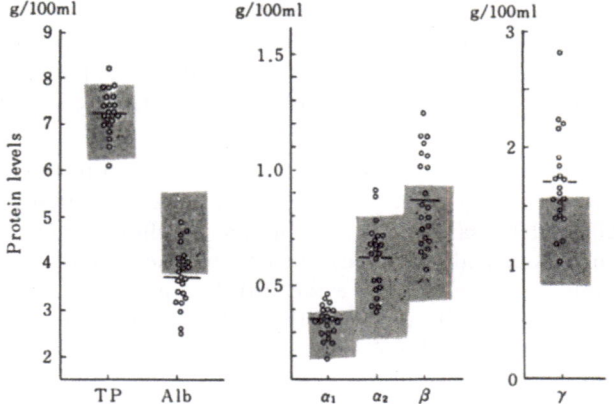

Fig. 175 Serum protein changes in sarcoidosis.
The serum protein fractionation was done on the patients with sarcoidosis by the filter paper electrophoretic technique.

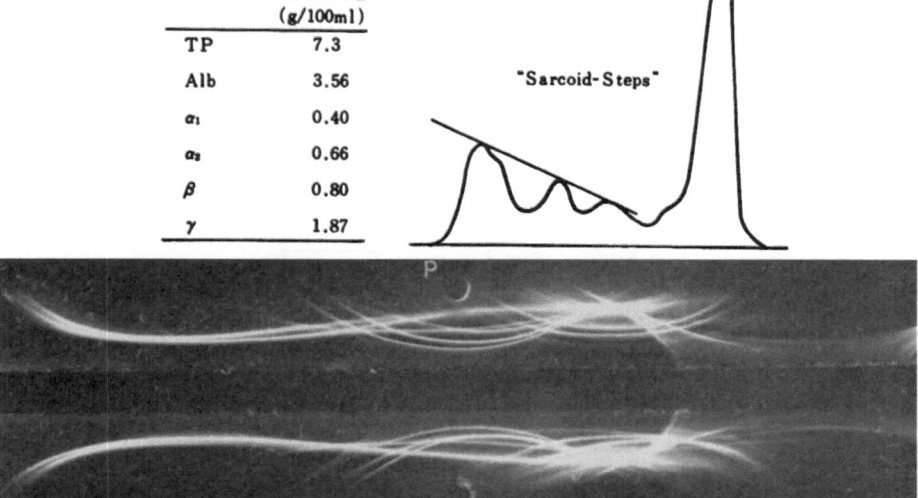

	(g/100ml)
TP	7.3
Alb	3.56
α_1	0.40
α_2	0.66
β	0.80
γ	1.87

"Sarcoid-Steps"

Fig. 176 Serum protein electrophoretic patterns in sarcoidosis.
K. S., 23 y.o., female, sarcoidosis.
Diagnosed as sarcoidosis because of bilateral hilar lymphadenopathy and pulmonary infiltration, and KVEIM reaction was positive. RA test (−), STS (−) and antinuclear factor (−).
On filter paper electrophoresis, noted are slight decrease in the albumin fraction and slight increase in the γ fraction. So-called "sarcoid-steps" is also recognized. IgG 1610 mg/100 ml, IgA 250 mg/100 ml, and IgM 160 mg/100 ml.

phoretic pattern of serum proteins in sarcoidosis is said to be of the chronic inflammatory type. SUNDERMAN et al.[371] once emphasized the characteristic increase in the α_2, β and γ fractions, representing a pattern called "sarcoid-steps" as shown in Fig. 176. These changes are again not peculiar to sarcoidosis, but are observed in other diseases. In the author's own experience, this was affirmed in 5 out of 22 cases studied with the filter paper method, and in 8 of 24 cases with the cellulose acetate method.

With regards to alterations in the serum immunoglobulin levels, they seem to be no characteristic change in sarcoidosis. The author has observed to be no characteristic change in sarcoidosis. The author has observed very similar changes to those seen in rheumatoid arthritis. As shown in Fig. 177, he has seen 10 cases where IgG, IgA, and IgM were all whithin the normal levels, and 5 cases where IgG alone showed a significant increase. In similar fashion, there have been other reports where there was a definite increase of IgA, somewhat like in other connective tissue diseases.[50,272]

There are reports stating that both the serologic tests for syphilis and the RA test often show false positivity.[50,166] However, in 24 cases studied by the author himself, the same results could not be obtained with the serologic test for syphilis, RA test, LE test, and the fluoresecent antibody technique for detecting the antinuclear factor. In only one case, RA test showed a weak positive result.

It is certainly interesting to note that the sarcoidotic patients in Japan have less tendency to show hypergammaglobulinemia, serological false positivity and hypercalcemia in comparison to those encountered in the western countries.

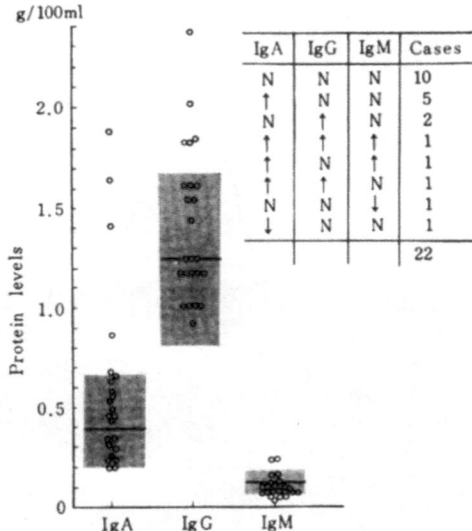

IgA	IgG	IgM	Cases
N	N	N	10
↑	N	N	5
N	↑	N	2
↑	↑	↑	1
↑	N	↑	1
↑	↑	N	1
N	N	↓	1
↓	N	N	1
			22

Fig. 177   Serum immunoglobulin levels in the patients with sarcoidosis.
IgG is increased significantly in a few patients, but IgA is increased markedly in some patients.

MALIGNANCIES

Malignant tumors can be divided into two classes: (1) non-reticular malignant tumors (principally, carcinoma) and (2) primary reticular malignant tumors.   Since the reticular tissue has a very intimate participation in the formation of immunoglobulins, it is natural to expect a difinite difference between the two groups, as far as the serum protein changes are concerned.   However, since there is the same pathophysiological background in the both groups basically, the plasma proteins other than the immunoglobulins show the changes consistent with the acute phase response pattern.

Non-Reticular Malignancies

The plasma protein abnormality in non-reticular malignancies is basically of the acute phase response type, and in a relatively advanced stage, it is accompanied by an increase of the γ fraction.   However, the changes are extremely variable among different patients. In other words, the multiplicity of plasma protein patterns as observed in cases of malignant tumor does not depend on the basic nature of cancer, but rather, on the differences in the pathophysiological backgrounds of the pateints; for instance, age, primary site, the type and extent of its expansion and metastasis, nutritional state, infections, bleeding and other such complications.   The plasma protein abnormalities observed in malignant tumors can be classified into the following four major groups.

1) Acute phase response pattern
2) Broad hypergammaglobulinemic pattern
3) M-proteinemic pattern
4) Malnutritional pattern.

Most of the reports have pointed out that the serum γ fraction in cases of malignant tumor were increased, particularly a prominent increase towards the last stages of cancer.[7]

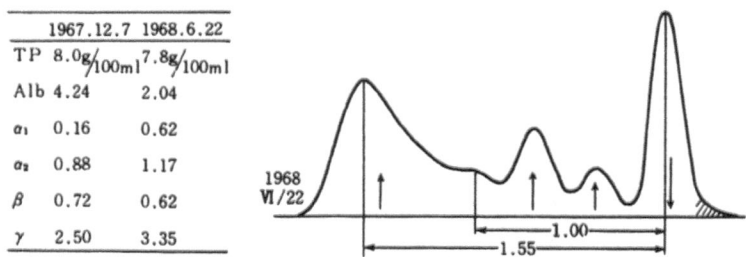

	1967.12.7	1968.6.22
TP	$8.0g/100ml$	$7.8g/100ml$
Alb	4.24	2.04
α_1	0.16	0.62
α_2	0.88	1.17
β	0.72	0.62
γ	2.50	3.35

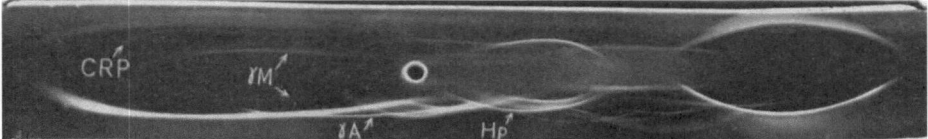

Fig. 178   Serum protein electrophoretic patterns in malignancy.
S. K., 65 y.o., carcinoma of the rectum.
In June, 1967, the adenocarcinoma of the rectum was surgically removed.   The postoperative course was uneventful, and the serum protein analysis showed slight increase in α_2 and γ fractions.
One year after the surgery, developed obstructive jaundice due to wide-spread metastasis. On cellulose acetate electrophoresis, the serum protein pattern was that of the chronic inflammatory type.   Because of marked icterus, a portion of the serum albumin was migrated faster (shaded area).   On immunoelectrophoresis, noted are a significant decrease in prealbumin, albumin and transferrin, and a marked increase in the acute phase reactants and all classes of the immunoglobulins.

However, according to the author's own experience, this is not necessarily the case. On studying the serum concentration of the γ fraction, it will decrease frequently at the last stage in association with malnutritional and cachectic states.   Generally, when the cancer originates on the body surface, the alterations are slight, but with visceral carcinomas, the γ fraction has a tendency to increase.   With primary hepatoma, because of its frequent combination with cirrhosis of the liver, the γ fraction increases significantly quite often.   The wide-spread metastasis also causes usually a significant increase of the γ fraction.   This tendency is even more evident when carcinomatous pleuritis and peritonitis are complicated.   But here also, the variability among the cases are very marked. In using the filter-paper and cellulose acetate electrophoresis, it is relatively rare to see cases where the serum γ fractions become more than 2 g/100 ml.   With visceral carcinomas other than primary hepatoma, the γ fraction level of over 2 g/100 ml may be sometimes encountered in cases of pulmonary carcinoma, and carcinomas of the colon, rectum, stomach, esophagus, kidney, uterine cervix and gall bladder, frequently complicated with peritonitis (Fig. 178, 179).   In addition, most of the cases mentioned above had a long survival of more than 3 years.   AKAI et al. also indicated that in cancer cases of long survival there exists a marked increase of the γ fraction[2].

On the other hand, working with cancer-bearing experimental animals, in practically all instances observed was a decrease in the γ fraction, regardless of whether the cancers had been developed spontanously or transplanted.   The reports could be summarized as follows: (1) the γ fraction in various experimental animals showed a decrease as the tumors proliferated; (2) in those where the cancer was cured or rejected, it tended to increase significantly.   The reason for such discrepancies in the clinical and experimental results is yet unknown.   But it might lie in the fact that the time element in the clinical cases are of relatively long endurance, whereas experimentally, the time factor is comparatively negligible.

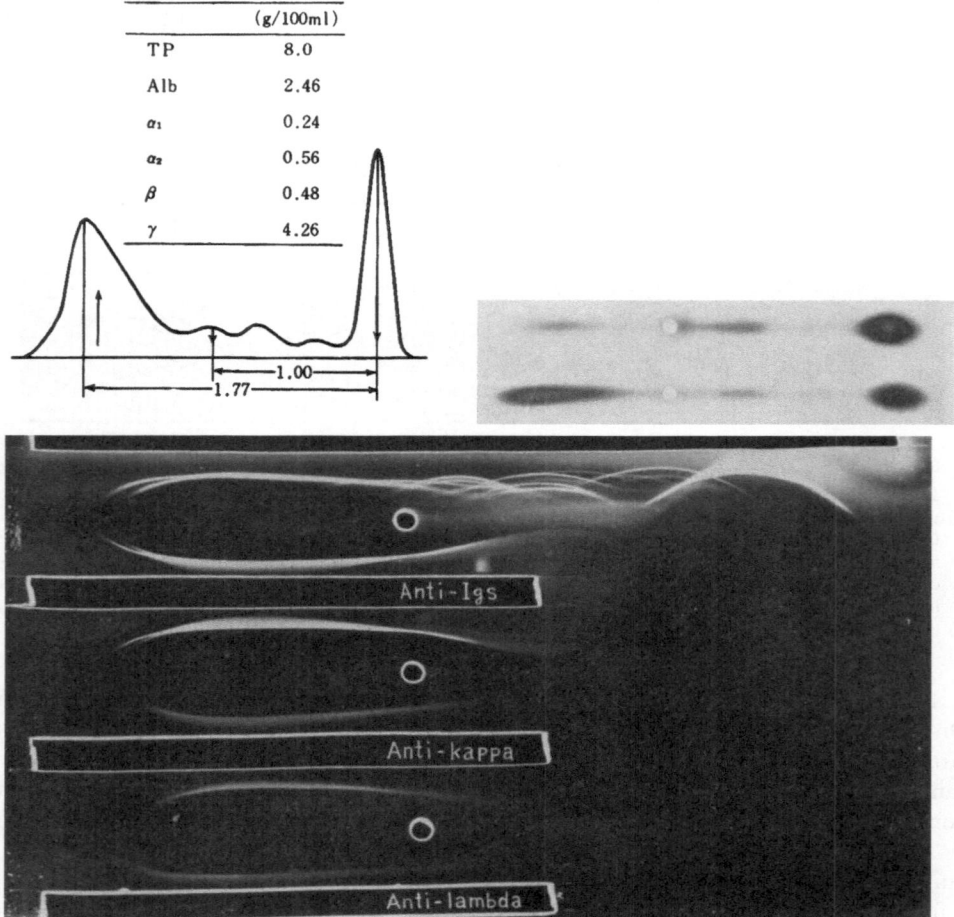

	(g/100ml)
TP	8.0
Alb	2.46
α_1	0.24
α_2	0.56
β	0.48
γ	4.26

Fig. 179 Serum protein electrophoretic patterns of the broad hypergammaglobulinemic type. K.H., 58 y.o., female, carcinoma of the kidney.
Complained of left abdominal pain, hematuria and lumbago for the past 3 months treated with chemotherapy extensively, but no improvement.
On cellulose acetate electrophoresis, the serum protein pattern shows a broad γ fraction. The modal mobility of the fraction is 1.77, abnormally slow on both the cellulose acetate membrane and the agar gel. On immunoelectrophoresis, noted are a marked increase in IgG (249% of normal) and IgA (140% of normal). The IgG line is split at the cathodal end. The line against the antilambda is particular elongated towards the cathode.

Working with immunoelectrophoresis, in the cancer patients having the increased γ fraction, there is also a corresponding increase in the serum levels of IgG, IgA and IgM, representing what is called the polyclonal hyperimmunoglobulinemic pattern. But as will be discussed later, occasionally monoclonal immunoglobulinopathy can be recognized. As illustrated in Fig. 179, there can be recognized the bifurcation of the anodal end of the IgG precipitation line.

It is not yet determined just what function the increase of the serum immunoglobulins in cancer patients serves, but it is probably to act as an immune reaction to cancerous tissue components. Histologically, it has been frequently pointed out that there is a marked infiltration of lymphocytes and plasma cells around cancerous lesions.   Recently, BURTIN and others found a large amount of immunologulins being produced in human pericancerous lymph nodes by immunofluorescent techniques.[52]   Also they stated that the amount of immunoglobulins produced appeared to have no relation to the presence or absence of metastatic cells in the lymph nodes studied, and that IgA was the predominant immunoglobulin found in the pericancerous lymph nodes.   Furthermore, in recent years, HIRAI, working with experimental tumors, was successful in refining and isolating cancer specific antigen, and actually crystallizing it.[147]   On the other hand, by the use of immunoelectrophoretic technique found are several abnormal antigenic components in sera of cancer pateints.   Since then the existence of such cancer-specific antigen is beyond doubt, it is not difficult to imagine that it in some way elicits an immune reaction, and contributes to the increase of the immunoglobulins in serum.   At the present level, with either humans or animals, it is thought to exist an immunological body response against malignant tumors.   Furthermore, it seems to depend mostly on cellular immunity and for lymphocytes to contribute an important role.[4]

Reticular Malignancies

Basically, in much the same fashion as other malignant tumors, there is seen the plasma protein abnormality of the acute phase response type, combined with various changes of the immunoglobulins.   They can be divided into five major groups.

1) Acute phase response pattern
2) Broad hypergammaglobulinemic pattern
3) Hypogammaglobulinemic pattern
4) M-proteinemic pattern
5) Malnutritional pattern

None of these patterns are specific for the reticular malginancy.   The changes of the serum immunoglobulins are also variable.   Even in normal individuals, we are still ignorant to understand completely the mechanisms of immunoglobulin production.   Therefore, with reticular malignancy, it is still unknown as to just what type of pathophysiological process is going on as far as immunoglobulin changes are concerned.   However, from clinical observation, it may be possible to show schematically as in Fig. 180.   The immunoglobulin level in serum is thought to be controlled by the following two factors;

1)   activation of the immunoglobulin-producing cells resulting from non-specific stimulation of the malignancy, and 2)   the extent of its invasion into the immunoglobulin-producing cells.

With the latter, there should be a great difference between the reticular and non-reticular malignancies.   In the reticular malignancy, the extent of its infiltration differs as well among different types of malignancy.   That is to say, with malignant tumors arising from non-reticular tissues, until there has been a wide-spread metastasis in the reticular tissues at the final stages, the immunoglobulin-producing cells are not directly affected.   On the other hand, with reticular tumors, it is recognized that, regardless of what cell system it owes its genesis to, there exists the possibility of its extending invasion into the reticular tissues from relatively early stages.   Thus, under non-specific stimulation of malignancy, there is a general tendency for the γ fraction to increase.   However, depending on the extent of destruction of the reticular tissues, especially the immunoglobulin-producing tissue, the serum level of the γ fraction should vary markedly.

In myeloproliferative disorders included are myelogenous leukemia, polycythemia, myelofibrosis, etc.   In all of these, the pathological process is limited in the myelogenous

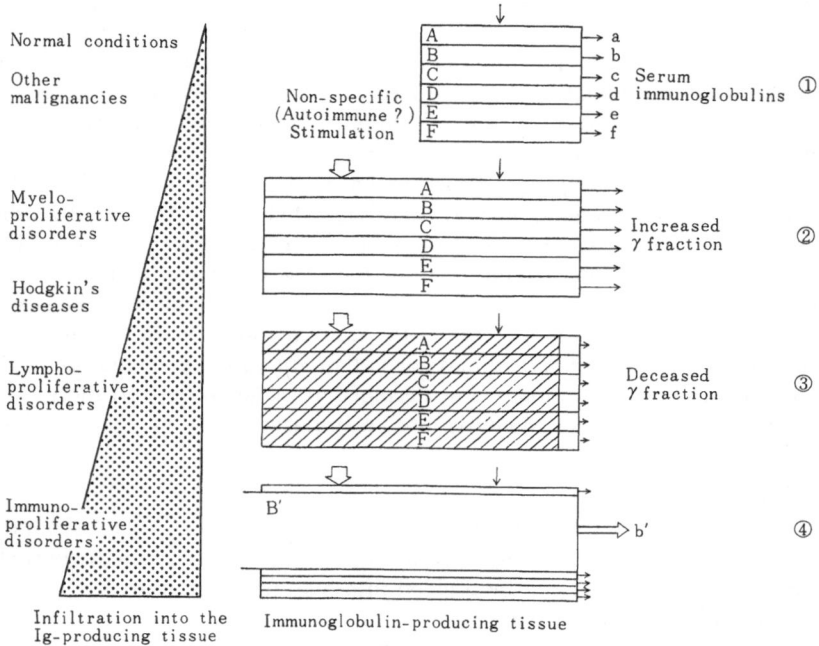

Fig. 180 A schematic representation on the pathogenesis of immunoglobulin changes in reticular malignancies.

① indicates the activity of the immunoglobulin-producing tissue in normality or the disease having no significant change in the immunoglobulin-producing tissue.

② indicates polyclonal stimulation of the immunoglobulin-producing tissue due to malignancy (autoimmunity?).

③ indicates the decreased activity of the immunoglobulin-producing tissue which has been mostly replaced by malignancy, even with increased stimulation.

④ indicates monoclonal stimulation of the immunoglobulin-producing tissue.

cell series, and thus the immunoglobulin-producing cells are not directly affected. As a result, until it has spread into a relatively wide area, the immunoglobulin production is not reduced, and in much the same fashion as with carcinomas, very often the γ fraction is increased.

In lymphoproliferative disorders included are lymphatic leukemia, lymphosarcoma, and reticulum cell sarcoma. It is a fairly well established fact that the lymphoid cells serve an important role, much like plasma cells, in the production of immunoglobulins. And where there is neoplastic proliferation of this type, a direct influence is brought to bear on the immunoglobulin-producing cells. As a result, from relatively early stages, they replace the immunoglobulin-producing tissue, causing a significant decrease in immunoglobulin production. Therefore, they are infrequently observed to have hyperimmunoglobulinemia.

HODGKIN's disease seems to fall between the myeloproliferative and the lymphoproliferative disorders.

Among the immunoproliferative disorders, in situations where there is monoclonal neoplastic proliferation, plasmocytoma (multiple myeloma), plasma cell leukemia, and WALDENSTRÖM macroglobulinemia are included. However, as yet, there have been no reports concerning polyclonal neoplastic proliferation of the immunoglobulin-producing cells. However, with the case discussed next, non-neoplastic or probably reactive proliferation of the reticular tissues may at times accompany polyclonal hyperimmunoglobulinemia.

(APPENDIX) NON-NEOPLASTIC PROLIFERATION OF THE RETICULAR TISSUES ASSOCIATED WITH HYPERGLOBULINEMIA

It is extremely difficult in some cases to make clear distinction as to whether reticular proliferation is neoplastic or non-neoplastic. But temporarily, with the exception, of course, of well-recognized leukemia, malignant lymphoma and HODGKIN's disease, any proliferation of a so-called unknown origin shall be regarded as "non-neoplastic".

When polyclonal hyperimmunoglobulinemia is coupled with the non-neoplastic proliferation of the reticular tissues, the pathogenesis is uncertain, and very often, there is difficulty with diagnostic terminology. In such situations, the cardinal clinical symptoms are just used as diagnosis in the following ways.

1. Lymphoreticulosis Hyperglobulinaemica

This is a case where marked polyclonal hyperimmunologbulinemia is coupled with generalized lymphadenopathy and hepato-splenomegaly. In Japan, there have been cases reported by SUGIYAMA et al.

> **Case** 70 y.o. female. Generalized lymphadenopathy, and hepatospelenomegaly noted. Not anemic, but leukocytosis (18,000) including 26.5% of plasmocytoid and lymphocytoid cells. Serum γ fraction 60.9%, mainly of IgG. On autopsy, the normal architecture remained in the lymph nodes, liver (1620 g) and spleen (400 g), containing many lymphocytes, lymphoid atypical cells, plasmocytoid cells and reticulum cells.

2. Splenomegalia Hyperglobulinaemica

This is where there is a marked increase in the γ-globulin fraction combined with splenomegaly. MIYOSHI and others have reported two such cases in Japan.

> **Case 1** 16 y.o. male. Susceptibility to infections and abdominal swelling complained for the past year. Anemic and malnutritional. Several small lymph nodes palpable at the inguinal region. Liver palpable 3 f.b. below the costal margin. Spleen markedly enlarged down to the navel. Sed rate 145 mm/hour. Hb 45% with SAHLI and WBC 1,200 (lymphocyte 69%). Plasmocytosis in bone marrow. γ fraction 59% or 4.95g/100 ml, mostly of 7S globulin. Histologically found marked infiltration of small lymphocytes, containing some plasma cells.

> **Case 2** 11 y.o. male. Susceptibility to infections with lymphadenopathy and hepatosplenomegaly for several years. Sed rate 55 mm/hour. Hb 37%. WBC 10,700 and 4,000; neutrophilia and later neutropenia. At the age of 10 years, splenectomy under the diagnosis of BANTI's disease. 4.4% plasma cells in bone marrow. Serum γ fraction 39.5% or 3.25g/100 ml, mostly of 7S globulin.

In the above two cases, since the reticular tissues were widely affected, it would be included with the former category, but an evident point of difference was the presence of marked splenomegaly.

3. Lymphadenosis Hyperglobulinaemica

As can be seen in detail in Fig. 181, it is a combination of polyclonal hyperimmunoglobulinemia and generalized lymphadenopathy. However, here, there is no abnormal cell

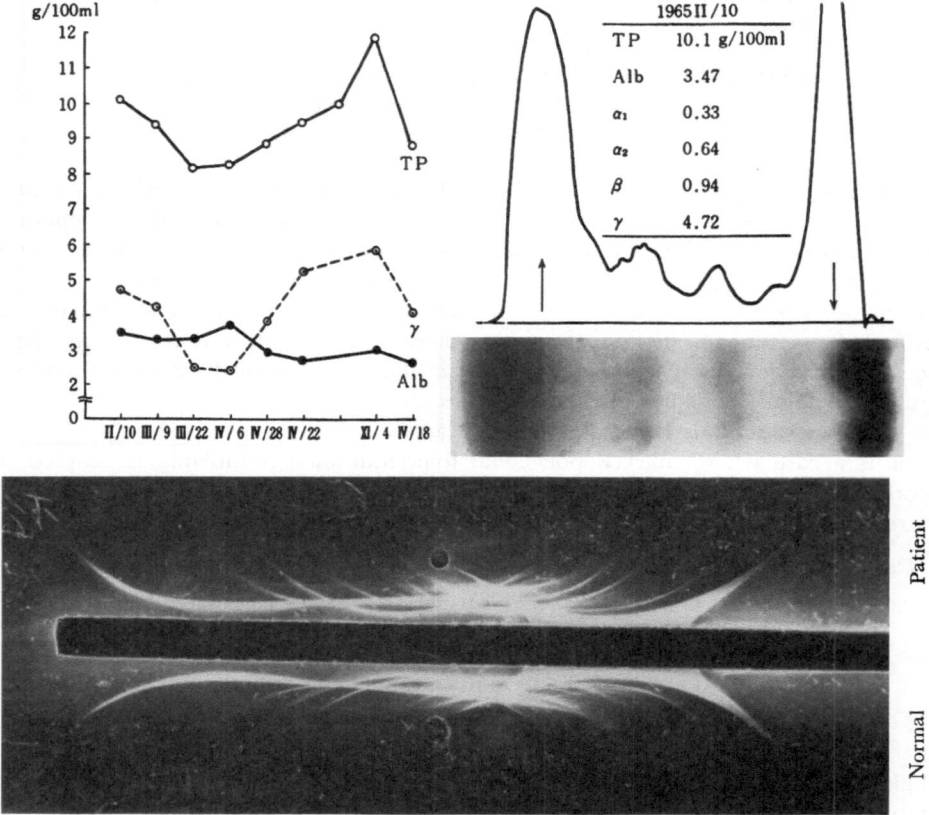

Fig. 181 (See also the next page).

infiltration in the bone marrow, hepatomegaly nor splenomegaly. It is presently being investigated as to whether this disease represents the first stage of lymphoreticulosis hyperglobulinaemica or is in itself a distinct disease entity.

AUTOIMMUNE DISEASES

The term "autoimmunity" denotes to an abnormal immune response resulting in the production of auto-antibodies and/or sensitized lymphoid cells capable of reacting with normal endogenous antigenic body constituents. The pathological conditions occuring on the basis of autoimmunity are called autoimmune diseases.[57,109] According to von PIRQUET's original concept, the term immunity implies the condition showing no pathological reactivity against repeated intrusion of foreign bodies, or at least the occurrence of some process strictly advantageous to the host organism. In contrast, the term *allergy* is used for a changed reactivity against foreign substance introduced repeatedly, resulting in some pathological condition. Therefore, some investigators prefer the term *autoallergic diseases* to the term autoimmune diseases.[57] Autoimmunity is also referred to by some as autosensitivity or autosensitization.[394] Multiplicity of the nomenclature given to the condition seems to be due to its undetermined pathogenesis.

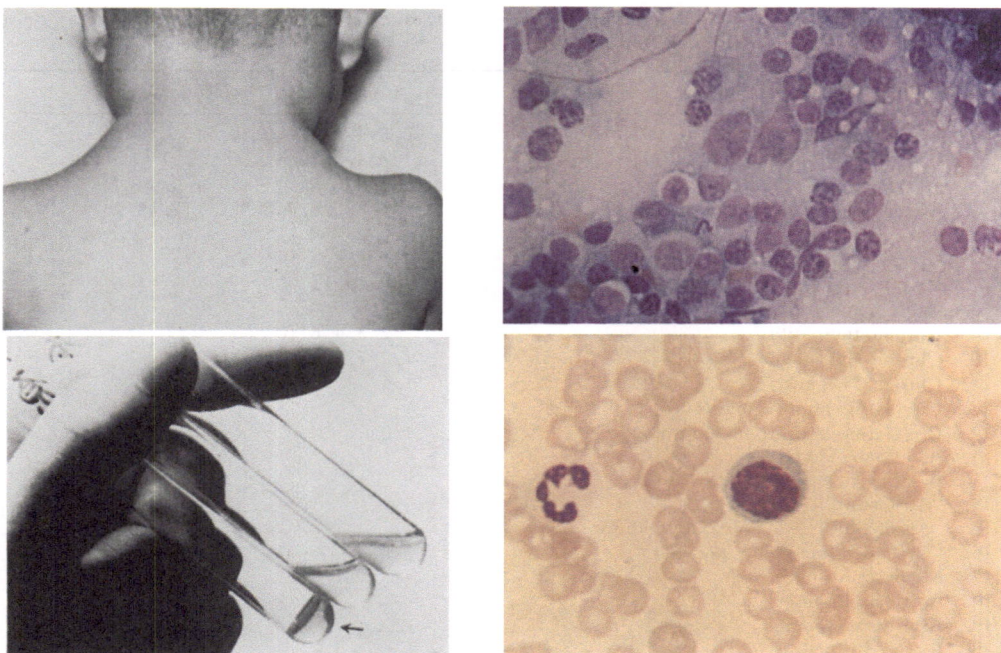

Fig. 181 Serum protein electrophoretic pattern of the broad hypergammaglobulinemic type. M. M., 4 y.o., male, lymphadenosis hyperglobulinaemica.
Since the age of 2 years, developed occasionally cervical lymphadenopathy (left upper). Small lymph nodes palpable also in the axillary and inguinal region. No hepatosplenomegaly nor hilar lymphadenopathy. Sed rate 141 mm/hr, Hb 68% (Sahli), WBC 8,100, having many atypical lymphocytes (16%), lymphocytes (46%) and monocytes (18%) (right lower). Bone marrow showing no abnormal cells. Lymph node showing reactive reticulum cell hyperplasia and prominent plasmocytosis (right upper). Heterophile antibody titer 7x, RA test (±), STS (−), CRP (−), ASO 333 Todd units.
The filter paper electrophoretic pattern of the serum protein shows slight decrease in the albumin fraction and a broad increase in the γ fraction. On immunoelectrophoresis, a significant increase of IgG, IgA and IgM was confirmed. On heating the patients serum at 60 °C for 30 minutes, jellification was noted (arrow), indicating pyroglobulinemia.
Clinical recurrence has been noted in many occasions although the patient responded temporarily to antibiotics and corticosterioids.

But in any event, in dealing with patients, whether it be autoimmune diseases or disorders akin to them, two major divisions are made.   (1) Organ-specific autoimmune diseases; and (2) Connective tissue disease (or collagen disease).   However, in reality, there are times when a clear distinction between the two groups is difficult.   For instance, a form of multiple immunopathy can be observed, as with SJÖGREN's syndrome, where there is a combination of two or more disorders.   WALDENSTRÖM, for example, calls the phenomenon of where a patient has contracted several disorders simultaneously, "clinical constellation"[399].   In situations where this is seen at a high rate of frequency, rather than assigning it to simple coincidence, such patients ought to be thought of as belonging to the same group.   Viewed from this point, in addition to organ-specific autoimmune diseases and connective tissue diseases, sarcoidosis and the conditions having circulating anticoagulants are included in the similar category (Table 48).

Table 48 Clinical constellations among autoimmune diseases and other related disorders (from WALDENSTRÖM).

Disease	"Constellations"
1. Chronic rheumatoid arthritis (RA)	2, 3, 4, 5, 6, 8, 9, 10, 11
2. Scleroderma diffusa (SD)	1, 3, 4, 9, 10
3. Dermatomyositis (DM)	1, 2, 4, 8
4. Systemic lupus erythematosus (SLE)	1, 2, 3, 5, 6, 7, 8, 9, 11, 12, 13, 20
5. Thrombotic thrombocytopenic purpura (TTP)	4
6. Idiopathic thrombocytopenic purpura (ITP)	4
7. Purpura hyperglobulinaemica	4, 8, 10, 11 ?
8. Acute hemolytic anemia	3, 4, 6, 9
9. Chronic thyroiditis	1, 2, 3, 4, 8, 10
10. Chronic sialoadenitis	1, 4, 7, 9, 11
11. Chronic "hepatitis"	1, 4
12. Circulating anticoagulants	4
13. Circulating antithromboplastin	4, 6
A. "Sarcoidosis"?	—
B. Sperm granuloma (Sterility)	—
14. "Adrenalitis"	—
15. "Adrenalitis" + hypoparathyroidism	—
16. "Gastritis" (pernicious anemia)	—
17. Mediastinitis fibrosa	—
18. Aortic arch syndrome	—
19. Arteritis temporalis	12
20. Polyangitis nodosa	4

Table 49 Organ-specific autoimmune diseases.

 I. Demyelinating Diseases
    Acute hemorrhagic encephalopathy
    Acute disseminated encephalomyelitis
    Multiple sclerosis
 II. Endogenous Uveitis
    Phacoanaphylaxis, or hypersensitivity to lens material
    Sympathetic ophthalmia
 III. Thyroiditis
    De QUERVAIN's acute granulomatous thyroiditis
    HASHIMOTO's thyroiditis
    RIEDEL's fibrous invasive thyroditis
    Myxedema
    Cretinism
 IV. Autoimmune Hemolytic Anemia
    Warm-type autoimmune hemolytic anemia
    Paroxysmal cold hemoglobinuria (PCH)
    Cold hemagglutinin syndrome
 V. Idiopathic Thrombocytopenic Purpura (ITP)
 VI. Thrombotic Thrombocytopenic Purpura (TTP)
 VII. Pernicious Anemia and Chronic Gastritis
 VIII. Liver Diseases (chronic hepatitis, liver cirrhosis, etc.)
 IX. Ulcerative Colitis
 X. Adrenalitis and Idiopathic ADDISON's disease
 XI. Orchitis and Aspermatogenesis (Infertility)
 XII. Myasthenia Gravis
 XIII. SJÖGREN's Syndrome*

* May be included in connective tissue disease.

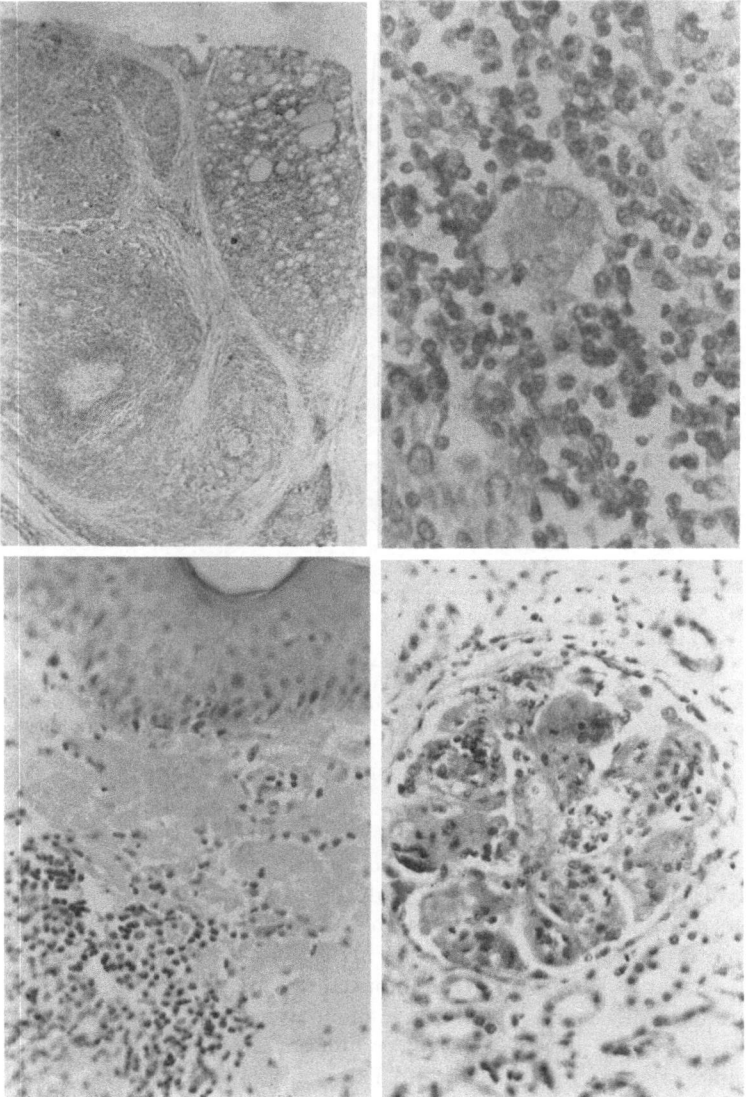

Fig. 182 Histopathological findings in autoimmune diseases. Upper: Chronic thyroiditis (HASHIMOTO's disease). The upper left is the lower magnification showing fibrosis and marked lymphocytic infiltration with lymph follicles. The upper right is the higher magnification showing many lymphocytes and a few plasma cells. Lower: Systemic lupus erythematosus. The lower left is the skin lesion showing fibrinoid degeneration of the collagen, lymphocytic infiltration and edema of the epidermal basal layer. The lower right is the glomerular lesion showing a typical "wire-loop lesion".

Organ-Specific Autoimmune Diseases

In this disease, it is brought about by an autoimmune phenomenon against antigenic components present in certain normal organs. However, pathological changes are limited to certain specific organs. In Table 49 are encompassed those diseases, which have been broadly accepted as auto-immune diseases.[10,303]

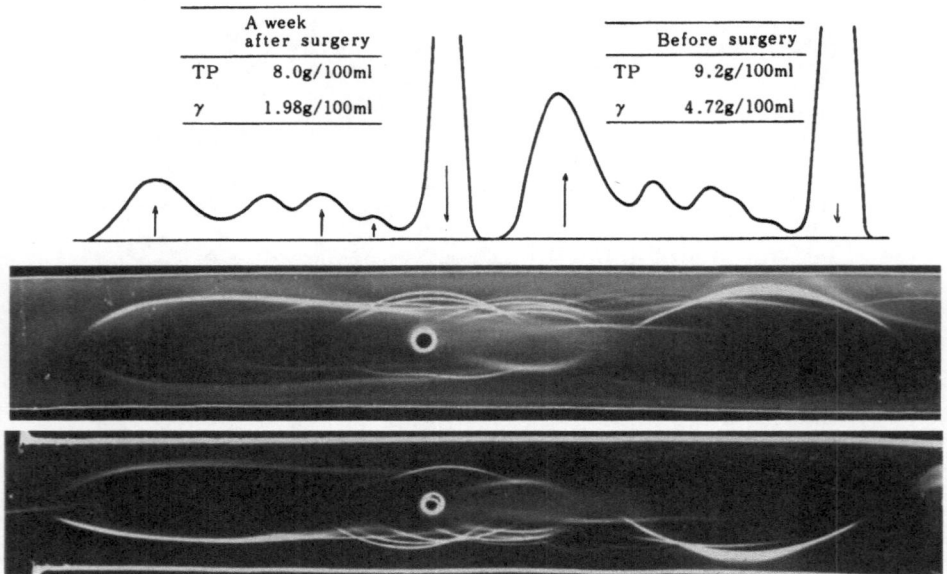

Fig. 183   Serum protein electrophoretic pattern of the broad hypergammaglobulinemic type.
H. U., 56 y.o., female, HASHIMOTO's disease.
Treated with corticosteroids under the diagnosis of subacute thyroiditis, but not improved.   BMR 7%,
Triosorb test 23.9%.   Serum total cholesterol 225 mg/100 ml.   Sed rate 98 mm/hr.   TA test positive.
Before thyroidectomy, the serum γ fraction was increased broadly, and the immunoelectrophoretic
pattern (upper) showed marked increase in IgG, IgA and IgM.   One week after surgery, although
the acute phase response pattern was recognized, the γ fraction became significantly lower than the
preoperative level, confirmed by the immunoelectrophoretic pattern (lower).

With these diseases, the organs involved show chronic inflammatory lesions characterized by marked lymphocytic and plasmocytic infiltration (Fig. 182).   In such disorders, there is no definite microbial infection, nor neoplastic proliferation of the lymphoid tissues. From this, one can gather that marked cell infiltration is the result of immune reaction. Consequently, as far as plasma protein abnormalities are concerned, they represent basically the chronic inflammatory plasma protein pattern.   In many instances, since the pathological changes are localized, when compared to systemic connective tissue disorders, the changes remain relatively mild.   For example, in many instances, there is no marked increase in the γ fraction, and frequently, various organ-specific autoantibodies are merely detected.   However, when pathological changes are severe, along with the appearance of acute phase reactants, there is an accompanying γ fraction increase.   This tendency is particularly true of both chronic thyroiditis (HASHIMOTO's disease) and ulcerative colitis.   Particularly with the former, it is not infrequent that a marked increase of the γ fraction is observed   (Fig. 186).   With regard to the serum levels of immunoglobulins, there are no particular characteristics, and at the present level, either diagnostically or pathophysiologically, there seems to be little reason for searching out any significance. Frequently, when there is an increase in IgG, at the same time, there is a parallel increase in both IgA and IgM.   In such instances with increased immunoglobulin levels, local lesions seem to play the primary role.   For example, as with the cases illustrated in Fig. 183, the immunoglobulin increase tends to return gradually to the normal level once the diseased organ has been removed.

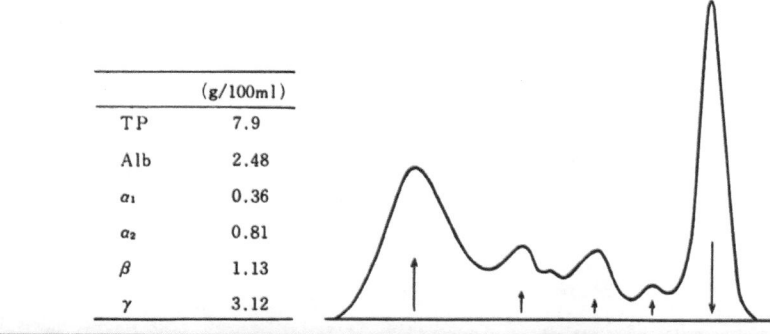

	(g/100ml)
TP	7.9
Alb	2.48
a_1	0.36
a_2	0.81
β	1.13
γ	3.12

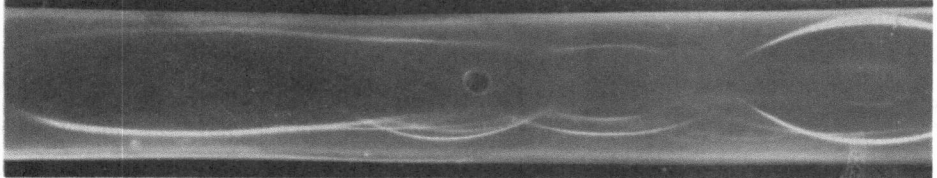

Fig. 184 Serum protein electrophoretic pattern of the broad hypergammaglobulinemic type. S. A., 29 y.o., female, aortitis syndrome.

For the past 4 years, suffered from circulatory disturbance of the left arm and occasional fever. No significant abnormality on hematological and urinary examinations. CRP (5+), ASO 500 Todd units. On cellulose acetate electrophoresis, the serum protein pattern was that of the chronic inflammatory type. On immunoelectrophoresis, IgG was markedly increased, and IgA was also increased slightly.

Autoimmune liver disease are discussed elsewhere, while autoimmune hemolytic anemia is handled under monoclonal immunoglobulinopathy.

Although not included in Table 50, the aortitis syndrome should be discussed here. In this disease, an increase in the serum γ fraction is found in about one third of the cases (Fig. 184), and anti-aorta antibodies are detected. The author himself has experienced one case with the presense of cryofibrinogen accompanying an increased fibrinogen concentration (see page 396).

Furthermore, it is known that the lungs, along with the skin, act as organs giving rise to very strong allergic reactions. However, pulmonary diseases thought to be included in the autoimmune diseases have yet to be uncovered, but it would be worthwhile drawing attention to the fact that there have been occasional reports linking pulmonary diseases with hyperimmunoglobulinemia (Fig. 185). Again, there can be seen some notable correlation between the γ fraction increase and the plasma cell infiltration in the pulmonary interstitium; and as mentioned previously, as with silicosis and berylliosis, participation of some immune mechanism can be easily surmised.

Table 50 Connective tissue diseases

1. Systemic lupus erythematosus (SLE)
2. Rheumatoid arthritis
3. Diffuse scleroderma, or progressive systemic sclerosis
4. Rheumatic fever
5. Dermatomyositis
6. Polyarteritis nodosa

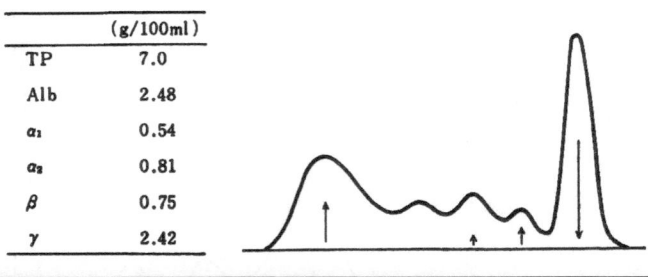

	(g/100ml)
TP	7.0
Alb	2.48
α_1	0.54
α_2	0.81
β	0.75
γ	2.42

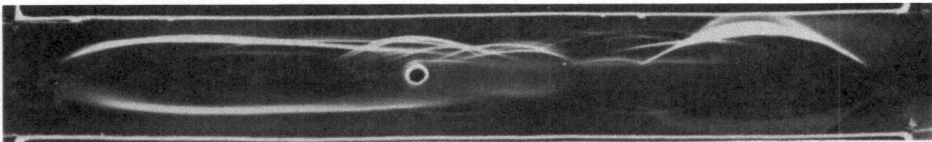

Fig. 185 Serum protein electrophoretic pattern of the broad hypergammaglobulinemic type. F. T., 66 y.o., female, chronic bronchitis and interstitial pulmonary fibrosis.
Developed asthmatic attacks for the past 6 months, and became worse recently. Eventually died of cardiac failure. Moderate hypochromic anemia (Hb 8.0 g/100 ml), leukocytosis with a shift to the left (WBC 21,800, stab. 24%). Sed rate 130–150 mm/hr. CRP (6+—12+), ASO 100 Todd units, RA test 2+.
On cellulose acetate electrophoresis, the serum albumin fraction was decreased markedly, and the α_1 and γ fractions were increased significantly. On immunoelectrophoresis, IgG, IgA and IgM are all increased, but β_{1C}-globulin within normal limits.

Connective Tissue Diseases

Unlike situations where pathological changes are localized to certain individual organs in the organ-specific autoimmune diseases, various organs and tissues in numerous combinations are damaged in the connective tissue diseases. It is characteristic for the autoimmune process to occur not only in one simple organ but in such components having much in common with organs and tissues as nucleoproteins and connective tissues (Fig. 182). The fact that there existed some disorders characteristically involving connective tissues was first cited by KLINGE in 1929. Later carried on by KLEMPERER, notions concerning "diffuse collagen disease" were widely accepted.[190] Subsequent to this, it has become known that it involves not simply the collagen fibres but damages connective tissues in a widespread fashion. As a result, presently there is a tendency to call it "connective tissue disease". The disorders encompassed within the connective tissue disease are listed in Table 50. Other than these one could include as well such disorders as BEHÇET's disease and WEGENER's granulomatosis, but in any of these cases the evidence for systemic connective tissue diseases is not firmly established.[10] Again, with SJÖGREN's syndrome there are severe chronic inflammatory changes in both the salivary and lacrimal glands, and from the fact that specific auto-antibodies against the organs have been detected, they seem rather to be included in the organ-specific autoimmune diseases. Over and above, since it appears very often as a complication of rheumatoid arthritis, it presents a very interesting case.[10]

Connective tissue diseases are thought to be a group of autoimmune diseases. However, as yet auto-antibodies have not been detected against connective tissue or any specific components of it. Nevertheless, the rheumatoid factor, the antinuclear factor (ANF) as well as the antibodies against cytoplasmic components have been detected in high frequency. Either these antibodies or antibody-like substances are by no means specific for individual disorders. In practically all cases of connective tissue diseases, not only have they been detected in varying titers but have been observed in the organ-specific autoimmune diseases as well.   As a result, considering it from this viewpoint, they could be called non-specific auto-antibodies.[10,318]   Therefore, as discussed previously, various diseases accompanying the hypergammaglobulinemic state other than the autoimmune diseases may have those non-specific auto-antibodies.[399]

The following laboratory abnormalities are frequently met with connective tissue diseases: namely, accentuated erythrocyte sedimentation rate, anemia, abnormal urinary findings (proteinuria, abnormal urinary sediment) and hypergammaglobulinemia.   The frequency of these findings varies according to individual diseases, but the author has presented them in model form in Fig. 186.   However, since they will not be of any direct

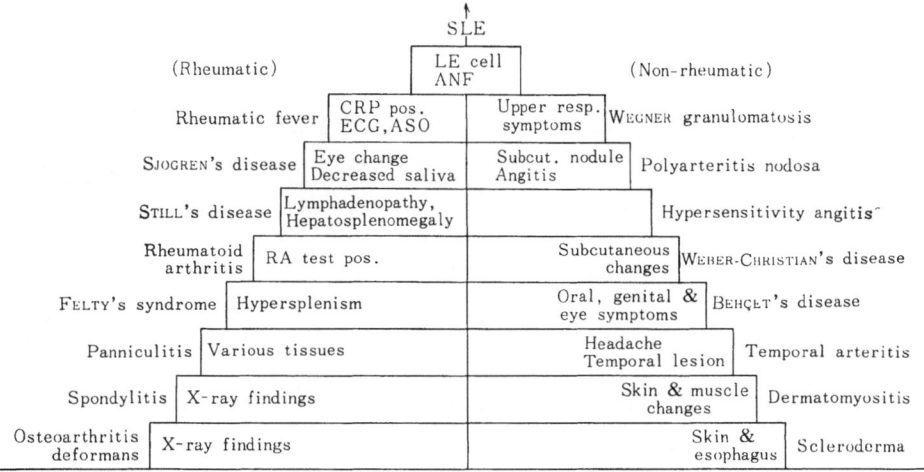

Fig. 186 A schematic representation of various connective tissue diseases and their related conditions (from KAWAI et al.).
The frequency of showing laboratory abnormalities such as increased sed rate, hypergammaglobulinemia and proteinuria tends to become high along the arrow shown in the center.   In the columns listed are most characteristic clinical symptoms and signs on each condition.

help for diagnosis of each specific disease, it goes without saying that other findings must be considered as well.   Even as regards plasma protein abnormalities, the same can be said. As a general tendency, the serum albumin decreases and the immunoglobulin concentrations increase significantly.   Therefore, the plasma protein abnormality in connective tissue diseases is summarized as the acute phase response pattern associated with varying degrees of polyclonal hyperimmunoglobulinemia.   This tendency has characteristics in common with all cases of connective tissue diseases, and here only rheumatoid arthritis and systemic lupus erythematosus will be discussed.

1.   Systemic Lupus Erythematosus (SLE)

As shown in Table 47 and Fig. 186, this disease exists at the very center of the clinical constellations that occur in autoimmune diseases, and very often gives the most serious problems clinically.   As can be seen in Fig. 187, in accord with patients' general condition and activity of the disease, the acute phase response pattern is observed in various degrees. Viewed from the study of many cases, the total serum protein concentration shows no constant tendency.   However, according to Fig. 188, when the total protein concentration shows an increase, the γ fraction increase plays a leading role.   In situations where it is

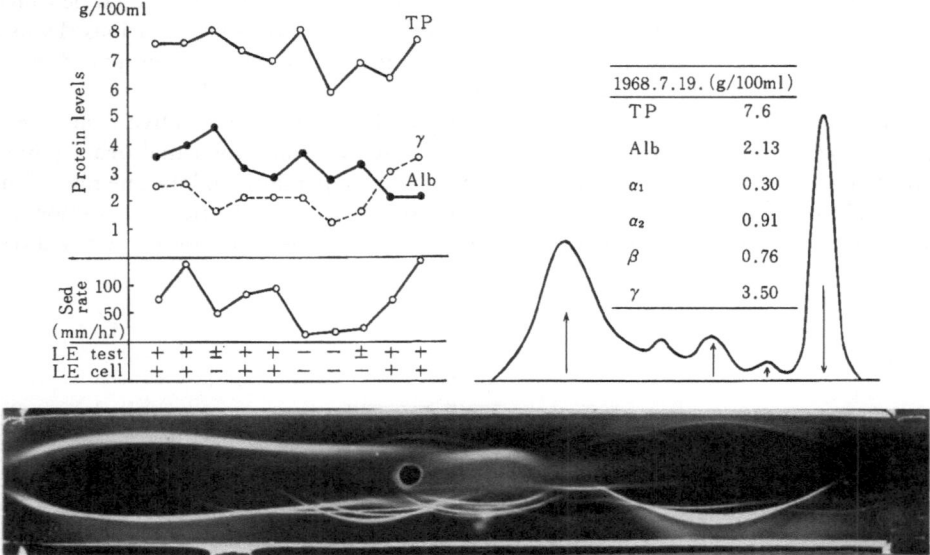

1968.7.19. (g/100ml)	
TP	7.6
Alb	2.13
α_1	0.30
α_2	0.91
β	0.76
γ	3.50

Fig. 187   Serum protein electrophoretic pattern of the broad hypergammaglobulinemic type.
M. S., 17 y.o., female, systemic lupus erythematosus.
Disgnosed as SLE because of typical skin lesion and positive serological tests.   Later, readmitted
with severe anemia, proteinuria and marked accentuation of the sed rate.   Eventually died 3 years
after the diagnosis was made.
The clinical course was well correlated with the sed rate values, also with the serum γ fraction levels.
Immediately before death, the serum protein electrophoretic pattern was that of the active chronic
inflammatory type.
On immunoelectrophoresis, IgG and IgA were markedly increased, and IgM was within normal limits.
β_{1C}-globulin was only faintly demonstrated.

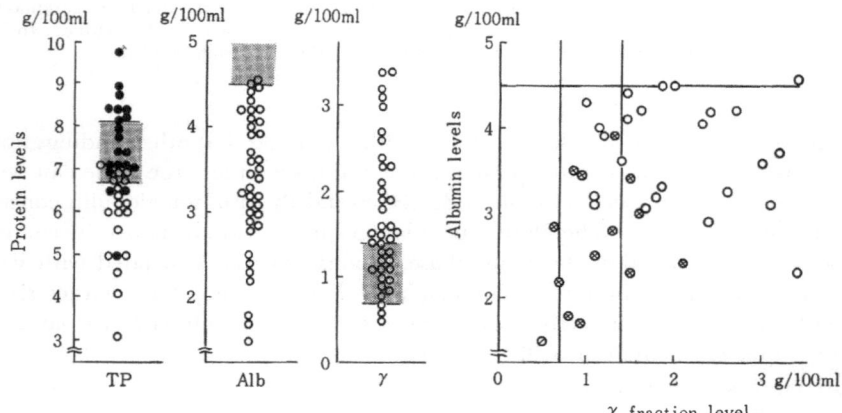

Fig. 188   Serum protein fractional values in systemic lupus erythematosus.
All the cases with hyperproteinemia showed an increase in γ fraction.   No constant
relationship was noted between the albumin fraction and the γ fraction.   In the
case where hypoproteinemia was associated with proteinuria, the albumin fraction level
was decreased in parallel to the γ fraction level.
● Cases having the γ fraction of more than 1.4 g/100 ml.
⊗ Cases having the serum total protein concentration of less than 6.7 g/100 ml.

decreased significantly, lupus nephritis is usually complicated with.   The protein-losing mechanism, as a result of the lupus nephritis, is rather characteristic for SLE among the connective tissue diseases.   Also, characteristically seen in SLE is a marked decrease in $\beta_{1C/A}$-globulin and serum complement titer (Fig. 187, 191)[128].   The albumin fraction shows a decrease in all cases, while the γ fraction shows rather an increase in about half the cases.   In those cases where either a normal or decreased level is noted, frequently there are complications of protein-losing hypoproteinemia.   No mutual relationship is seen between the albumin and γ fractions, but in situations where the albumin fraction is markedly decreased through an evident protein loss, the serum γ fraction tends to be decreased also.   However, in cases not associated with urinary protein-losing condition, it is thought that decreased albumin fraction and increased γ fraction reflect different pathophysiological processes.

2.   Rheumatoid Arthritis

With this disease, unlike SLE, there is very seldom to see complications such as protein-losing process.   Therefore, it is considered as a condition showing the alterations charac-teristic of connective tissue diseases.   On filter-paper electrophoretic fraction ation of serum proteins, as seen in Table 51, there is a slight decrease in the albumin fraction and an increase in the γ fraction.   However, depending on the degree of the activity of the disease process, an increased values can be seen in the a_1, a_2, and β fractions, representing

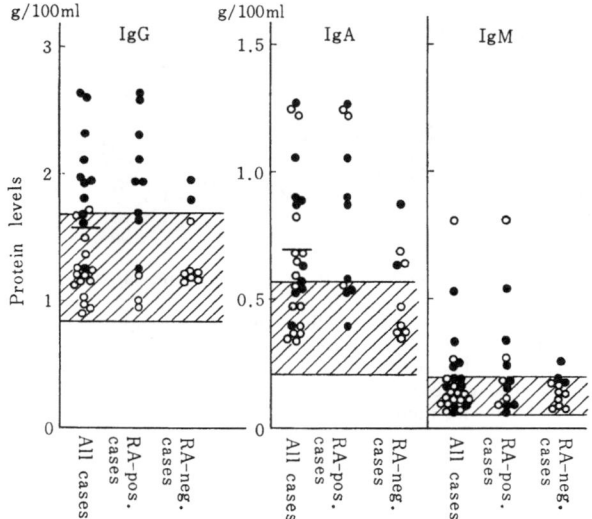

Fig. 189 Serum immunoglobulin levels in rheumatoid arthritis. The black dots indicate the cases showing the positive CRP test. Increased IgG level was recognized more frequently in the cases showing the positive CRP test. Increased IgM level was recognized frequently in the RA-positive cases, but there was no constant rela-tion between the IgM level and RA-positivity. In general, IgA tends to be increased most frequently.

either the subacute or chronic inflammatory patterns.   On the average, the γ fraction increase is more evident in the RA test-positive group than the RA test-negative group. As shown in Fig. 189, the serum level of IgG tends to increase slightly and IgA to increase moderately.   The increase of IgA was affirmed in 52% of the cases studied, and an increase

Table 51   Filter-paper electrophoretic fractionations of serum proteins in rheumatoid arthritis.

	RA-positive group	RA-negative group
Total serum protein	7.36±1.109	6.90±0.387
Albumin fraction	3.38±0.616	3.39±0.318
α_1 fraction	0.36±0.102	0.35±0.236
α_2 fraction	0.61±0.160	0.65±0.195
β fraction	0.74±0.319	0.82±0.114
γ fraction	2.20±0.940	1.89±0.339

(mean±S. D., unit: g/100 ml)

of IgG in 36%, and of IgM in 20%.   The increase of IgG was overwhelmingly high in the CRP-positive group, and in the same fashion as SLE, it roughly paralleled to the activity of the disease.   In many cases with positive RA test, the serum IgM level was increased, but no definite relationship could be necessarily found between the two.   However, ordinarily in the synovial fluid of patients suffering from this disease, the greater part of IgM has the rheumatoid factor activity (Fig. 190).

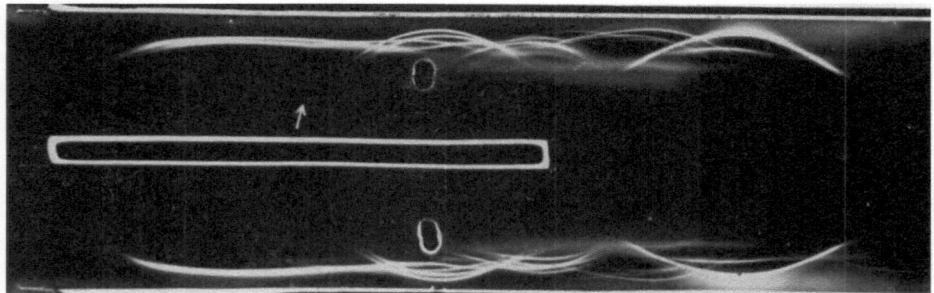

Fig. 190   Immunoelectrophoretic pattern of the synovial fluid in rheumatoid arthritis.
The upper well contains the synovial fluid showing strongly positive RA test, and the lower well contains the supernatant synovial fluid which has been absorbed by the RA-test reagent.   The middle trough contains the specific anti-IgM.   Although the IgM line is clearly recognized before the absorption, IgM is not demonstrated for the absorbed synovial fluid.

3.   Increase in Serum IgA

IgA in serum tends to increase significantly in connective tissue diseases.[20,62,223] Even in SLE, with the immunoelectrophoretic technique, this same tendency has been often observed.   It is not infrequent to find such an increase in serum IgA to a marked degree even without the presence of serum IgG increase (Fig. 191).   The IgA precipitin line is often extended towards the anode, and represents its polyclonal increase.   In this fashion, polyclonal IgA increase, either associted with or without the increase of other immunoglobulins, seems to be rather characteristic of connective tissue disease. The same finding has been experienced not infrequently in the author's laboratory in cases with dermatomyositis, rheumatic fever, diffuse seleroderma, BEHÇET's disease, WEGE-NER's granulomatosis (Fig. 192) and FELTY's syndrome as well as rheumatoid arthritis and SLE.   The similar isolated increase in serum IgA is encountered occasionally also in such organ-specific autoimmune disease as chronic thyroiditis, ulcerative colitis, mya-

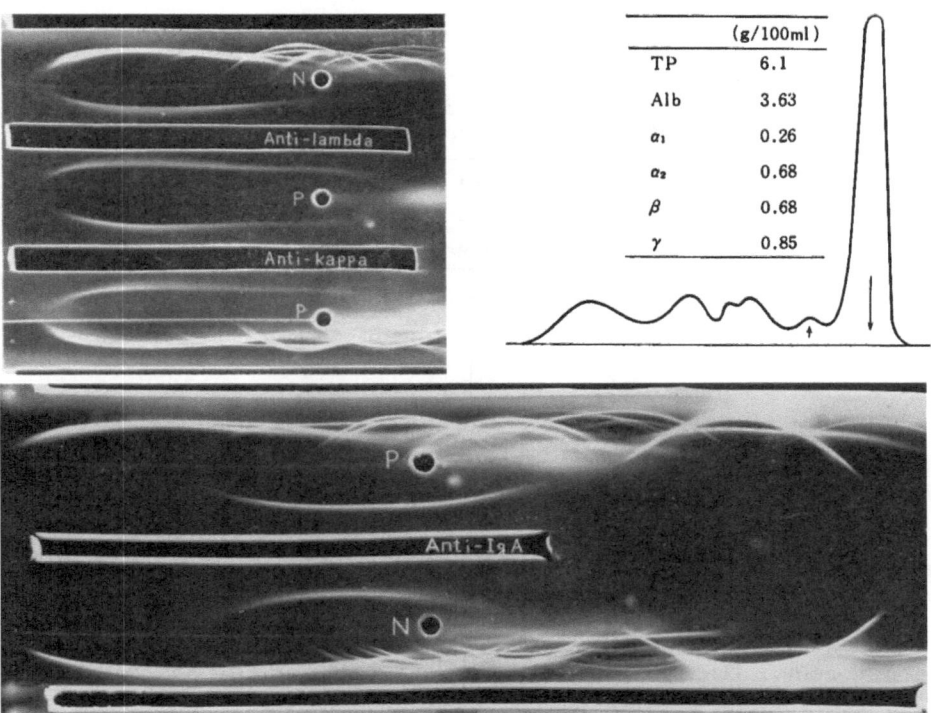

	(g/100ml)
TP	6.1
Alb	3.63
α_1	0.26
α_2	0.68
β	0.68
γ	0.85

Fig. 191 Serum protein electrophoretic pattern in hyperimmunoglobulinemia A.
Y.Y., 26 y.o., female, SLE and lupus nephritis.
Suffered from SLE for the past 6 years, developing multiple episodes of clinical recurrence. Slight anemia (Hb 10.8 g/100 ml), marked thrombocytopenia (3,000/mm³), bleeding time 7 min., sed rate 130 mm/hr. Serum total cholesterol 375 mg/100 ml, serum urea-N 16 mg/100 ml, marked proteinuria (6 g/day), PSP excretion in 15 min. 9%.
LE cell (+), antinuclear factor (+) CRP (−) RA test (−), STS (−), direct Coombs test (−).
Cellulose acetate electrophoresis of the serum protein resulted in hypoporteinemia and hypoalbuminemia.
The γ_1 zone is rather full, but no M-protein band is seen.
On immunoelectrophoresis, IgA and LDL were markedly increased, and β_{1C}-globulin was markedly decreased. The IgA line elongated towards the anode, reacting against both the anti-kappa and the anti-lambda immune sera.

sthenia gravis, and others. Presently, even though the clinical significance of the isolated IgA increase is not certain, the finding may warrant to search for the possiblity of connective tissue disease.

ESSENTIAL HYPERIMMUNOGLOBULINEMIA

The author defines essential hyperimmunoglobulinemia to be polyclonal hyperimmunoglobulinemia of unknown etiology and that not associated with any distinct neoplastic and reactive proliferations of the reticuloendothelial tissues. The condition is represented by purpura hyperglobulinaemica which was reported by WALDENSTRÖM in 1948.

This disease is of a benign, chronic process and as indicated previously, not only is there lacking complication of lymphadenopathy, hepatomegaly, or splenomegaly but clinically, reticular neoplastic and reactive proliferations are equally ruled out. Nevertheless, in the bone marrow there is only a mild proliferation of plasma cells and reticulum

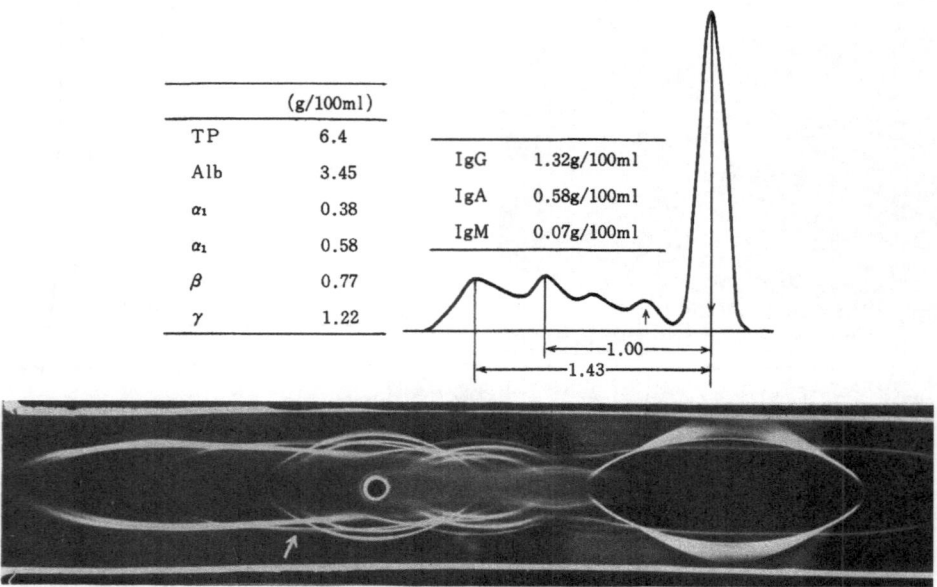

	(g/100ml)
TP	6.4
Alb	3.45
α_1	0.38
α_1	0.58
β	0.77
γ	1.22

IgG	1.32g/100ml
IgA	0.58g/100ml
IgM	0.07g/100ml

Fig. 192 Serum protein electrophoretic pattern showing markedly increased IgA. R. T., 46 y.o., male, WEGENER'S granulomatosis.
Sed rate 45 mm/hr. WBC 6,700 with a shift to the left (stab 25%). Proteinuria ($\pm$)—(2+), but normal renal function tests. CRP 2+, RA test (–), STS (–), cold hemagglutinin (–).
On cellulose acetate electrophoresis, the serum protein showed slight hypoalbuminemia. The modal mobility of the γ fraction was decreased, being 1.43, and the γ_1 zone was full. On immuno-electrophoresis, the IgA line is extremely prominent, but the IgM line is barely seen.

cells, if any. However, it is an important fact that it is not associated with the various diseases known to be complicated with hyperimmunoglobulinemia, and this disease, apart from a marked hyperglobulinemia, must be diagnosed by exclusion based entirely on negative findings. There is absolutely no indication as to the cause of the disease. As for the clinical symptoms, to be discussed later, they are not of a serious nature, taking a benign, chronic course, and it is difficult to establish the fact that the patient's immune mechanism shows any abnormal change. The author feels that there might be an abnormality in the feed-back mechanism to control the serum immunoglobulin level, as indicated in Fig. 172.

There are various diagnostic terms given to the similar cases. For example, a few would be purpura hyperglobulinaemica (WALDERSTRÖM), calcinosis hyperglobulinaemica (MIYOSHI), dysproteinemia associated with neurological symptoms, dysproteinemia associated with osteoporosis, and the like. All these diagnostic terms are assigned to the most prominent of clinical symptoms found. Not that these symptoms hold any peculiarities as regards the various disorders, but simply that they represent clinical manifestations resulted from hyperglobulinemia. As illustrated on Fig. 193, it is possible to group the clinical manifestations that arise as a result of a marked increase in serum immunoglobulins, but especially the IgG-globulin. They would be as follows: (1) an increased tendency for the rouleaux formation of the erytherocytes; (2) increased serum viscosity; (3) pathologically increased affinity to various blood constituents. The four principal clinical symptoms thought to arise from the participation of these three are rarely recognized individually, but they are frequently encountered in varying combinations.

Encephalopathia hyperglobulinaemica: With the hyperviscosity syndrome, various neurological symptoms can be recognized as a result of increased resistance of peripheral circu-

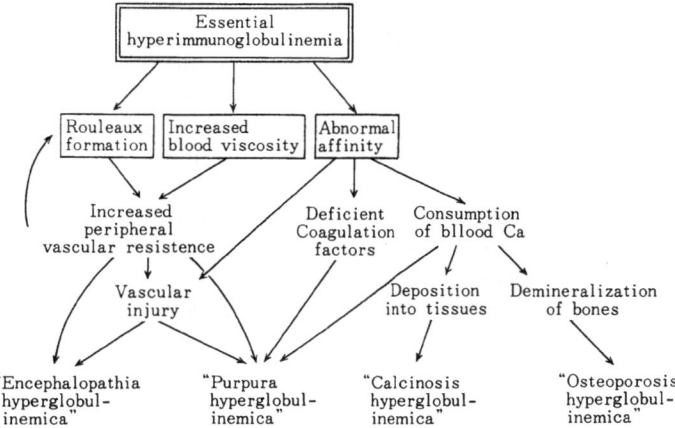

Fig. 193 Possible mechanisms for cardinal symptoms in essential hyperimmunoglobulinemia.

lation. These would be, for example, abnormal electroencephalograms, vertigo, nystagmus, ataxia, paresis, convulsions, visual disturbance, and the like. The case experienced in Japan involved an 18 year-old female, showing a broad increase of the γ fraction in the serum, up to 4~5.2 g/100 ml. Other than complicated neurological symtpoms, there were no remarkable clinical findings.

Purpura hyperglobulinaemica: Since WALDENSTRÖM first reported it, there have been a number of worldwide cases reported under the same diagnostic name.[379] A total of about 30 cases have been reported in Japan. The condition, called by this diagnostic term, is divided into primary and secondary forms. WALDENSTRÖM himself has reported that this syndrome is very often combined with SJÖGREN syndrome, SLE and others, or eventually develop various connective tissue disease.[379] However, the author finds difficult to understand that the cases combined with any connective tissue disease are included in the category of essential hyperimmunoglobulinemia. In these cases, purpura can be understood to occur symptomatically due to hyperglobulinemia which has been resulted by connective tissue diseases. Therefore, the author admits the occurrence of essential hyperimmunoglobulinemia only of the primary form.

In this disorder, the purpuras are detected mostly in the lower extremity, and further there is no hemorrhage in other organs or tissues. In rare instances, they have been detected in the arms, trunk, and abdomen; and even in situations where the purpuras are not in the upper extremity, the RUMPEL-LEEDE test frequently appears positive, leading up to suspect a strong postural factor. WALDENSTRÖM thinks that these clinical features are characteristic for the disease. The author himself would prefer to think that, with this disorder, the characteristic postural distribution of the petechiae basically fulfills the role of increased peripheral circulatory resistance, rather than the systemic disturbance of the capillary walls. For example, it is common knowledge, that for most of us who spend our days either standing or sitting, blood stagnation of the legs is brought on. Capillary blood vessels in the skin are likely to be damaged both extrinsically and intrinsically in comparison to those embedded in the visceral organs. Circulatory congestion will cause relative hemoconcentration, subsequently resulting in increased tendency for rouleaux formation of the erythrocytes and increased relative viscosity of the plasma. In this fashion, the abundance of purpuras in the skin of the lower extremities can be easily agreed upon.

The author finds it hard to accept such symptoms to be characteristic of purpura hyperglobulinaemica, but would rather think of them in terms of being one of the clinical symptoms associated with essential hyperimmunoglobulinemia, instead.

There are cases where immunoglobulinopathy is accompanied by bleeding tendency. There are still many unsettled problems as regards the causes for this, but the fact that a relative deficiency state is engendered by blood coagulation factors (in particular, Factor II, V, VII, etc.) is thought of as being one possibility.[359]

Calcinosis hyperglobulinaemica or Osteoporosis hyperglobulinaemica: In reference to both monoclonal and polyclonal hyperimmunoglobulinemia, the pathologically increased immunoglobulins can bind with calcium in the blood plasma.[117] As a result, the Ca^{++} ion concentration in the plasma is reduced. Then, in order to maintain the physiological equilibrium represented by the formula

$$[Ca^{++}] \times [HPO_4^{--}] = K$$

there may occur excessive release of the calcium ion from the bones, causing secondarily osteoporosis.[117]

On the other hand, the pathological immunoglobulins carrying abundant calcium circulate throughout the body, and result in calcification of various tissues. Nephrocalcinosis has been frequently observed[615], and MIYOSHI has reported a case of calcinosis hyperglobulinemica in which calcification of the lymph nodes was verified diffusely[386]. In general, pathologic calcification is encountered in the following two ways: (1) dystrophic calcification and (2) metastatic calcification[11]. The former represents cases where it occurs in necrotic foci and degenerated tissues. With the latter, where it happens in normal tissues due to the increase of plasma Ca^{++} level, usually involving the kidneys, gastric mucosa and lungs. In other words, in these latter three organs, because of secreting acid substances (phosphoric, hydrochloric, carbonic acids), there is a tendency for the tissues to become relatively alkali and the solubility of Ca^{++} is lowered. Furthermore, since necrotic tissues tend to be alkali,dystrophic calicification will be brought about. Therefore, when an excess of calcium combined with pathological immunoglobulins is circulating in great quantity, then metastatic calcification may occur subsequently.

In whatever instance taken, rather than thinking of the above symptoms as characteristic features for individual conditions, they should be understood as peculiar symtoms secondarily induced by pathological properties of the immunoglobulins. Therefore, the author has included them as a single clinical entity under the heading "essential hyperimmunoglobulinemia."

Chapter 23

Plasma Protein Changes in
M-proteinemic Type

DEFINITION AND SYNONYMS OF M-PROTEINEMIA

Each class of normal immunoglobulins is a more or less continuous spectrum of molecular variability, showing a characteristic electrophoretic heterogeneity. As a general rule, a single clone of the immunoglobulin-producing cells produces only a single variant of heavy chain and light chain, and they are commonly called as "monoclonal immunoglobulin". A single monoclonal immunoglobulin should be electrophoretically separated as a homogeneous, compact band or a discrete peak (Fig. 194). Monoclonal immunoglobulin is also called as M-protein, M-component, or M-fraction. At the beginning, "M" was understood to mean "myeloma" or "macroglobulinemia".[307] However, the identical immunoglobulin has been encountered in other clinical conditions, and thus "M" denotes "monoclonal" according to WALDENSTRÖM.[399] M-protein is also designated by WHO (1964) as "pathological immunoglobulin" (P.Ig), and defined as "abnormally homogeneous populations of immunoglobulin molecules".[338]

Monoclonal or pathological immunoglobulins may be recognized by the following two characteristic features. First, a sharp, discrete peak is observed on zone electrophoresis, and a localized distortion of the immunoglobulin lines (M-bow formation) is shown on immunoelectrophoresis. Secondly, they belong to a single immunochemical population involving both the heavy and light chains.

Many of the M-proteins are composed of both heavy and light chains, forming a complete molecular unit. However, they may be molecular fragments of either H chain or L chain. Properties of individual M-proteins will be discussed later.

M-proteinemia is defined as the pathological conditions where M-protein appears in blood and other body fluids. The term monoclonal gammapathy or gammopathy has been widely used since WALDENSTRÖM called as such.[399] Monoclonal gammopathy is frequently used interchangeably with monoclonal hypergammaglobulinemia. However, both terms should not be used interchangeably in a strict sense. Not all of M-proteinemias are associated with an increase in the serum immunoglobulin concentration. This condition is also refered to plasma cell dyscrasia,[280] paraimmunoglobulinopathy,[84] or many other terms.

CHARACTERISTICS OF THE M-PROTEINEMIC SERUM PROTEIN ELECTROPHORETIC PATTERN

Appearance of M-protein Band

On electrophoretic separation of serum proteins, the albumin fraction is typical of a homogeneous protein band. Therefore, any abnormal protein band must be compared

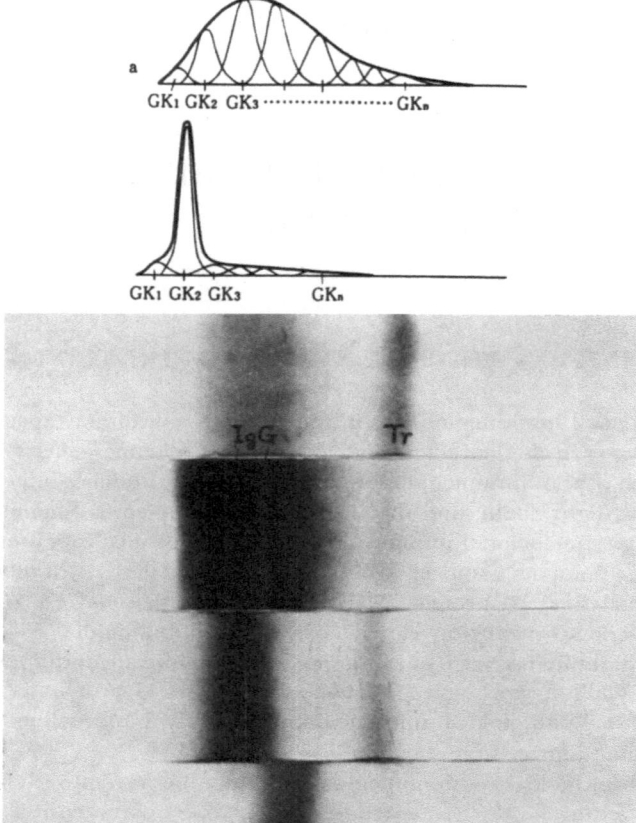

Fig. 194 Electrophoretic hemogeneity of M-protein.
In normal and polyclonal hyperimmunoglobulinemic sera, the IgG molecules are composed of those produced from multiple clones (GK$_1$, GK$_2$, ...) (a). However, in monoclonal hyperimmunoglobulinemic sera, only the single clone (GK$_2$) is stimulated to produce an increased amount of the corresponding molecule (b). On filter paper electrophoresis, even the normal IgG shows a broad protein band in comparison to a discrete protein band of transferrin (Tr.) In polyclonal hyperimmunoglobulinemia (C), the IgG band is broader than the normal (N). In the sera from IgG type M-proteinemia (1, 4), an extremely compact band of the IgG is seen, each having different electrophoretic mobility. The electrophoretic patterns were obtained on Rivanol-treated sera (mainly containing IgG and transferrin).

with the albumin fraction in order to see if it be homogeneous. A compact band similar to M-protein band may be recognized in the prealbumin and α_1 areas, resulted from para-albumin, alloalbumin, prealbumin, α_1-lipoprotein and α_1-fetoprotein. The latter protein bands should not be confused with M-protein band.

On cellulose acetate electrophoresis, M-protein band may be clearly recognized when the serum concentration of M-protein is more than 0.3 g/100 ml. However, M-protein of less than 0.1 g/100 ml serum may be recognized if it migrates at the γ_1 area or the γ fraction is decreased markedly.

M-proteins of the IgG type are the most homogeneous and often are observed as very compact protein bands (Fig. 194). On the contrary, other M-proteins, that is, the IgA type, the IgM type, the BENCE JONES type, and the heavy chain fragments usually all migrate rather broadly in comparison to the IgG type. So in many cases it may be difficult to distinguish a monoclonal increase from a polyclonal increase unless other analytical procedures are used.

Increase of Globulin Fractions

As described in the previous section, when M-protein exists in a great amount, it can be observed as a distinct M-protein band. However, M-proteins other than the IgG type having almost the same mobility as those of normally identified globulin fractions may be difficult to be identified as M-protein bands. Therefore, it is recommended to use agar gel electrophoretic methods, in order to detect a small amount of M-proteins. Agar gel is appropriate for the observation of the γ and slow- γ areas, while agarose gel is appropriate for the observation of the γ_1, β and α_2 areas.

Even if the M-protein is present in the serum concentration of above 0.1 g/100 ml, they sometimes escape observation when they migrate overlapping the β fraction. It is particularly like to occur with M-proteins of the IgA, IgM, and IgD as well as the BENCE JONES protein type. Therefore, when the presence of M-proteins is clinically suspected, immunoelectrophoretic analysis must be done.

Changes in the Serum Protein Fractions Other than the M-protein Band

The serum protein fractions other than M-fractions shows fundamentally the change of acute phase response type. Serum total protein concentration increases in most cases of IgG-and IgA-type myeloma, reaching sometimes up to 14 g/100 ml (Table 52). The albumin fraction decreases in every case and the α_1 and α_2 fractions tend to increase in

Table 52 Frequency of various types of M-proteinemia.

A total of 177 cases with M-proteinemia is listed here, all of which have been studied in Central Clinical Laboratories of Nihon University Hospitals, from 1966 to 1970.

	No. of cases	Frequency (%)	Light Chain (K : L)	BENCE JONES Protein found
Multiple Myeloma				
IgG type	26	14.6	13 : 10	5/21 (23.8%)
IgA type	11	6.2	4 : 6	3/8 (37.5%)
IgD type	3	1.7	0 : 3	3/3 (100%)
Bence Jones type	6	3.4	0 : 5	6/6 (100%)
Atypical	1	0.6		0/1
Primary Macroglobulinemia	4	2.3	3 : 1	2/3 (66.7%)
Cold Agglutinin Disease	1	0.6	0 : 1	
Essential M-proteinemia				
IgG type	83	46.9	42 : 33	4/18 (22.0%)
IgA type	21	11.9	13 : 7	0/6
IgM type	9	4.4	3 : 3	0/2
Mixed type IgG+IgA	1	0.6	0 : 1	
IgD+BJP	1	0.6	0 : 1	
Undetermined				
IgG type	10	6.2	2 : 6	0/2
IgM type	1			
Total	177		80 : 76	24/72 (33.3%)

most cases.   This is the same as in the cases of malignant tumors.   The β fraction increases remarkably when the M-proteins have their mobility at the β zone, but in other cases it mostly shows the tendency to decrease.   The γ fraction is influenced by the mobility of M-proteins as in the case of the β fraction.   When M-proteins are present in the globulin fractions other than the γ fraction, or in M-proteinemia of the minor anomaly type, the γ fraction mostly decreases.   Positive CRP-test is observed in about 30% of all cases.

Changes in the Serum Immunoglobulins Other than the M-protein

As shown in Fig. 195, normal immunoglobulins other than M-proteins generally tend to decrease, and the tendency is notable in the patients of multiple myeloma.[54,68,283]   It is observed that the greater the amount of M-proteins, the stronger the tendency to decrease of normal immunoglobulins (Fig 196).   In some cases of essential M-proteinemia, normal immunoglobulins may be increased instead.

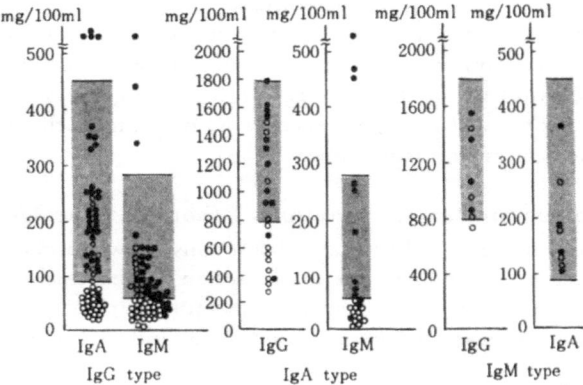

Fig. 195   Serum immunoglobulin levels in M-proteinemia.
The white dots indicate the values from the patients with multiple myeloma or primary macroglobulinemia, while the black dots indicate those from essential M-proteinemia.
The normal immunoglobulin components are decreased more prominently in myeloma than in essential M-proteinemia. No significant decrease of the normal immunoglobulin components is seen in macroglobulinemia.

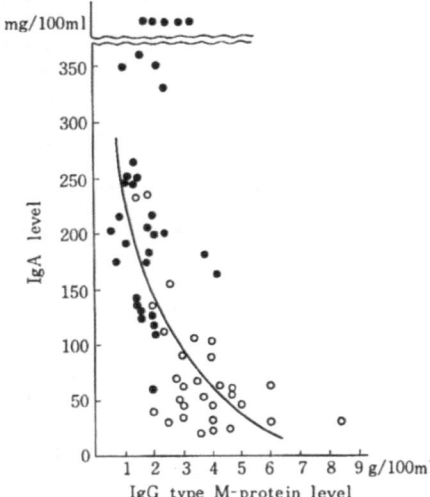

Fig. 196   Relation between the M-protein level and the IgA level in IgG type M-proteinemia.
The black dots indicate the values from the patients with essential M-proteinemia, and the white dots indicate those from multiple myeloma.

Changes of Serum Lipids in the M-Proteinemias

In M-proteinemia, serum lipids generally remain normal or rather tend to decrease. Especially this can be said of cases where M-protein concentration is remarkably high and the synthesis of other serum protein components is decreased. However, it was reported that in rare cases M-proteinemia is accompanied by marked hyperlipidemia[99]. There have been several cases reported in which multiple myeloma is accompanied by hyperlipidemia or xanthoma, but except for the cases studied by NEUFELD et al.[267], no detailed analysis has been made available. The author also experienced two cases of M-proteinemia combined with hyperlipidemia (Fig. 197, 198). Two mechanisms may be possible to explain the association of hyperlipidemia: 1) M-proteinemia and hyperlipidemia occur in complete independence of each other, 2) both diseases result from a single pathological condition. The former case is thought to be very rare, if any. The author's second case seems to belong to this category, but detailed analysis is not available. At least, the IgG-type M-protein was confirmed not to carry lipids.

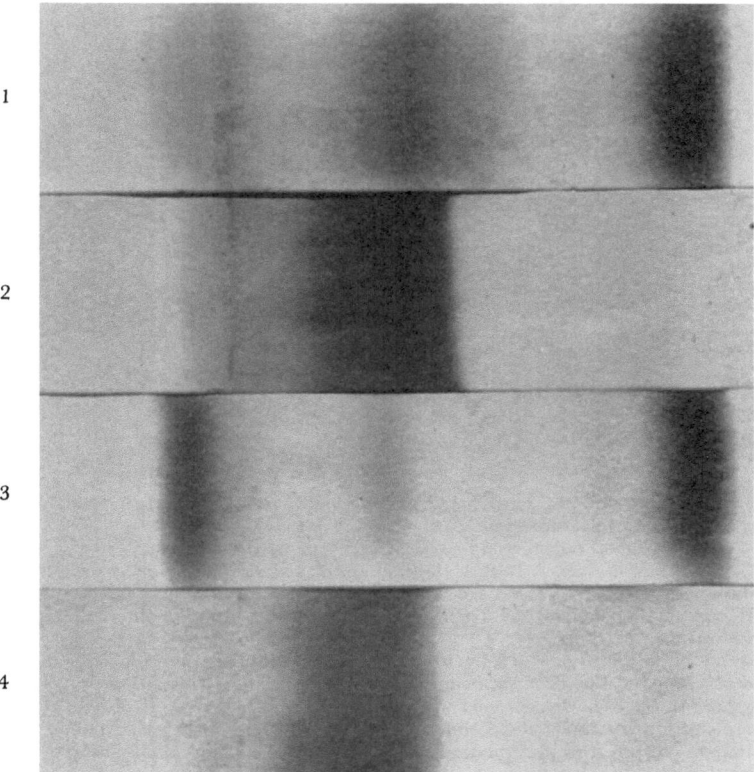

Fig. 197 Filter paper electrophoretic patterns in 2 cases of essential M-proteinemia associated with hyperlipidemia.
Case 1 (S. H. 56 y.o., female): Patterns 1 and 2, also refer to Fig. 198.
Case 2 (T. S. female): Patterns 3 and 4.
The IgG type M-protein is migrated at the slow-γ zone. LDL is increased markedly, the serum total cholesterol being 442 mg/100 ml. Slight plasmocytosis in the bone marrow. No anemia, and excellent general condition at the time of protein analysis. The patterns 1 and 3 were stained with BPB dye for protein moiety, and the patterns 2 and 4 were stained with lipophilic dye (Sudan black) for lipid moiety.

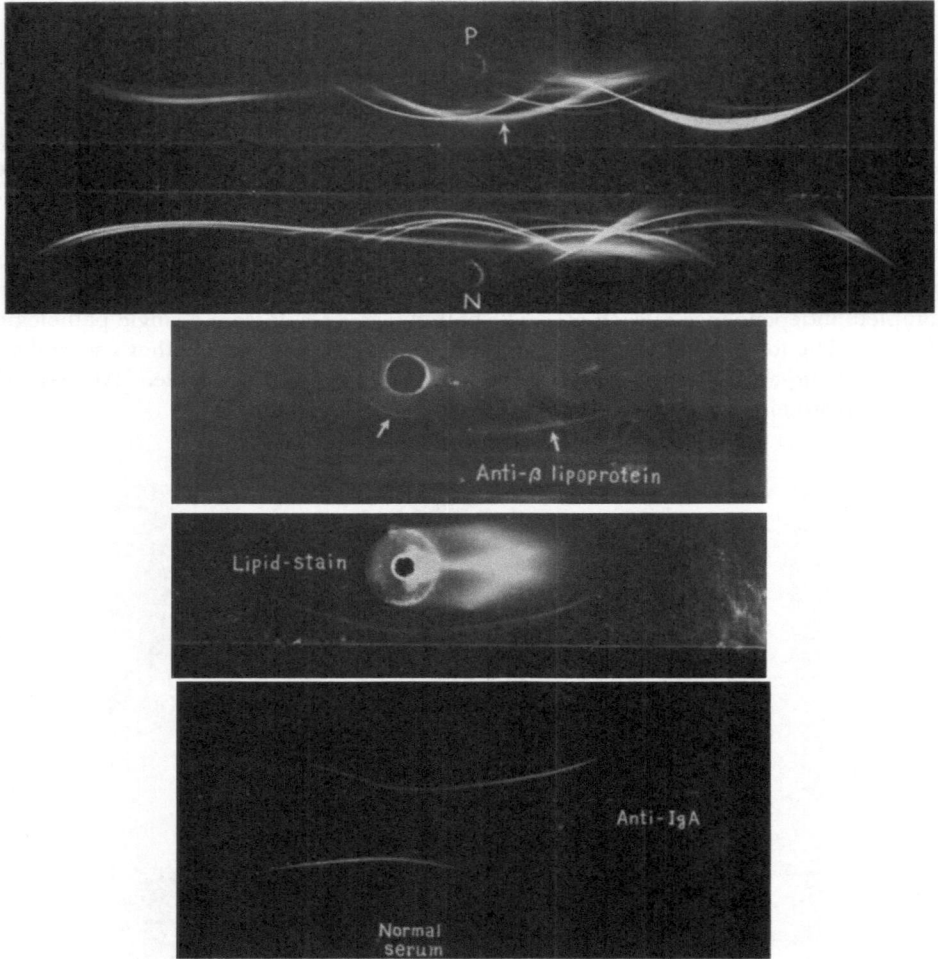

Fig. 198 Immunoelectrophoretic patterns of serum proteins in essential IgA type M-proteinemia
associated with hyperlipidemia.

S. H., 56 y.o., female, hyperlipidemia and essential IgA type M-proteinemia.

Suffered from hypertension since the age of 18 years, developed xanthoma at the age of 49 years, and
hepatomegaly at the age of 52 years.   No familial occurrence.   Glycosuria noted.   Serum neutral
fat 925 mg/100 ml, total cholesterol 754 mg/100 ml (esterified, 66%), phospholipid 435 mg/100 ml, free
fatty acid 1144 mEq/L.

Serum total protein concentration was 8.1 g/100 ml, measured with the micro-biuret method.   A broad
protein band was noted at the β zone, amounting to 36.1% or 2.9 g/100 ml (note Fig. 197).   On immunoelectrophoresis, the abnormal protein was identified as the IgA type M-protein.   The IgA line
formed a characteristic M-bow at the β zone, tailing towards the anode.   Using the specific anti-β-
lipoprotein immune serum, two precipitation lines were formed.   The lipid staining of the immunoelectrophoretic patterns was helpful to indentify the complex formation of the IgA type M-protein
and β-lipoprotein.

The latter case is of great interest to the author from the pathophysiological stand point.
The case reported by NEUFELD et al. is best investigated.   The first case that the author
experienced (Fig. 198) resembles multiple myeloma very much, although it was not
diagnosed as such. Both belong to the IgA type M-proteins migrating at the β zone.
NEUFELD et al. supposed that pathological β-globulins which contain great amounts of

lipids are synthesized in plasma cells. However, as is the same in the author's case, it can not be completely excluded the possiblity that the IgA type M-proteins after their synthesis may cause inter-molecular aggregation with lipoproteins. On the other hand, in some cases, M-proteins bind with heparin in the blood and therefore PHLA (postheparin lipolytic activity) decreases, causing serious hyperlipidemia.[99] At any rate, it is rare that M-proteinemia is accompanied by hyperlipidemia.

DIFFERENTIATION OF THE M-PROTEINEMIC PATTERN

Differentiation from the Polyclonal Hypergammaglobulinemic Pattern

If one reads only the numerical results obtained by the zone electrophoretic technique on each serum protein fraction, M-proteinemia will not be detected. Therefore, the electrophoretic strips should be observed carefully. At least, before excluding the possibility of M-proteinemia, a careful observation of the electrophoretic strip or of the densitometric pattern is mandatory.

Differentiation from Artificial Discrete Protein Peaks

When one observes electrophoretic strips, especially densitometric patterns, one has to exclude "discrete peaks" caused by the following artificial effects (see p. 188). Sometimes apparently sharp peaks appear due to the presence of degenerated proteins and chylomicron at the point of serum application. However, in this case, if one observes the electrophoretic strip itself, one can easily distinguish the degenerated proteins adsorbed to the surface of the strips.

Differentiation from Sharp Protein Peaks of the Normal Globulin Components

Some of normal plasma protein components including low-density lipoproteins, hemoglobin, β_{1C}-globulin and fibrinogen may be observed as discrete protein bands on electrophoretic fractionation.

The low-density lipoproteins are observed as a linear band or a very sharp peak when cellulose acetate membrane (especially Separax) is used. This can be easily distinguished according to the mobility and the stained pattern. That is, the peak is observed as a line from the α_2 zone to the β_1 zone. On immunoelectrophoresis, an increase of LDL will be easily distinguished.

Hemoglobins migrate to the β zone and they are very difficult to be differentiated from M-protein bands by zone electrophoretic separation alone. Of course, they can be distinguished easily by the red color of the protein band before staining. They may be definitely identified by immunoelectrophoresis and by benzidine stain.

β_{1C}-globulin band is sometimes observed at the β_2 zone as a separate protein band when fresh serum sample is applied for analysis. However, it disappears because of its conversion into β_{1A}-globulin after stored in a refrigerator for more than three or four days.

Fibrinogen is found at the γ_1 zone as a rather discrete peak. On cellulose acetate and filter-paper electrophoresis, it forms a protein band of comparatively characteristic appearance. Therefore, once one gets accustomed to it, it can be easily distinguished. In order to identify it definitely, immunoelectrophoresis using anti-fibrinogen is helpful.

Sometimes, the hemoglobin-haptoglobin complex may be observed as the α_3 fraction. Also, α_{2M}-globulin may be observed to be somewhat separated from other α_2 globulins.

Unidentified Protein Bands not to be Neglected

Unidentified protein bands must not be ignored even when there is no abnormality on the immunoelectrophoretic patterns.   They might be some proteins coming from various destroyed tissues, but one should rather suspect the presence of IgD, IgE and BENCE JONES proteins.   On immunoelectrophoresis, only the protein components which correspond to the antibodies contained in the immune sera used are detected.   Therefore, the author always sees the agar gel electrophoretic patterns obtained under the same electrophoretic conditions as in the immunoelectrophoretic fractionation, when the immunoelectrophoretic patterns are to be examined.

REPRESENTATIVE DISEASES ASSOCIATED WITH THE M-PROTEINEMIC PATTERN

It is known that there are many kinds of M-proteins.   The properties of each kind of M-protein is described later.   The immunological type of the various M-proteins does not directly correlate with clinical pathological patterns.   Up to the present, M-proteinemia has been encountered in various diseases and conditions listed in Table 53.

Table 53   Representative diseases associated with M-proteinemia.

I.	Multiple myeloma		
	A. IgG type	E.	IgM type
	B. IgA type	F.	IgE type
	C. BENCE JONES type	G.	Mixed type
	D. IgD type	H.	Atypical
II.	Primary macroglobulinemia WALDENSTRÖM		
III.	Essential M-proteinemia		
	A. IgG type	D.	IgD type (?)
	B. IgA type	E.	BENCE JONES type
	C. IgM type		
IV.	7S IgM disease (SOLOMON-KUNKEL disease)		
V.	Half-molecule IgG type plasmacytoma (HOFFS-JACOBS disease)		
VI.	Heavy chain disease		
	A. γ chain disease (FRANKLIN disease)		
	B. α chain disease		
	C. μ chain disease		
V.	Amyloidosis		
VI.	Chronic cold hemagglutinin syndrome		
VII.	Autoimmune diseases with monoclonal autoantibodies		

PATHOGENESIS OF M-PROTEINEMIA

M-proteins are not only electrophoretically homogeneous, but also in many respects including physical, chemical, immunochemical and biological properties.   Therefore, one would expect each M-protein to be the product of a single clone of cells.   Also, in general, the antibody molecules of single cell line are directed against only a single antigen.   In fact, a normal antibody immunoglobulin of a monospecific nature is quite similar to M-proteins in many respects.   While the M-proteinemias are themselves pathological, there is no convincing evidence that such M-proteins are abnormal.

What is the reason for appearance of monoclonal immunoglobulins in some cases, while a single antigenic stimulation may usually cause polyclonal hyperimmunoglobulinemia.   At least in multiple myeloma, a single clone of immunoglobulin-producing cells is thought to show a neoplastic proliferation of unknown etiology. However, in other conditions accompanying M-proteinemias, many uncertain points have been left to be solved. First, let us see what kinds of clinical facts have been observed.

1. Transition from a Polyclonal Proliferation to a Monoclonal Proliferation

As indicated by the statistical observation in many cases, it is very rare that polyclonal hyperimmunoglobulinemia changes to monoclonal proliferation, and even if they exist, they may be thought to be very special cases.[133,281,399] That is, there is only one such case among 300 cases of M-proteinemia which the author has experienced (Fig. 199). In this case with HODGKIN's disease, polyclonal hyperimmunoglobulinemia was seen at the beginning and later hypogammaglobulinemia appeared due to diffuse involvement of the lymphoreticular tissues, followed by M-proteinemia before death. Similar cases have been reported by KRAUSS and SOKAL, and they state that the appearance of M-proteinemia leads to unfavorable prognosis.[197] Therefore, in such cases, rather than to conclude that the polyclonal abnormality changes directly to the monoclonal, the author prefers to think that the fundamental natures of their pathology are entirely different. In other words,

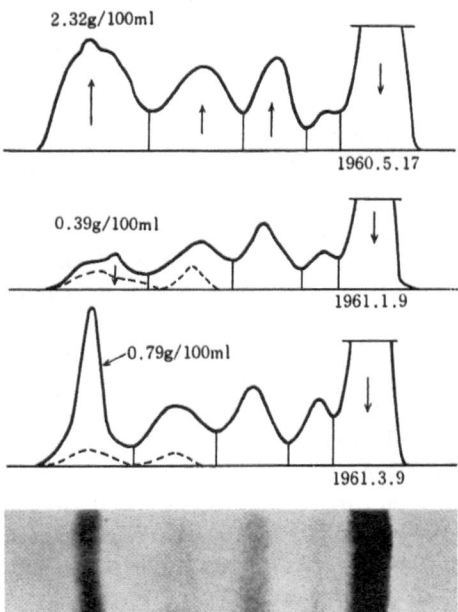

Fig. 199 A transition from polycolonal to monoclonal immunoglobulinopathies.
13 y. o., male, HODGKIN's disease.
In May, 1960, diagnosed as HODGKIN's disease through biopsy. No significant response to X-irradiation, but lymphadenopathy improved with Endoxan and Mitomycin. Two months later, lymphadenopathy became worse, and Endoxan was given. Later developed cachexia, icterus and marked lymphadenopathy. A discrete peak on serum protein electrophoretic fractionation, immediately before death.
On immunoelectrophoresis, IgG, IgA and IgM were all increased at the beginning. Only IgG line was vaguely seen when hypogammaglobulinemia was recognized. At the end, no immunoelectrophoretic analysis was possible, but the M-protein peak was removed completely with Rivanol precipitation, suggesting the possibility that the IgM type M-proteinemia was present.

immunoglobulin-producing tissues which show an almost normal reactivity during the poly-clonal proliferating period are destroyed to the point of losing their control mechanism with regard to clonal distribution in its final stages, thus developing monoclonal proliferation. Similar pathological patterns are found in the exhaustion stage of experimental protracted sensitization, examined by OKABAYASHI et al. and also in experimental myeloma in mice.[277]

With the exception of the special cases cited above, there is no transition between the polyclonal and monoclonal types and the two pathological conditions are thought to come from different pathophysiological backgrounds. However, there are findings which sup-port indirectly this theory; that is, in M-proteinemia, serological false positive reactions occur very rarely (serological tests for syphilis, rheumatoid factor, and antinuclear factor). The author experienced only 2 cases which show positive RA test among about 300 cases. One of the two was of chronic hepatitis in which IgG type M-protein is present in associa-tion with polyclonal increase in IgG and IgM. In the other case, a positive reaction to the serological tests for syphilis was noted, but this was not biological false positivity.

2. Predominance of the L-type IgG-Globulin

As shown in Fig. 179 of the previous chapter, the author observed 3 cases in which IgG increases broadly, but predominantly of L-type IgG increase. In each case as shown in Fig. 179, the cathodal end of the IgG precipitin line exhibits crossing or bifurcation, when the particular horse immune serum is used.

All of this was observed in cancer patients, and their serum and urine did not contain free light chain components. This phenomenon may serve as an intermediate condition between polyclonal and monoclonal disorders.

3. Disappearance of M-protein

ANDERSON and FERRIMAN reported that in a case of WALDENSTRÖM's macroglobulinemia M-protein disappeared after the administration of nitrogen mustard.[8] All the other cases reported belong to essential M-proteinemia and most of them are of the IgG type. A few cases of the IgM type M-proteinemias have been observed. In the two cases ob-served by the author (see Fig. 200, 238), M-proteins disappeared after treatment. Cases which have been documented up to the present are mostly the diseases accompanying proliferation of the immunoglobulin-producing cells, such as autoimmune diseases, infec-tions, and cirrhosis of the liver. Moreover, M-proteins disappear upon recovery from or improvement of the primary disease. Also as listed in Table 59, the diseases which are combined with essential M-proteinemia often are accompanied by polyclonal hyperim-munoglobulinemia. This fact suggests that a similar pathogenesis is shared partly in both polyclonal and monoclonal abnormalities

4. Frequent Occurrence of M-proteinemia in Aged Persons.

Essential M-proteinemia is prone to occur in an advanced age. HÄLLÉN reported that it was found in 3% of the individuals of over 70 years of age. The author et al. also con-firmed in Japaneses of over 60 years of age; 8 out of 395 males (2.03%) and 4 out of 414 females (0.96%) were found to have M-proteinemias, including 5 cases of IgG-K type, 2 cases of IgG-L type, 3 cases of IgA-K type and 2 cases of IgA-L type. Relatively frequent occurrence of the essential M-proteinemia in an advanced age may be explained in that aging would cause instability of the reticular tissue in some way.

5. Familial Occurrence of M-proteinemias

There are rare reports indicating the familial occurrence of M-proteine-mia.[15, 33, 339, 409] The author has experienced a case of macroglobulinemia (Fig. 209) in which the father of the patient had died from plasma cell leukemia associated with

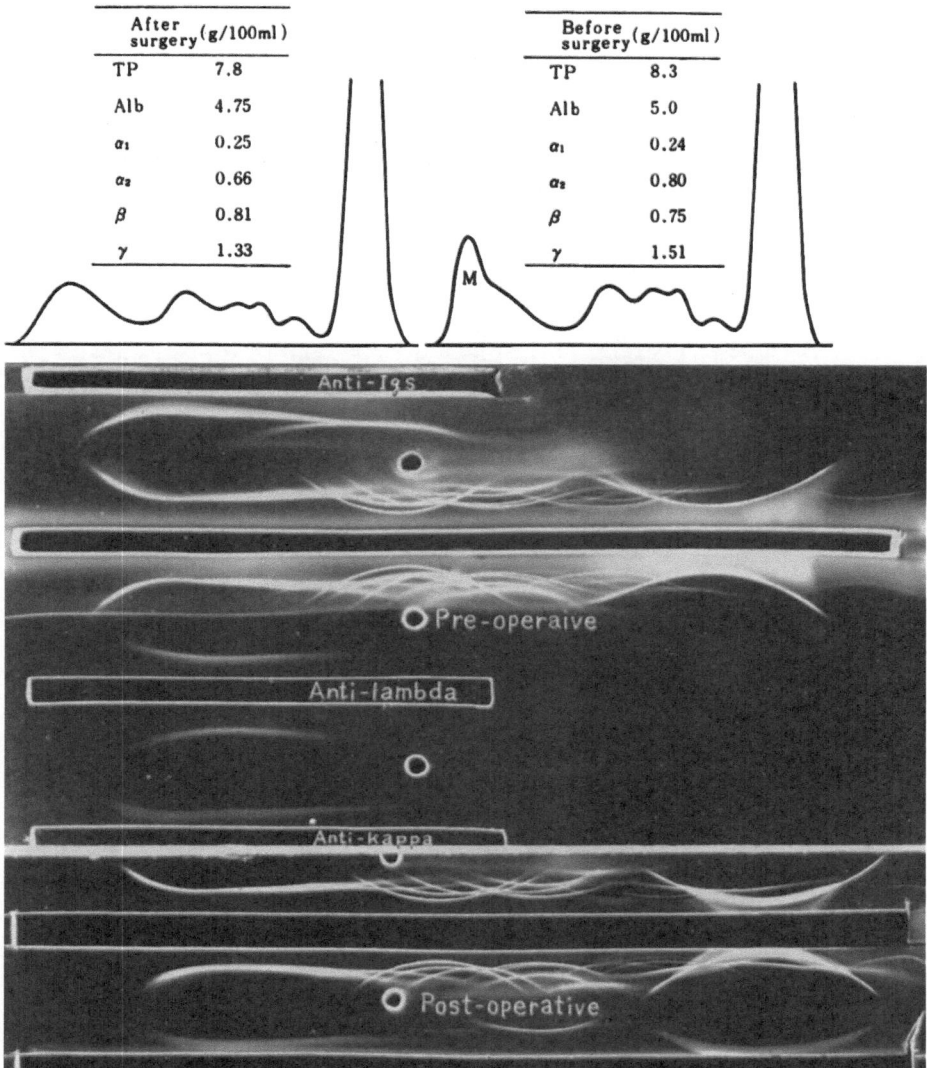

After surgery (g/100ml)	
TP	7.8
Alb	4.75
α_1	0.25
α_2	0.66
β	0.81
γ	1.33

Before surgery (g/100ml)	
TP	8.3
Alb	5.0
α_1	0.24
α_2	0.80
β	0.75
γ	1.51

Fig. 200   Serum protein electrophoretic patterns in essential M-proteinemia.
S. M., 58 y. o., female, HASHIMOTO's disease.
Complained of struma for the past 2 years, and total thyroidectomy done.   The removed thyroid weighed 77.5 g, and was diagnosed as HASHIMOTO's thyroiditis histologically.
Before surgery, the serum protein electrophoretic pattern showed a minor spike at the slow-γ zone, which was confirmed to be the IgG-K type M-protein.   The M-protein did not disappear even after the serum was absorbed by the reagent for TA test.   Two months after surgery, the M-protein band disappeared completely, and its immunoelectrophoretic pattern failed to show any abnormality.

IgM-type M-proteinemia.   On the other hand, AXELSSON and HÄLLÉN investigated a total of 6,995 inhabitants being above 25 years old in a particular district of Sweden and observed M-proteinemia in 64 cases and reported familial occurrence of M-proteinemia in 3 instances.[15]   These cases were found to be normal hematologically and diagnosed as essential M-proteinemia, all of them being aged persons from 50 to 80 years old. This occurrence does not seem to be merely accidental.   Among these families, no poly-

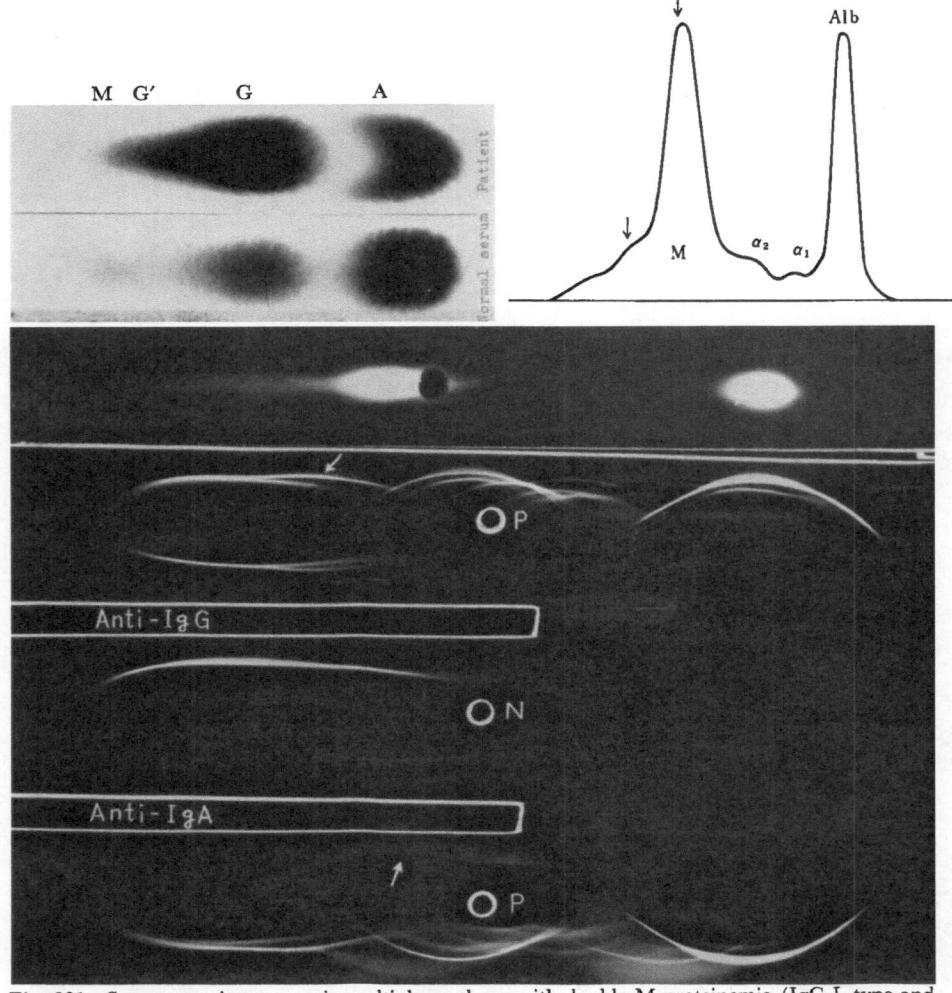

Fig. 201 Serum protein patterns in multiple myeloma with double M-proteinemia (IgG-L type and IgA-L type).

T. K., 52 y.o., male, multiple myeloma

Admitted with lumbago, anemia (Hb 7.5 g/100 ml) and marked accentuation of the sed rate (148 mm/hr). On X-ray examinations, the lumbar vertebrae showed osteolytic lesion. The bone marrow contained markedly atypical plasma cells in 24%. On fluorescent antibody techniques, the IgA-producing cells were much more predominant than the IgG-producing cells, and there was no definite morphological difference between the two different groups of the plasma cells. The serum total protein was 10.5 g/100 ml. On cellulose acetate electrophoresis, a major M-protein band was recognized at the β zone, amounting to 53% or 5.6 g/100 ml, and a minor M-protein band at the β_2 area. On thin-layer gel filtration, marked decrease in the A and M fractions was noted, while the G fraction was increased and the G' fraction was also recognized. On immunoelectrophoresis, identified were a large amount of the IgG-L type M-protein at the β_2 zone.

clonal hyperimmunoglobulinemia was found. This also indicates that polyclonal proliferation and monoclonal proliferation are essentially different pathological conditions.

6. Appearance of More Than 2 Kinds of M-Proteins

There are times when, in M-proteinemia, two M-peaks are observed (Fig. 201, 205). This can be interpreted as two different phenomena; that is, (1) it is caused by one kind of M-protein, but, because of strong molecular aggregation, giant molecular aggregates are separated from monomeric M-protein, (2) two different kinds of M-proteins are present. In the former case, the phenomenon is merely the result of physico-chemical molecular interaction, but in the latter case it is essentially different.[66,375,392] FATEH-MOGHADAM, et al. called "Doppelparaproteinaemie" in which two kinds of M-proteins appear, and observed in 0.4% of all M-proteinemias, occurring in the following combinations: IgG-K+IgA-K, IgG-K+IgA-L, IgA-K+IgA-L, IgG-K+IgG-L.[87] Until the present, there has been no definitive evidence as to whether, in these cases, two kinds of M-proteins are produced from different clones. In rare cases, more than two kinds of M-proteins are encountered in the same patient. Sometimes, multiple minor spikes may be observed at the γ area on agarose gel electrophoresis, particularly in diseases such as auto-immune diseases and malignant lymphoma which are frequently associated with marked destruction of the immunoglobulin-producing tisseus. These facts may lead one to think that multiple clones are proliferated incidentally at the same time. However, the author prefers to think that appearance of multiple minor spikes are the result of the instability of the immunoglobulin-producing tissues.

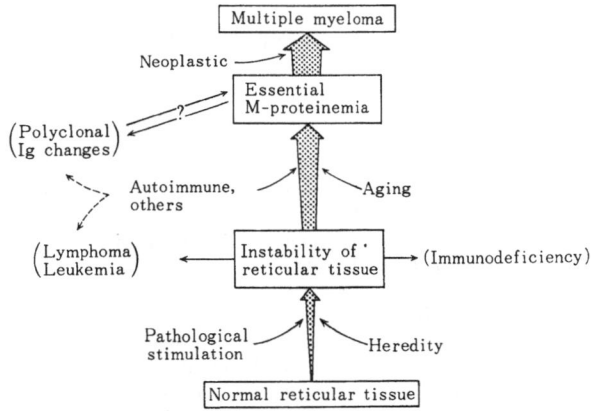

Fig. 202 Hypothetical diagram on the pathogenesis of M-proteinemia.

7. Transition from the Essential Form to the Neoplastic Form

Essential M-proteinemia is not basically the result of the neoplastic proliferation of immunoglobulin-producing cells. It may last for a long period of time without any significant change and it even disappears in rare occasions. However, there are few reports of essential M-proteinemia which changed into multiple myeloma eventually after more than 10 years, and some researchers think that essential M-proteinemia is a potentially malignant condition.[152, 204] But considering the fact that in recent years a great number of cases of essential M-proteinemia has been observed, the change into multiple myeloma occurs so rarely that it may be considered just as an incidental combination of the two conditions. Therefore, the author does not necessarily agree with the idea of pre-malignancy.

Taking these clinical facts into consideration, the author has attempted to summarize the possible mechanisms involving the pathogenesis of M-proteinemia (Fig. 202). It may be thought that there are persons who congenitally have "unstable" reticular tissues, while some may acquire "unstable" reticular tissues through certain pathological conditions involving the reticular tissues. At any rate, the instability of the reticular tissues is likely to precede, causing abnormal clonal distribution of the immunoglobulin-producing cells. Various inducing factors may then result in reticular malignancies in some patients, while immunodeficiencies in some other patients.   M-proteinemia may be produced due to an excessive proliferation of a certain clone or due to suppression of all clones except for one left to be proliferating, when certain inducing factors are added to "unstable" reticular tissues or "ready-to-go" state.   The inducing factors could be aging which may be of an autoimmune process, or the pathological conditions causing polyclonal hyperimmunoglobulinemia such as malignancies, autoimmune diseases and others.   Essentially, as long as certain factors necessary for neoplastic transformation are not added, the transition from essential M-proteinemia to multiple myeloma may not occur.   In essential M-proteinemias, it seems that the feed-back mechanism of controlling the serum immunoglobulin concentration is maintained on a different level from the normality.   Even if there may not be a direct relation between essential M-proteinemia and polyclonal hyperimmunoglobulinemia, the inducing factors for the two conditions seem to have much in common.

CLASSIFICATION AND CHARACTERISTICS OF M-PROTEINS

Various types of M-proteins are listed in Table 54.   All of these are, in a broad sense, immunoglobulins or their fragments.

Table 54   Various Types of M-proteins.

I.   M-proteins composed of both H chain and L chain.
    A.   IgG type M-protein
    B.   IgA type M-protein
    C.   IgM type M-protein (or 19S IgM type)
    D.   IgD type M-protein
    E.   IgE type M-protein
    F.   7S IgM type M-protein
    G.   Half-molecule IgG type M-protein (composed of
        one H chain and one L chain)
II.   M-proteins composed of immunoglobulin fragments
    A. γ type M-protein
    B. α type M-protein
    C. μ type M-protein
    D.   BENCE JONES protein

IgG Type M-proteins

1.   Electrophoretic Characteristics

On electrophoresis, IgG type M-proteins migrate as a very compact band or a sharp peak, and their mobility ranges from the α_2 zone to the slow-γ zone.   On cellulose acetate electrophoresis using Separax or Oxoid membrane, they sometimes form a characteristic wavy band.   The mobility of M-proteins differs in different cases.   However, among the cases with essential M-proteinemia of IgG type, their frequency distribution is almost identical to that of normal IgG molecules (Fig. 203).   In multiple myeloma of IgG type, the slow-γ mobility is recognized predominantly.

Since the TISELIUS method and the filter-paper method do not have the molecular sieve effect, the relative mobility of M-proteins is comparable to the true mobility.   However, by the use of starch gel and polyacrylamide gel electrophoresis which have significant molecular-sieve effect, the true mobility cannot be assumed.   By the agar gel method, M-proteins and agar gel interact in the buffer of low ionic strength and sometimes cause white precipitation resulting in a remarkable variation of their mobility. The phenomenon is frequently observed with the IgG type M-proteins of the slow-γ mobility, and the M-proteins stay around the point of serum application. M-proteins are observed sometimes as two or more M-protein bands in starch gel electrophoresis and rarely in agar gel electrophoresis. In this case, it is not necessarily implied that more than two kinds of M-proteins appear but as explained later, it may be that some M-proteins cause molecular aggregation and a portion of the M-protein exists in the living body as giant molecular aggregates (Fig. 205).

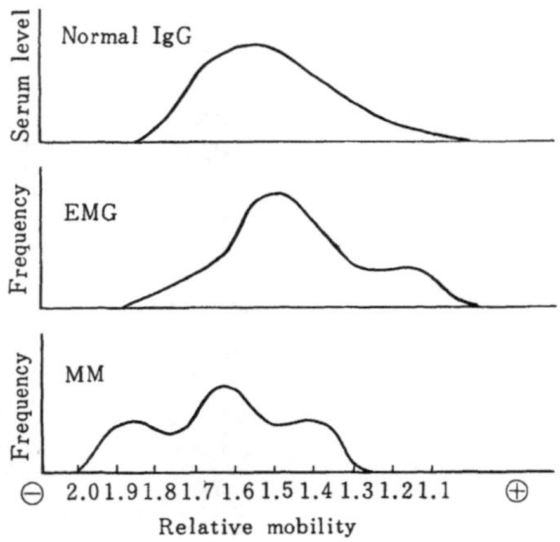

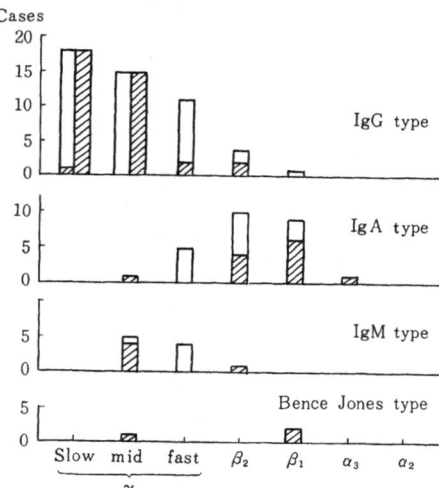

Fig. 203 Comparison between the clonal distribution of the normal serum IgG and the relative mobility of IgG type M-proteins.

The relative mobility of the M-proteins was expressed in comparison to the distance between the albumin and β-fractions. In essential M-proteinemia, the distribution of the relative mobility among the M-proteins is quite comparable to the clonal distribution of the normal serum IgG. However, in IgG type multiple myeoma, the slow-γ type M-proteins appear rather frequently.

Fig. 204 Distribution of the relative mobility among different types of M-proteins.

The shaded areas indicate the cases with multiple myeloma or primary macroglobulinemia. The other empty areas indicate the cases with essential M-proteinemias.

2. Physical Characteristics

Usually, the IgG-type M-proteins have a sedimentation constant of about 7S and are observed in the G-fraction on the ultracentrifugal and thin-layer gel filtration patterns. However, since molecular aggregation sometimes occurs among M-proteins, they sometimes form giant molecular aggregates of about 12 S.[261, 350] By using the thin-layer gel filtration technique, also, the author and his collaborators classify into 3 different patterns: 1) patterns in which M-proteins are observed only in the G-fraction, 2) patterns in which M-proteins are observed mainly in the G-fraction but extend a little toward the M-

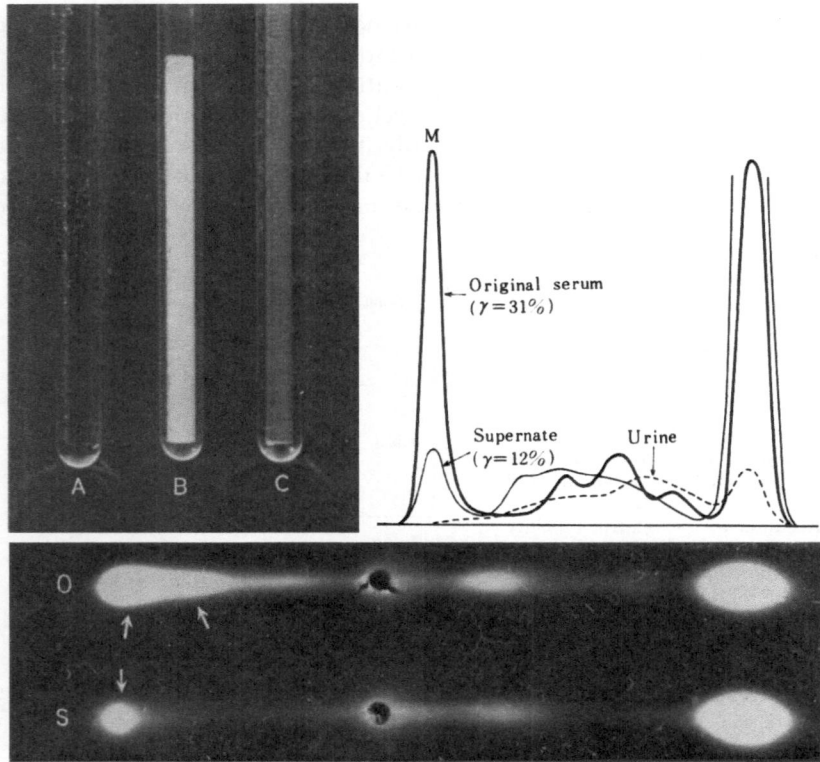

Fig. 205 Serum protein patterns in pyroglobulinemia.

E. I., 58 y.o., male, multiple myeloma.

Complained of lumbago and loss of appetite.   Atypical plasma cells, 12% in the bone marrow.   Sed rate 125 mm/hr.   Slight anemia with marked rouleaux formation of the erythrocytes.   ZTT 20 units, TTT 12.3 units.

The serum showed white coagulation after heating at 56 °C for 30 minutes (upper left).   The pyrocrit value after centrifugation was 73% (B).   (A) is the non-heated pateint's serum, and (C) is the normal serum heated at 56 °C.

The serum total protein was 10.0 g/100 ml, and the relative viscosity was 3.35 (normal, 1.6–1.9).   On cellulose acetate electrophoresis, the wavy M-protein band was seen at the slow-γ zone (the relative mobility, 1.74), amounting to 31% or 3.1 g/100 ml.

After heating at 56 °C, the M-protein fraction decreased to 12%, and did not show the wavy band. On agar gel electrophoresis (lower patterns), the original serum (O) showed biphasic M-protein bands, while the heated serum (S) showed only a single M-spot at the slow-γ zone.

On immunoelectrophoresis (upper patterns, next page), the M-protein was identified as the IgG-L type.   With the original serum, the IgG precipitation line was biphasic.   The heated serum (supernate) showed only a single M-bow at the slow-γ zone.   IgA and IgM are not seen.

On thin-layer gel filtration the A fraction was 32%, the G fraction 55%, the G′ fraction 10%, and the M fraction 3%.   However, after heated, the G′ fraction (about 10 S) disappeared.   On immuno-gel filtration (lower patterns, next page), the IgG precipitation line was biphasic, extending to the G′ fraction.   After heated, the IgG line is mostly around the G fraction, showing a minor tailing towards the G′ fraction.

It is apparent, therefore, that the pyroglobulin is composed of aggregated IgG-L type M-protein molecules, being around 10–14S.

Minimal proteinuria was noted, but no BENCE JONES protein was found.   Only a trace amount of the IgG-L type M-protein (P) was excreted into urine.

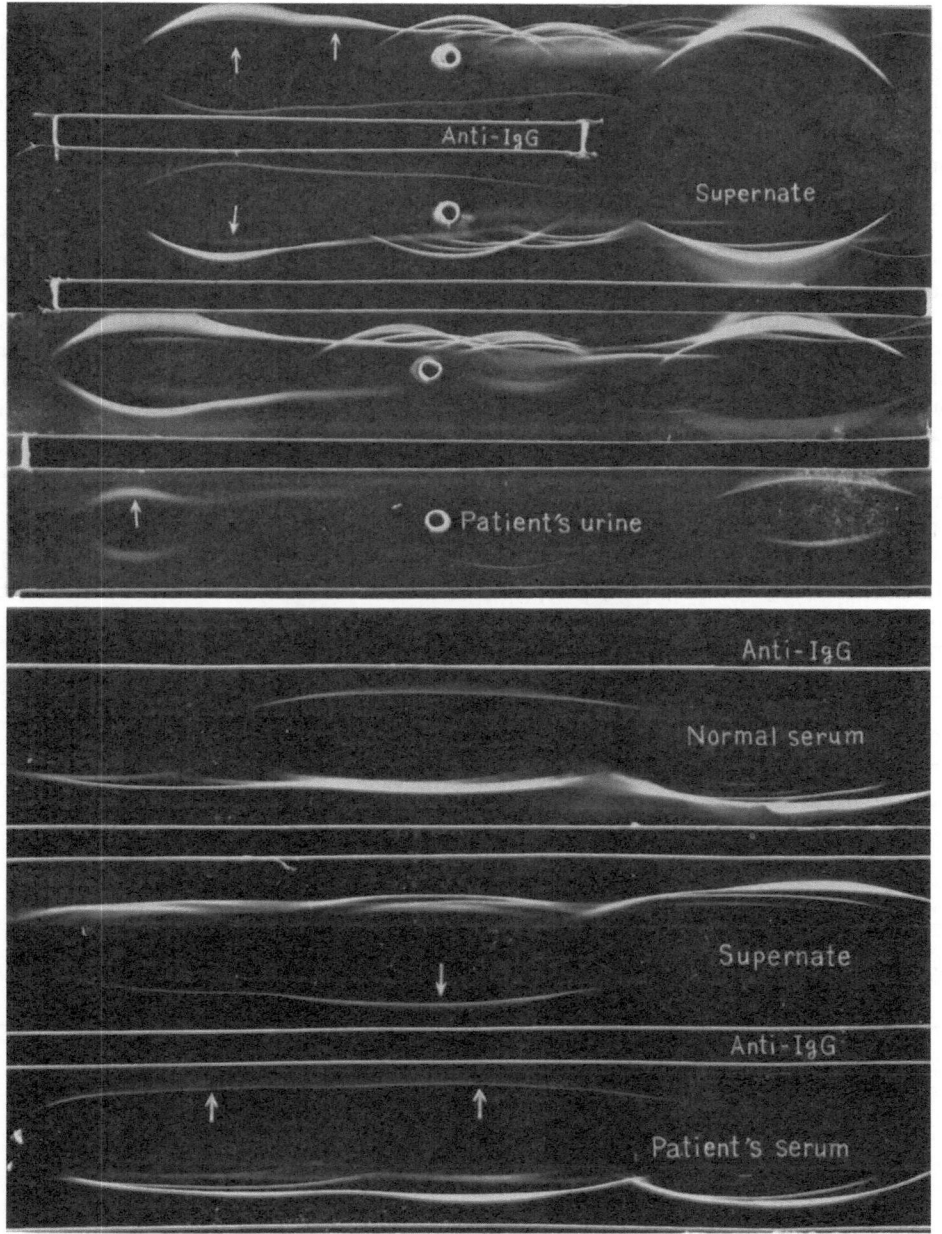

fraction, and 3) patterns in which M-proteins are extended markedly toward the M-fraction and become biphasic in the immuno-gel filtration pattern.   Among 15 cases of IgG-type myeloma, two cases belonged to the third type and seven to the second type. On the oth. r hand, among 35 cases of essential M-proteinemia, 7 cases belonged to the second type, but no cases are found to belong to the third type.   In other words, M-proteins in myeloma have a stronger tendency toward molecular aggregation.

3.   Immunological Characteristics

As shown in Table 52, IgG-type M-proteins are the most frequently encountered.

On immunoelectrohphoreisis, the characteristic M-bow is observed in the IgG line of precipitation, reflecting electrophoretic homogeneity (Fig. 206).

When the serum concentration of IgG-type M-proteins is remarkably high, the characteristic M-bow may not be observed because of antigen excess.   Therefore, when the M-protein concentration is high, the serum should be diluted for immunoelectrophoresis. Occasionally the IgG line splits into M-protein arc and normal residual IgG component separately (Fig. 206).

The light chains of IgG-type M-proteins belong to either the L or the K type, never in both.   However, in rare cases, it consists of two M-peaks and moreover, IgG-K type M-protein and IgG-L type M-protein sometimes coexist.[84,193]

In the cases found in Europe and U.S.A., the K:L ratio of IgG-type M-proteins is observed as 7: 3, but in Japan, L-type is comparatively numerous, the K: L ratio being 55: 43 (Table 52).   In order to determine the types of the light chains, anti-lambda and anti-kappa specific antisera must be used (Fig. 200).

Table 55   Subclasses of IgG and their frequency in IgG type myeloma found in U.S.A. (from WHO report).

Subclass	Synonyms	Frequency
IgG1	We, γ_{2b}, C	70–80 %
IgG2	Ne, γ_{2a}	13–18 %
IgG3	Vi, γ_{2c}, Z	6–8 %
IgG4	Ge, γ_{2d},	3 %

Table 56   Frequency of IgG subclasses in IgG type myeloma found in Japanese and other races (from TAKATSUKI).

Subclass	Japanese	Caucasian	Negro
C type (IgG1)	100	113	45
Z type (IgG3)	1	12	3
C-Z- (IgG2, 4)	34	14	4
Total	135	139	52

IgG-type M-proteins are unique to each patient.   Therefore, after the anti-serum obtained through immunization by a single IgG-type M-protein is absorbed with normal IgG, the absorbed anti-serum reacts only to the M-protein which have been used for the immunization.   According to KORNGOLD[193], none of 110 cases of IgG-type myeloma has the same antigenicity.   Therefore, each M-protein has an antigenic determinant differing slightly from the others, even if they belong to the same IgG class.   However, the classification into four subclasses depending on the antigenicity of the γ chain is possible, including γ_1, γ_2, γ_3 and γ_4.[161, 381]   All the IgG-type M-proteins belong to either one subclass

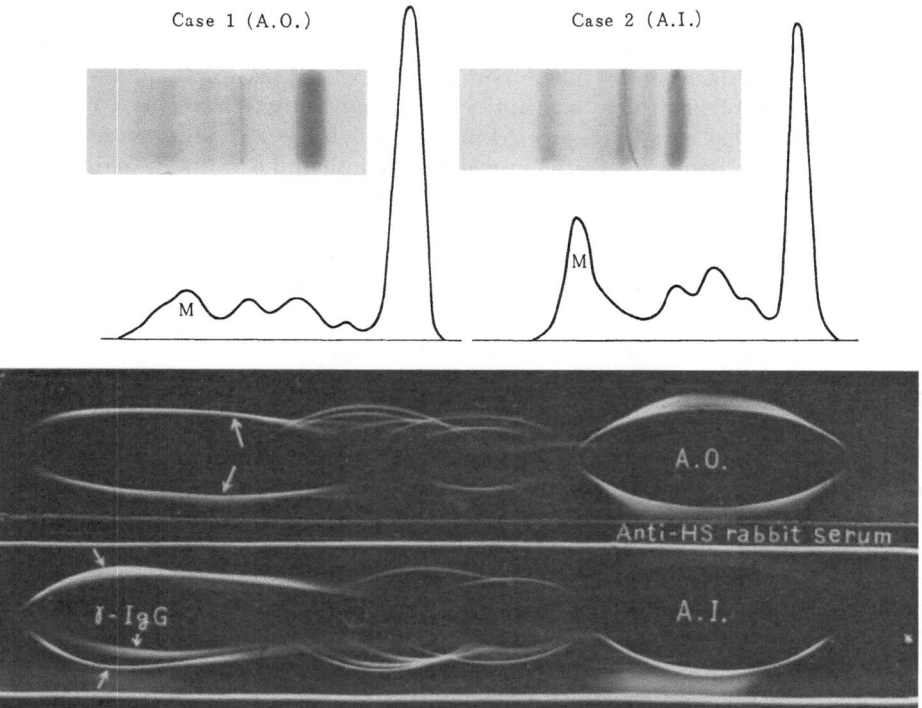

Fig. 206 Serum protein electrophoretic patterns in IgG type M-proteinemia.
Case 1 (A. O.): 65 y.o., male, cerebral apoplexy.
Case 2 (A. I.): 77 y.o., female, pneumonia and pleuritis.
On cellulose acetate electrophoresis, the M-protein bands showed no wavy appearance. The serum
total protein and the γ fraction were 7.1 g/100 ml and 6.4 g/100 ml, 0.78 g/100 ml and 2.05 g/100 ml,
respectively.
On immunoelectrophoresis, the IgG lines showed characteristic M-bow (arrows). In Case I, a small
amount of the M-protein can be demonstrated more clearly with the rabbit immune serum than with
the horse immune serum. In Case 2, the residual IgG (r-IgG) line is demonstrated separately from
the M-protein line. In this case, the residual IgG line can be noted more clearly with the horse
immune serum.

of the four. The frequency distribution of the subclasses studied in U.S.A. is shown in
Table 55.[130] There seem to be racial differences also (Table 56). It is reported in
U. S. A. that IgG_1 has a K: L ratio of 2: 1, while IgG_2 and IgG_3 have a K: L ratio of
1: 1.[130,378] But among the Japanese, irrespective of the subclasses of the γ chain, the K: L
ratio is almost 1: 1.

    4.   Chemical Characteristics

Through the observation using various M-proteins, the following facts have been
reported[94]: 1) in one IgG molecule, the peptide loop which consists of about 60 amino
acid residue forms a fundamental "pseudosubunit" and twelve of these subunits form
one γ chain; 2) among these 12 loops, 4 pairs, that is, one pair of light chains, three pairs
of heavy chains are common to various IgG type M-proteins. The other variable loops
are characteristic for each individual M-protein.[94]

    5.   Biological Characteristics

Formerly, M-proteins were considered so-called pathological "dummy proteins" which
do not have antibody activity. However, recently, some of the M-proteins have been

known to have some kind of antibody activity.[399] ZETTERVALL reported that among IgG type M-proteins, there are those which have remarkably high antistreptolysin O activity and those which have anti-staphylolysin activity.[420,421] Also, BLAUMONT reported that IgG type M-proteins which have anti-β lipoprotein activity have been observed.[44] In Japan, also reported is the IgG type M-protein which has anti-albumin activity.

EISEN found in one patient where IgG type M-protein had the antibody activity against 2, 4-dinitrophenyl-hapten although this patient had not been immunized by DNP.[82] This antibody activity has been proved to exist in Fab fragments and to be highly specific. The reason why antibody activity is not confirmed in many of the M-proteins may be that the corresponding antigens are not available at present. Although research on immunoglobulins has developed rapidly and incessantly, it must be admitted that knowledge in regard to antibody globulins is still in the primitive stage.

IgA Type M-proteins

1. Electrophoretic Characteristics

On electrophoresis, the IgA type M-proteins migrate in the areas ranging from the α_2 and to the γ zones (Fig. 204), most frequently in the γ_1 and β zones. When IgA type M-proteins are present in large amounts, they can be observed easily. Generally, electrophoretic homogeneity of the IgA type M-proteins is not as prominent as that of the IgG type (Fig. 207). Therefore, when a small amount of IgA type M-proteins is present around β_1 zone, it becomes very difficult to be distinguished because they are buried in the β fraction. Moreover, as explained later, the IgA type M-proteins have a strong tendency toward inter-molecular aggregation, and therefore, on agar gel and cellulose acetate electrophoresis, they sometimes are separated to be biphasic or triphasic. Also, on starch gel electrophoresis, they not only migrate widely but also are often observed as several protein bands.

2. Physical Characteristics

In many cases, IgA-type M-proteins exist as various polymers. Since there are M-proteins which show a sedimentation constan of 7S, 10S or 14S, IgA type M-proteins are often observed as the Z-fraction (Zwischen-fraktion) in ultra-centrifugal analytical patterns (Fig. 207).[35] Formerly, they were also known as atypical macroglobulins, in contrast to the fact that IgG-type M-proteins appear mainly in the 7S fraction, but later they were clearly separated immunologically.

By the thin-layer gel filtration technique, similar results are obtained as the IgG type M-proteins, and the G'fraction is observed between the G and M fractions. However, among the six cases the author experienced, two cases showed the G" fraction close to the M fraction. It is probably assumed that the G' fraction is a spot caused by M-proteins of about 10S and the G" fraction is a spot caused by M-proteins of about 14S (Fig. 207).

3. Immunological Characteristics

On immunoelectrophoresis, a prominent precipitin line appears in the γ_1 zone, but since the electrophoretic homogeneity of the IgA type is not as prominent as that of the IgG type, it is not common that the line is observed as a definite M-bow. When it makes the typical M-bow formation or is observed as a biphasic precipitin line, normal residual IgA sometimes shows a vague precipitin line (Fig. 208). However, when the IgA line spreads in a rather wide range, it is necessary to distinguish it from the polyclonal one through studying on antigenicity of the light chain.

Among the IgA-type M-proteins, the L-type is a little more frequent than the K-type in Japan. In the cases found in the U.S.A and Europe, the K: L ratio is said to be 59: 49, while in Japan, it is 27: 32. Sometimes it is difficult to determine the light chain type

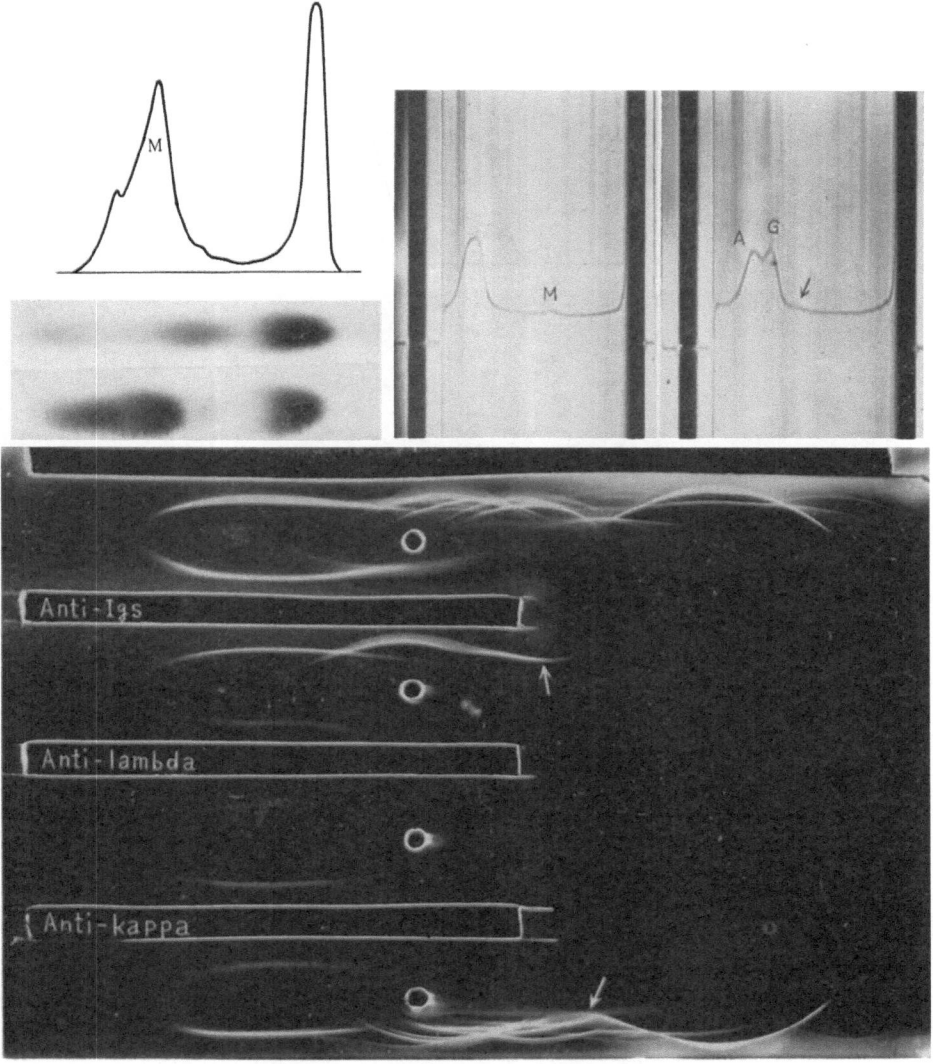

Fig. 207 Serum protein patterns in IgA type M-proteinemia.

T. U., 40 y.o., male, multiple myeloma.

Complained of general malaise and abdominal pain for the past one year, and admitted with markedly accentuated sed rate (150 mm/hr) and severe anemia (Hb 7 g/100 ml). Osteolysis noted radiologically, and 59% plasma cells in sternal bone marrow. The plasma cells showed marked dilatation of the rough-surfaced endoplasmic reticulum on electron microscopy (note Fig. 220).

The serum total protein was 10.2 g/100 ml, and the serum relative viscosity was markedly increased, being 4.5 at 25 °C. On filter paper electrophoresis, a relatively broad M-protein band was seen at the β zone, amounting to 44% or 4.5 g/100 ml. On cellulose acetate electrophoresis, the M-protein band showed a pseudo-wavy pattern. On immunoelectrophoresis, the M-protein was confirmed to be the IgA-L type. No Bence Jones protein was found in both the serum and the urine. The IgA precipitin line was noted to extend towards the anode, and was found to form a complex with the serum albumin (arrows). On thin-layer gel filtration (uppermost pattern), the IgA type M-protein was noted as the G′ fraction. On ultracentrifugal analysis (upper right), Z fraction (Zwischen-fraktion, intermediate fraction) was demonstrated. The IgA type M-protein has a tendency to form various polymers.

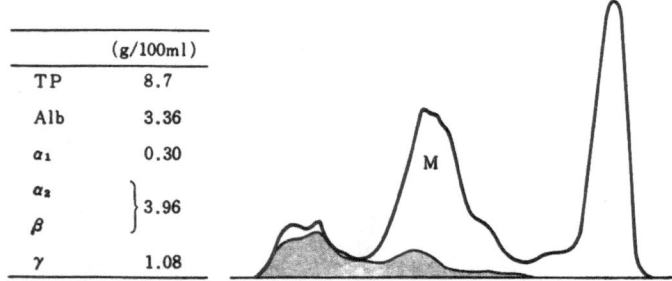

	(g/100ml)
TP	8.7
Alb	3.36
α_1	0.30
α_2	
β	3.96
γ	1.08

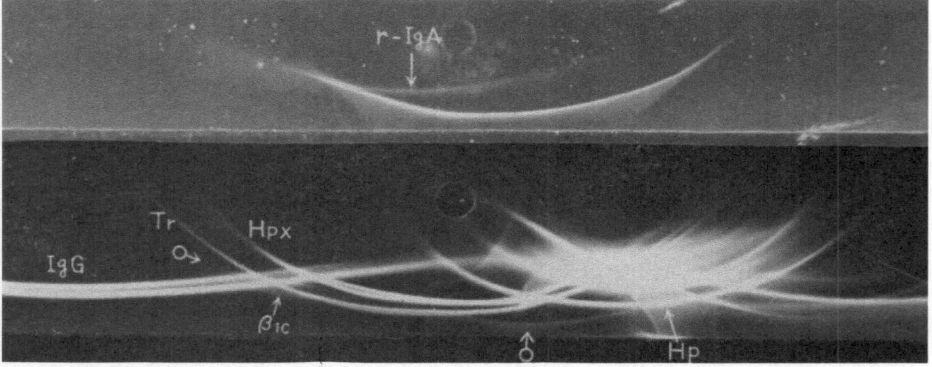

Fig. 208 Serum protein electrophoretic patterns in IgA type M-proteinemia.
S. O., 50 y.o., female, multiple myeloma.
Complained of lumbago for the past 2 years. The plasma cells 40% in the bone marrow, and osteolytic
lesions radiologically. No BENCE JONES proteinuria.
Serum relative viscosity was slightly increased, being 2.20. On filter paper electrophoresis, a relatively
broad M-protein band was noted around the α_2 and β zones. On immunoelectrophoresis, only faint abnormal precipitin line was noted between transferrin and haptoglobin (arrow). Using the specific
anti-IgA immune serum, the M-protein was identified as the IgA type, migrating faster than the normal
residual IgA (r-IgA). The shaded area at the upper pattern indicates the disappearance of the M-
protein after treated with Rivanol precipitation.

of IgA-type M-proteins, even by the use of anti-λ and anti-$\varkappa$ immune sera. This is thought
to be due to polymerization of the IgA-type M-proteins. Therefore, it may be necessary
that the IgA-type M-proteins are made to react against anti-λ and anti-$\varkappa$ after splitting
them into monomeric units by mercaptoethanol treatment.

Sometimes IgA-type M-proteins are bound to serum albumin in the blood. In such
cases, the cathodal end of the albumin precipitation line is often observed to be extended
toward the cathode (Fig. 207, 227). It is said that when a careful examination is made
by using the anti-albumin serum, the binding of M-proteins to albumin is confirmed in
about 53–64% of IgA myeloma cases.[234,332] Besides these, haptoglobin, β-lipoproteins,
α_1-glycoproteins and other serum proteins are bound to IgA type M-proteins. These
complexes are sometimes formed non-covalent bonds, but may be with intermolecular
disulfide bonds.[234] At least, two subclasses have been confirmed with the α chain.[392]

IgM-Type M-Proteins

1. Electrophoretic Characteristics

IgM type M-proteins appear within the β_2 and mid-γ zones on electrophoresis, most
frequently seen around the fast-γ zone (Fig. 204). The electrophoretic homogeneity of
the IgM-type M-proteins tends to be less prominent than IgG-type, but more prominent

than the IgA type.   The IgA-type M-proteins are often observed as rather sharp protein peaks, but because they are macro-molecules and they have high intrinsic viscosity, often they stay at the origin in agar gel.   On starch gel electrophoresis, they sometimes show two or three protein bands.   However, the IgM-type M-proteins are generally observed as a narrow protein band than the IgA-type M-proteins on cellulose acetate electrophoresis (Fig. 209).

2.   Physical Characteristics

IgM-type M-proteins exist as macroglobulins which have a sedimentation constant of about 19S.   Therefore, on thin-layer gel filtration patterns, they show a remarkable increase of the M-fraction.   As explained later, sometimes the 7S IgM molecules may coexist in the serum.

3.   Immunological Characteristics

On immunoelectrophoresis, the IgM-type M-proteins usually show a characteristic M-bow.   However, especially when the buffer solution of low ionic strength is used in agar gel, the IgM-type M-proteins precipitate around the origin and in many cases a clear precipitin line may not be recognized.   Moreover, even when the precipitin line is clearly seen, it appears often in a completely different position and in a different shape from that of the normal.   So monospecific anti-serum may be necessary for final identification.   It is not rare that IgM line is observed to be biphasic or triphasic because of the coexistence of 7S IgM, or various polymers (Fig. 209).

Usually, as described previously, the presence of IgM-type M-proteins can be identified easily.   In order to determine the light chain type, mercaptoethanol treatment of the serum sample may be necessary before they are reacted with anti-$\varkappa$ and anti-λ (Fig. 237).

There are at least two subclasses of μ chain of IgM-type M-proteins.[404]   WELTON et al. reported a case of IgM type M-proteinemia accompanied by bone destruction and plasmacytosis, in which the IgM type M-protein has the antigenicity that is not shared with normal IgM.[404]   IgM-type M-proteins may be bound with serum albumin in some cases of macroglobulinemia.

4.   Chemical Characterisics

IgM-type M-proteins are split from 19S to 7S through chemical reduction, so that the IgM type M-proteins are the polymeric aggregates of the molecular units composed of $\mu_2 L_2$.[35]

5.   Biological Characteristics

In many cases, IgM-type M-proteins do not have any antibody activity demonstrated but there are reports indicating the IgM type M-proteins with a certain antibody activity.[338]   As described later, a typical example is cold hemagglutinin.   Some of the IgM type M-proteins were proved to have the activity similar to the rheumatoid factor.

The IgD Type M-proteins

1.   Electrophoretic Characteristics

IgD type M-proteins migrate electrophoretically around the β_2 and fast-γ zones, but slight differences in mobility are noted among the different cases.   It is rare that the M-protein concentration becomes more than 3 g/100 ml, and a rather broad M-protein band is observed between the β and the γ fractions (Fig. 210).

2.   Physical Characteristics

It is reported that the IgD type M-protein isolated from serum has a sedimentation constant of about 6.5–7.0S by ultracentrifugal analysis.   However, it may contain also the

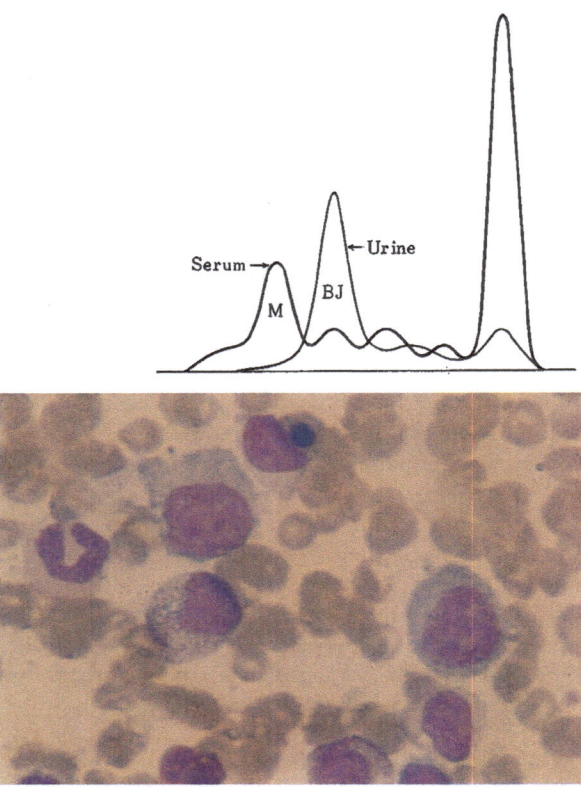

Fig. 209 a   Serum protein patterns in macroglobulinemia.

S. A., 27 y.o., male, primary macroglobulinemia.

Six months before admission, complained of back pain, left abdominal pain and heavy feeling at the epigastrium.   Small lymph nodes (less than one cm in diameter) palpable in many areas, superficially, and a large hard tumor probably at the retroperitoneum.   The bone marrow contained 5% plasma cells, 18% basophilic monocytoid cells (shown in the photograph) and 10% small lymphocytes.   Lymph node biopsy revealed no neoplastic change, but reticulum cell proliferation with the disappearance of the normal structure (note Fig. 222).

By treating with 6 MP and predonine, the abdominal mass disappeared rapidly, and the patient's general condition was much improved one month later.   However, he developed later thrombocytopenia, leukopenia and fever, and died of pneumonia 2 months after admission.

Sed rate 17 mm/hr.   TTT 0.4 units, CCF (−).

The urine contained BENCE JONES protein of the L type, migrating at the β zone.

The serum total protein was 8.0 g/100 ml, and SIA water dilution test was negative.   On cellulose acetate electrophoresis, the M-protein band was at the γ_1 zone, amounting to 20% or 1.6 g/100 ml, and showing pseudo-wavy appearance.   On immunoelectrophoresis, the M-protein was identified to be the IgG-L type.   Using the specific anti-IgM immune serum, the IgM line showed biphasic M-bow.   The serum relative viscosity at 25°C was increased, being 3.15.   The immunochemical quantitation of the immunoglobulins resulted as IgG 85% of normal, IgA 96% and IgM 950%.

The **patient's father**, 70 y.o., had primary macroglobulinemia associated with plasma cell leukemia. He was treated in other hospital.   Admitted with marked emaciation, severe anemia, and coma, and died one week later with coma paraproteinemicum.   WBC was 28,700 with 59% plasma cells.   The bone marrow contained 25% plasma cells.   On autopsy, hepatosplenomegaly and hyperplastic bone marrow being infiltrated diffusely by plasma cells.   No osteolytic lesion, radiologically.   Serum total protein 17.4 g/100 ml, γ fraction 69%.   On ultracentrifugation, 19 S fraction was 65.2%, and 23S 5.6%.   On immunoelectrophoresis, the IgM-K type M-protein was identified.   The serum relative viscosity was extremely high, being 27.4 at 20°C.   IgG and IgA in serum were decreased markedly. SIA water dilution test was strongly positive.   No BENCE JONES protein in urine.

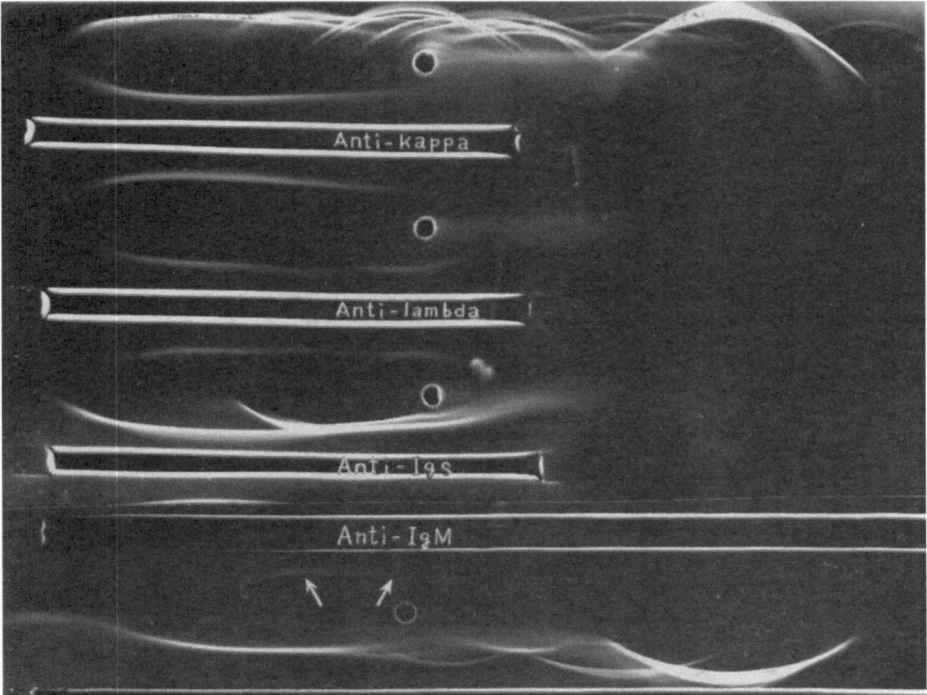

Fig. 209 b Serum protein immunoelectrophoretic patterns in macroglobulinemia

component of 11.7S in approximately 4.5%.[314] When gel filtration technique is used, ROWE et al. noted that normal IgD appears as an almost symmetrical single protein peak between the 7S fraction and the 19S fraction.[314] In 3 cases in which thin-layer gel filtration technique was employed, the author observed that all the IgD type M-proteins were found in the G' fraction between the G and M fractions. The thin-layer gel filtration pattern of the IgD type M-proteins is very similar to that of the IgA type M-proteins (Fig. 210). Moreover, when immuno-gel filtration technique are used, the IgD type M-proteins are found in a rather wide area. Therefore, it seems that they exist *in vivo* as various polymers.

3. Immunological Characteristics

By immunoelectrophoretic technique, a comparatively long precipitin line is observed and barely forms a distinct M-bow. Also, since most anti-human whole serum immune sera do not contain a significant amount of anti-IgD, it is necessary to use the specific anti-IgD immune serum to indentify the presence of IgD type M-proteins. The particular lot of the anti-human whole serum horse immune serum used routinely in the author's laboratory contains significant amount of the antibody against the δ chain, and the IgD type M-proteins can be observed on routine immunoelectrophoretic analysis (Fig. 210). With the use of this antiserum, the IgD line shows a partial identity with the IgG line, and the cathodal end of the IgD line always fuses into the IgG line. Therefore, a common antigenicity seems to be shared partly between the δ and γ chains.

With regard to the types of the light chains in the IgD type M-proteins, the L-type is much more frequent than the K-type.

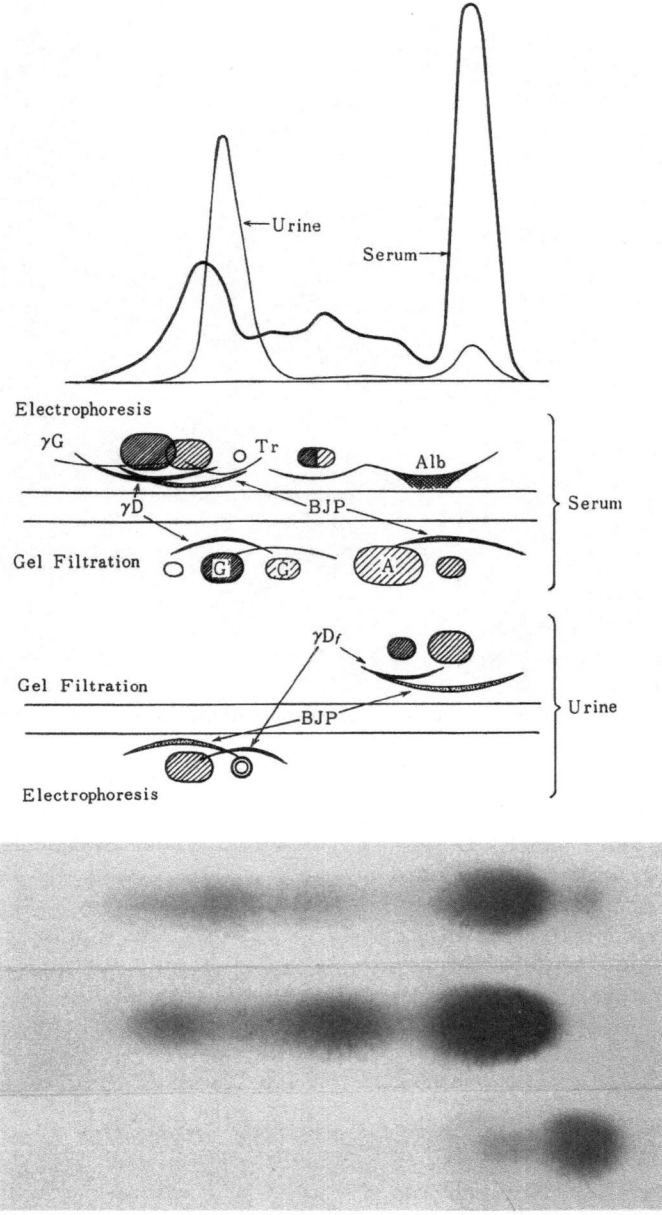

Fig 210 a

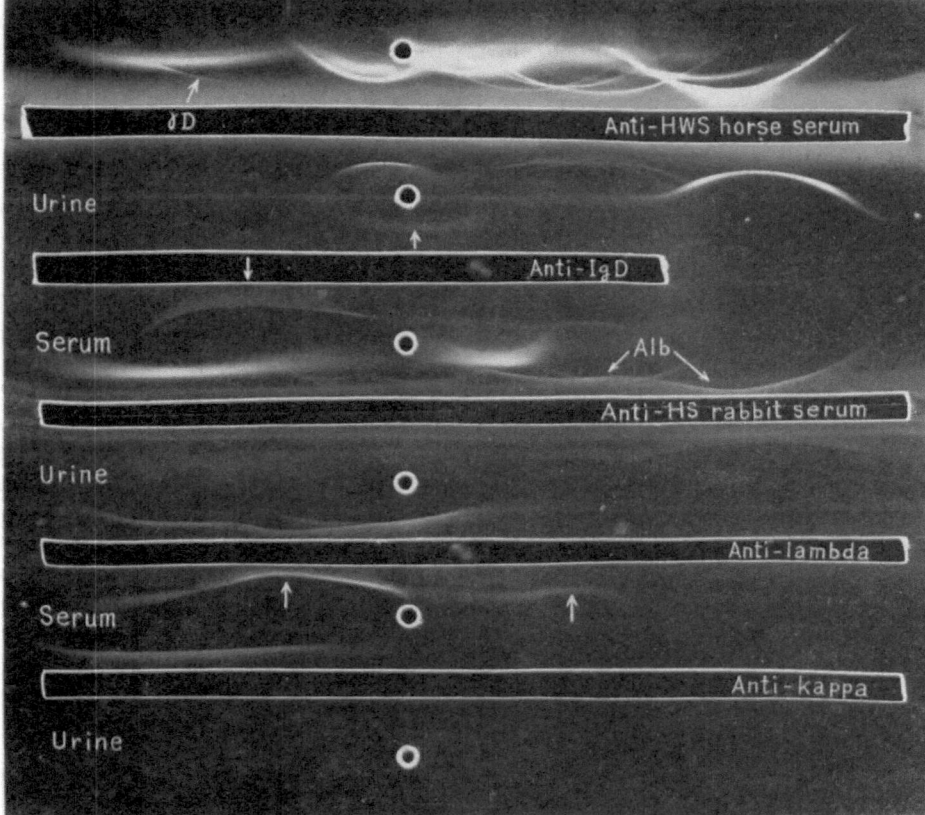

Fig. 210 b   Serum and urine protein patterns in IgD type M-proteinemia.

N. H., 62 y.o., male, IgD type multiple myeloma.

Complained of lumbago and chest pain for the past one year, and admitted with fever and anuria. Severe anemia (Hb 5.4 g/100 ml), sed rate 165 mm/hr., serum urea N 130 mg/100 ml and serum creatinine 11.3 mg/100 ml.   Proteinuria (0.8 g/day) and the urinary sediment containing a few erythrocytes and leukocytes.   Multiple "punched-out" lesions in the skeletal system, and the bone marrow containing 30% atypical plasma cells.   Although the renal failure improved rapidly with corticosteroids and Exdoxan, eventually died of recurrent renal failure.

The serum total protein was 6.4 g/100 ml.   On cellulose acetate electrophoresis, the γ fraction was 25% or 1.6 g/100 ml, and a relatively broad M-protein band was recognized at the fast-γ zone.   Agar gel electrophoretic patterns showed two M-spots at the γ_1 area.   On immunoelectrophoresis, the fast-moving M-protein spot was identified as the L type BENCE JONES protein, and the slow-moving spot as the IgD-L type M-protein.   On thin-layer gel filtration, BENCE JONES protein was separated as the post-A fraction, and the IgD type M-protein was identified as the G' spot between the G and M fractions.   A portion of the IgD type M-protein was formed a complex with the serum albumin, and the complex was migrated at the α_1 zone.

The urine contained the L type BENCE JONES protein, migrating at the β_2 zone and amounting to about 85% of the urinary protein.   On thin-layer gel filtration, a minor spot was at the A fraction and a major spot was at the post-A fraction.   In addition to the BENCE JONES protein, found was a trace amount of IgD$_f$, which reacted against both the anti-lambda and the anti-IgD, having a molecular size of approximately 4.5 S.   This IgD fragment was not identified in the serum, and seems to be a product resulting from degradation of the IgD type M-protein.

On the previous page, the electrophoretic and thin-layer gel filtration findings on the serum and urine are schematically shown.

Of six cases reported in Japan, all were of the L-type. IgD type myeloma is frequently associated with BENCE JONES proteinemia. In one case studied by the author, the IgD type M-protein is bound with serum albumin, the complex migrating at the a_1 zone (Fig. 210).

IgE Type M-proteins

There is only two cases of IgE type myeloma reported in the world. The IgE type M-protein in the first case migrates at the fast-γ zone and its light chains belong to the L-type. The sedimentation constant was about 8S and the carbohydrate content was 10.7%. Also, the IgE type M-protein has been proven to have biologically human reaginic skin senstizing activity.[358]

7S IgM Type M-proteins

19S IgM molecules exist as the polymers of five IgM-monomeric units. However, 7S IgM molecules have been found occasionally in primary macroglobulinemia, ataxia telangiectasis and SLE, also rarely in liver cirrhosis. It is found that 7S IgM have the activity of incomplete isohemaglutinin.[53,364]

There is only one case reported in which 7S IgM type M-protein was detected in serum.[352] In this case, 7S IgM type M-protein was observed as a M-protein band in the β_2 zone and its serum concentration was 0.3 g/100 ml. By immunoelectrophoresis, the abnormality of the IgM precipitin line was observed and it was of the L-type. The increase of 19S component was not observed by ultracentrifugal analysis. 7S IgM does not precipitate in water, thus not belonging to the euglobuling group. It is not yet known whether the M-protein is produced through degradation of ordinary 19S IgM.

γ Type M-Proteins

1. Electrophoretic Characteristics

The γ type M-proteins migrate in the β_2 zone, and are observed as relatively broad M-protein bands. However, the γ type M-proteins almost always show the same mobility among different cases.[96,282]

2. Physical Characteristics

When ultracentrifugal analysis is used, γ type M-proteins have a sedimentation constant of 3.6–3.8S. The molecular weight is from about 52,000 to 55,000. They are easily passed into urine through the glomerular basement membrane.

3. Immunological Characteristics

On immunoelectrophoresis, the γ type M-proteins are observed as an almost symmetrical line close to the antibody trough in the γ_2 zone. They react with anti-IgG, but not with the antibodies against light chains. They have common antigenicity with the Fc fragments of the γ chain, but they seem not to be identical to the Fc fragments. The γ type M-proteins lack Fab fragments. They belong to either one of the four subclasses of the γ chains.[282]

α Type M-Proteins

Up to the present, there has been only a few cases reported in which the α type M-proteins were observed. The α type M-proteins migrate in the β zone, and are observed as a relatively broad protein band. They do not have the light chains and have common antigenicity with IgA or α chain. They belong to one of the two subclasses related to the α chains. They seem to be very similar to the Fc fragments of the α chains.[340]

μ Type M-Proteins

The μ type M-protein has been recognized only in the sera of the patients with μ-chain disease, mostly accompanied with BENCE JONES proteins. The μ type M-proteins migrate in the $\alpha_2-\beta$ zones, and are usually not recognized on cellulose acetate electrophoresis. Immunoelectrophoretically it is revealed to react with an antiserum to μ chains but not with antisera to $\varkappa$ and λ light chains. It represents a fragment of the μ chain having a molecular weight of about 55,000.[95] Both the μ type M-protein and the light chains fail to assemble presumably because of a structural delection involving the region of the Fd fragment which contains the cysteine residue linking the heavy chain to the light chain.

BENCE JONES Proteins

1. Heat Precipitability

It is characteristic of typical BENCE JONES proteins to precipitate when heated above 56 °C, but when boiled above 100 °C they redissolve. Such heat precipitability is strongly dependent on pH and also influenced by the other various physicochemical conditions.

Therefore, as reported by PUTNAM, et al, the test must be carried on under a fixed condition by the use of an acetate buffer of pH 4.9. Pseudo–BENCE JONES proteins which do not redissolve below 110° also are known to exist.[35] Moreover, when considerable amounts of albumin and other serum protein components coexist in urine and serum, the dissolution of BENCE JONES proteins cannot be observed even when heated 100 °C. More than 600 mg per day of BENCE JONES proteins must be excreted for the heat precipitation test to become positive in urine.[222] Therefore, heat precipitability does not have crucial significance for diagnosis of BENCE JONES proteins.

Clinically, the appearance of white precipitation by heating below 60 °C is sifficient to suspect the presence of BENCE JONES protein. In order to confirm the finding, electrophoretic and immunoelectrophretic analyses must be performed.

2. Electrophoretic Characteristics

BENCE JONES proteins show various electrophoretic mobility from the α_2 to the γ zones, and migrate as a rather broad M-protein band. Usually, they are monophasic, but sometimes they are biphasic on cellulose acetate or/and agar gel electrophoresis. The biphasic electrophoretic patterns indicate that the BENCE JONES proteins appear in both monomeric and dimeric forms (Fig. 211). By the use of starch gel electrophoretic technique they often show many bands. In BENCE JONES proteinemia, BENCE JONES proteins are pre sent usually in the form of dimers.

3. Physical Characteristics.

In most cases, BENCE JONES proteins have the sedimentation constant of 3.6S and a molecular weight of about 45,000, and they are dimers of the free light chains. However, the sedimentation constant sometimes ranges from 1.8S to 5.5S. Free light chains which have the sedimentation constant of 1.8S and the molecular weight of about 22,000 are present as monomers in some cases. Moreover, those with the sedimentation constant of 5.5S and a molecular weight of about 90,000 are considered to be tetramers of the light chains. Similar findings have been observed by thin-layer gel filtration technique. 3.6S dimers are observed as an abnormal spot present at the post-A fraction. Moreover, the monomers are sparated clearly at a slower rate than the dimers (Fig. 211).

Dimers and monomers both show similar heat-precipitability, but in the case of cryo-BENCE JONES proteins, only dimers show cryoprecipitation.

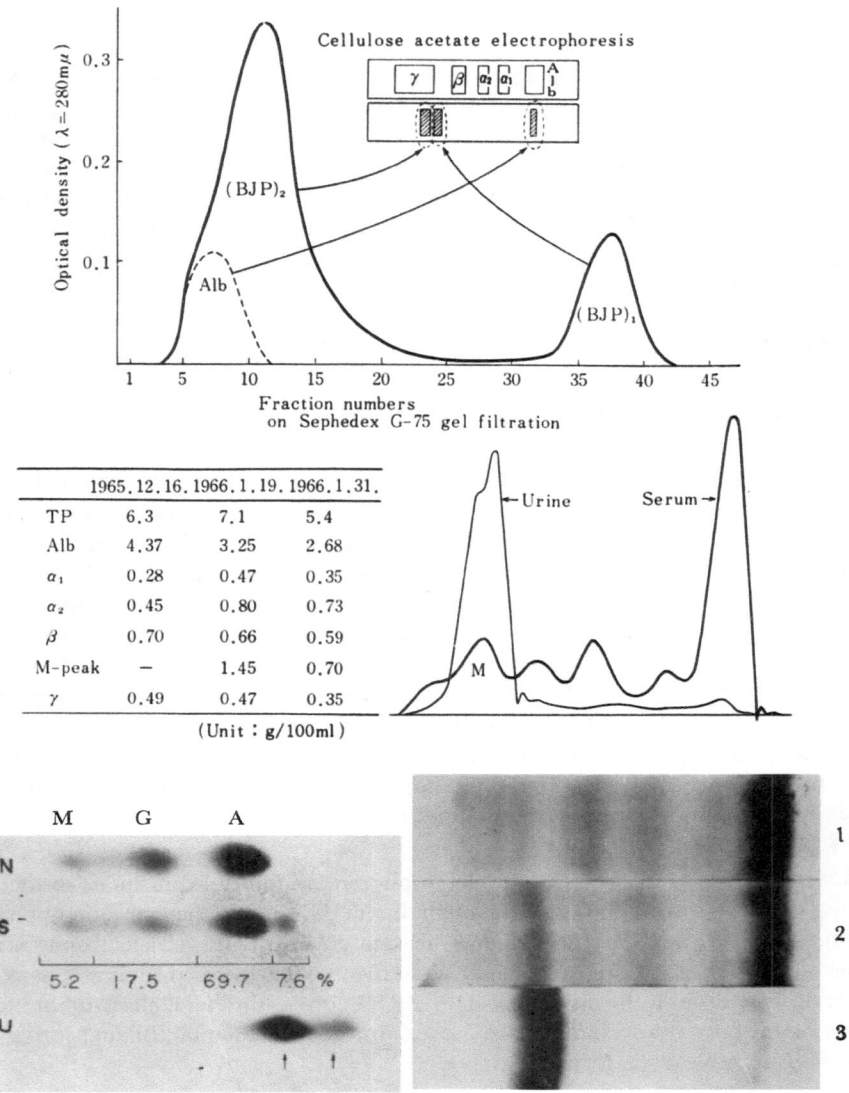

Fig. 211a

4. Immunological Characteristics

BENCE JONES proteins are observed, on immunoelectrophoresis, usually as a faint, relatively flat arc very close to the antibody trough.   They are located most frequently between IgG and transferrin, and their lines do not cross with that of the IgG, but it shows partial identity with IgG.   Therefore, sometimes, it is difficult to distinguish those proteins from IgA type M-proteins, IgG type M-proteins migrating in the γ_1 zone, IgM type M-proteins, and the IgD type M-proteins and other proteins having γ_1-mobility. Furthermore, many lots of commercially available anti-human whole serum do not contain a sufficient amount of the antibodies against the light chain, and the specific antisera against

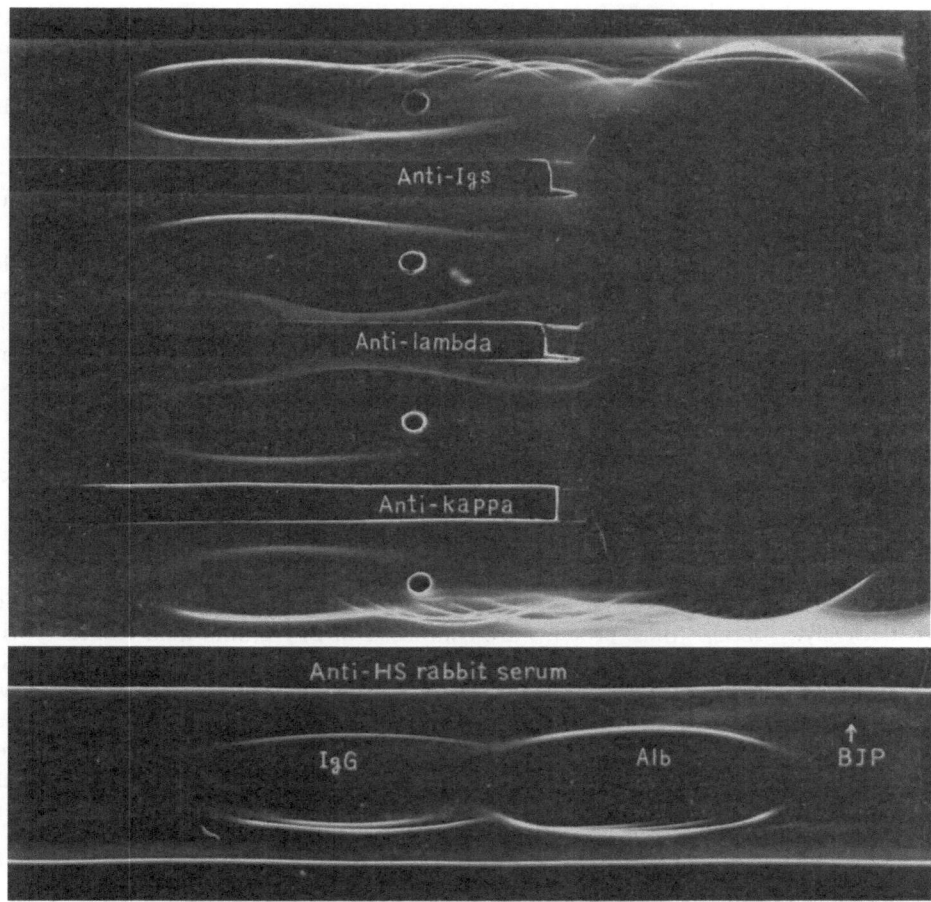

Fig. 211b Serum and urine protein patterns in BENCE JONES proteinemia.
M. T., 50 y.o., male, multiple myeloma.
Admitted with proteinuria. The sternal bone marrow containing 60% atypical plasma cells, and no osteolytic lesion on X-ray examinations. Died of renal failure (urea N, 141 mg/100 ml). On autopsy, the bone marrow was diffusely infiltrated by plasma cells, and a plasma cell tumor, 1 cm in diameter, was found in the 4th lumbar vertebra. The kidneys showed a characteristic histological finding of "myeloma kidney".
On filter paper electrophoresis, the serum showed no M-peak and the γ fraction was decreased significantly (1). At the time when a renal failure developed, the L type BENCE JONES protein was identified in the serum, migrating at the β_2 zone (2). A large amount of the BENCE JONES protein was found in urine (3), and it tended to show double peaks.
On thin-layer and column gel filtrations, the BENCE JONES protein in the serum was of the dimer molecule, and the urinary BENCE JONES proteins were composed of both the monomer and the dimer molecules. On cellulose acetate electrophoresis of the urinary protein, the monomer BENCE JONES protein migrated faster than the dimer.
The dimer BENCE JONES protein in urine was identified as the cryo-BENCE JONES protein, showing reversible precipitation at the temperature of lower than 15°C in the protein concentration of more than 4 g/100 ml (note Fig. 225). The monomer BENCE JONES protein did not show any cryo-precipitation.

the light chain must be used to detect them. Concentrated urine specimens containing a large amount of BENCE JONES proteins may not demonstrate clear precipitin lines because of antigen excess.

BENCE JONES proteins are divided into the K-type which has common antigenicity with normal $\varkappa$ chains and the L-type which has common antigenicity with normal λ chains. Moreover, in most cases, BENCE JONES proteins should belong to one of the two types. However, the primary structure of the constant portion shares a common antigenicity between the two types.   It is also known that such a chemical structure of light chains is under genetic control.   However, in rare cases, BENCE JONES proteins of both types may coexist.[84]   The K:L ratio among BENCE JONES proteins is about 7:8.   In Japan, the ratio is about 7:12; the L type being overwhelmingly numerous.

BENCE JONES proteins originally exist in the serum, but because of their low molecular weight, they are found in urine in large amounts.   Therefore, with careful examinations BENCE JONES proteins may be almost always found in the serum even if they are in small amounts when they exist in urine.   Also, BENCE JONES proteins are often found together with other M-proteins in the same patient.   When this happens, the type of BENCE JONES proteins and the light chain type of the M-proteins are likely to be the same (Fig. 210).

Also BENCE JONES proteins may be further classified into various sub-types, even if they belong to the same K or L type.   It seems that a variable portion of each BENCE JONES protein is difierent from the others.[35]

5.   Chemical Characteristics

BENCE JONES proteins are able to be collected in a large amount and at the same time in an almost pure state.   So the chemical structure of BENCE JONES proteins has been studied most extensively.   BENCE JONES proteins are composed of 214 amino acid residues, with a variable portion consisting of 106 amino acids from the N-terminus and with an invariable portion consisting of 107 amino acids from the C-terminus.   Formerly it was thought that the K-type BENCE JONES proteins and the L-type BENCE JONES proteins had completely different antigenicity and had no common antigenic determinants.   However, the primary structure of the invariable portion shares a common antigenicity between the two types.   It is also known that such a chemical structure of light chains is under genetic control.

CHANGES IN THE BLOOD AND THE URINE CAUSED BY THE PRESENCE OF M-PROTEINS

M-proteins have certain properties which are fairly differnt from those of the normal immunoglobulins.   Therefore, when they are present in a large amount in the serum or urine, they may cause characteristic changes in serum and urine of M-proteinemic patients.

Changes in the Blood Mainly due to Abnormal Aggregation of the M-protein

Among various plasma proteins, immunoglobulins have a strong tendency of molecular aggregation.   Some of the M-proteins have a tremendous tendency to show molecular aggregation.   Typical changes caused by the increased molecular aggregation of M-proteins are: 1) Rouleaux formation of erythrocytes, 2) Increased erythrocyte sedimentation rate, 3) Increased serum viscosity, 4) Appearance of a wavy M-protein band on cellulose acetate electrophoresis, and 5) Cystic dilatation of the rough-surfaced endoplasmic reticulum in the plasma cells.

1.   Rouleaux Formation of Erythrocytes

Rouleaux formation of erythrocytes is a characteristic aggregation recognized in certain protein solutions (Fig 212), being quite different from immunological hemagglutination. Increased roluleaux formation is always associated with marked acceleration of erythrocyte

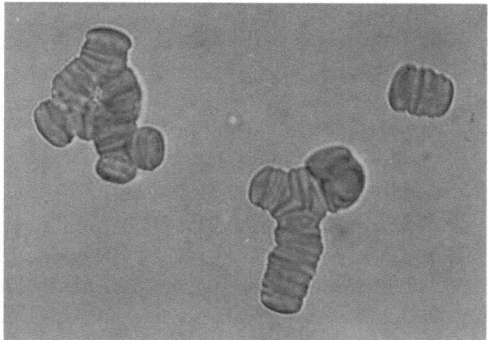

Fig. 212 Rouleaux formation of the erythrocytes.

sedimentation rate. Its characteristic pattern is recognized on peripheral blood smears. Rouleaux formation of erythrocytes occurs in the blood having abnormal protein compositions, particularly in severe M-proteinemia and occasionally in polyclonal hyperimmunoglobulinemia. The phenomenon may be recognized sometimes when the serum low-density lipoproteins are markedly increased.

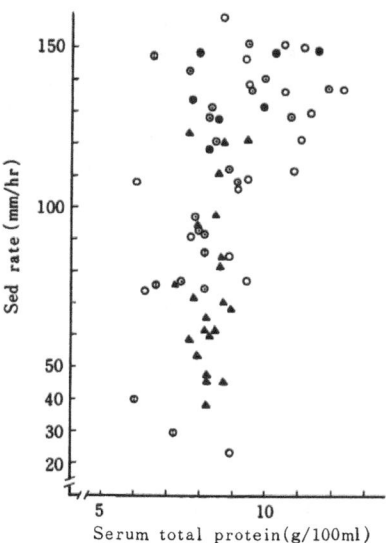

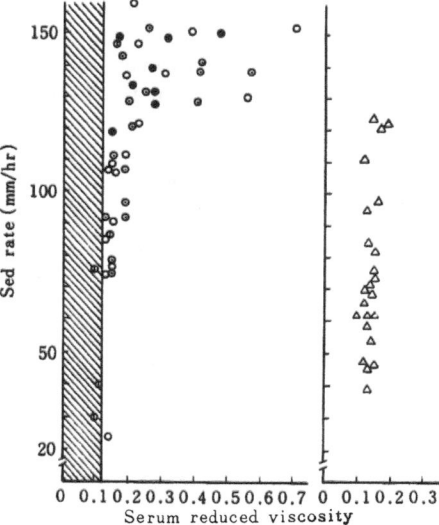

Fig. 213 Relation between the sedimentation rate and the serum total protein concentration in various conditions.

No definite correlation can be recognized, but it is apparent that a marked accentuation of the sedimentation rate is noted in all cases showing the serum total protein of more than 10 g/100 ml.

Fig. 214 Relation between the sedimentation rate and the serum reduced viscosity in various conditions.

When the reduced viscosity is more than 0.2, a marked accentuation of the sedimentation rate is recognized always.

○ IgG type myeloma
● IgA type myeloma
⊙ Macroglobulinemia
◍ BENCE JONES type myeloma
△ Chronic inflammations

2. Increased Erythrocyte Sedimentation Rate

Increased erythrocyte sedimentation rate is generally associated with infections, inflammations and wide-spread tissue destruction, and seems to parallel to positive CRP test. The following conditions should be suspected when increased sedimentation rate without positive CRP test is recognized: anemia, pregnancy, nephrotic syndrome and M-proteinemia (monoclonal hyperimmunoglobulinemia).

In almost all cases with significant M-proteinemia, the erythrocyte sedimentation rate is increased significantly, as shown in Fig. 213. There is no constant correlation between the erythrocyte sedimentation rate and the total serum protein concentration. However, a marked increase in the sedimentation rate is always noted if the total serum protein level exceeds 10 g/100 ml. Fig. 214 shows the relation between the erythrocyte sedimentation rate and the reduced specific viscosity of serum which is closely related to the molecular aggregability of M-proteins. There is no constant correlation between the two. In other words, the erythrocyte sedimentation rate is more closely related to the molecular properties of the M-protein rather than its serum concentration.

3. Increased Serum Viscosity

As shown in Fig. 217, in M-proteinemias of the IgG, IgA and IgM types, there are cases in which a marked increase in serum viscosity occurs. This tendency is pronounced in cases of myeloma and primary macroglobulinemia. That is, about 23 of the myeloma cases of IgG and IgA types show a definite increase in serum viscosity; about 30% of the cases showing a marked increase of more than 3. All cases of primary macroglobulinemia showed a significant increase, because the intrinsic viscosity of 19S IgM is great. In BENCE JONES proteinemia, however, the serum viscosity is within the normal range. Fig. 216 shows the relation between the reduced specific viscosity of M-protein and relative

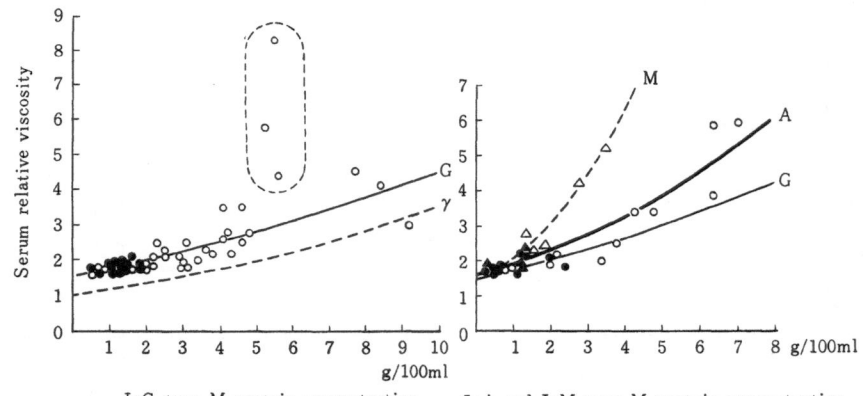

IgG type M-protein concentration   IgA and IgM type M-protein concentration

○ IgG and IgA type multiple myeloma
● IgG and IgA type essential M-proteinemia
△ Primary macroglobulinemia
▲ IgM type essential M-proteinemia

Fig. 215 IgA and IgM type M-protein concentration.
γ-curve is obtained by measuring relative viscosity on buffered saline solution of normal COHN-Fraction II (mainly composed of IgG). G-, A- and M-curves are for IgG, IgA and IgM added to the agammaglobulinemic serum. In the IgA and IgM type M-proteinemias, each value is lined along the A- and M-curves. However, in the IgG type M-proteinemia, most of the cases are lined along the G-curve, but 3 cases circled with the dotted line showed a marked increase in their serum relative viscosity.

mobility on filter-paper electrophoresis. The IgG type M-proteins with a high reduced specific viscosity tend to migrate at the slow-γ zone, and thus the slow migrating IgG type M-proteins seem to have an increased molecular aggregation.

The serum viscosity is dependent on the serum concentration of a particular M-protein present. As the M-protein concentration increases, the serum viscosity also increases ex-

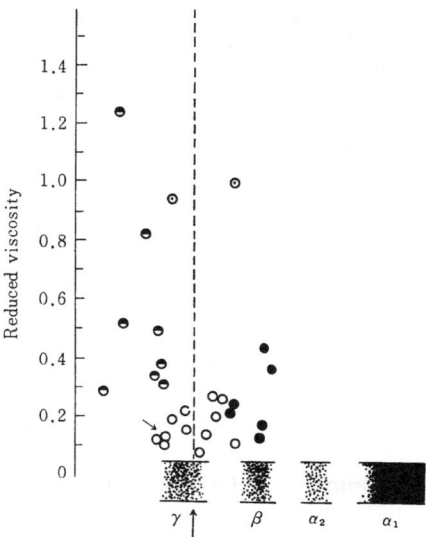

Fig. 216 Relation between the reduced viscosity and the relative mobility on the filter paper electrophoresis.
In the cases where the reduced viscosity was more than 0.29, the wavy M-protein band was noted on cellulose acetate electrophoresis. In the IgG type M-proteinemia, the ones migrating at the slow-γ zone tend to show the wavy band, while the fast-migrating ones of the IgA type M-proteins tend to show an increased viscosity.

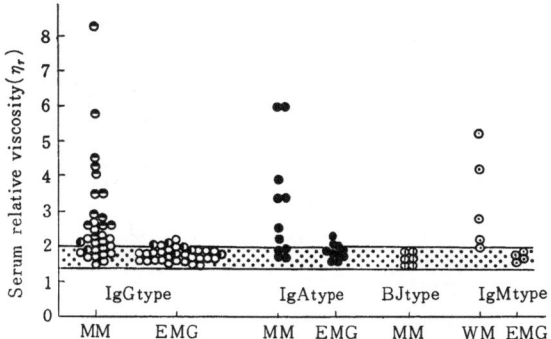

Fig. 217 Serum relative viscosity in various forms of M-proteinemia.
MM: Multiple myeloma
EMG: Essential M-proteinemia
WM: Macroglbulinemia WALDENSTRÖM
◗ IgG type M-proteins showing the characteristic wavy pattern.

ponentially. Furthermore, the higher the reduced specific viscosity of M-proteins, the more the serum viscosity shows the concentration-dependence. In general, IgM type M-proteins show marked protein concentration-dependence.[170,302] As shown in Fig. 215, in cases of the IgA and IgM type M-proteinemia, the serum relative viscosity increases in proportion to M-protein concentration. Though in most cases of the IgG type M-protein serum relative viscosity and M-protein concentration keep a definite proportion along the curved lines, labeled as G, a few exceptions are noted. These exceptional cases showed tremendous high reduced specific viscosity and temperature-viscosity index, being almost comparable to those of IgM type M-proteins. This indicates that in rare cases molecular aggregation of IgG-type M-proteins is very prominent and it exists already in living bodies as macro-polymers.[191]

Also, the temperature-dependence of serum viscosity is more distinct in M-proteinemia and the temperature-viscosity index (TVI) runs parallel to reduced specific viscosity. However, TVI is not related to M-protein concentration. That is, temperature-dependence of serum viscosity has a close relation to the molecular aggregation of M-proteins rather that to M-protein concentration (Fig. 218). This point is related to the cyro-precipitability of M-proteins, which will be discussed later.

When serum relative viscosity increases markedly, "hyperviscosity syndrome" may result clinically, and some of the important symptoms are listed in Table 57. In the begnning, they were thought to be symptoms characteristic for macroglobulinemia, but the same symptoms are sometimes seen in cases of myeloma.[350] In rare cases, it causes "coma paraproteinemicum"

Macroglobulinemia or myeloma should be suspected when serum relative viscosity is more than 3, and macroglobulinemia is suspected when it is more than 6. But very high serum relative viscosity is rarely noticed in myeloma. Besides M-proteinemia, serum relative viscosity increases occasionally up to about 3 in hypergammaglobulinemic cases of chronic inflammatory diseases, collagen diseases and liver cirrhosis.

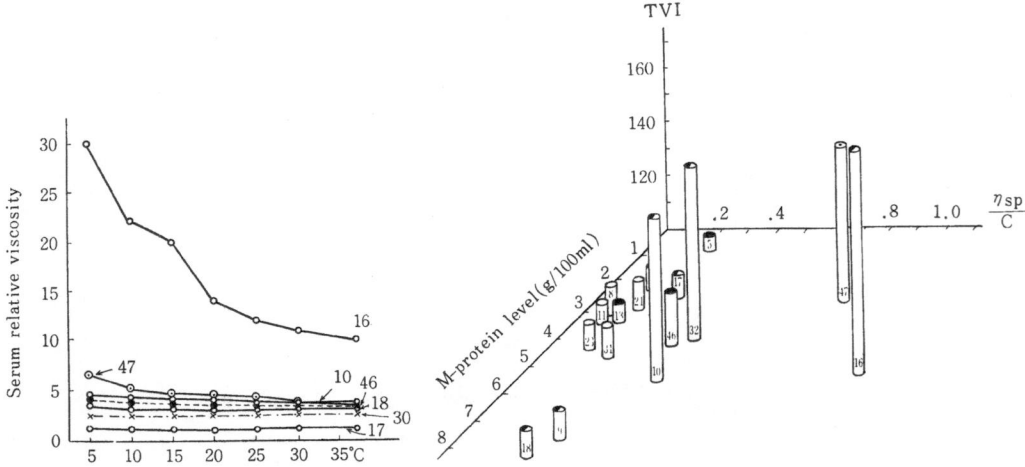

Fig. 218 Temperature dependency of the serum viscosity in M-proteinemia.
η_{sp}/c: Reduced viscosity, TVI: Temperature-viscosity index.

Case 47 is of WALDENSTRÖM macroglobulinemia, cases 5, 13, 46 are of the IgA type myeloma, and cases 8, 9, 10, 11, 16, 17, 18, 21, 23, 31, 32 are of the IgG type myeloma. Case 47, 10, 16, 32 showed an abnormally high TVI, indicating a strong tendency to form molecular aggregation.

Table 57 Cardinal clinical symptoms in hyperviscosity sndrome (from BARTH).

Symptoms	Frequency	
Hemorrhage	often	Nasal and mucosal bleeding without thrombocytopenia.
Retinal change	often	Sausage-like dilatation of veins, occasionally associated with hemorrhage and papilloedema.
Neurologic signs	often	Abnormal EEG, vertigo, nystagmus, abnormal gait, convulsion.
Cardiac signs	rare	Congestive heart failure
Complaints	often	Anorexia, easy fatigability

WALDENSTRÖM proposed the temperature-viscosity index (TVI) defined as follows:

$$TVI = \frac{\eta_r \text{ at } 13°C}{\eta_r \text{ at } 37°C} \times 100$$

Normally, TVI is calculated to be 100 ± 10. When TVI is more than 120, macroglobulinemia is to be suspected. However, as WALDENSTRÖM himself pointed out later, even myeloma occasionally shows TVI of more than 120. Macroglobulin molecules are cleaved to 7S molecules through treatment with penicillamine, mercaptoethanol and cystein, thus, the serum viscosity being reduced significantly.[336] On the contrary, in myeloma serum having increased viscosity, the relative viscosity does not show a marked change even after the treatment.

4. Wavy M-protein Band on Cellulose Acetate Electrophoresis

On cellulose acetate electrophoresis, some of the M-proteins may often appear as a wavy band (Fig. 219). The typical wavy bands are most frequently encountered with the use of Separax and Oxoid cellulose acetate membranes. Millipore and Separaphore III membranes rarely demonstrate the typical wavy bands.

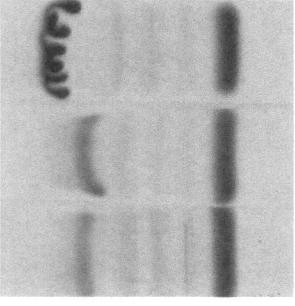

Fig. 219 Wavy appearance of the M-protein band on cellulose acetate electrophoresis.
The upper pattern shows a typical "wavy band" in the IgG type multiple myeloma. The middle pattern shows a pseudo-wavy band in macroglobulinemia. The lower pattern shows no tendency of the wavy appearance in the IgG type essential M-proteinemia.

The factors influencing to form the typical wavy bands are summarized in the following. First, the electrophoretic conditions are not directly related to the phenomenon. Secondly, the serum samples must contain an abnormal protein or M-protein having a strong tendency for intermolecular aggregation. Generally, the IgM type M-proteins have a high intrinsic viscosity and tend to show marked intermolecular aggregation. Some of the M-proteins of both IgG and IgA types may have the similar characteristic, particularly the IgG type M-proteins migrating at the slow-γ zone. When the electrophoretic separation has completed, the M-protein fraction must have a high protein concentration.

Therefore, the serum concentration of M-proteins is to be high and also the electrophoretically separated M-protein band is to be sufficiently compact or sharp. Since the IgG type M-proteins are generally more homogeneous than the IgA or IgM type M-proteins, the wavy pattern is encountered predominantly with the IgG type M-proteins. Thirdly, the phenomenon depends largely on the property of the supporting media used for electrophoretic separation. The porosity of the supporting media must be sufficiently small, and also irregular in their size. For example, the phenomenon does not occur with the agar gel medium, although the porosity is extremely small. In addition, the M-protein must move from the point of serum application in a certain distance in order to develop the wavy band.

The typical wavy band is exclusively encountered in IgG type myeloma. It is found in approximately 50% of the IgG type myeloma cases, in which the reduced specific viscosity is calculated to be over 0.29 and hyperviscosity syndrome is likely to develop clinically. "Pseudo-wavy" bands, which show an incomplete wavy appearance, are frequently encountered in most of the cases with macroglobulinemia and the IgA type myeloma having the serum viscosity of more than 3.0. Also occasionally found is the "pseudo-wavy" pattern in essential M-proteinemia and polyclonal hyperimmunoglobulinemia with a markely increased IgG.

5.   Cystic Dilatation of Rough-Surfaced Endoplasmic Reticulum in Plasma Cells

It has been pointed out that in myeloma the plasma cells producing M-proteins often show a marked cystic dilatation of the rough-surfaced endoplasmic reticulum (Fig.

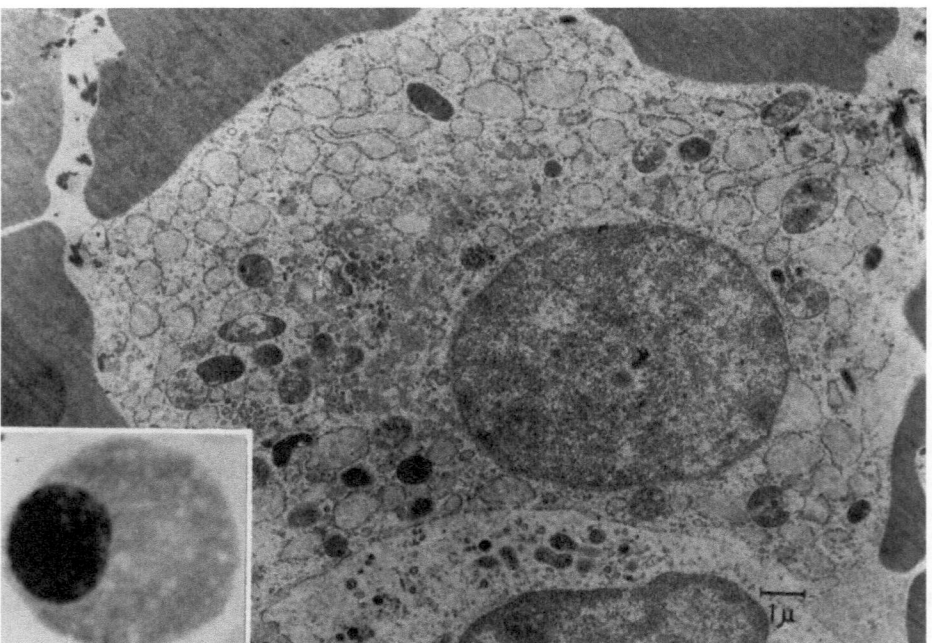

Fig. 220   Electron-photomicrogram of the myeloma cell showing dilated rough-surfaced endoplasmic reticulum.
The plasma cell of the bone marrow taken from the case of the IgA type myeloma listed in Fig. 207. The intraplasmocytic protein-viscosity index (I. P. P. V. I.) was 5.27. The IgA type M-protein was migrated at the β zone, amounting to 3.7 g/100 ml. The sternal marrow contained 59% plasma cells.

220, 221). There has been no widely-accepted opinion as to its clinical significance. The author has proposed the possible mechanism of the phenomenon from the rheological veiwpoint. The M-protein molecules synthesized in the polysomes move mainly through the cysterna of the rough-surfaced endoplasmic reticulum, accumulates in the Golgi area, and eventually are discharged from the cells. Cystic dilatation of the endoplasmic reticulum may occur either due to obstruction of the outlets of the cell membrane through which M-proteins are discharged, or the increased viscosity of the M-protein solution in the endoplasmic reticulum, resulting in stagnation of viscous M-proteins. The former malformation may occur accidentally even in normal plasma cells, and in fact, its appearance is only rarely noticed in normal bone marrow smears. Certainly, abnormal neoplastic plasma cells may have the malformation more frequently.

The intraplasmocytic protein-viscosity index (I.P.P.V.I.) is introduced by the author as an attempt to express quantitatively the rheological resistance in the endoplasmic reticulum.

Therheological resistance of a M-protein in the endoplasmic reticulum increases when theM-protein synthesized in each plasma cell tends to show increased molecular aggregation (qualitative factor), and also when the M-protein is synthesized more in each plasma cell (quantitative factor).

$$\text{I. P. P. V. I.} = \binom{\text{Reduced Viscosity}}{\text{of M-protein}} \times \frac{\text{M-protein level in serum}}{\text{Plasma cells (\%) in bone marrow}} \times 100$$

In the cases with myeloma of both IgG and IgA types, I.P.P.V.I. values are correlated

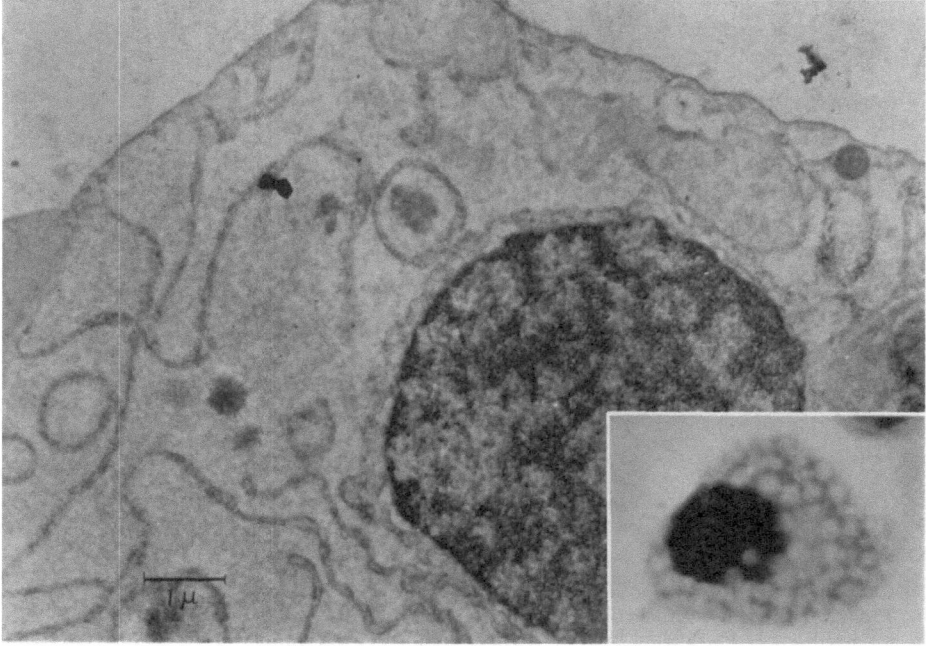

Fig. 221 Electron-photomicrogram of the myeloma cell showing dilated rough-surfaced endoplasmic reticulum.
50 y.o., male, IgG type myeloma. M-protein was 2.0 g/100 ml at the γ zone. I.P.P.V.I. was 5.85. 12.3% plasma cells in the sternal marrow.

with the electron microscopic findings of the plasma cells.   Regardless of the types of myeloma, no cystic dilatation of the endoplasmic reticulum is recognized when the calculated value is less than 2.   In contrast, when the calculated value is more than 5, most of the myeloma cells in all cases show significant dilatation of the rough-surfaced endoplasmic reticulum.   In the cases of myeloma excreting only BENCE JONES proteins, I.P. P.V.I. cannot be calculated because the serum does not contain a significant amount of the BENCE JONES protein.   However, the reduced specific viscosity of the BENCE JONES proteins is always low, and none of the cases show a significant dilatation of the endoplasmic reticulum.

Accordingly, although the possibility of morphological malformation in the myeloma cells cannot be completely neglected, the rheological conditions in the endoplasmic reticulum of the myeloma cells may have an important contribution for the cystic dilatation of the endoplasmic reticulum.

Changes in the Blood Mainly Due to Abnormal Solubility of the M-Protein

1.   Abnormalities in Serum Colloidal Reactions

Serum colloidal reactions are mainly influenced by changes in the serum A/G ratio, but there may be abnormal results in cases with M-proteinemia.

Generally, in cases of polyclonal hyperimmunoglobulinemia, almost all serum colloidal reactions show a high value.   This is especially so with the commonly-used thymol turbidity test (TTT) and the zinc sulfate turbidity test (ZTT).   In IgG type M-proteine-

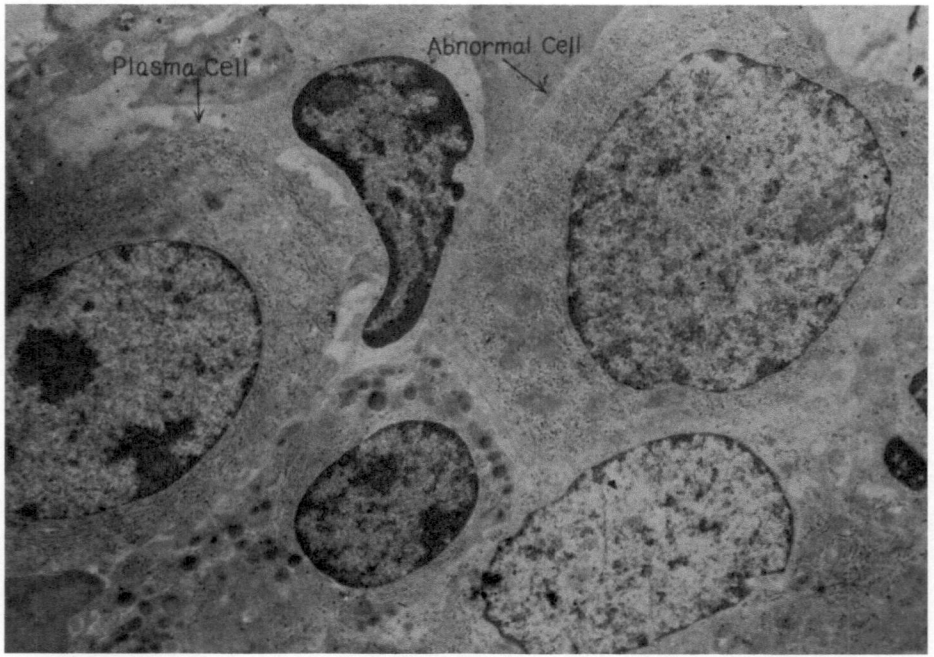

Fig. 222   Electron-photomicrogram of the abnormal cell in primary macroglobulinemia. The case was listed in Fig. 209.   Abnormal cell 18.5%, plasma cell 5% in the sternal marrow.   I.P. P.V.I. for the abnormal cell was 9.2, while I.P.P.V.I. for both the abnormal cell and the plasma cell was 7.25.   On light microscopy, the abnormal cells had the monocytoid nucleus and the large stongly basophilic cytoplasm.   On electron microscopy, the abnormal cell contains no endoplasmic reticulum, but aggregates of the ribosomes.   The mature plasma cell does not show dilatation of the endoplasmic reticulum.

mia, ZTT shows usually an extremely high value, but there are some cases where TTT shows normal or only slightly elevated values despite of high ZTT values (Fig. 223).   In the BJ, IgA and IgM types of M-proteinemia, TTT and ZTT show a low value although the serum M-protein levels are fairly high.   In the IgG type M-proteinemia TTT values tend to be extremely high when the slow-migrating M-protein exist in the serum.   IgA type myeloma rarely shows a high value of ZTT (Fig. 223).

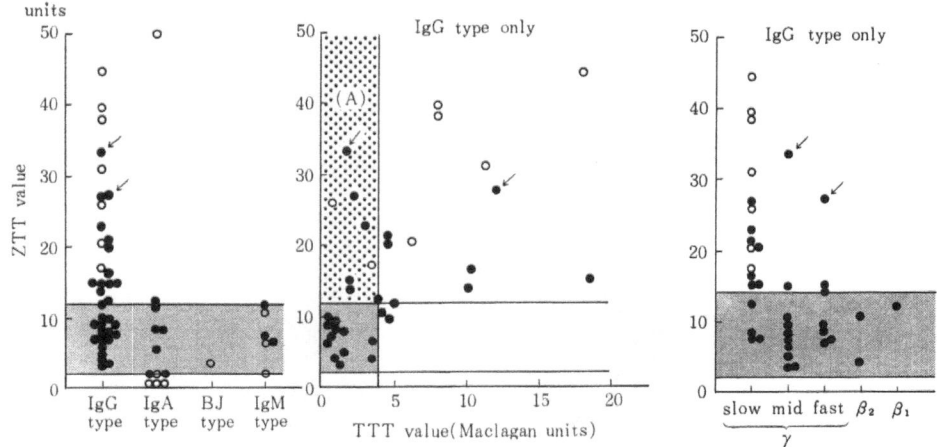

Fig. 223 Serum thymol turbidity test and zinc turbidity test in M-proteinemia.
(left) ZTT increased in all cases of the IgG type myeloma and about a half of the cases of the IgG type EMG, and more prominent in myeloma cases.   In the IgA and IgM types, ZTT was within normal limits in almost all cases.
(middle) In many of the cases with IgG type M-proteinemia, both ZTT and TTT were elevated.   In 7 cases included in the zone (A), the M-proteins migrated at the slow-γ zone, showing a discrepancy between ZTT and TTT values.
(right) Among the IgG type M-proteins, an elevated ZTT value was mostly encountered with the ones migrating at the slow-γ zone.

2. Abnormalities in Serum Gel Reactions

M-proteins frequently show abnormal solubility with different inorganic and organic solvents.   Especially, hydrochloric acid gelation, formol gel reaction, chromate gelation and urethan gel reaction are sometimes useful for screening the presense of M-proteinemia.

3. Positive Water Dilution Test

Water dilution test is commonly called as SIA test, based on the fact that serum euglobulins are insoluble in water.[346]   The test should be called more correctly as BRAHMACHARI–SENN test, because they reported on the test already in 1923.[332]

In the test, a few drops of serum sample are droped into the test tube containing approximately 5 ml of water from which carbon dioxide gas has been sufficiently removed through boiling.   The appearance of coarse flocculation is registered as a positive result.

The positive test is observed in approximately 50% of the cases with macroglobulinemia,[212] and also occasionally in chronic infections, rheumatoid arthritis, liver cirrhosis and IgG type myeloma.   The test is a non-specific reaction, and slight turbidity may be observed even in nephrotic syndrome where α_2-macroglobulin is significantly increased.

When the serum containing a significant amount of the M-proteins of an euglobulin character is submitted to agar gel electrophoresis using the buffer of a low ionic strength, these M-proteins are precipitated around the point of serum application.   In addition,

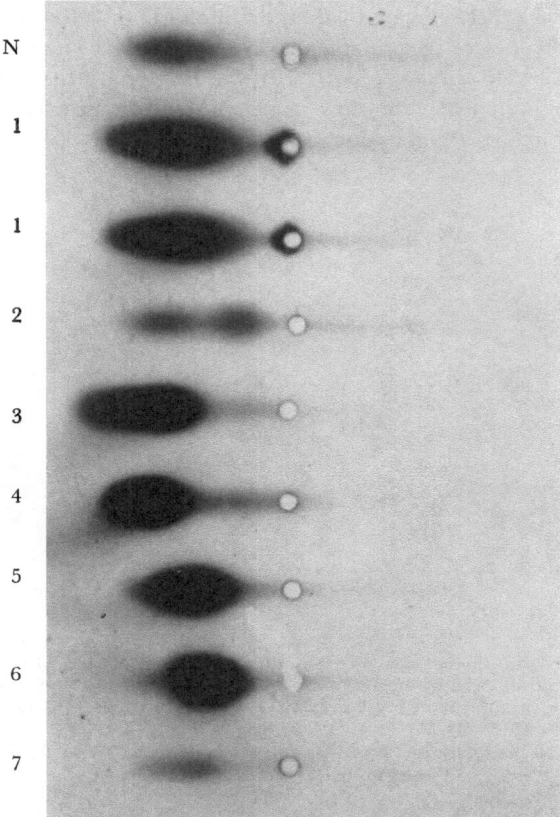

Fig. 224 Positive water dilution tests in agar gel electrophoretic patterns.
After agar gel electrophoresis, the agar plate was placed in distilled water without being fixed. After 24 hours' washing in distilled water, the plate was stained with Amido Black 10 B.
Even in normal serum (N), a small amount of euglobulins is identified in the γ fraction. Cases 2, 7 are of the IgA type M-protein, showing a very little amount of euglobulins. Case 1 had both the IgG and IgM type M-proteins, the final diagnosis being uncertain. Cases 3, 4 are of the IgG type, showing a strongly positive water dilution test. Cases 5, 6 are of the IgM type, showing a strongly positive water dilution test.

white euglobulin spots are identified easily if the agar gel plate is left in distilled water after routine agar gel electrophoretic separation (Fig. 224).

4. Cryo-Precipitation and Cryoglobulins

Normal serum does not show any noticeable macroscopic change when it is kept at low temperature. But in some cases of M-proteinemia, the serum shows cryoprecipitation or cryogelation when cooled to $0°–4°C$, which will re-dissolve on heating up to $37°C$. This phenomenon was first noticed by WINTROBE and others in 1933, and LERNER et al. later named it cryoglobulin. Table 58 shows the plasma proteins which are confirmed as showing cryoprecibitability. Among them, cryofibrinogen is mentioned in Chapter 24 (page 371). The other cryoproteins are all abnormal immunoglobulins, and are in general, referred to as cryoglobulin. Among immunoglobulin fragments, the cryo-BENCE JONES protein is known to show cryo-precipitation, but there is no report of other fragments showing this property.

Table 58 a   Various Cryoproteins.

A. Cryoglobulins
    1. IgG type cryoglobulin
    2. IgM type cryoglobulin
    3. IgA type cryoglobulin
    4. IgG.IgM mixed type cryoglobulin
    5. IgG.IgA mixed type cryoglobulin
    6. IgG.IgA.IgM mixed type cryoglobulin
    7. IgG.IgM.complement mixed type cryoglobulin
    8. IgG.IgM.α_{2M} mixed type cryoglobulin
    9. Ig.Lp mixed type cryoglobulin
    10. Cryo-BENCE JONES protein
B. Cryofibrinogen

Ig: immunoglobulins,   Lp: lipoproteins

Table 58 b   Clinical conditions and diseases accompanying cryoglobulinemia.

I.   Essential cryoglobulinemia
II. Symptomatic cryoglobulinemia
     1. Multiple myeloma
     2. Primary macroglobulinemia
     3. Malignant lymphomas
     4. Chronic lymphatic leukemia
     5. Reticulosis
     6. Periarteritis nodosa
     7. Systemic lupus erythematosus
     8. Rheumatoid arthritis
     9. SjÖGREN syndrome
     10. Liver cirrhosis
     11. Sarcoidosis
     12. Ankylosing spondylitis
     13. Hemolytic anemia
     14. Subacute bacterial endocarditis
     15. Leprosy
     16. Syphilis
     17. Cytomegalovirus mononucleosis
     18. Infectious mononucleosis
     19. Acute glomerulonephritis
     20. Kala azar
     21. Purpura

Cryoglobulinemia is the state where cryoglobulin appears in the blood.   It is classified into essential or idiopathic cryoglobulinemia and symptomatic or secondary cryoglobulinemia.   It is most often encountered in M-proteinemias, such as multiple myeloma, primary macroglobulinemia, reticular malignancies, cancers, autoimmune diseases, connective tissue diseases, etc. (Table 58).   In these cases, a portion or a majority of the M-protein carry a characteristic cryoprecipitable character.   In mixed types, however, the one immunoglobulin component carries a M-protein character, while the other is not usually of monoclonal.   In cases of polyclonal hyperimmunoglobulinemia (liver cirrhosis, endocarditis, purpura, SLE, rheumatoid arthritis, kala azar, etc.), the cryoglobulin contains both the K and L types immunoglobulins.   Cryo-BENCE JONES proteins are of either the L type or the K type (Fig. 211), but predominantly of the L type.[7, 184]

The cryo-precipitability is not lost when heated at 56 °C for 30 minutes, nor does it show any variation when subjectd to repeated cryo-precipitation. But when it is purified cryo-precipitability is easily lost. Cryoprecipitation is certainly dependent on the protein concentration. A monomeric unit of the cryo-BENCE JONES protein does not show a typical cryoprecipitation, and only the dimer shows cryoprecipitation at the protein concentration higher than 4 g/100 ml. Cryo-precipitability varies in different pH, ionic strength, and urea concentration. In cases of mixed-type cryoglobulin, a single protein component alone usually does not show cryo-precipitability. It shows precipitation when both components coexist. MANOR suggested the presence of "a cryo-precipitating factor", and he was able to demonstrate that this factor gives cryo-precipitability even to normal serum proteins. The relation between temperature and the viscosity of the cryo-BENCE JONES protein is shown in Fig. 225, and a marked increase in the relative viscosity is noted, followed by gelation and white precipitation successively.[184,191]

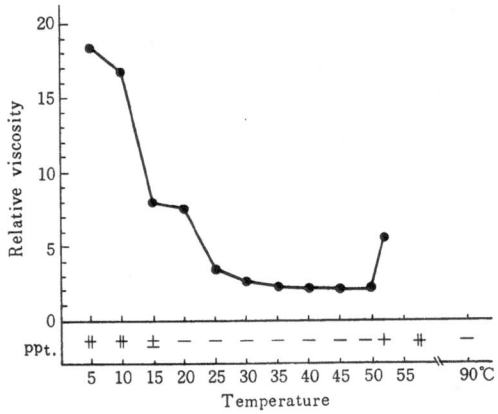

Fig. 225 Temperature dependency of the cryo-BENCE JONES protein.
The dimer BENCE JONES protein presented in Fig. 211 is dissolved in the phosphate buffer of pH 7.4 to make a protein concentration of 7 g/100 ml. ppt. indicates the macroscopic appearance of the protein solution, indicating the extent of white precipitation. On lowering the temperature, a marked increase in viscosity was recognized before precipitation appeared.

Thus, this phenomenon seems to occur when cryoglobulin molecules increase molecular aggregation as the temperature decreases, and the subsequent changes in surface potential, conformation and hydrophilic character may result in cryoprecipitation (Fig. 226).

But the physico-chemical and immunological difference between cryoglobulins and other M-proteins has not yet been established. Also, details are not yet available as to why such characteristic cryo-precipitability exists.

Recently, some of the cryoglobulin complexes have been found to have specific antibody activities such as rheumatoid factor, antinuclear antibody, anti-cytomegalovirus, cold hemagglutinin and anti-complement activity[21]. Therefore, the cryoglobulin, particularly the mixed type cryoglobulin, is suggested to be immune complex[21]. In fact, speaking of the immunological precipitation reaction, more precipitates are formed at a low temperature than at 37 °C. However, the complement components, β-lipoproteins, α_2-macroglobulin, nucleo-proteins are well known to be adsorbed to protein precipitates

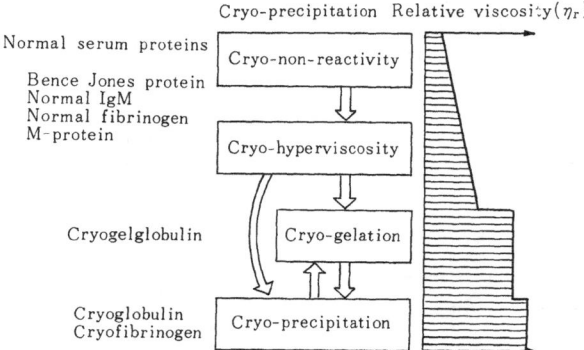

Fig. 226 Relation between cryo-precipitation and viscosity. On lowering the temperature, all the serum protein components show an increase in their intrinsic viscosity. However, since water itself increases its intrinsic viscosity the relative viscosity of normal serum does not change significantly on lowering the temperature. Normal IgM and fibrinogen which have extremely high intrinsic viscosity show slight increase in their relative viscosity on low temperature. Almost all the M-proteins also show slight increase in their viscosity on low temperature.

Cryo-gelation is the state of a solution showing an extremely high viscosity without losing hydrophilic property. Thus, cryo-gelation may change to cryo-precipitation easily once hydrophilic property is lost by adding salts, changing pH or other physico-chemical factors.

non-specifically. Therefore, it has to be studied further to conclude that all the cryoglobulins are of immune complex.

Cryoglobulinemia may cause clinically RAYNAUD's symptom or cryopathy, resulting in purpura, arthralgia and anemia. However, not all the cases with cryoglobulinemia are associated with RAYNAUD's symptom, and some patients may be entirely asymptomatic.

5. Heat-coagulation and Pyroglobulins

Pyroglobulin, named by MARTIN et al.[237], is a pathological immunoglobulin which is easily coagulated by heating. On inactivation of the serum containing pyroglobulin, white jelly-like coagulation takes place irreversibly. In general, the name pyroglobulin is given to that pathological globulin which shows jelly-like coagulation when heated at 56 °C for 30 minutes.[47,90,237,287,353] In four cases of pyroglobulinemia studied by the author, the temperature at which jelly-like coagulation began ranges from 54° to 60 °C. With normal sera, heating at 60 °C for 30 minutes does not produce any significant change in macroscopic appearance nor in relative viscosity. For this reason, the author defines pyroglobulin as that pathological globulin which shows irreversible coagulation when heated at 37 °C–60 °C.

Since pyroglobulin shows jelly-like coagulation on heating, the "pyrocrit" value after centrifugation is usually around 80%, and only a small amount of the supernatant serum is obtained. The "pyrocrit" value itself does not indicate the quantity of the pyroglobulin present in serum, but it seems to be structurally similar to agar gel.

Pyroglobulins usually migrate at the γ area, most frequently at the slow-γ zone, and demonstrate characteristic wavy protein bands on cellulose acetate electrophoresis. Pyroglobulins are mostly of the IgG type, but occasionally of the IgA and IgM types.

In pyroglobulinemia, the serum relative viscosity is markedly increased, probably due to an increased molecular aggregation of the pyroglobulin molecules in the blood. As

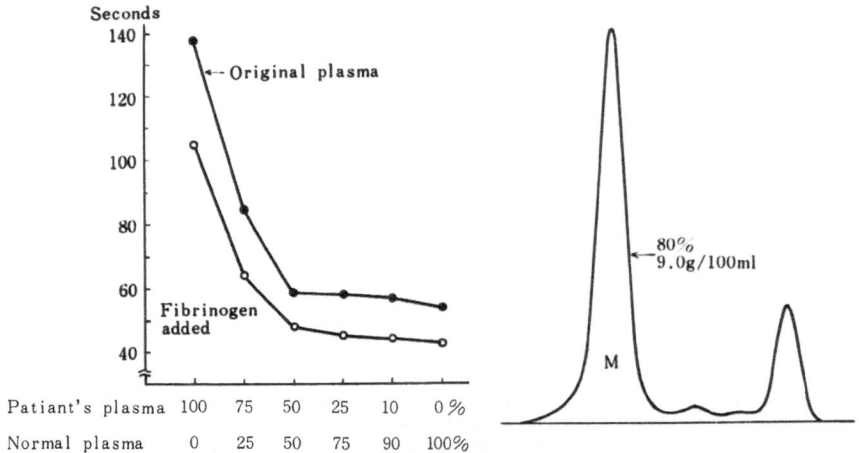

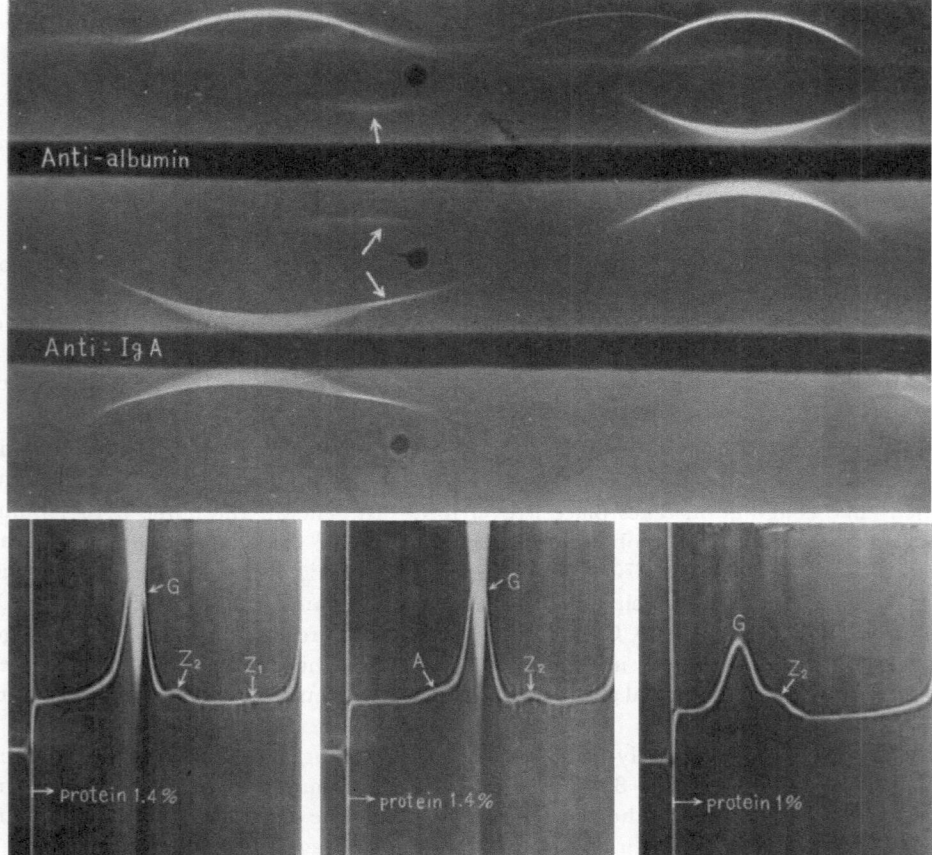

Fig. 227 Serum protein pattern in IgA type myeloma with abnormal blood coagulation.
T. N., 67 y.o., female, multiple myeloma.
Admitted with general malaise and anemia (Hb 4.5 g/100 ml). "Punched-out" lesion in the skeletal system and bleeding tendency noted. Sed rate 176 mm/hr. Blood coagulation studies resulted as below, indicating marked hypofibrinogenemia and deficiency of other blood coagulation factors. No
(continue to the next page)

shown in Fig. 205, a portion of the M-protein showing the pyroglogulin character is proved to be composed of the aggregated M-protein molecules.

Pyroglobulins are not different immunologically and biologically from the M-proteins lacking the pyroglobulin character.[335] The mechanism of demonstrating the characteristic heat-coagulability has not been well understood.

Pyroglobulinemia is encountered most frequently in multiple myeloma, but rarely in essential M-proteinemias or in the cases having no M-proteinemia.[225] Hyperviscosity syndrome may appear in some cases of pyroglobulinemia.

Changes in the Blood Mainly Due to Abnormal Interaction of the M-proteins with Other Plasma Protein Components

As indicated previously, there are some cases where M-proteins, especially those of the IgA and IgM types, combine with normal serum protein components such as serum albumin, haptoglobin, low-density lipoproteins and a_1-glycoproteins. They combine also with some of the blood coagulation factors (fibrinogen, prothrombin, factor V, factor VIII, etc.), often causing blood coagulation disorders (Fig. 227).[341] The complements may also be combined with M-proteins, demonstrating the anticomplementary effect.

Not only the protein components mentioned above are bound to some of the M-proteins, but also with non-protein constituents such as heparin, calcium ion, copper and so on. When they combine with heparin in the circulating blood, the post-heparin lipolytic activity (PHLA) becomes lower, resulting in hyperlipidemia. When they combine with a large amount of calcium or copper, hypercalcemia or hypercupremia may result.[120]

It has been well known that the BENCE JONES proteins may combine with some of the serum protein components, but also show a strong affinity to various tissues. In myeloma kidneys, the BENCE JONES proteins are precipitated out in the renal tubules, causing a marked epithelial degeneration. It was also reported that the complex composed of M-protein and low-density lipoprotein is precipitated on the glomerular basement membrane,

---

anticoagulants were traced (upper left). Soon after drawing the blood, the erythrocytes sedimented extremely rapidly because of marked rouleaux formation. In several minutes after the blood was drawn, the whole blood became jelly-like, giving a difficulty for separation of the serum. This peculiar phenomenon may be due to abnormal fibrin clotting and poor clot retraction which was resulted possibly from thrombocytopenia and markedly increased viscosity (6.70).

The serum total protein was 11.7 g/100 ml, and on cellulose acetate electrophoresis, a M-protein band was seen at the β zone, amounting to 80% or 9.0 g/100 ml. On immunoelecrophoresis, the M-protein was identifed to be the IgA type. The IgA line showed a significant tailing towards the anode, and the complex between the IgA and the albumin was identified in a small amount (arrows). Other classes of the immunoglobulins were almost completely failed to be demonstrated. On ultracentrifugation, two small spikes (Z_1, Z_2) were seen in addition to the G fraction.

Blood coagulation studies:
Bleeding time 13 minutes (prolonged)
Thrombocytopenia 3,000/mm³
Plasma fibrinogen 112 mg/100 ml (decreased)
Euglobulin lysis time 1 hr. 45 min. (decreased)

	Original plasma	Fibrinogen added	Normal plasma
PTT Platelin	154″	113.4″	120″
Act. cephaloplastin	137″	105.0″	45″
Plasma prothrombin time	27.3″	21.8″	14.1″
Plasma thrombotest	23.5%		
Correction for prothrombin time			
1. Ba-plasma	22.0″	19.4″	
2. Stored serum	24.2″	19.0″	
3. Stored plasma	21.4″	19.4″	

causing typical nephrotic syndrome.[297]   As discussed in detail later, amyloid may be formed through the interaction between the tissue glycoproteins and the M-proteins (most likely BENCE JONES proteins).

MULTIPLE MYELOMA

Definition and Diagnostic Criteria of Multiple Myeloma

Multiple myeloma may be defined as "multiple, progressive bone destruction due to neoplastic proliferation of the plasma cells".   Since the plasma cells potentially have the ability to synthesize immunoglobulins, one can find in the serum or other body fluids a detectable amount of pathological immunoglobulins or M-proteins released from the neoplastic plasma cells.   At times, the M-proteins are excreted into urine.   However, very rarely encountered are the cases of atypical multiple myeloma where no M-protein is detected in the serum or urine.   Atypical myeloma is said to occur in 2–10% of the myeloma cases, but only 3 such cases have been reported in Japan (Fig. 228). Occasionally, plasma cell leukemia develops in multiple myeloma.

Solitary myeloma has been reported occasionally.   In two cases of so-called solitary myeloma studied by the author, the M-protein has not disappeared completely even after the tumor was removed (Fig. 229).   Generally, many investigators are reluctant to admit the presence of solitary myeloma accompanying M-proteinemia.   If one could examine the bones systematically and extensively, one may find myeloma lesions other than the notable solitary myeloma lesion.   However, it is practically impossible to

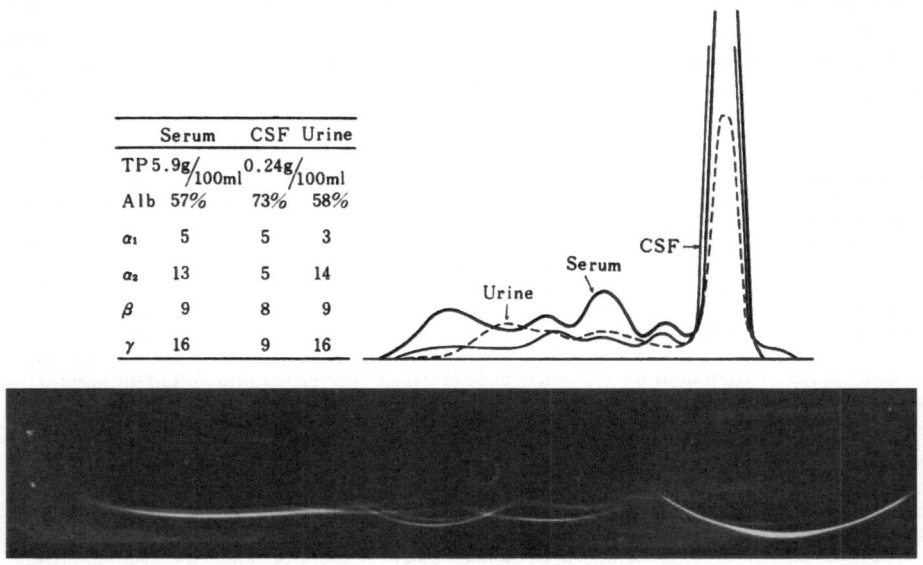

	Serum	CSF	Urine
TP	5.9g/100ml	0.24g/100ml	
Alb	57%	73%	58%
α_1	5	5	3
α_2	13	5	14
β	9	8	9
γ	16	9	16

Fig. 228   Solitary myeloma without M-proteinemia.
Y. S., 42 y.o., male, solitary myeloma.
Admitted with the osteolytic lesion in the 12th thoracic vertebra, which was confirmed to be plasmocytoma histologically.   The sternal marrow was normal, and no other skeletal lesion found.   No BENCE JONES proteinuria.
The serum protein electrophoretic patterns show no abnormality, and the cerebrospinal fluid proteins are also normal.
Discharged 2 months later.   Readmitted 5 months later with increased pain in the lower extremities. The osteolytic lesion was still localized, and no M-protein was demonstrated.

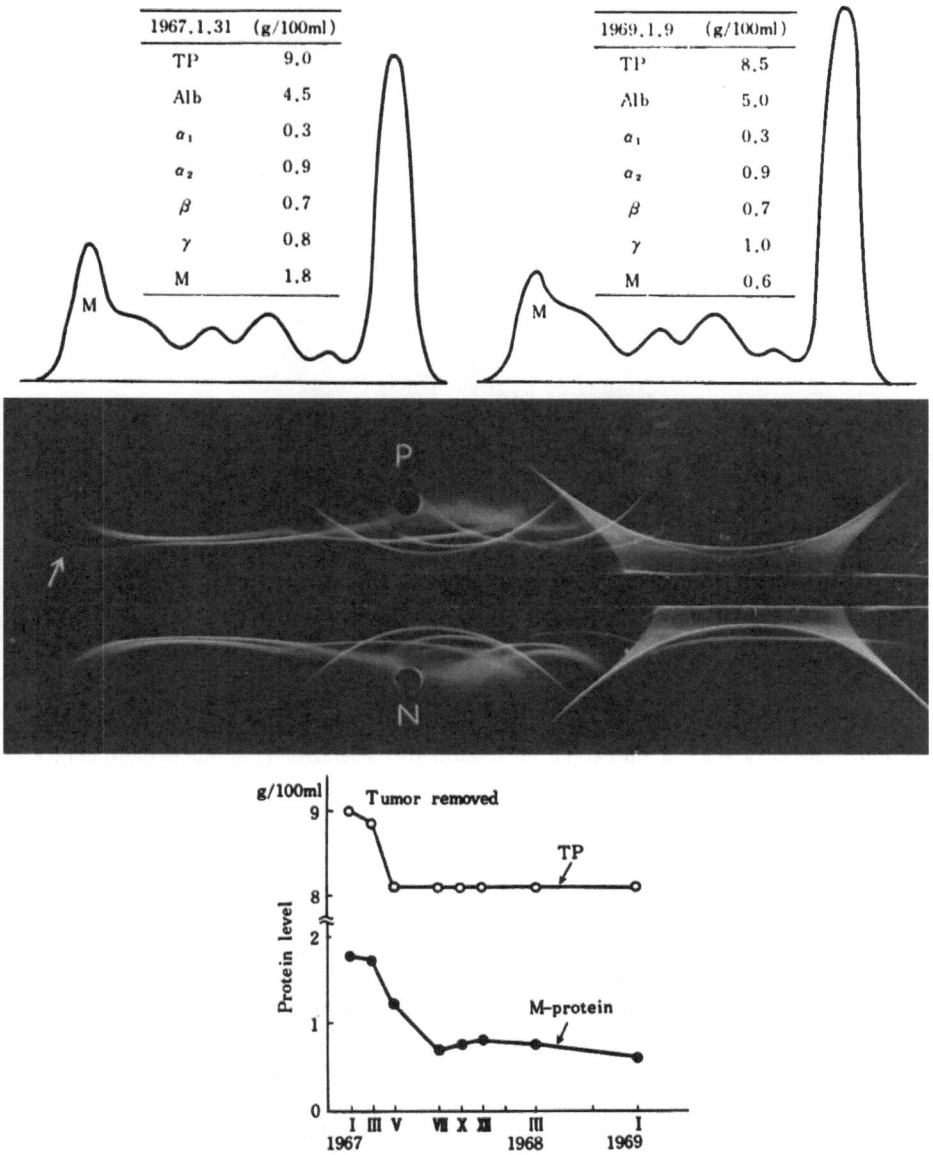

1967.1.31	(g/100ml)
TP	9.0
Alb	4.5
a_1	0.3
a_2	0.9
β	0.7
γ	0.8
M	1.8

1969.1.9	(g/100ml)
TP	8.5
Alb	5.0
a_1	0.3
a_2	0.9
β	0.7
γ	1.0
M	0.6

Fig. 229 Serum protein patterns in solitary myeloma.

K. M., 48 y.o., male, solitary myeloma of the right femur.

Admitted with pain in the right leg. The tumor was found radiologically in the right femur, and was resected the upper one third of the femur. The tumor was diagnosed as plasmocytoma histologically, and the tumor was found at the surgical margin of the femur. No other skeletal lesion was found. Endoxan was given continuously after surgery.

Before surgery, a small amount of the IgG type M-protein was found at the slow-γ zone. Two years after surgery, no subjective nor objective abnormality was found.

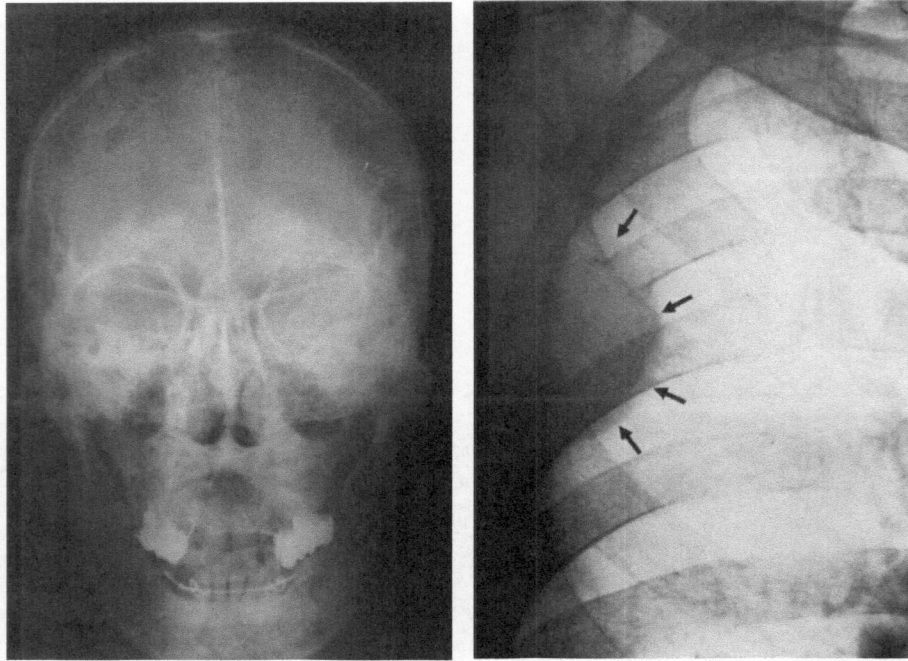

Fig. 230 Skeletal lesions in multiple myeloma.
Radiologically recognized are the two kinds of the skeletal abnormalities in multiple myeloma; 1) osteolytic lesions such as "punched-out" lesion (upper left) and osteoporosis, and 2) tumorfaction (upper right). Occasionally, the pathological fracture of the involved bones is encountered.
Histopathologically, the two types of the involvement are recognized; 1) diffuse infiltration by plasma cells, showing no apparent macroscopic change, and 2) tumorfaction.

examine the entire skeletal system completely.   Or one may assume that multiple myeloma would eventually develop if the cases with solitary myeloma be followed for a sufficiently long period of time.

Diagnostic criteria for multiple myeloma have changed in recent years mainly because of a significant progress in the analytical procedures.   In very early days, it used to be diagnosed solely by histopathological findings, and then mainly by hematological findings. In the past, the presence of myeloma proteins in blood or urine was considered to be one of the most diagnostic features in multiple myeloma.   However, at the present, because of a marked improvement of clinical, biochemical and immunological analyses, M-proteinemia itself is not helpful for diagnosing multiple myeloma.   Therefore, for diagnosis of multiple myeloma, the author emphasizes to find evidences of bone destruction due to plasma cell proliferation.   Radiologically most cases of multiple myeloma show evidences for bone destruction such as punched-out lesion or osteoporosis, and the bone destruction should be proved to be due to proliferation of atypical plasma cells through biopsy or aspiration of the bone marrow lesions.   If no osteolytic lesion is recognized radiologically, diffuse proliferation of plasma cells must be noted histologically through bone biopsy or necropsy. (Fig. 230)

Even though the above-mentioned criteria are not fully observed, multiple myeloma is strongly suggestive if the following findings are obtained: 1) atypical plasma cells in bone marrow aspiration being more than 20%, 2) appearance of definite wavy M-protein band on cellulose acetate electrophoresis, and 3) increased serum relative viscosity of more than 3.0.

M-Proteins Recognized in Multiple Myeloma

According to their immunologic types, M-proteins in multiple myeloma cases appear in the order of frequency, as follows: IgG type, IgA type, BENCE JONES type, IgD type, mixed type (more than two kinds of M-proteins), IgM type, atypical type, and IgE type. Among these various types, the IgG type myeloma makes up about 60% of all cases; the IgA type, 20%; and the BENCE JONES type, 10%. The IgD type myeloma has been recognized in approximately 50 cases, and the IgE type is only two cases in the world. In approximately one half of the cases of the IgG and IgA types, BENCE JONES proteins are found also. In the IgD type myeloma, almost all cases are associated with BENCE JONES proteinuria, predominatly of the L type.

IgM Type Myeloma

In several cases of IgM type M-proteinemia, a progressive bone destruction and predominant plasma cell proliferation are associated with.[240, 338, 404] These cases are referred to as "macroglobulinemia with bone destruction" "macroglobulinemia associated with myeloma" or "atypical macroglobulinemia". However, the author has proposed to call this particular condition as the IgM type myeloma. In the case reported by WELTON, the M-protein isolated from the patient's serum reacts only with anti-IgM, and it has the extra-antigenicity which is not recognized in normal IgM molecules.[404] The author also has experienced three such cases, and they have been followed up carefully.

IgM molecules are normally synthesized in certain lymphoid cells, but it is also recognized through fluorescent antibody techniques that certain plasmocytoid cells contain IgM. Therefore, it is assumed that a clone or clones of plasma cells are potentially capable of producing IgM. Thus, IgM type myeloma is understood to be a neoplastic proliferation of the IgM-producing plasma cells.

PRIMARY MACROGLOBULINEMIA

Macroglobulinemia is the condition where IgM type M-protein appears in serum, and is classified into primary and secondary.

Primary macroglobulinemia is also called as macroglobulinemia WALDENSTRÖM. The differentiation between the primary and the secondary forms is extremely difficult in some cases.

As pointed out by SELIGMANN, BASCH and HÄLLEN, no definitive diagnostic evidence is found in primary macroglobulinemia, including clinical findings, serum protein changes, prognosis and cellular morphology.[133,240,338] However, the definition given by HÄLLEN seems to be quite reasonable, clinically. First of all, reticular malignancies such as maligant lymphoma, leukemia and HODGKIN's disease have to be ruled out completely through various clinical studies. In addition to IgM type M-proteinemia, an abnormal proliferation of atypical lympho-plasmocytoid cells must be shown in the bone marrow and other reticuloendothelial organs. Atypical cell proliferation is mostly of lymphoid. Plasma cell infiltration may be associated with, but it should not be the major feature (Fig. 209, 222). In contrast to what is recognized in multiple myeloma, bone destruction is not seen in primary macroglobulinemia.

ESSENTIAL M-PROTEINEMIA

The condition has been variously called, but it includes principally all the M-proteinemias with the exception of myeloma, primary macroglobulinemia, heavy chain diseases.

Frequency of Essential M-Proteinemias

It is very difficult to obtain a true incidence of essential M-proteinemias. It would be necessary to examine all the inhabitants of a given district, as AXELSSON et al. did in Sweden.[15] However, with the recent progress of clinical laboratory examinations, many samples may now be examined in a short time with the resulting increase of its incidence reported year after year. Upon examination of all the samples submitted to the author's laboratories, essential M-proteinemia was found in 115 cases among 177 cases showing M-proteinemias, being approximately two thirds of all the cases. This is twice the incidence of multiple myeloma and primary macroglobulinemia. In the author's laboratories, all the routine serum protein electrophoretic patterns are reviewed carefully by either clinical pathologists or senior technologists to find even a very small spike. This incidence is very close to the incidence of M-proteinemias reported by WALDENSTRÖM.[133]

Diseases Associated with the Essential M-Proteinemias

As shown in Table 59, various diseases are associated with essential M-proteinemias, and they can be classified into 3 groups. The first group includes the diseases primarily involving the reticular tissues or organs; the second group includes the diseases which are usually associated with polyclonal hyperimmunoglobulinemia; and the third group includes the diseases which are not related usually to the reticuloendothelial tissues. The second group is most frequently encountered, occupying approximately 60% of all the cases with essential M-proteinemias. In the third group, the occurrence of M-proteinemia may have been quite accidental, and it seems to be not directly related to the basic diseases.

Nearly all the cases of multiple myeloma occurred after the age of 40 years, but the age distribution among the cases with essential M-proteinemias ranges widely from 4 months

Table 59   Various diseases associated with essential M-proteinemia.

The First Group Diseases with primary reticular abnormalities.	
Lymphoproliferative disorders (reticulosarcoma, HODGKIN's disease, etc.)	6
Myeloproliferative disorders (myelogenous leukemia, myelofibrosis, etc.)	3
Immunodeficiencies (GITLIN's syndrome)	1
	(10)
The Second Group Diseases usually associated with polyclonal hyperimmunoglobulinemia.	
Carcinomas (stomach, esophagus, rectum, etc.)	25
Autoimmune diseases (rheumatoid arthritis, HASHIMOTO's disease, etc.)	13
Respiratory tract infections (tuberculosis, pulmonary gangrene, etc.)	13
Hepato-biliary diseases (liver cirrhosis, hepatitis, etc.)	12
Renal diseases (chronic nephritis, chronic pyelonephritis, etc.)	5
Systemic amyloidosis	2
WEIL's disease	1
	(71)
The Third Group Diseases usually not related to reticular abnormalities.	
Cardiovascular diseases (hypertension, apoplexy, myocardial infarction, etc.)	20
Diabetes mellitus	5
Digestive tract diseases (peptic ulcer, gastritis)	4
Blood diseases (iron-deficiency, thrombocytopenia)	2
Polyp of vocal cord	1
Prostatic hypertrophy	1
Osteomalacia	1
	(34)
Total	115

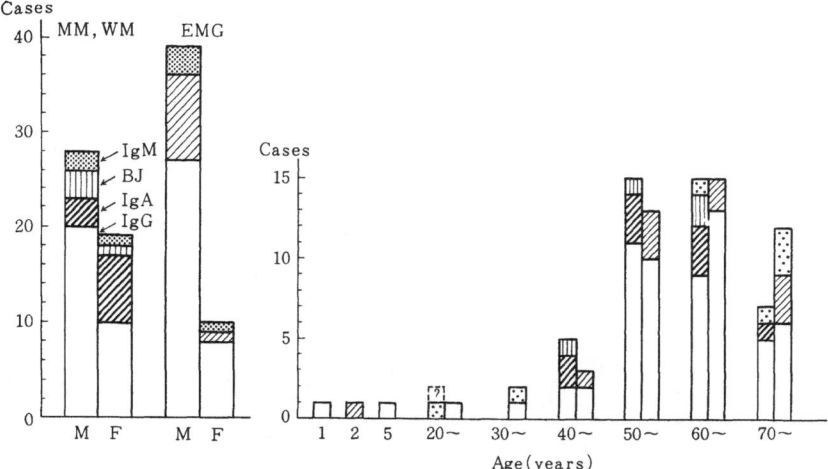

Fig. 231 Frequency distribution of M-proteinemia in different ages and sexes.
The thick solid lines are for multiple myeloma (MM) and primary macroglobulinemia
(WM), while the thin solid lines are for essential M-proteinemia (EMG).? indicates
a 27 y.o. male having solitary myeloma.
M-proteinemia occurred more frequently in males than in females, particularly so in
essential M-proteinemia.   Multiple myeloma occurred in the pateints of more than 40
years, the mode being in the fourth decade.   Essential M-proteinemia occurred in all
age groups, the mode being in the 6th decade.

to 89 years (Figs. 232, 238), although it may be said that the condition is more prevalent
to occur in older age (Fig. 231).   Essential M-proteinemia in infancy and childhood occurs
only in the first and second groups.   The average age of the patients belonging to the first
group is 35.8 years, and they occur in any age.   The second group may occur in infancy,
but mostly in adults; their average age being 58.3 years.   None of the pediatric patients
are seen in the third group, and their average age is 62.9 years.

Serum Protein Changes in the Essential M-Proteinemias

In all the cases with essential M-proteinemias, a varying amount of M-protein is detected in serum, being immunologically of the IgG type, IgA type, IgM type and BENCE
JONES type.   The M-proteins found in essential M-proteinemias are fundamentally similar
to those found in multiple myeloma and primary macroglobulinemia, but in general they
are found in a lesser amount and show less abnormal physico-chemical properties (Fig.
215).   In essential M-proteinemias, BENCE JONES protein is found less frequently in
urine.[133]   In almost all the cases with essential M-proteinemias, the serum concentration of M-protein may not change significantly for a long period of its observation
(Fig. 233).   However, the author has experienced a case of HASHIMOTO's disease in which
the M-protein disappeared after thyroidectomy (Fig. 200).   The serum concentrations
of normal immunoglobulins may not be significantly decreased in most of the cases (Figs.
195, 196).

Differentiation of the Essential M-Proteinemias

1. The IgM Type M-proteinemias
Essential M-proteinemia of the IgM type should be differentiated from primary macro-

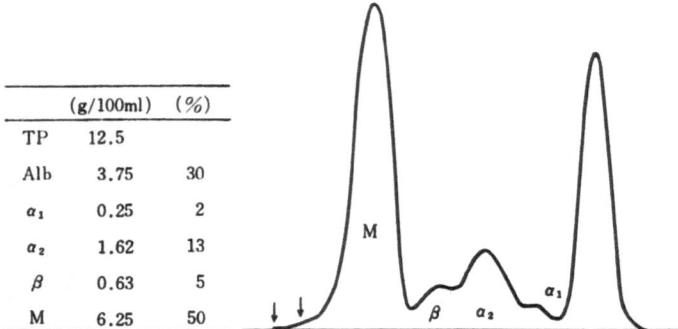

	(g/100ml)	(%)
TP	12.5	
Alb	3.75	30
a_1	0.25	2
a_2	1.62	13
β	0.63	5
M	6.25	50

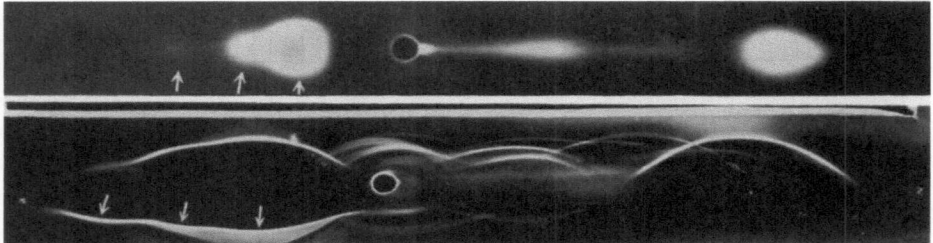

Fig. 232 M-proteinemia in a female infant.

Y. G., 1 y.o., female, lymphatic leukemia.

Admitted with fever, lymphadenopathy and leukocytosis. No skeletal abnormality radiologically.
Slight proteinuria without Bence Jones protein. Sed rate 110 mm/hr. CCF 3+, TTT 1.7 units,
ZTT 33.6 units, icterus index 2, total cholesterol 284 mg/100 ml, LDH 730 units, GOT 36 units. Died
of Klebsiella septicemia 2 months after the admission.

On autopsy, no acute pneumonia was found. Diffuse leukemic cell infiltration throughout the internal
organs.

On cellulose acetate and agar gel electrophoreses, the M-protein band was found at the γ zone, demonstrating three spots or triphasic precipitin line (arrows). The M-protein was identified as the IgG
type. On thin-layer gel filtration, the M-protein was found mostly at the G fraction. G′ fraction
was not clearly demonstrated.

globulinemia and IgM type myeloma. In IgM type myeloma, the plasma cells are predominantly proliferating and their infiltration accompanies significant bone destruction.
Therefore, the differentiation from IgM type myeloma is usually easy. However, the differentiation from primary macroglobulinemia is quite difficult clinically. In addition, in
primary macroglobulinemia, the normal immunoglobulins (IgA and IgG) are usually
normal or only slightly decreased, and this finding is not different from that found in essential M-proteinemia (Fig. 195). Although both the conditions tend to be different in
many clinical laboratory findings as shown in Table 60, definitive diagnostic differentiation is not easy. Therefore, as stated previously, primary macroglobulinemia should
be diagnosted only when the primary diseases involving reticuloendothelial tissues are
ruled out completely.

2. The IgG Type and IgA Type M-Proteinemias

Essential M-proteinemias of both the IgG and IgA types should be differentiated from
multiple and "solitary" myelomas. In solitary myeloma, as shown in Fig. 229, M-
protein may decrease significantly after the removal of the tumor lesion. The tumor
lesion may localize for a long period of time, and most investigators in this field believe
that the lesion may progress to multiple myeloma eventually. Therefore, the differentia-

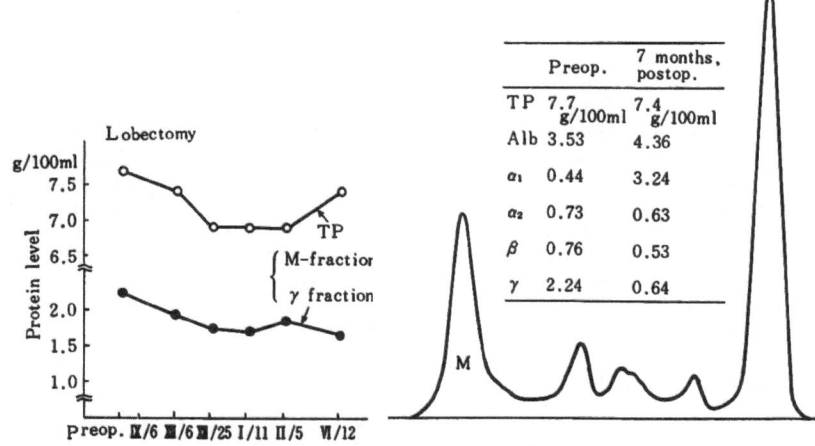

	Preop.	7 months, postop.
TP	7.7 g/100ml	7.4 g/100ml
Alb	3.53	4.36
α_1	0.44	3.24
α_2	0.73	0.63
β	0.76	0.53
γ	2.24	0.64

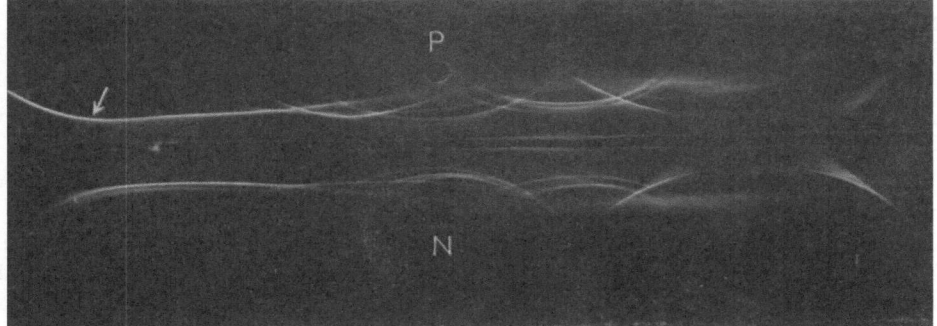

Fig. 233 Serum protein patterns in essential M-proteinemia.
Y. S., 50 y.o., female, middle lobe syndrome.
Complained of cough, sputa and general malaise, and diagnosed as the middle lobe syndrome. No malignancy noted on cytological studies. Sed rate 57 mm/hr. Sternal marrow contained 9.6% plasma cells, showing no atypism. No skeletal lesion noted radiologically. No proteinuria. ZTT 21 units, TTT 4.4 units.
Preoperatively, the cellulose acetate electrophoresis revealed a M-protein band at the slow-γ zone. The M-protein was identified to be of the IgG type. IgA was 200 mg/100 ml and IgM was 340 mg/100 ml. Postoperatively, the M-protein concentration did not change significantly.

tion of essential M-proteinemia from non-progressive solitary myeloma and extramedullary plasmacytoma must depend on histopathological findings.

The diagnosis of multiple myeloma is usually easy when typical punched-out lesions are evidenced roentgenologically. However, when diffuse infiltration causes the least osteoporotic changes, its differentiation may be quite difficult. Even though a definite pathological fracture is present, it may occur also in various diseases other than myeloma. As a matter of fact, essential M-proteinemias are frequently associated with malignancies, and their bone metastases may certainly give us a diagnostic confusion. Therefore, in order to make a definite diagnosis of multiple myeloma, a given bone destruction must be proved to be due to plasmocytic infiltration. Further, in the cases where no definite bone destruction can be recognized roentgenologically, the diagnosis must depend on bone biopsy or necropsy.

As shown in Table 61, 62, there are many laboratory data which are useful for the differentiation, and the following points are to be emphasized.

Table 60   Useful laboratory data for the differentiation of the IgM type M-proteinemias.

	Abnormal Data	MW	IgM-EMG
1.	Hyperproteinemia (over 7.0 g/100 ml)	7.4 g/100 ml (75.0 %)	6.5 g/100 ml (55.0 %)
2.	Hypoalbuminemia (less than 3.0 g/100 ml)	2.4 g/100 ml (75.0 %)	3.5 g/100 ml (22.2 %)
3.	Hypobetaglobulinemia (less than 0.55 g/100 ml)	0.59 g/100 ml(50.0 %)	0.52 g/100 ml(44.4 %)
4.	Hypogammaglobulinemia (less than 1.0 g/100 ml)	1.07 g/100 ml(75.0 %)	0.97 g/100 ml(66.7 %)
5.	Serum M-protein (more than 1.5 g/100 ml)	2.25 g/100 ml(75.0 %)	0.84 g/100 ml(0 %)
6.	Serum IgG (less than 1.0 g/100 ml)	0.63 g/100 ml(100.0%)	1.25 g/100 ml(11.1 %)
7.	Serum IgA (less than 0.17 g/100 ml)	0.16 g/100 ml(50.0 %)	0.2 g/100 ml (44.4 %)
8.	Serum IgM (more than 1.2 g/100 ml)	3.43 g/100 ml(100.0 %)	0.53 g/100 ml(0 %)
9.	Increased serum viscosity (more than 2.0)	2.56 (100.0 %)	1.63 (22.2 %)
10.	Serum TTT (more than 4.0)	2.55 (0 %)	5.20 (50.0 %)
11.	Serum ZnTT (more than 11.0)	10.8 (25.0 %)	11.5 (33.3 %)
12.	Anemia (less than 10.0 g/100 ml)	8.6 g/100 ml (75.0 %)	12.6 g/100 ml (25.0 %)
13.	Sed rate (more then 65 mm/l hr)	66 mm/1 hr (50.0 %)	61 mm/1 hr (25.0 %)
14.	Proteinuria	(100 %)	(57.1 %)
15.	Positive CRP test	(66.7 %)	(33.3 %)
16.	Positive RA test	(0 %)	(66.7 %)

Note 1:   MW indicates primary macroglobulinemia.
Note 2:   See Note 2 in Table 61.

1)   Multiple myeloma occurs at the age of more than 40 years, while essential M-proteinemia occurs at any age (Fig. 231).

2)   Unless there is no basic disease causing anemia, essential M-proteinemia itself does not cause significant anemia.

3)   Essential M-proteinemia is quite probable, if only mature plasma cells are found on bone marrow smears.   However, atypical plasma cells are not diagnostic for multiple myeloma, and do not rule out the possibility of essential M-proteinemia.   If the bone marrow smears show the plasma cells in less than 10%, essential M-proteinemia is suggestive, but a small number of plasma cells may be seen also in some cases with multiple myeloma.   Multiple myeloma is quite probable, if the rough-surfaced endoplasmic reticulum is cystically dilated on electron microscopy.   Histochemically, esterase and acid phosphatase activities may be increased more in myeloma cells.

4)   With the total serum protein concentration of more than 9.0 g/100 ml, multiple myeloma is more likely to be present.   Only in rare instances of essential M-proteinemia, the serum total protein becomes over 9.0 g/100 ml (Fig. 232).   The serum albumin concentration tends to be lower in multiple myeloma, but it varies greatly depending on the basic diseases.

5)   The serum concentration of M-protein is lower in essential M-proteinemias, rarely being more than 2.0 g/100 ml.   In contrast, in multiple myeloma, the serum M-protein exceeds a level of 2.0 g/100 ml in more than 80% of the cases.

6)   The wavy M-protein band on cellulose acetate (Oxoid or Separax) electrophoresis is highly suggestive of multiple myeloma of the IgG type.   The pseudo-wavy pattern may appear in both the conditions.

7)   Increased serum viscosity of more than 3.0 is likely to be due to multiple myeloma or macroglobulinemia.   In essential M-proteinemia, it rarely exceeds 2.5.   Even in multiple myeloma, it may be less than 3.0 (Fig. 217).

8)   If the serum concentrations of normal immunoglobulins are not decreased, the possibility of essential M-proteinemia is highly suggested (Fig. 195).

9)   Significant BENCE JONES proteinuria indicates the possible presence of multiple

Table 61 Useful laboratory data for the differentiation of IgG type M-proteinemias.

Abnormal Data	IgG-MM	IgG-EMG
1. Hyperproteinemia (over 8 g/100 ml)	8.5 g/100 ml (73.1 %)	7.1 g/100 ml (16.9 %)
2. Hypoalbuminemia (less than 3.5 g/100 ml)	3.5 g/100 ml (46.2 %)	3.6 g/100 ml (39.7 %)
3. Hypobetaglobulinemia (less than 0.55 g/100 ml)	0.53 g/100 ml (50.0 %)	0.61 g/100 ml(45.7 %)
4. Hypogammaglobulinemia (less than 0.3 g/100 ml)	0.16 g/100 ml (84.5 %)	0.61 g/100 ml(34.9 %)
5. M-protein (more than 2 g/100 ml)	3.65 g/100 ml (84.5 %)	1.24 g/100 ml(11.0 %)
6. Serum IgG (more than 3 g/100 ml)	4.3 g/100 ml (70.0 %)	2.3 g/100 ml (18.8 %)
7. Serum IgA (less than 0.2 g/100 ml)	0.1 g/100 ml (72.0 %)	0.26 g/100 ml(47.5 %)
8. Serum IgM (less than 0.08 g/dl)	0.06 g/100 ml (43.5 %)	0.12 g/100 ml(12.5 %)
9. Increased serum viscosity (more than 2.0)	2.30 (65.0 %)	1.82 (17.8 %)
10. Serum TTT (more than 5.0)	5.8 (50.0 %)	4.6 (27.8 %)
11. Serum ZnTT (more than 20.0)	28.8 (73.3 %)	14.6 (25.4 %)
12. Decreased serum iron (less than 80 μg/100 ml)	102.3 μg/100 ml(33.3 %)	78.8 μg/100 ml(61.6 %)
13. Anemia (less than 11 g/100 ml)	10.5 g/100 ml (52.2 %)	12.4 g/100 ml(26.0 %)
14. Sed rate (more than 80 mm/1 hr)	95 mm/1 hr (55.6 %)	40 mm/1 hr (17.0 %)
15. Proteinuria	(52.4 %)	(47.8 %)

Note 1: IgG-MM indicates the IgG type multiple myeloma, while IgG-EMG indicates the IgG type essential M-proteinemia.

Note 2: The values are of the average, and those in parenthesis indicate the incidence of each abnormal data.

Table 62 Useful laboratory data for the differentiation of IgA type M-proteinemias.

Abnormal Data	IgA-MM	IgA-EMG
1. Hyperproteinemia (over 8 g/100 ml)	9.6 g/100 ml (90.9 %)	7.1 g/100 ml (19.0 %)
2. Hypoalbuminemia (less than 3.5 g/100 ml)	3.0 g/100 ml (72.7 %)	3.6 g/100 ml (33.3 %)
3. Hypobetaglobulinemia (less than 0.4 g/100 ml)	0.15 g/100 ml (81.8 %)	0.61 g/100 ml (9.5 %)
4. Hypogammaglobulinemia (less than 0.6 g/100 ml)	0.37 g/100 ml (81.8 %)	0.95 g/100 ml (28.6 %)
5. Serum M-protein (more than 2.0 g/100 ml)	5.0 g/100 ml (90.9 %)	0.61 g/100 ml (9.5 %)
6. Serum IgG (less than 1.0 g/100 ml)	0.65 g/100 ml (80.0 %)	1.44 g/100 ml (14.3 %)
7. Serum IgA (more than 1.5 g/100 ml)	2.27 g/100 ml (70.0 %)	0.77 g/100 ml (0 %)
8. Serum IgM (less than 0.1 g/100 ml)	0.05 g/100 ml (90.0 %)	0.18 g/100 ml (33.3 %)
9. Increased serum viscosity (more than 2.0)	4.12 (70.0 %)	1.69 (4.8 %)
10. Serum TTT (more than 4.0 units)	0.63 units (0 %)	5.73 units (39.3 %)
11. Serum ZnTT (less than 8.0 units)	9.9 units (20.0 %)	7.6 units (56.3 %)
12. Serum iron (less than 100 μg/100 ml)	127 μg/100 ml (0 %)	67 μg/100 ml (100 %)
13. Anemia (less than 10.0 g/100 ml)	7.9 g/100 ml (100 %)	13.0 g/100 ml (5.3 %)
14. Leukopenia (less than 7000/mm³)	5400/mm³ (70.0 %)	7800/mm³ (38.9 %)
15. Increased sed rate (more than 100 mm/1 hr)	138 mm/1 hr (100 %)	57 mm/1 hr (14.3 %)
16. Proteinuria	(62.5 %)	(20.0 %)
17. Positive RA-test	(20.0 %)	(62.5 %)

myeloma, but occasional cases of essential M-proteinemias may be associated with BENCE JONES proteinuria.

HEAVY CHAIN DISEASES

Heavy chain diseases are the conditions in which a single kind of the heavy chain or its fragment appears in serum and/or urine. At the present, three kinds of heavy chain diseases have been encountered clinically; γ chain disease, α chain disease and μ chain disease. Other kinds of heavy chain diseases related to the δ and ε chains may be found in future.

Mechanism of the Appearance of Free Heavy Chain

The mechanism of immunoglobulin production may be assumed as follows.[180] In the immunoglobulin-producing cells, there seem to be two kinds of the polysomes; a large polysome producing the heavy chains and a small polysome producing the light chains. These two polypeptide chains are produced under different genetic controls. The light chain is synthesized in the small polysome, is discharged into the "free" light chain pool and is taken up onto the H-polysome where the combination of the light and heavy chains occurs to produce a complete immunoglobulin molecule, which is then discharged into the rough-surfaced endoplasmic reticulum. Fc fragments of the heavy chain are independently synthesized in the H-polysome, and Fd fragments seem to be synthesized only in the presence of the corresponding light chain. Therefore, in the immunoglobulin-producing cell, there seem to be complete immunoglobulin molecules and a small amount of "free"

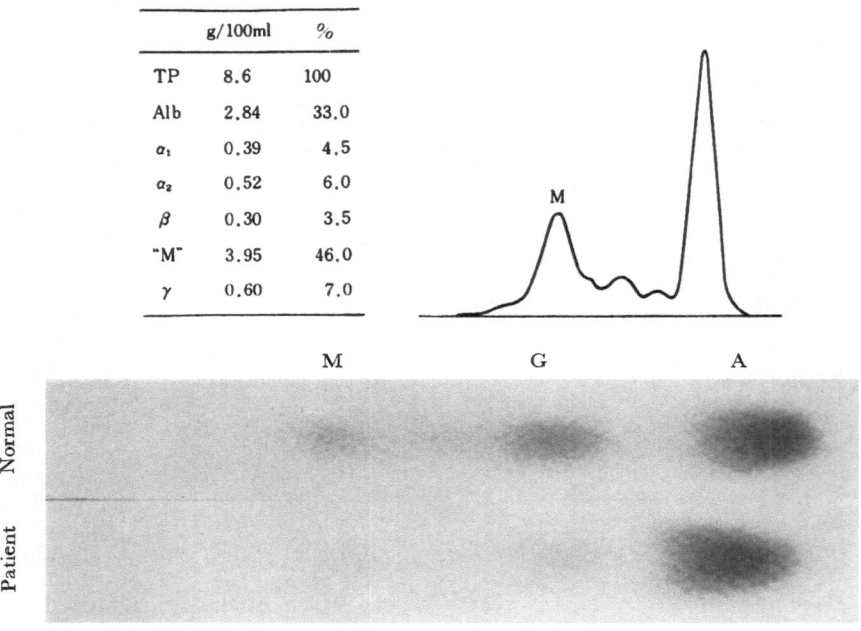

	g/100ml	%
TP	8.6	100
Alb	2.84	33.0
α_1	0.39	4.5
α_2	0.52	6.0
β	0.30	3.5
"M"	3.95	46.0
γ	0.60	7.0

Fig. 234 Serum protein patterns in γ chain disease.

H. Y., 37 y.o., female, γ chain disease. (Okayama University Hospital)
Complained of stupor, high fever and dyspnea and treated as pulmonary tuberculosis, 5 years prior to the present admission. Since then, developed high fever of unknown etiology, occasionally. Admitted presently with vomiting, right hypochondriac pain, fever and dyspnea. On laparotomy, found were hepato-splenomegaly and lymphadenopathy around the stomach. Slight anemia. Mild proteinuria, but no BENCE JONES protein. Plasmacytoid cells 36.4% in the bone marrow, but no bone destruction on X-ray examination.
On cellulose acetate electrophoresis, a rather broad M-band was found at the fast-γ zone, amounting approximately 46%. On immunoelectrophoretic analyses, the M-protein reacted against the anti-IgG and the anti-Fc. On thin-layer gel filtration, the abnormal protein was found in the A fraction. On ultracentrifugal analysis, the M-protein was calculated to be 3.7S.

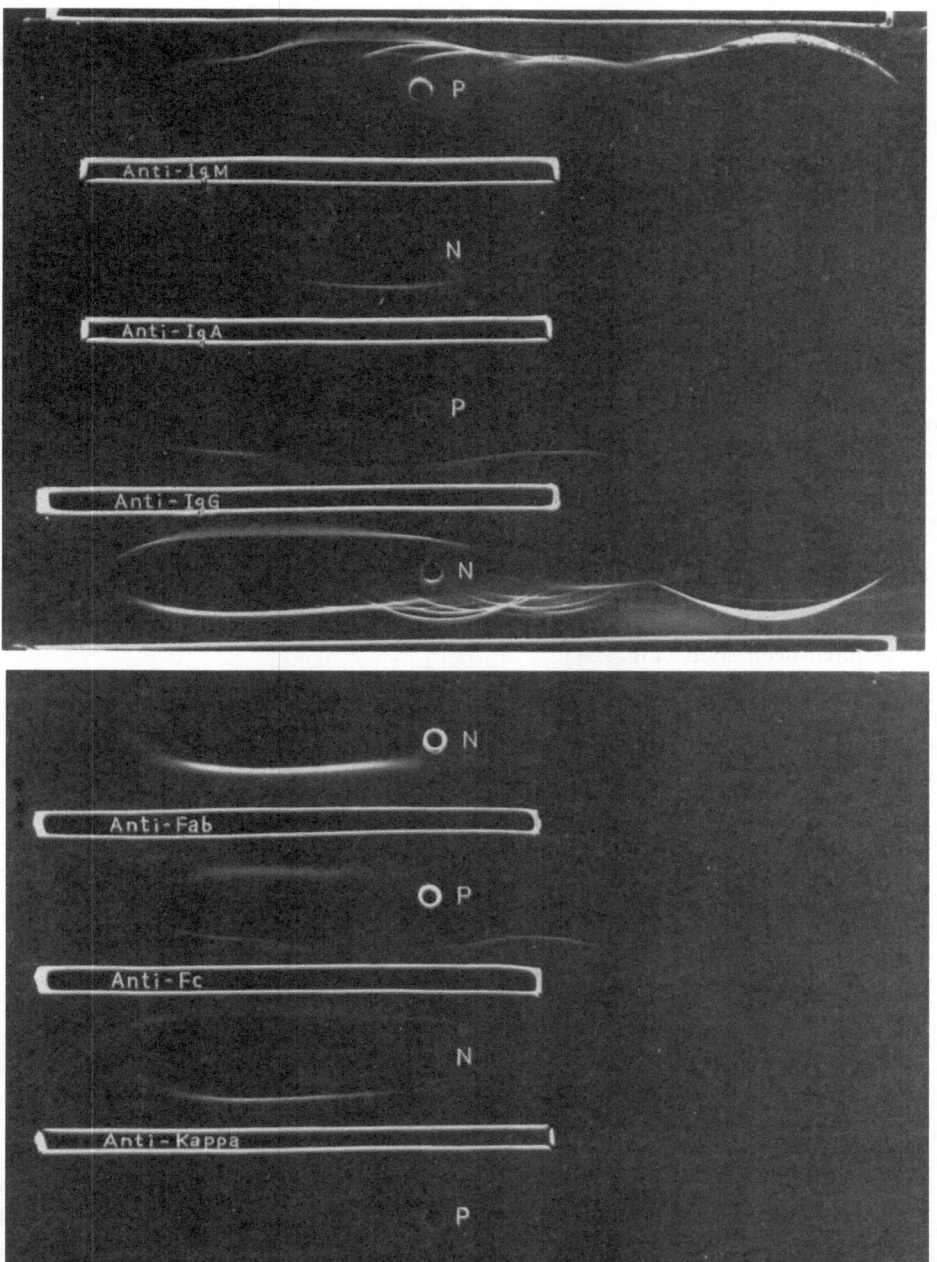

Fig. 234 b Serum protein immunoelectrophoretic patterns in γ chain disease.
P: patient's serum N: normal adult serum

light chain. In normal conditions, the production of both the heavy and light chains occurs in a balanced state. Although a small amount of "free" light chain in the cell may not be secreted outside the cell, degeneration or destruction of the cell may cause "free" light chain to be discharged outside. This may explain the fact that a minute amount of the light chain could be detected in normal urine.

Supposed that the syntheses of both the heavy and light chains are not synchronized in abnormal cells, it is easily understood that immunoglobulin fragments increase in serum or urine. In addition, impairment of the combination between the heavy chains and the light chains may also lead to the appearance of "free" heavy chains in serum or urine.

It should be emphasized that immunoglobulin-fragmenturia shown in Fig. 169 is to be differentiated clearly from the heavy chain diseases. In the former, various immunoglobulin fragments are found only in urine, but not significantly in serum. Therefore, the condition described as immunoglobulin-fragmenturia must be fundamentally different from the heavy chain disease which is one of the monoclonal gammopathies. The possible mechanism of producing immunoglobulin-fragmenturia is an excessive destruction of immunoglobulin molecules in the kidney tissues.

γ Chain Disease

The first report of the condition was made by FRANKLIN et al., and it has been frequently called as FRANKLIN disease.[96,194,282,338] In Japan, at least three cases of γ chain disease have been sited (Fig. 234).

Clinically, the condition is encountered in the age of more than 40 years. Chief clinical symptoms include generalized lymphadenopathy and hepatosplenomegaly, but no bone destruction is noted. In almost all cases, pancytopenia is noted, and lympho-plasmocytic proliferation is seen in the bone marrow and lymph nodes.

The M-protein of γ type is relatively broad on electrophoresis, usually migrating at the β_2 zone, and it is proved to be quite similar to the Fc fragment of γ chain both immunologically and physicochemically. The identical M-protein is usually found also in urine. The corresponding light chain is not detected in various body fluids and also in the abnormal cells with fluorescent antibody techniques.

α Chain Disease

The condition was first reported by SELIGMANN et al. and many additional cases have been reported.[340] Only one case has been found in Japan. The M-protein of α type has a common antigenicity with the α chain, and does not have the light chain. It usually migrates around β zone in a relatively wide area, and is found in both serum and urine. It contains a large amount of carbohydrate, and is thought to be quite similar to the Fc fragment of α chain.

The condition is encountered usually in Syrian females, suffering from abdominal lymphoma and severe malabsorption syndrome. The small intestines are widely infiltrated by lymphocytes and plasma cells. The skeletal system is not destroyed, but contains aggregates of the abnormal cells which are composed of plasma cells and large lymphocytes. The corresponding light chain is not found in body fluids nor in the abnormal cells.

μ Chain Disease

The condition was first reported by BALLARD et al.[17] It was seen in a 59-year-old man having the following major clinical and laboratory features: bone pain, a lymphoproliferative disorder, amyloidosis, bilateral carpal-tunnel syndrome, BENCE-JONES protein in urine and marked plasma-cell and lymphocytic bone-marrow infiltration. Being quite different from the other heavy chain diseases, both the μ chain fragment and BENCE JONES protein were detected in the same patient and also radiographic evidence of skeletal involvement was noted. The serum protein electrophoretic pattern does not reveal an M-component, but the immunoelectrophoretic pattern shows an anomalous heavy chain fragment of IgM. Urine does not contain the μ chain fragment but the K type BENCE JONES protein.

AMYLOIDOSIS

Amyloidosis was termed for the first time by VIRCHOW, because "amyloid" was thought to be consisted primarily of carbohydrates. On light microscopy, amyloid is observed as amorphous, eosinophilic, starch-like material, depositing diffusely in the interstitial tissues of the parenchymal organs or the connective tissues (Fig. 235). On electron microscopy, the amyloid deposits show the characteristic fibrillar pattern. It has been proved histochemically to be the complex material consisting of proteins and polysaccharides.

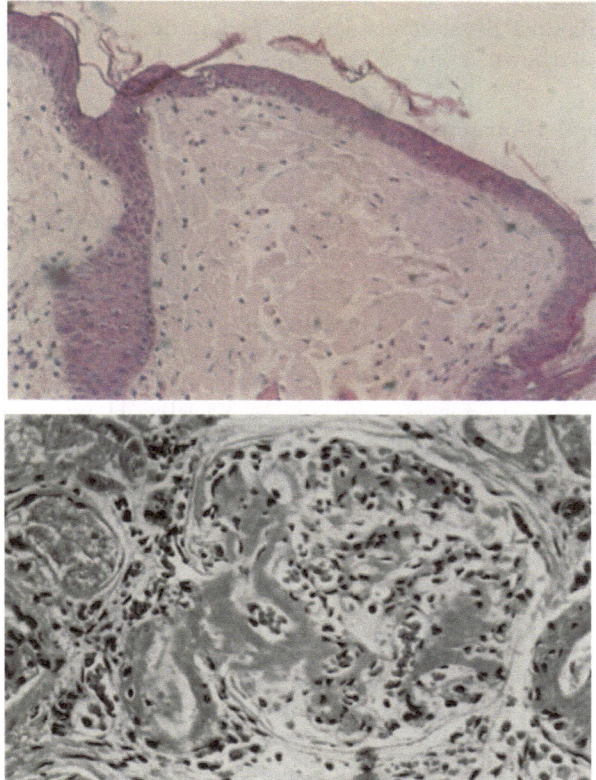

Fig. 235 Histopathological finding in amyloidosis.
The upper photomicrograph shows typical amyloid deposits in the skin (the case shown in Fig. 236).
Shown above is the glomerular lesion seen in secondary amyloidosis associated with chronic tuberculosis. Amorphous amyloid deposits are seen in the wall of the afferent arterioles and the basement membrane.

Pathogenesis of Amyloidosis

Amyloidosis is classified clinically and pathologically into the following 4 groups: 1) secondary amyloidosis, frequently combined with chronic inflammatory or necrotizing lesions such as tuberculosis, osteomyelitis, lepra, rheumatoid arthritis, HODGKIN's disease, carcinoma, etc., involving the spleen, kidneys, liver and adrenals; 2) primary amyloidosis, chiefly involving the heart, tongue, gastrointestinal tract, nervous tissues, joints and skin; 3) amyloidosis associated with multiple myeloma, occurring in approximately 5–10% of the cases with multiple myeloma.[12, 280]   In addition, local amyloidosis or amyloid tumor is recognized in the lungs, tongue, pharynx, thyroid, etc.   Also, in the pancreatic islets of patients, amyloid-like deposition is frequently seen.   Local amyloid deposition seems to be fundamentally different from systemic amyloidosis.

However, the above-mentioned classification of amyloidosis is not necessarily clear-cut. As stated later, OSSERMAN et al. emphasized the frequent association of BENCE JONES proteinuria in amyloidosis, and has considered amyloidosis to be included "plasma cell dyscrasia".[279]   Therefore, primary and secondary amyloidosis are considered to belong to BENCE JONES type essential M-proteinemias, while amyloidosis associated with multiple myeloma to be a variant of BENCE JONES type multiple myeloma.   Only when the particular BENCE JONES proteins having strong tissue affinity are present, amyloidosis may appear as a clinical entity.   The tissue affinity may or may not be derived from the auto-antibody activity against tissue components.   The pathogenesis of amyloidosis may be similarly thought as shown in Fig. 202.

Physicochemical Characteristics of Amyloid

Amyloid shows a characteristic metachromasia on staining with Congo Red, iodine, Methyl Violet, Gentian Violet and so on.   When Congo Red is given parenterally to the patient with amyloidosis involving the liver, its excretion will be markedly disturbed.

Amyloid is not a simple "starch-like" material, but is the complex consisting of protein and polysaccharides.   Among the polysaccharides included in amyloid, the uronic acids seem to be important for amyloid formation and its metachromasia.[328]   Amyloid contains approximately 42% of the proteins to be insoluble, and the soluble protein components include the complements and various plasma protein components such as immunoglobulins, lipoproteins, fibrinogen, etc.[57,58,250,328,331]   However, these components usually combine easily with various tissue components and plasma proteins.   The role of BENCE JONES protein or immunoglobulins has been in dispute, although the author believes that BENCE JONES protein or monoclonal immunoglobulins may contribute essentially to amyloid deposition.

Serum and Urine Protein Changes in Amyloidosis

There is no diagnostic serum protein electrophoretic pattern consistent with amyloidosis.   As stated previously, when nephrotic syndrome appears in amyloidosis, the serum protein electrophoretic pattern becomes of the nephrotic type.   Naturally, various patterns may appear, depending on different pathophysiological conditions occurring in the patients.   One of the most interesting findings related to serum protein changes is a frequent association of BENCE JONES protein.

In 1931, MAGNUS–LEVY first commented on the frequent association of amyloidosis with BENCE JONES proteinuria.   Later, OSSERMAN and his associates presented extensive evidence for a significant association between amyloidosis and monoclonal immunoglo-

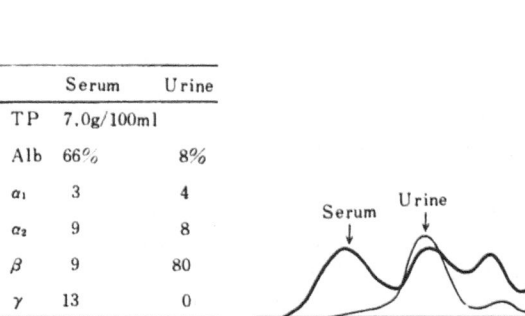

	Serum	Urine
TP	7.0g/100ml	
Alb	66%	8%
α_1	3	4
α_2	9	8
β	9	80
γ	13	0

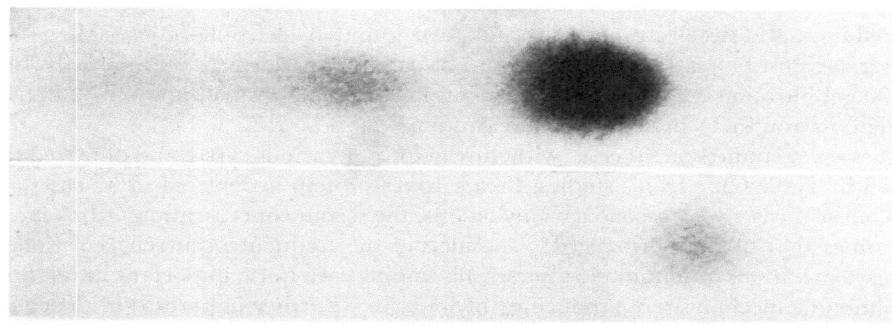

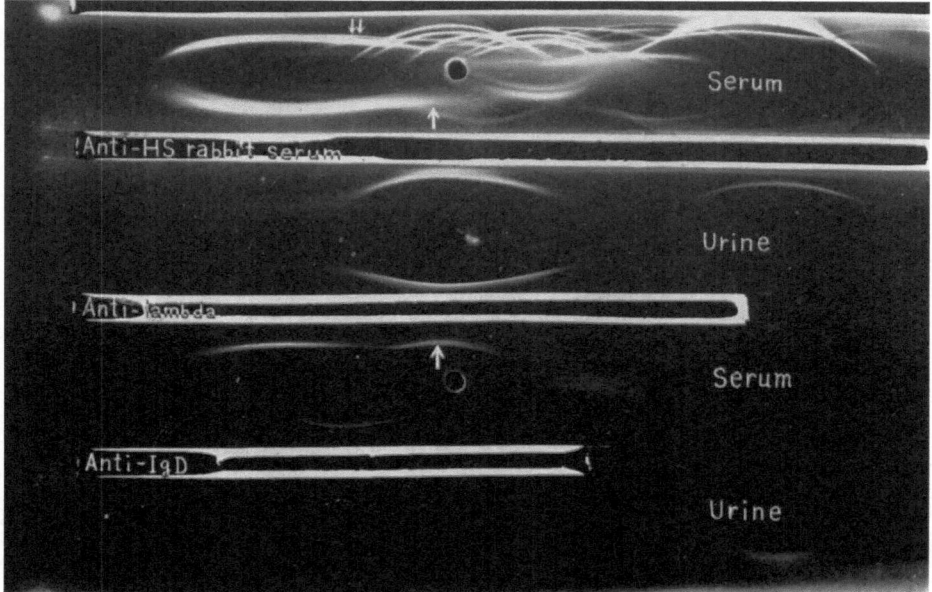

Fig. 236 Serum protein patterns in amyloidosis.

T. A., 48 y.o., female, generalized amyloidosis.

Developed reddish papules around the eyes for the past 6 years, and spread throughout the body. The skin lesion was diagnosed as amyloidosis (see Fig. 235). Slight anemia, but no other chemical abnormalities. Atypical plasma cells 13% in the bone marrow.

On cellulose acetate electrophoresis, no significant abnormality was found. On immunoelectrophoresis, an increased concentration of IgD was confirmed. A small amount of the L type BENCE JONES protein was identified in the serum, and also in the urine (single arrow). On thin-layer gel filtration, the urine showed a post-A spot which was confirmed as the BENCE JONES protein immunochemically.

bulins, particularly BENCE JONES proteins.[279)]   However, many other investigators failed to substantiate this relationship.

In order to detect a small amount of BENCE JONES protein in amyloidosis, the following points should be kept in mind.   As a matter of fact, it is the author's experience that amyloidosis is usually associated with the presence of a small amount of BENCE JONES protein, if any.   When nephrotic syndrome is combined with amyloidosis, marked albuminuria will cause a significant turbidity on boiling and thus a typical heat precipitability of BENCE JONES protein will not be recognized.   Therefore, the positive test on sulfosalycilic acid method is warrant to analyze the urine sample further, if BENCE JONES proteinuria is suspected.   Usually, the cellulose acetate electrophoresis of urine proteins will demonstrate a discrete peak indicating the presence of BENCE JONES protein.   However, in nephrotic syndrome, a small amount of BENCE JONES protein will not be detected on cellulose acetate electrophoresis, and then immunoelectrophoretic analysis using the antisera against BENCE JONES proteins must be carefully performed.   Frequently, Sephadex G-200 gel filtration technique is quite useful for detecting a small amount of BENCE JONES protein, as routinely performed in the author's laboratories.

OSSERMAN studied on 40 cases with amyloidosis of various types, and obtained the result shown in Table 63.   In his studies, BENCE JONES protein is detected in 37 cases, or 93%. In 3 out of 4 cases with secondary amyloidosis, the serum concentration of IgA is increased 2–4 times that of the normal.[279)]   Considering the frequent occurrence of isolated IgA increase in various autoimmune diseases, his findings are quite important in relation to the autoimmune mechanism of producing amyloidosis.   Further important evidence has been reported by GLENNER et al.[116)]

Table 63   Detection of BENCE JONES protein in amyloidosis (from OSSERMAN et al.)

	M-proteins in serum			BENCE JONES Protein			Incidence of BJP
	IgG	IgA	IgM	K	L	?	
Group I (14 cases)	5	2	0	9	5	0	14/14
Group II (11 cases)	0	1	0	5	4	2	11/11
Group III (11 cases)	2	0	1	1	5	4	10/11
Group IV (4 cases)	0	1	0	1	0	1	2/4

GroupI: Combined with multiple myeloma having bone destruction.
Group II: Primary without nephrotic syndrome.
Group III: Primary with nephrotic syndrome.
Group IV: Secondary.

AUTOIMMUNE HEMOLYTIC DISEASES

Autoimmune hemolytic diseases are the conditions in which autoantibodies against the patient's own erythrocytes are detected in serum, and they include the warm type autoimmune hemolytic anemia, the cold type autoimmune hemolytic anemia (or cold agglutinin disease) and the paroxysmal cold hemoglobulinuria (PCH).   Autoimmune hemolytic anemia due to drug allergy may be included in the category.   Autoimmune hemolytic diseases are classified into the idiopathic and the secondary types, and the latter is accompanied with leukemia, malignant lymphoma, myeloma, macroglobulinemia, myeloproliferative syndrome, malignancies, allergic diseases, and viral diseases.   Paroxysmal cold hemoglobulinuria is classified into the syphilitic and the non-syphilitic (idiopathic).

The erythrocyte autoantibodies are known to include the following three kinds; the warm autoantibodies, the cold autoantibodies and the DONATH–LANDSTEINER antibodies. Recently, these autoantibodies have been studied immunologically and physicochemically, and some of them are known to be related closely to M-proteinemia.[187]

Idiopathic Chronic Cold Hemagglutinin Disease

It is also known as the cold type autoimmune hemolytic disease, and the disease process is usually chronic, associated with mild anemia. Upon exposure to cold, cyanosis usually appears in the fingers, toes, face and ear lobes. The clinical symptoms tend to become worse in winter, and as shown in Fig. 237, the condition usually progresses consistently. Hematologically, slight lymphocytosis is recognized occasionally, but the bone marrow always shows an increased number of small lymphocytes.[187] SCHUBOTHE considers the condition to be a variant of primary macroglobulinemia.[187]

1. Serological Characteristics of the Cold Hemagglutinin

Cold agglutinin reacts most prominently at a low temperature (0–5 °C), and does not react at 37 °C. In most cases, it does not react at room temperature, but it may react up to 33 °C if the titer is significantly high. Most of the cold hemagglutinins show complement-fixing activity. They are usually eluted from erythrocytes when warmed to 37 °C, but the complements tend to be remained on the erythrocyte surface. In a few occasions, the cold agglutinins do not show complement-fixing activity, and they do not cause hemolysis. The cold agglutinin titer usually ranges from 1,024 to 64,000 dilutions, but rarely it may rise over 1 : 540,000.[187]

The direct antiglobulin test generally shows the positive reaction even at 37 °C. Cold hemagglutinins are not related to any of the blood typing systems, but they have mostly the I–specificity. That is, the most of the cold hemagglutinins have the anti-I activity, reacting with adult erythrocytes and only slightly with fetal erythrocytes. In very rare occasions, as shown in Fig. 237, I-specificity is not recognized with the cold hemagglutinin. According to TÖNDER et al., the individual cold hemagglutinin reacts differently to rabbit erythrocytes, showing serological heterogeneity.[383] In the author's case, also, it does not react with Human O-erythrocytes at 37 °C, but does react significantly with rabbit erythrocytes with the titer of 1 : 512.

In general, the cold hemagglutinin titer in primary viral pneumonitis and mycoplasma pneumonitis is relatively low, showing no I–specificity. These symptomatic condition seems to be fundamentally different from chronic cold hemagglutinin disease.

2. The Monoclonal Nature of the Cold Hemagglutin

With the exception of the case reported by ANGEVINE et al. where the IgA type cold hemagglutinin was detected, all the cold hemagglutinins detected so far belong to the IgM. According to SCHUBOTHE,[187] all 100 cases with chronic cold hemagglutinin disease that he experienced had the IgM type cold hemagglutinins, the light chains of which belonged all of the K type. In the case reported by COOPER and HARBOE[65,134] and the author's case, the IgM-L type cold hemagglutinin was detected, and it did not have the I–specificity. Therefore, the I–specificity seems to be dependent upon the K-type light chains of the IgM cold hemagglutinins. Further, the specific activity of the cold hemagglutinin varies among different patients.

The IgM type M-proteins having the cold hemagglutinin activity usually do not show compact bands on electrophoresis, but it certainly may form a compact band as shown in the author's case. The IgM type M-protein in chronic cold hemagglutinin disease is usually absorbed completely by the addition of human O-erythrocyte , but in some cases only 40% of the M-protein may be absorbed.[187] Thus, serologically non-specific M-protein

	(g/100ml)	(%)
TP	8.3	
Alb	4.50	55
a_1	0.17	2
a_2	0.32	4
β	0.50	6
M	2.49	30
γ	0.32	4

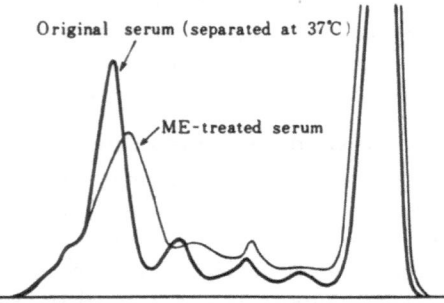

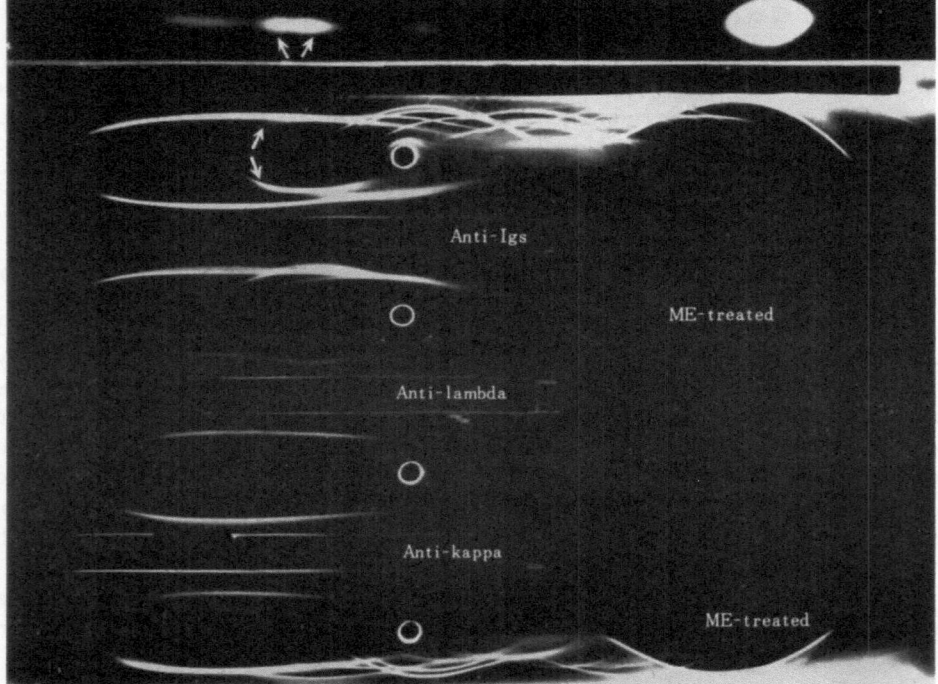

Fig. 237 Serum protein patterns in cold hemagglutinin syndrome.

K. K., 70 y.o., female, cold hemagglutinin syndrome.

Admitted with poor appetite and loss of weight. No definite abnormality on physical examination. Moderate normochromic anemia (Hb 8.2 g/100 ml) with reticulocytosis (1.5%). Bone marrow containing 0.8% plasma cells, and no other abnormal cells. Urinary urobilinogen (↑). Serum total bilirubin 1.65 mg/100 ml with the direct 0.35 mg/100 ml. No other abnormality in blood chemistry.

Blood type, A, CcDEe. Cold hemagglutinin, 800,000 x, showing marked temperature dependency (upper center). No I-specificity found (upper left). The positive direct Coombs test was due to the presence of the complements (upper right).

Using the serum separated at 37 °C, the electrophoretic fractionation resulted in a relatively broad M-protein band at the fast-γ zone. On thin-layer gel filtration, the serum showed a marked increase of the M (19 S) fraction (middle left). On ultracentrifugation, a marked increase of the 19 S (M_3) fraction, as well as 2 other minor spikes (M_2, M_1). On agar gel electrophoresis, the M-protein was noted as double spots at the γ_1 area, and it was found to be the IgM type (arrows). IgG and IgA were within normal limits. The light chain type was determined as the L type only after ME treatment of the serum. After the serum was absorbed with human O cells, the serum IgM disappeared almost completely (middle center).

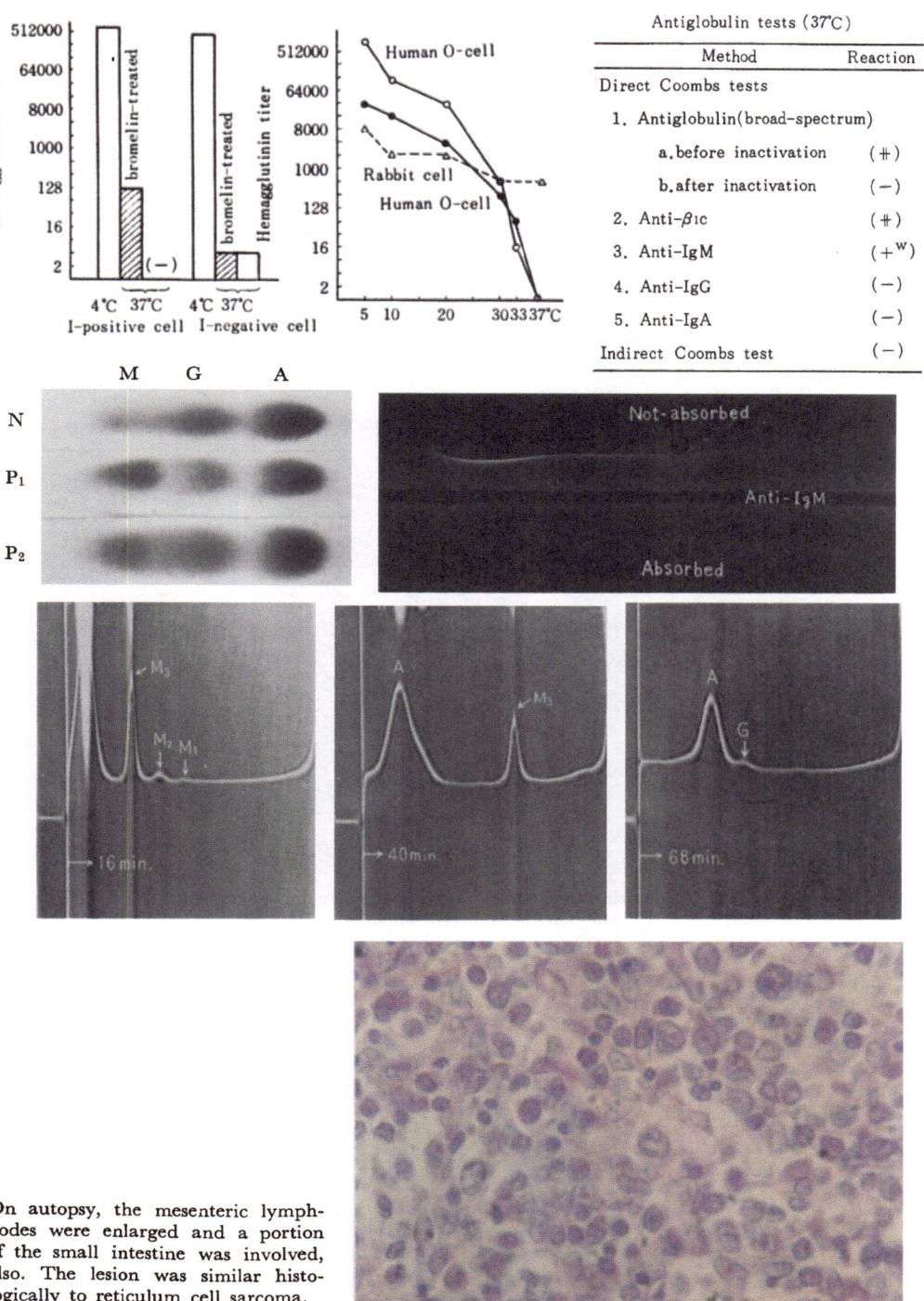

On autopsy, the mesenteric lymph-nodes were enlarged and a portion of the small intestine was involved, also. The lesion was similar histologically to reticulum cell sarcoma.

Antiglobulin tests	
Direct Coombs test	(+)
Indirect Coombs test	
Anti-globulin(broad-spectrum)	(+)
Anti-IgG	(+)
Anti-kappa	(+^{w})
Anti-lambda	(+^{w})

Before treatment	(g/100ml)
TP	7.3
Alb	4.7
α_1	0.2
α_2	0.7
β	0.7
γ	1.0

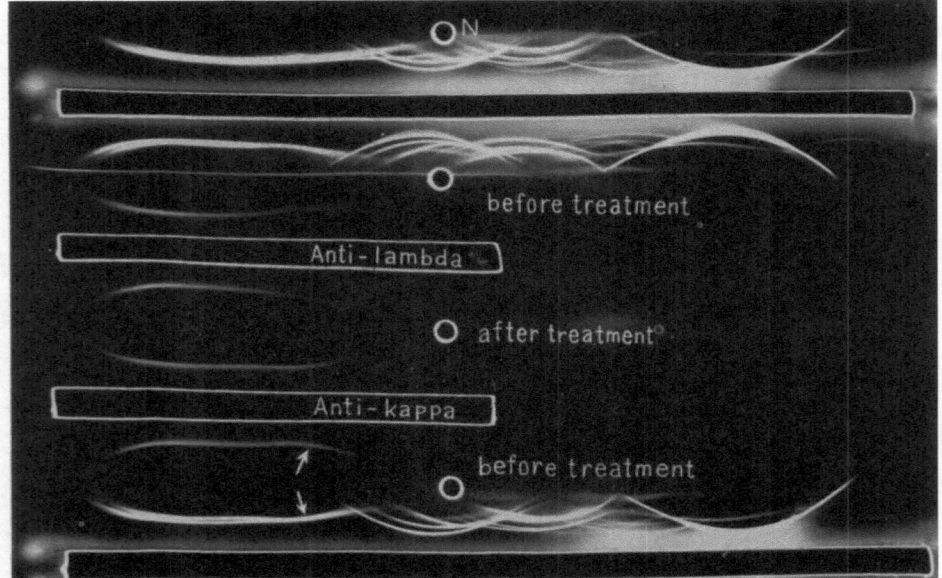

Fig. 238   Serum electrophoretic patterns in autoimmune hemolytic anemia.
K. S., 4 mo., male, warm type autoimmune hemolytic anemia.
Physiological neonatal jaundice appeared one week after birth, and continued for about one month.
Anemia (Hb 8.1 g/100 ml) noted at the age of 2 months.   Reticulocytosis (3.6%), color index 0.91,
WBC 16,350 (lymphocyte, 54%).   Blood type, O, Rh$_0$ (+).   Serum iron 144 μg/100 ml, serum total
bilirubin 1.6 mg/100 ml (direct 0.8 mg/100 ml), s-GOT 75 units, s-GPT 46 units.   Urinary urobilinogen
(+) and bilirubin (−).
On cellulose acetate electrophoresis, a minor M-protein band was noted at the fast-γ zone (arrow).   On
immunoelectrophoresis, the M-protein was found to be the IgG-K type.   The warm type autoantibody
was identified to be the IgG type, reacting to both the anti-kappa and the anti-lambda.
After treatment with predonine, the direct Coombs test became negative, and the M-protein band was
not recognized on cellulose acetate electrophoresis.   On immunoelectrophoresis, however, a trace
amount of the M-protein was still noted.   Thus, not all the IgG-K type M-protein molecules belong
to the autoantibody.

may be also synthesized along with the monoclonal cold hemagglutinin. This phenomenon
seems to be identical to that of essential M-proteinemia presented in Fig. 200, 202.   Further, it may be considered to be the similar phenomenon to "status hyperglobulinemica"
frequently recognized in chronic infections, as presented in the chapter on polyclonal
hyperimmunoglobulinemia.

The increased titer of the cold hemagglutinin recognized symptomatically in vaious respiratory diseases does not accompany with M-proteinemia, and both the IgM-K and IgM-L types are mixed.

Warm Type Autoimmune Hemolytic Anemia

Warm type autoimmune hemolytic anemia is recognized clinically in both acute and and chronic forms, and is generally associated with marked jaundice and severe anemia.

The warm autoantibodies usually react most prominently at 37 °C, having the serological features of the incomplete antibody. They may rarely belong to the complete antibody, reacting even in saline. In approximately one third of the cases, the warm autoantibodies may show a specificity to Rh antigens.[109] Using antiglobulin reactions, they are classified into the IgG type, the complement-fixing IgG type, and the complement type.[187] In Japan, the complement-fixing IgG type warm autoantibodies are found most frequently. In about a half of the cases having the IgG type autoantibody, they are of monoclonal type, but the rest of the autoantibodies are both of the K type and of the L type. However, this does not rule out the possibility that warm autoantibodies have the monoclonicity, but they may be a mixture of different monoclonal autoantibodies.

The case shown in Fig. 238 suffers from warm type autoimmune hemolytic anemia associated with IgG type M-proteinemia. In this case, however, the M-protein was not absorbed by the addition of human erythrocytes, and M-proteinemia was recognized even after the direct antiglobulin test become negative.

Paroxysmal Cold Hemoglobinuria

The condition occurs infreqnently, and is recognized in both the acute and chronic forms. The direct antiglobulin test becomes positive only when the actute attack occurs. The DONATH–LANDSTEINER antibodies combine to erythrocytes at low temperature, accompanied with complement fixation. However, hemolysis occurs only when the sensitized cells are warmed to 37 °C, again. The DONATH–LANDSTEINER antibodies belong to the IgG, showing probably the anti-Tj^a activity. According to WORLLEDGE and ROUSSO, in 11 cases, the D–L antiboes reacted with P_1 and P_2 erythrocytes, but did not react with PP and P^k erythrocytes. Thus, the D–L antibodies seem to be specific autoantibodies against the patient's own cells. The monoclonicity of the D–L antibodies has not been proved definitely.[187]

Chapter   24

Abnormal Plasma Proteins

CLASSIFICATION AND DEFINITION OF
THE ABNORMAL PLASMA PROTEINS

"Abnormal plasma protein" is defined as the protein that cannot be detected in normal plasma but appear in plasma only in pathological conditions. APITZ first used the term 'paraprotein' in 1940, with the meaning of abnormal proteins not found in normal plasma. However, APITZ used the term in relation to pathological immunoglobulins found in cases of multiple myeloma, and in accord with this meaning, WUHRMANN and WUNDERLY classified plasma protein abnormalities into (1) dysproteinemia and (2) paraproteinemia.[413] The former indicates a quantitative abnormality, while the latter a qualitative one. Subsequent to this, in Japan, the term 'paraproteinemia' has become widely used, being understood to mean principally the appearance of pathological immunoglobulins in blood. Since pathological immunoglobulins found in myeloma and other monoclonal immunoglobulinopathies cannot be proven to be absolutely non-existent in normal plasma, the use of the term "abnormal proteins" itself has been criticized. For instance, WALDENSTRÖM and others have never used the term 'paraproteinemia' and there are many other investigators in aggreement with him. Since the two terms 'paraprotein' and 'paraproteinemia' have uncertain meanings, the author has chosen rather to avoid their use.

When pathological changes are seen in the plasma protein components (putting aside the existence of minor components known only through their biological activity), regardless of whether or not these abnormalities are essential for patients' illness, they are called "plasma protein abnormalities". If essential for patients' illness or very important, the author will refer to them as "plasma protein disease". When the term 'abnormal proteins in plasma' is used, reference is to a situation where the proteins not found in normal plasma appear pathologically in plasma. Or again, in using the term 'abnormal plasma protein', as in the former case it refers to a plasma protein having its molecular structure clearly different from the normal. Those recognized as hereditary having the normal original function are of course handled as hereditary types and are not included in the abnormal plasma proteins. Furthermore, as regards many of the plasma proteins, there still remains a number of unclarified points, and the distinction between, for instance, "hereditary type", "para-type", and the "allo-type" is not too clear. But based on what is known to date, they have been compiled in Table 64. Among the abnormal plasma proteins, the C-reactive protein and the abnormal immunoglobulins are discussed elsewhere.

ABNORMAL FIBRINOGENS

The terms parafibrinogen and parafibrinogenemia were first used by BECK.[24]   Hold-

Table 64   Abnormal proteins in plasma.

I. Abnormal Plasma Proteins
    1. Abnormal fibrinogens
        a. Parafibrinogens, congenitally encountered.
        b. Parafibrinogen, acquiredly encountered.
    2. Paralbumin
    3. Paraimmunoglobulins (?), including various pathological
        immunoglobulins encountered in M-proteinemia.
    4. C-reactive protein (?)
II. Tissue Proteins and similar protein components
    1. Hemoglobins and non-hemoglobin stromal proteins of erythrocyte
    2. Myoglobin
    3. α_2-globulin appearing on renal transplantation
    4. Cancer-specific protein and α_1-fetospecific protein
    5. Miscellaneous proteins

---

ing with the same meaning there are times also when it is called dysfibrinogenemia.[45] The author prefers to use the term 'parafibrinogenemia' following BECK's nomenclature, since there is no function of normal fibrinogen and also from the fact that the molecular structure differs.   The author intends to use the term dysfibrinogenemia in relation to quantitative changes in fibrinogen.

The ones that are included in abnormal fibrinogen are listed in Table 64.   Among these, with cryofibrinogen as with cryoglobulin, there is no sufficient proof as to its abnormal structure, but for the sake of convenience it will be handled in this chapter.   Up to the present, for the sake of reference, there have been reports of eight families with congenital parafibrinogenemia, and there has been one acquired case in association with primary hepatoma.

BECK suggested the following findings to suspect the presence of an abnormal finbrinogen: (1) familial hypofibrinogenemia; (2) prolonged thrombin clotting time; (3) friable clots without evidence of acquired hypofibrinogenemia or accentuated fibrinolysis; (4) increased anticoagulant activity interfering with clotting of normal fibrinogen by thrombin; and (5) abnormal fibrin stabilization.   Of course, the final diagnosis of parafibrinogenemia must depend on further appropriate studies including the structure of fibrinogen molecules and other properties.

Cryofibrinogen

If cryofibrinogen is preserved at 4 °C, white precipitation is brought about, but when heated to room temperature or 37 °C, it becomes redissolved.   One thing that should be noted here is that the detected amounts of cryofibrinogen will alter using different anticoagulants.   This is particularly true of heparinization, where in acute diseases, transiently, a small amount of cryofibrinogen (a portion of the total fibrinogen) is detected.[169,351]   The situation where cryofibrinogen appears in blood is called cryofibrinogenemia.   It is extremely rare and the author has run across only one case and a few in Japan (Fig. 239).   Results are not yet conclusive on the chemical composition of cryofibrinogen.   It is still unclear whether or not it is identical to normal fibrinogen.   However, at least with the author's case, normal coagulability is seen and no marked differences were observed as regards immunological antigenicity as well as electrophoretic properties.   In fact, all normal plasmas happen to show the similar phenomenon.   Namely, when normal plasma are once frozen and redissolved in a refrigerator, white floccula always appear and they disappear when warmed to 37 °C.   The floccula occupy approximately one third of the total fibrinogen

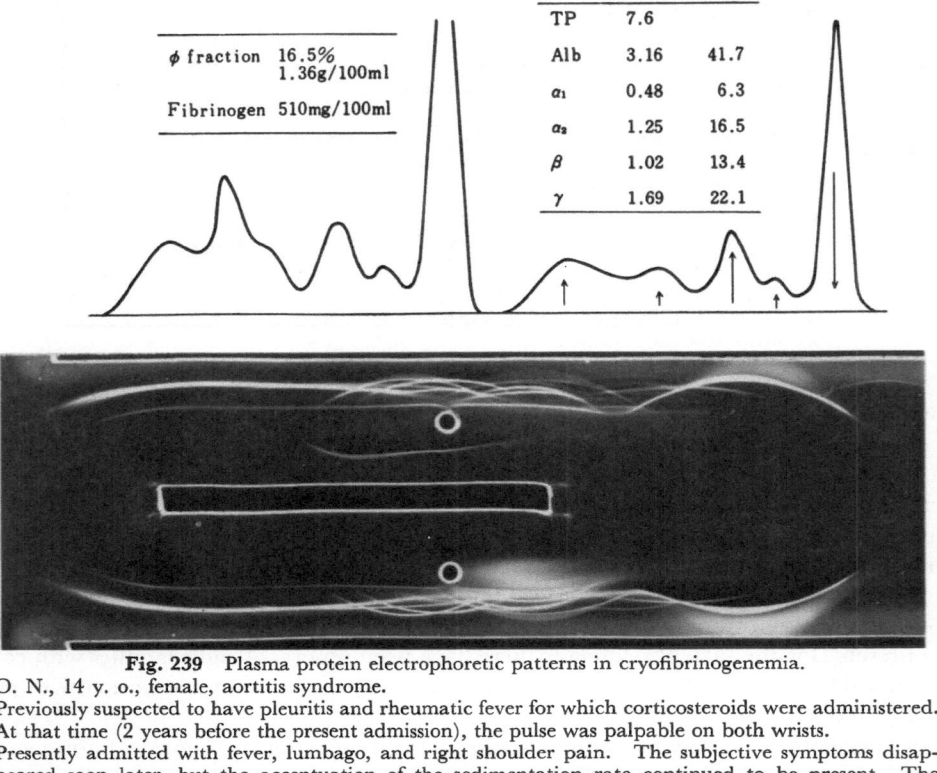

	(g/100ml)	(%)
TP	7.6	
Alb	3.16	41.7
α_1	0.48	6.3
α_2	1.25	16.5
β	1.02	13.4
γ	1.69	22.1

ϕ fraction 16.5%
1.36g/100ml

Fibrinogen 510mg/100ml

Fig. 239 Plasma protein electrophoretic patterns in cryofibrinogenemia.

O. N., 14 y. o., female, aortitis syndrome.

Previously suspected to have pleuritis and rheumatic fever for which corticosteroids were administered. At that time (2 years before the present admission), the pulse was palpable on both wrists.

Presently admitted with fever, lumbago, and right shoulder pain. The subjective symptoms disappeared soon later, but the accentuation of the sedimentation rate continued to be present. The pulse was not palpable on the left arm. No cryopathy was clinically apparent. Sed rate 115 mm/hr. Moderate hypochromic anemia (Hb 9 g/100 ml). Serum iron 20–40 μg/100 ml, iron binding capacity 305 μg/100 ml. No hemorrhagic tendency. CRP 6+, RA test negative.

The serum protein electrophoretic pattern was of the subacute inflammatory type, showing slight increase in the γ fraction. On electrophoresis of the plasma, the ϕ fraction was 16.5 %, and the immunochemical quantitation of the plasma fibrinogen was 510 mg/100 ml. On immunoelectrophoresis, the acute phase reactants were increased as well as the low-density lipoproteins and β_{1C}-globulin. The oxalated plasma showed marked reversible white precipitation in a refrigerator.

White precipitate formed at low temperature was composed only of fibrinogen. The cryofibringoen occupied approximately one third of the total fibrinogen. The heparinized plasma showed the most prominent precipitation.

Table 65 Various clinical conditions accompanied with cryofibrinogenemia.

I. Idiopathic or essential cryofibrinogenemia

II. Secondary cryofibrinogenemia
  A. Malignancy
    1. Carcinomas of the lung, stomach, ovary and prostate.
    2. Sarcomas: fibrosarcomatosis
    3. Reticular malignancies: multiple myeloma, chronic lymphatic leukemia
  B. Autommune diseases
    1. Rheumatic fever (acute)
    2. Ulcerative colitis (severe)
    3. Aortitis syndrome

in normal plasma, and we assume that they represent the fibrinogen portion 'ready to clot', possibly of fibrin monomers or like. The floccula become insoluble at 37 °C after repeated washing and cryoprecipitation.

With cryofibrinogenemia, as shown in Table 65, very often there are complications of maligant tumors and autoimmune diseases. Infrequently, however, idiopathic ones are observed as well.[169,179,306] In order to understand the pathogenesis of the abnormality, however, it is interesting to know that it is very often complicated with reticular tissue abnormalities such as malignancy and autoimmune diseases. Almost all cases showing cold sensitivity (cryopathy) belong to primary idiopathic cryofibrinogenemia. The secondary or symptomatic cryofibrinogenemia usually does not accompany clinical manifestations of cryopathy, although the case reported by JAGER was accompanied with lymphatic leukemia and demonstrated significant ischemic changes in toes and fingers.[169]

Parafibrinogens

To date, in all reported cases of para-fibrinogenemia, there were seen blood clotting abnormalities, and the physiological formation of fibrin seems to be deficient.[24,25,93,136,164, 248,395]

Disorders of fibrin formation may be classified as follows: (1) deficiency involving the formation of fibrin monomer, or abnormal release of fibrinopeptides by thrombin from α and β chains of the fibrinogen molecules (fibrinogen Zurich, Detroit, Cleveland, St. Louis, Los Angeles): (2) deficiency involving the formation of urea-insoluble fibrin-i from fibrin-s catalyzed by factor XIII activity.

Parafibrinogenemia was also reported to occur in acquired form.[396] It was discovered in a 24 year-old female, with no clinical signs of hemorrhagic tendency. The liver was almost completely replaced by primary hepatocellular carcinoma, weighing 3,000 g. No similar abnormality was found in any member of the family. Both the prothrombin and thrombin times were prolonged and polymerization of the fibrin monomer was deficient. It is thought that the abnormal fibrinogen was formed by neoplastic cells in the liver.

ABNORMAL ALBUMIN

As is discussed elsewhere (page 11), various names have been given to alloalbumins, which are the hereditary molecular variants of serum albumin, and paralbumin is one of those. However, in this book, based on the opinion of the Society of Electrophoresis, we would prefer to use the terms "paralbumin" and "paralbuminemia", not for the hereditary molecular variants but rather as being recognized transiently as pathological, or having amino acid composition different from that of normal albumin.

Abnormal albumin was reported by GABL and HUBER.[105] It was discovered in an 11 month-old male infant suffering from slight anemia and ascites. Both in the serum and ascites found was a fast-moving albumin along with normal albumin. The total serum protein concentration was 5.0 g/100 ml, with the albumin fraction constituting 42% of it. Approximately one third of the albumin was migrated faster. Further detailed results were unclear. However, the infant's general condition improved, and the abnormal albumin disappeared.

A similar case was experienced in Japan. In this instance, it involved a 60 year-old male, who was hospitalized complaining of abdominal distension.[376] On peritoneoscopy, the liver was found to be diffusely granular. Upon biopsy, slight cellular infiltration in

the GLISSON's sheath as well as hepatocellular degeneration were recognized.   In the
serum and ascites, biphasic albumin fractions were found, and the fast-moving abnormal
albumin occupied only one fifth of the normal albumin.   The abnormal albumin did not
have the BPB-binding capacity, but on immunoelectrophoresis, it was found to be identical to the normal albumin.   During the hospitalization ascites disappeared rapidly,
and then the abnormal albumin could no longer be detected.

As for the cause of transient paralbuminemia, it is thought that the formation of albumin is abnormal due to liver damage.[105,332,376)   However, the albumin molecules bind
normally with various metabolites and exogenous substances, and its bound form would
be fast-moving.   Therefore, the above possibility should be excluded carefully before
paralbuminemia is diagnosed.

TISSUE PROTEINS PATHOLOGICALLY APPEARING IN PLASMA

A detailed discussion of these components deviates from the purpose of this book, but
they shall be touched on very lightly.

Since the molecular biology of myoglobin and hemoglobin is a rather famous issue,
discussion here will be avoided.   Furthermore, on page 170, non-hemoglobin erythrocyte
stromal proteins have already been covered.   Conseuqently, discussion will be limited to
simply 2 or 3 components that in recent years have attracted some attention.

a_2-Globulins in Renal Transplantation

When a renal transplantation is performed, the plasma protein abnormality of the
acute phase response type can be observed post-operatively, showing the decreased albumin fraction and the increased a_1 and a_2 fractions.   However, after passing through a

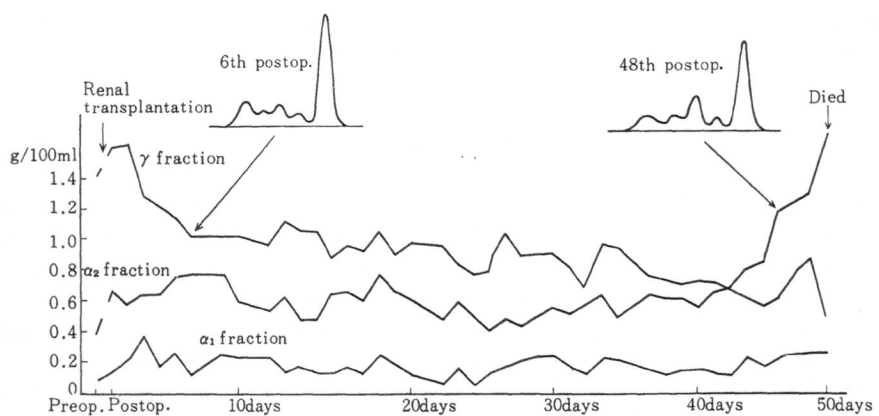

Fig. 240 Serum protein electrophoretic patterns on renal transplantation.
T. I., 38 y.o., male, chronic glomerulonephritis and uremia.
Treated for chronic glomerulonephritis for the past 10 years, but developed uremia
and severe anemia.   Received hemodialysis 31 times in the past one year.   Bilateral
nephrectomy done, followed by renal transplantation 40 days later.   Died 50 days
after the renal transplantation.
On cellulose acetate electrophoretic analysis of the serum proteins, the acute phase
response pattern was recognized and the γ fraction continued to decrease after the
renal transplantation.   From the 4th postoperative week on, the a_2 fraction started to
increase prominently and showed a tremendous increase for the week immediately
before death.

post-operative transient stress reaction, when the rejection of the transplanted homologous kidney occurs, the increase of the a_2 fraction becomes significant, reaching to more than twice of the pre-operative level (Fig. 252). This tremendous increase of the serum a_2 fraction usually appear a little after the renal function is started to be impaired. As long as the rejection continues, so does the increase in the a_2 fraction. In the event where the transplanted kidney is removed, once again it tends to return to normal.[290] The same findings were verified, upon experiments with dogs. By the use of immunoelectrophoresis, the extreme increase of the a_2 fraction could not be verified through the existence of the serum proteins, but at least, with our experience through such experiments, it is felt that it may originate from the necrotic renal tissues. Based on the results of RIGGIO's clinical cases as well, principally, there is an increase of glycoproteins, but not of haptoglobin, a_{2M}-globulin, ceruloplasmin, and orosomucoid.[305] Presently, it is not certain to decide whether the increased a_2 globulins are characteristic for the rejection, or they are of non-specific reaction from the renal tissue necrosis.[305]

Other than this, at the time of the renal transplantation, the excretion of the L chain in urine is measured.[301] In fatal cases, the L chain is excreted continuously in urine in a high concentration. In those who survived for a long period after renal transplantation, the L chain becomes not detected within 7–25 days after the operation. It might be useful then to measure the urinary concentration of the L chain after the operation, in order to assume the beginning of the rejection.

Cancer-Specific Proteins and a_1-Fetoprotein

Experimentally, HIRAI was able to isolate a specific antigen from cancer cells, and thus, the cancer-specific tissue proteins are easily thought to appear in blood.[147] Recently, in the serum of pateints suffering from primary hepatoma, a_1-fetoprotein has been detected in a high frequency. PURVES and others,[300] examining 133 cases of patients with primary hepatoma, discovered 100 cases or 75% to show the positive result.

In the author's laboratory, by using the counter electrophoretic technique, a_1-fetoprotein was found in approximately 90% of the patients with primary hepato-cellular carcinoma. It was also found in occasional cases of fulminant hepatitis, particularly those showing distinct regeneration of the liver cells. In neonatal hepatitis and embryonal testicular tumor found frequently is a_1-fetoprotein in the patient's serum. A small amount of it is found also in sera of pregnant women. Therefore, the protein is not specific for primary hepatoma.

In pregnant women, a_1-fetoprotein is probably transfered from the fetal circulation. However, the reason for the appearance of this protein in the patients with hepatoma and hepatitis is entirely unknown. It is assumed that drastic hepatocellular proliferation of either reactive or neoplastic nature may cause a functional de-differentiation of the liver cells.

The carcinoembryonic antigen, similar to a_1-fetoprotein, has been found also in the patients with carcinoma of the colon.[129]

Chapter   25

Defect Dysproteinemias

CHRACTERISTICS OF THE DEFECT-DYSPROTEINEMIC SERUM PROTEIN ELECTROPHORETIC PATTERNS

Pathophysiologically, defect dysproteinemias represent the situations where certain plasma protein components show isolated deficiency.   Therefore, they include mostly the congenital forms of plasma protein deficiencies.   For the purposes of clinical differentiation, some acquired or secondary protein deficiency states will be briefly discussed.   Congenital protein deficiency is a condition where the protein synthesis within the cells is deficient, due to various reasons that are discussed elsewhere (page 105).   Some of the important plasma protein deficiencies are listed in Fig. 241.   Among these plasma pro-

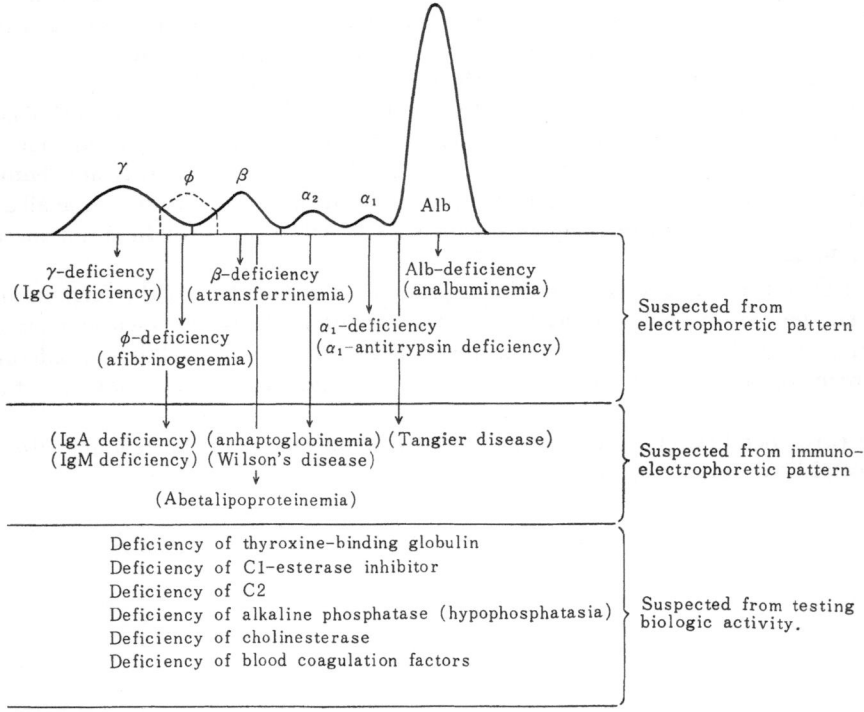

Fig. 241   Classificasion of congenital deficiency of plasma proteins.

tein deficiencies, the one which causes a significant hypoproteinemia is congenital analbuminemia. Complete deficiency of other plasma protein components may not necessarily
accompany hypoproteinemia.

Even on electrophoretic fractionation of serum proteins, only two conditions may result
in significant changes of the electrophoretic pattern: analbuminemia and animmunoglo-
linemia of IgG type. In other words, only the fractions which are composed mostly of a
single plasma protein component may reflect the severe deficiency of certain protein components. For example, the albumin and γ fractions are markedly decreased on each
deficiency state. The α_1, α_2 and β fractions are composed of many protein components,
and they will not be decreased markedly even when a single protein component is completely lacking. Nevertheless, on cellulose acetate electrophoresis, α_1-antitrypsin deficiency and a transferrinemia may be suggested to be present if the α_1 and β fractions are
markedly decreased (Fig. 245).

None of the serum protein fractions separated electrophoretically are composed of a
single protein component. Therefore, each deficiency is necessary to be confirmed by the
immunoelectrophoretic, immunochemical or biological techniques.

ANALBUMINEMIA

Analbuminemia is recognized as an extremely rare congenital disorder, and is a condition where there is almost an absolute lack of albumin in the serum. In Japan, there
has been only one reported case. Since BENNHOLD's report, 4 family reports have been
added.[26,29,124,258,343] Other than showing either generalized or localized edema, no
obvious symptoms are recognized clinically.

Characteristics of the Analbuminemic Pattern

The total serum protein concentration is decreased, usually within a range of from
4.5 to 6.0 g/100 ml. On serum protein electrophoresis, not only is the albumin fraction
almost absolutely deficient, but all the other globulin fractions tend to increase. In
particular, α_1 and α_2 fractions ordinarily show a significant increase (Table 66). In
cases where the γ fraction showed a marked increase, it was complicated with rheumatoid
arthritis, and did not seem to have a direct relationship with analbuminemia.

Table 66 Electrophoretic fractionation of the serum proteins in analbuminemia.

Reported by	Age (yrs)	Sex	TP	Alb	α_1	α_2	β	γ
BENNHOLD (1954)[29]	31	F	4.8	0.1	0.48	0.85	1.47	1.86
BENNHOLD (1954)[29]		M	5.3	0.1	0.70	1.25	1.73	1.65
GORDON et al. (1959)[124]	27	F	4.6	0.79	0.41	0.81	1.37	1.30
BECK et al. (1959)[26]	29	M	4.5	0.15	0.41	1.36	1.73	0.91
SHELTAR et al. (1959)[343]	61	M	5.7	0.06	0.31	1.10	1.02	3.27
MONTGOMERY et al. (1962)[258]	20	F	4.7	0.29	0.43	1.07	1.23	1.73

unit: g/100 ml.

Differentiation of the Analbuminemic Pattern

A differentiation may be necessary from the non-selective protein-losing pattern.
Usually the differentiation of the two patterns is not difficult. In the serum protein pat-

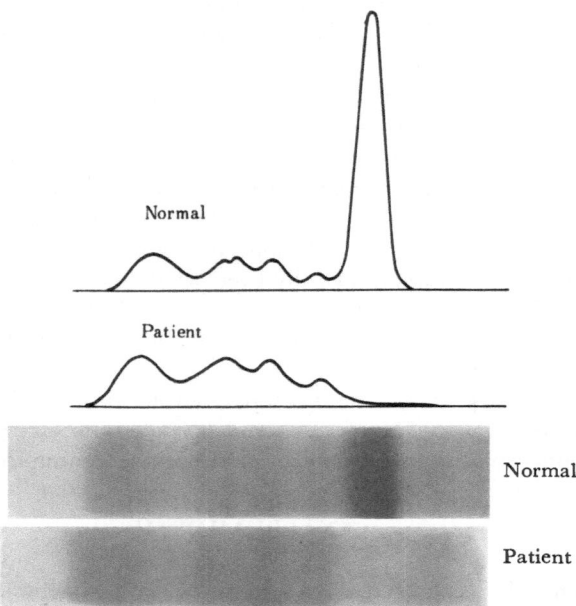

Fig. 242 Serum protein electrophoretic patterns of analbuminemia.
21 y.o., male, analbuminemia (Okayama Univ. Hospital).
Found incidentally, and showed no abnormality on physical examination. Serum
lipoproteins 918 mg/100 ml, serum total cholesterol 346 mg/100 ml, and NEFA abnormally low. The serum total protein was 5.0 g/100 ml. On cellulose acetate
electrophoresis, the albumin fraction was completely absent. The α_1 fraction was
11.6%, the α_2 fraction 31.9%, the β fraction 21.7% and the γ fraction 34.8%. The
serum albumin was completely lacking even on single radial immunodiffusion method.
The liver biopsy was normal, demonstrating no albumin. The half-life of the serum
albumin, using RISA, was prolonged to be over 38 days.

tern of the non-selective protein-losing type, the albumin fraction does rarely drop below
a serum level of 0.5 g/100 ml. If, however, as a protein-losing type, the albumin fraction shows a conspicuous drop below 0.5 g/100 ml, nephrotic syndrome has to be
suspected, where a marked increase of the α_2 fraction and a decrease in the β and γ
fractions are accompanied simultaneously. Thus, the differentiation becomes relatively
easy. Again, in the proteinlosing pattern, hypercatabolism should be recognized with
albumin metabolic studies.

Pathophysiological Background in Analbuminemia

In BENNHOLD's cases, the albumin fraction is hardly detected, and its immunochemical
quantitation shows a level of 1.6 mg/100 ml. With the TISELIUS electrophoresis, the albumin fraction shows a level of 0.4%, and is thought to be composed of seromucoids or
prealbumin.[319] WALDMANN et al., studying the albumin metabolism in this disease,
found that the half-life of RISA was remarkably prolonged, being about 55 days as compared to 13–20 days in normal individuals.[401] If 350 g of albumin is intravenously injected into the patients with this disease, the half-life is 42–48 days, showing a very little
improvement. Therefore, it is thought that in this disorder, along with a decreased
albumin synthesis, there exists also an abnormality in the albumin catabolism.[401] The
intravenous injection of 25g of albumin in each week is needed to maintain the normal
serum level of albumin in this patient.

In these cases, the albumin concentration is only 0.05% of the normal, and the serum colloid osmotic pressure is about 1/2 of the normal level. However, it seems to be compensated by the increase of various globulins.[113] According to MONTGOMERY,[258] the transferrin level is 2–3 times that of normal, α_2-macroglobulin 3.5 times, β-lipoproteins 3 times, and the IgM 3 times. The erythrocyte sedimentation rate is significantly increased.

With other serum components, the electrolytes are normal, as well as the aldosterone. As for serum lipids, triglyceride is normal, and the phospholipid shows an increase.[113] It is thought that the increase of the serum globulins and lipids is a secondary change due to the decrease of albumin. In other words, with an intravenous injection of 350 g of albumin, a decrease of 50% is brought about in serum concentrations of cholesterol, phospholipids, fibrinogen, γ globulin fraction and the iron-binding proteins. The γ-globulin (mainly composed of IgG) shows a normal half-life, and no significant change follows even after albumin injection. It seems, therefore, that γ-globulin has a metabolism completely independent of that of albumin.[381]

α_1-ANTITRYPSIN DEFICIENCY

α_1-antitrypsin deficiency is considered either congenital or familial, and represents a condition where the α_1-antitrypsin in the serum is gravely deficient. It was ERIKSSON[86] who pointed out that there is a relationship between α_1-antitrypsin deficiency and obstructive lung diseases, and in particular, pulmonary emphysema. Subsequently, its familial occurrence has been reported by other investigators, as well.[106,162,247]

This disorder is thought to be of the autosomal recessive type, and even with a heterozygous abnormality, the serum α_1-antitrypsin is recognized to be lower than the normal level (Fig. 243).[162]

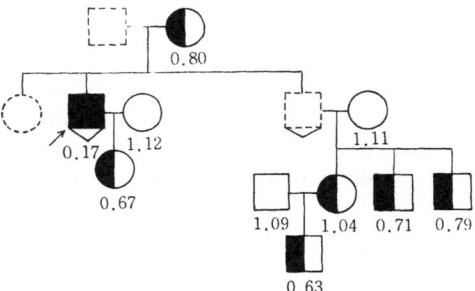

Fig. 243 A family study of α_1-antitrypsin deficiency (from HUMTER et al.).

Due to a lack of the α_1-antitrypsin, there should be an increase of trypsin activity in the blood and tissues, resulting in juvenile emphysema due to increased destruction of the pulmonary connective tissues.

Characteristics of Hypo-α_1-Globulinemic Pattern

Since 70–80% of the α_1 fraction are comprised of α_1-antitrypsin in normal serum, α_1-antitrypsin deficiency may cause a marked decrease of the α_1 fraction on serum protein electrophoresis. Without complications, the total serum protein concentration as well as other serum protein components show no significant change.

Differentiation of the Hypo-a_1-Globulinemic Pattern

Differentiation from the severe hepato-degenerative pattern with severe hepato-cellular damage, where the a_1-fraction may be decreased, there is also an increase in the γ fraction and decrease in a_2, β and albumin fractions.

With immunoelectrophoresis, often, because of antigen excess, the precipitation line of a_1-antitrypsin may become indistinct.

THYROXINE-BINDING GLOBULIN DEFICIENCY

Familial cases of thyroxine-binding globulin (TBG) deficiency are encountered infrequently.[269,377]

Ordinarily, with TBG deficiency, the thyroxine binding activity of serum is decreased. Although the Triosorb test (^{131}I–T3 resin sponge uptake) results in an extremely high value, there are no clinical symptoms suggesting hyperthyroidism.

Even with the lack of TBG, the serum protein electrophoretic pattern shows no significant alteration, unless polyacrylamide gel electrophoresis is used.

There have been reported cases where a hereditary increase of the serum TBG is recognized.[28,92]

ANHAPTOGLOBINEMIA AND CERULOPLASMIN DEFICIENCY

a_2 fraction is composed of many minor protein components, and the deficinecy of a single a_2 globulin component does not cause a singificant decrease of the a_2 fraction on electrophoresis. Therefore, in order to confirm a deficiency of a_2-globulin components it is necessary to employ immuno-electrophoresis or other special techniques. Haptoglobin and ceruloplasmin are the two representative a_2-globulins which may be deficient congenitally.

Anhaptoglobinemia

Congenital anhaptoglobinemia is relatively rare. By definition, it is a condition where there is either an absolute lack of haptoglobin in the serum or existing only in very small quantities. Clinically, no particular symtoms are forthcoming, but it may accompany slight anemia. Presently, about 30% of Nigerian Negroes are thought to be suffering from this disorder.[5]

Clinically, it is necessary to differentiate from the acquired deficiencies, such as hemolytic diseases (Fig. 244), neonate and liver damage. The total serum protein concentration is normal, the serum level of the a_2 fraction may be slightly decreased, and no significant change is usually recognized with this disorder. With immunoelectrophoresis, the precipitation arc of haptoglobin cannot be recognized.

According to a genetic analysis of families with anhaptoglobinemia, in all cases there is the possibility of the presence of Hp^2 allele. In other words, with the patients suffering from this disease, the phenotype of haptoglobin, as far as can be determined, has been found to be Type 2–1, or 2–2.[332] Nevertheless, the hereditary form is not clear and there are some who hypothesize that Hp^0 allele is present.

Ceruloplasmin Deficiency

WILSON's disease is well known to have ceruloplasmin deficiency. In this disorder, it gives rise to copper deposition in all organs of the body, particularly evident in the liver

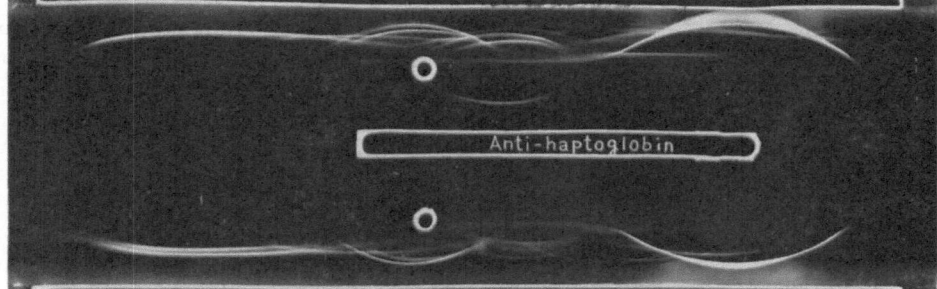

	(g/100ml)	(%)
TP	6.7	
Alb	3.86	58
α_1	0.50	7
α_2	0.40	6
β	0.67	10
γ	1.27	19

Fig. 244 Serum protein electrophoretic patterns in secondary anhaptoglobinemia.
K. M., 34 y.o., female, autoimmune hemolytic anemia.
Warm type autoantibody (IgG) was found.   Marked anemia (Hb 2.9 g/100 ml) and slight jaundice.
Slight plasmocytosis and marked erythroid hyperplasia of the bone marrow.
On cellulose acetate electrophoresis, the serum showed slight increase in α_1 fraction and slight decrease
in α_2 fraction.   On immunoelectrophoresis, the haptoglobin was not demonstrated in the patient's
serum (lower well).

Table 67 Serum concentration of ceruloplasmin in the patients with WILSON's disease and its
carriers (from STERNLIEB and SCHEINBER).

Ceruloplasmin concentration (mg/100 ml)	Homozygotes (WILSON's dis.)		Heterozygotes (Carriers)	
	Cases	Percentage	Cases	Percentage
< 1	27	24.4 %	0	0 %
1– 4.9	31	27.9	1	1.3
5– 9.9	32	28.8	1	1.3
10–14.9	10	9.0	2	2.7
15–19.9	7	6.3	11	14.7
20 <	4	3.6	60	80.0
Total	111	100	75	100

Normal: 20–35 mg/100 ml.

and brain.   Again, when there is copper deposition in the cornea, the KAYSER–FLEISCHER
ring is recognized clinically.   If the copper deposits over a long period of time, it causes
reactive fibrosis, resulting in impairment of the hepatic and cerebral functions.   Thus,
it is called also by another name, "hepatolenticular degeneration".[180,325]

The diagnostic criteria for WILSON's disease include ceruloplasmin deficiency, the
KAYER–FLEISCHER corneal ring, increased urinary excretion of copper (over 100 μg), and
increased copper content in tissues (over 150 μg per 1 g dry weight of the liver).   Therefore, ceruloplasmin deficiency may strongly suggest the possibility of WILSON's disease.

This disease is of the autosomal recessive type. The parents of patients with WILSON's disease have heterozygosity, and as can be seen on Table 67, the serum ceruloplasmin level is decreased in 20%. In contrast, the patients with the disease show ceruloplasmin deficiency in more than 95%. However, there is no direct relation between the ceruloplasmin level and the severity of the disease.[362]

The deficiency of ceruloplasmin itself does not give rise to a significant alteration of the serum protein electrophoretic pattern, and it is difficult to distinguish the deficiency even on routine immunoelectrophoresis. Therefore, the diagnosis must be confirmed with immunoelectrophoresis using the specific antiserum, oxidase activity, or copper quantitation.

Congenital ceruloplasmin deficiency has to be differentiated from the following conditions: (1) stored serum samples; (2) new-born infants; (3) copper malabsorption, such as sprue, intestinal scleroderma, etc. (4) abnormality in protein metabolism, such as infantile hypoproteinemia, iron deficiency state, Kwashiorkor, nephrotic syndrome, protein-losing gastoenteropathy, etc.

ATRANSFERRINEMIA

Atrasferrinemia is a condition where transferrin is almost absolutely lacking in the serum, and its congenital form is very rare.[124,196] Clinically, due to the lack of serum transferrin, the transport of iron is impaired, resulting mainly in iron-deficiency anemia and hemosiderosis. This disease is inherited as the autosomal dominant form (Table 68).[196]

A transferrinemia itself does not show any obvious changes on filter-paper electrophoresis. For example, in the case in Japan, the total serum protein was 7.0 g/100 ml, alb fraction 52%, α_1 3.5%, α_2 11.4%, β 13.4%, and γ 19.7%. This is because the β fraction is composed of many other globulin components such as hemopexin, β-lipoproteins, complements, etc. Thus, even with a complete lack of transferrin, there may be no definite decrease of the β globulin fraction (Fig. 245). On the other hand, with cellulose acetate and agar gel electrophoresis, atransferrinemia may show a notable decrease of the β fraction (Fig. 246). It is because most of the β-lipoproteins migrate in α_2 fraction, and transferrin becomes a major component of the β fraction.

Even on filter-paper electrophoresis combined with the RIVANOL precipitation technique, atransferrinemia can be easily diagnosed. Because of the β globulins other than transferrin are mostly precipitated out (Fig. 245). On immunoelectrophoresis, the transferrin

Table 68 Serum transferrin concentrations in the patient with congenital atransferrinemia and her family (from KOZURU).

	Serum concentration of transferrin (mg/100 ml)	Serum total iron binding capacity (γ/100 ml)
Normal	240–280	270 – 370
Father	171.8	188
Mother	264.2	
Brother (S)	304.1	
Propositus	9.0	50
Sister (N)	109.6	135
Aunt (T)		377
Aunt (M)		361
Uncle (Y)		285

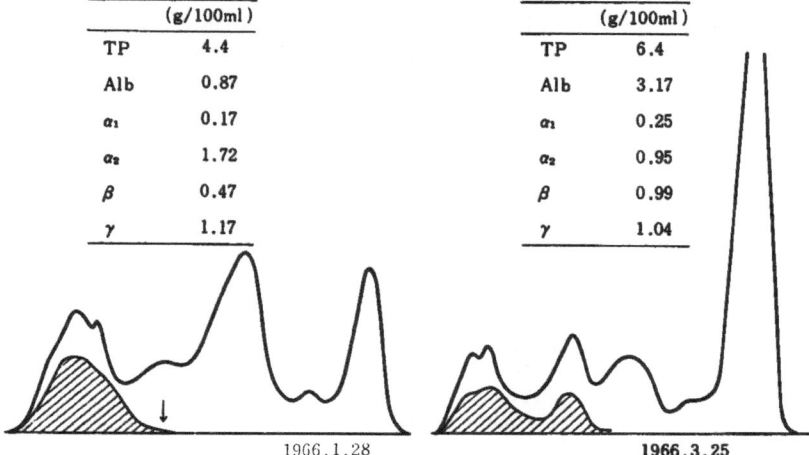

	(g/100ml)
TP	4.4
Alb	0.87
α_1	0.17
α_2	1.72
β	0.47
γ	1.17

	(g/100ml)
TP	6.4
Alb	3.17
α_1	0.25
α_2	0.95
β	0.99
γ	1.04

1966.1.28   **1966.3.25**

Fig. 245 Serum protein electrophoretic patterns of the hypobetaglobulinemic type.
E. K., 4 y.o., female, nephrotic syndrome and secondary atransferrinemia.
Marked generalized edema, proteinuria (8 g/day), total cholesterol 630 mg/100 ml. The shaded areas are the densitometric patterns of the serum proteins after Rivanol precipitation. Before the administration of corticosteroids, the β fraction was decreased and Rivanol-precipitated pattern showed a complete lack of the transferrin peak, being only 20 mg/100 ml immunochemically. In 3 days after the treatment the serum protein pattern (right) rapidly improved, and the serum transferrin was 280 mg/100 ml.

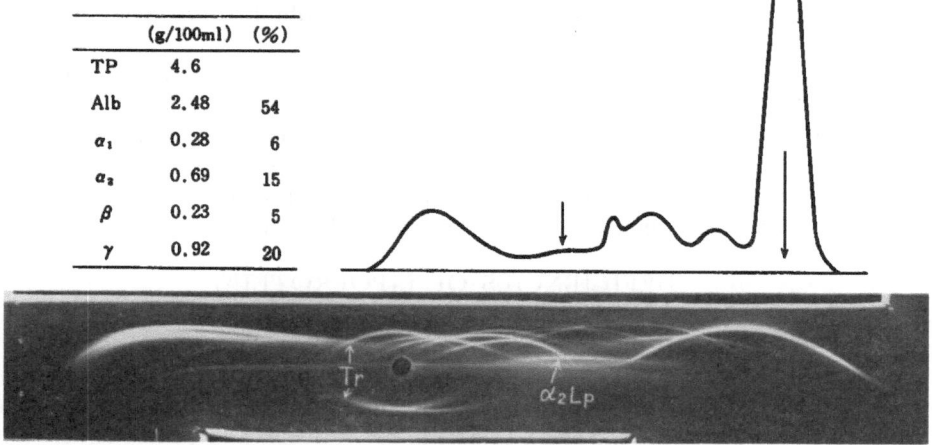

	(g/100ml)	(%)
TP	4.6	
Alb	2.48	54
α_1	0.28	6
α_2	0.69	15
β	0.23	5
γ	0.92	20

Fig. 246 Serum protein electrophoretic patterns of the hypobetaglobulinemic type.
S. M., 2 m.o., female, septicemia (Klebsiella).
Complained of diarrhea and fever for two weeks, and Klebsiella cultured from blood. Hypochromic anemia (Hb 9.5 g/100 ml), and leukocytosis (WBC 28,200) with leukemoid reaction. Serum iron 78 μg/100 ml, serum transferrin level 45 mg/100 ml.
On cellulose acetate electrophoresis, β fraction was markedly decreased. On immunoelectrophoresis, noted were an increase in the LDL and a marked decrease in transferrin. Haptoglobin, normal. In this case, latent decrease in transferrin synthesis during infancy seemed to be combined with increased metabolism of transferrin in acute infection.

line cannot be recognized or can be seen very weakly, if any.   With this condition, it is necessary to be confirmed by using the specific antiserum since it is often difficult to be differentiated from hemopexin.   On immunochemical quantitation, the serum transferrin level was only 1/100 or 1/30.   Congenital atransferrinemia must be differentiated from the acquired types as shown on Table 69.[141]

Table 69   Pathogenesis of atransferrinemia
(from HEILMYER).

A.   Decreased synthesis in the liver
      Congenital atransferrinemia
      Infections
      Liver diseases

B.   Increased catabolism
      Infections
      Malignancies
      Connective tissue diseases

C.   Increased transfer to infected lesions

D.   Increased external loss
      Nephrotic syndrome
      Protein-losing gastroenteropathy

DEFICIENCIES OF THE COMPLEMENT COMPONENTS

Acquired deficiencies of the complement components are recognized in acute glomerulonephritis, SLE, lupus nephritis, autoimmune hemolytic anemia, severe liver damage, etc.   Congenital deficiencies have been found on the inhibitor of Cl esterase, Cl, C2, C3 and C4.[74,137,332,347,384]

The inhibitor of Cl esterase as a congenital deficinecy has been reported also in Japan. With the lack of the Cl esterase inhibitor, Cl component becomes activated, resulting in the formation of Cl esterase which causes excessive destruction of both C4 and C2 components.   Furthermore, Cl esterase itself is considered to be the cause of accelerating vascular permeability.   As a result, hereditary angioneurotic edema is observed clinically in association with the deficiency of the Cl esterase inhibitor.

DEFICIENCIES OF LIPOPROTEINS

High-Density Lipoprotein Deficiency

In 1960, FREDRICKSON reported on familial high density lipoprotein deficiency (a-lipoprotein deficiency or analphaliporoteinemia) which was discovered in the Island of Tangier among two brothers, 5 and 6 years old, respectively, and consequently called the 'Tangier disease.'   Subsequent to this, three more families were discovered.[98]

With this particular disease, there is seen an almost absolute lack of high-density lipoproteins, but no evident changes observable in either the total serum protein concentration or serum protein electrophoretic pattern.   For the diagnosis of the disease, both immunoeleetrophoretic and immunochemical techniques are employed.   Since either obstructive jaundice or severe liver damage can cause high-density lipoprotein deficiency, the Tangier disease must be carefully differentiated from the acquired ones.   With the disease, both the total cholesterol and phospholipid concentrations in serum are decreased amounting to 70 mg/100 ml (50 – 130 mg) and 100 mg/100 ml (70 – 140 mg), respectively. However, neutral fat in serum shows a slight increase after meals. A clear deposition of the cholesterol ester to the reticuloendothelial system occurs clinically, and the tonsils show a characteristic orange colored swelling. At times, it causes hepatomegaly, splenomegaly or lymphadenopathy as well. Furthermore, the deposition of lipids can be seen in such areas as skin, cornea, blood vessels and rectal mucosa.

Since α-lipoprotein deficiency is hereditary, in the heterozygous abnormality, the serum α-lipoproteins show a decrease, but no typical tissue deposition of the lipids are recognized.[97] The homozygote of the abnormal genes may result in α-lipoprotein deficiency. However, with homozygous and heterozygous forms, the Tangier α-lipoprotein which is slightly different antigenically from that of the normal individual, is found in serum.[99] In the Tangier disease, the low-density lipoproteins appear normal on immunoelectrophoresis. However, in postprandial states, the characteristic electrophoretic pattern of serum lipoproteins is recognized, characterized by a broad β-lipoprotein band and no recognizable pre-β lipoprotein band. The β-lipoprotein contains a large amount of neutral fats and has the specific gravity of less than 1.006. Therefore, the β-lipoproteins would seem to be necessary for a pre-β lipoprotein band to be identified clearly.[218]

Abetalipoproteinemia and Hypobetalipoproteinemia

Familial or congenital deficiency of low-density lipoproteins can be divided into two major groups, abetalipoproteinemia and hypobetalipoproteinemia. They seem to be under different genetic control.

1. Abetalipoproteinemia

Since the first two cases were reported in 1950, more than 30 cases have been reported so far. Since BASSEN and KORNZWEIG noted first abnormal erythrocyte shapes and characteristic neurologic symptoms, it has become known as "congenital acanthocytosis', and in 1960, after SALT had pointed out the absolute lack of low-density lipoproteins, it has come to be called abetalipoproteinemia.[23]

Clinically, steatorrhea and abdominal distension are observed from the time of infancy, and neurological disorders such as muscular hypotonia, nystagmus, degeneration of the postero-lateral and cerebellar tracts will develope later. Furthermore, both retinal degeneration and visual disturbance may develop further, and the patients usually die at a relatively early age. Ceroid is deposited in various tissues, but the content of phospholipids is low. The serum lipids are extremely low, the total cholesterol concentration being 20–90 mg/100 ml, the phospholipids being 35–95 mg/100 ml, and the triglyceride being only 10–20 mg/100 ml.

On electrophoretic separation of the serum lipoproteins, there is a complete absence of the low-density lipoproteins. The α-lipoproteins are normal immunochemically, but their density ranges from 1.019 to 1.063, and they contain a great quantity of lipids. The lipids appear to be absorbed from the intestines in the usual manner, but chylomicron is not formed at all. Furthermore, there is an excessive storage of triglyceride in the liver. Thus, the α-apoproteins are considered as having the important function of transporting exogenous and endogenous triglycerides from the cells.[99,295]

2. Hypobetalipoproteinemia

In hypobetalipoproteinemia, the serum low-density lipoproteins decrease to 10–50% of the normal level. The condition is clinically associated with abnormal erythrocyte shapes and neurologic disorders, and the patients may not survive to adulthood. However, unlike abetalipoproteinemia, the intestinal absorption of fats and the formation of chylomicrons are recognized.[220,287] It is necessary to differentiate familial hypobetalipoproteinemia from secondary hypobetalipoproteinemias accompanied with cachexia, acute infections, malabsorption syndromes, and the like.

DEFICIENCIES OF FIBRINOGEN

Hemorrhagic tendency is likely to occur clinically when the plasma concentration of fibrinogen falls below 150 mg/100 ml or in hypofibrinogenemia. Afibrinogenemia is diagnosed when fibrinogen is not detected in plasma both hematologically and chemically. However, by the use of immunochemical procedures, the level of below 10 mg/100 ml may be detected. Therefore, afibrinogenemia can be further subdivided into absolute afibrinogenemia and severe hypofibrinogenemia.

Fibrinogen deficiency can be divided into the congenital and the acquired forms. Furthermore, the condition may occur with the following three mechanisms: 1) decreased synthesis in the liver, 2) increased consumption in the body (defibrinating syndrome) and 3) increased degradation in the body (accentuated fibrinolysis).

Characteristics of the Plasma Protein Electrophoretic Pattern of ϕ-Deficient Type

On electrophoretic fractionation of the plasma proteins, characteristically recognized is the lack of ϕ fraction which appears normally between the β and γ fractions. Generally, the fibrinogen concentration of less than 50 mg/100 ml will not show a recognizable band or peak on filter-paper or cellulose acetate electrophoresis (Fig. 247). IgA, IgM, IgD and IgE are migrated at the γ_1 zone, and their significant increase may result in appearance of an abnormal protein band similar to the ϕ fraction. Of course, the differentiation is easy, when the serum sample is analyzed electrophoretically. Also, the immuno-electrophoretic and immunochemical analyses are quite helpful for the differentiation.

In addition, the deficiency of fibrinogen can be easily suspected if a small fibrin clot or no clot is formed during blood coagulation. Furthermore, when the plasma sample is heated at 56 °C for 20 minutes, no white precipitate or a very little precipitate is formed.

Table 70 Clinical findings in the case of congenital afibrinogenemia experienced by the author.

K. K.   2 yrs. 6 mos.   female
Chief Complaint "easy to bleed"
Past History Blood transfusion because of bleeding from the umbilical cord after the birth.
Family History No consanguineous marriage
Coagulation Studies
    Sed rate   1 mm/30 min.: 1.5 mm/hour.
    Platelet count   16.2 10^0/mm^0
    Rumpel-Leede test   negative
    Bleeding time   15 minutes
    Prothrombin time (Quick) ∞
    TGT   normal
    No circulating anticoagulants
    Fibrinolysis   normal
Studies on Fibrinogen
    No clot formation on plasma thrombin time
    No precipitation on heating 56 °C, 20 min.
    No precipitation with 25 % saturated ammonium sulfate
    No fibrinogen with both chemical and immunochemical measurements
    No anti-fibrinogen detected in serum

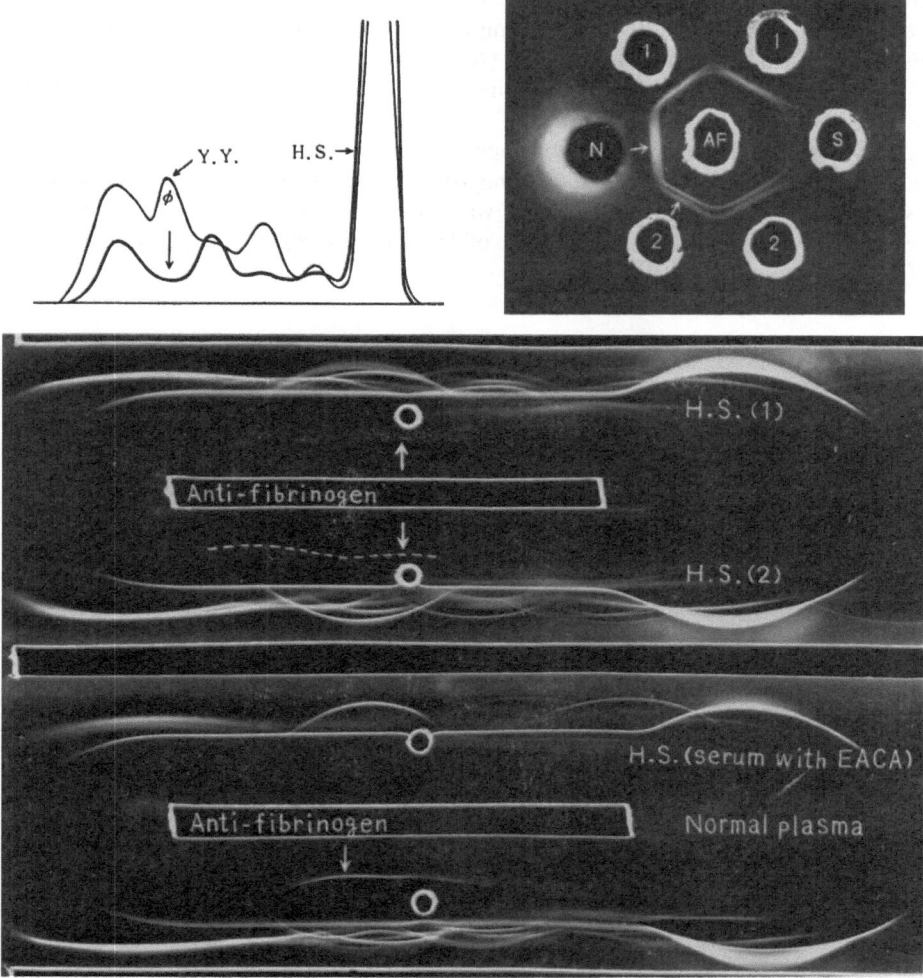

Fig. 247 Plasma protein electrophoretic patterns in afibrinogenemia.

H. S., 14 y.o., male, carcinoma of the colon, autoimmune hemolytic anemia and acquired afibrinogenemia.

Resection of the colon with carcinoma of the ascending colon, one year ago. Admitted again with bleeding tendency and autoimmune hemolytic anemia.

H. S. (1): on cellulose acetate electrophoresis, the ϕ fraction was not demonstrated, in comparison to the clear ϕ fraction demonstrated for the normal (N, or Y. Y.). Immunochemically, the fibrinogen concentration was 16.01 mg/100 ml biochemically. H. S. (2): the patient's plasma drawn 4 days after H. S. (1). Fibrinogen was zero, biochemically. Only a trace of fibrinogen was identified with the OUCHTERLONY double diffusion method (arrow). Also identified was the fibrinogen degradation product. Acquired afibrinogenemia was thought to have developed due to accentuated fibrinolysis, but the patient died before the laboratory studies were completed.

S indicates the pateint's serum separated from the blood which was drawn by adding EACA.

Congenital Afibrinogenemia

Since congenital afibrinogenemia has first been reported in 1920 by RABE and SALOMON,[310] there have been to date over 80 such cases cited. In Japan, up to 1967, 14 cases have been reported (Table 70). It is encountered in both sexes and takes on a recessive hereditary form.   Of the cases in Japan, there are 4 males and 10 females involved.   The cases are not necessarily limited to consanguineous marriages, and there are many occasions of umbilical bleeding after birth.   Generally, spontaneous bleeding is infrequent. But in times of trauma, there are many instances where the bleeding is difficult to be controlled.

In many cases, no trace of plasma fibrinogen is to be found, and there are some instances, where a small amount of it (below the 10 mg/100 ml.) is detected immunochemically.   In all cases reported to date, the fibrinolysis was either normal or low.   Fibrinogen is not synthesized in the liver, and the half-life of infused fibrinogen is normal, roughly 3–6 days (Fig. 248).

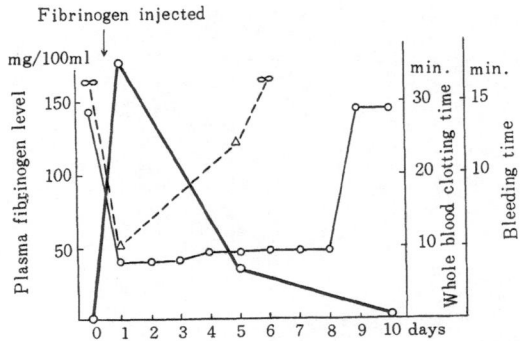

Fig. 248   Intravenous injection of fibrinogen in congenital afibrino-genemia.
The thick solid line indicates the plasma fibrinogen levels, showing the half-life to be approximately 4 days.   The thin solid line shows the whole blood clotting times, and the dotted line for the bleeding times.   Interestingly, the bleeding time remained normal till the 8th day, even though the blood coagulation failed to appear on the 6th day after the injection.

Acquired Afibrinogenemia

1.   Decreased Synthesis

Since fibrinogen is a plasma protein synthesized in the liver, severe liver damage and severe malnutrition may result in hypofibrinogenemia.

2.   Increased Consumption (Defibrination Syndrome)

When the blood coagulation proceeds excessively in the body, fibrinogen is consumed, thus resulting in hypofibrinogenemia.

Defibrination syndrome is most frequently encountered in some of the complications of pregnancy such as premature separation of the placenta, retained dead fetus, amniotic fluid embolism and infected abortions.[241, 260, 262]   No definite theory is available for explaining the mechanism underlying the defibrination in these disorders.   However, the most widely accepted explanation is that the over-utilization of fibrinogen occurs due to the release into the maternal circulation of a very potent placental or fetal thromboplastic material.[138]   On autopsy, however, intravascular fibrin deposits cannot be recognized in the cases dying of abruptio placentae, and this fact may be against the above-mentioned hypothesis.   In experimental animals receiving an excess amount of thromboplastin intravascularly, the similar changes are recognized.[290]

Fibrinolysis has also been blamed for the defibrination syndrome. In fact, accentuation of fibrinolysis is evidenced in the cases with amniotic embolism and infected abortion.[287]

Hypofibrinogenemia and thrombocytopenia are also encountered in a hypercoagulable blood state which is occasionally associated with thoracic surgery, open-heart surgery, incompatible blood transfusion and others. In these situations, a hypercoagulable state is brought about by a release of clot-promoting and heparin-neutralizing substances from destroyed tissues or red blood cells.[138, 262]

3. Increased Degradation (Increased Fibrinolysis)

Hypofibrinogenemia may occur due to accentuation of fibrinolysis since an increased plasmin activity causes an excessive destruction of both fibrin and fibrinogen in the body. In these instances, a significant amount of fibrin-degradation products are detected in the blood. In severe cases, fibrinogen may sometimes lack completely.

In accentuated fibrinolysis, marked hemorrhagic tendency may result frequently due to hypofibrinogenemia and the anticoagulant effect of the fibrin-degradation products such as fragments X and Y.[138, 262]

Accentuated fibrinolysis is clinically observed in the conditions listed in Table 5.[381]

IMMUNODEFICIENCY SYNDROMES

General Survey and Classification of the Immunodeficiencies

Since 1952, when BRUTON reported the case of an 8 year-old boy suffering from repeated infections and a lack of serum γ-globulin, there have been many reports describing the similar disorders.[49] The condition was previously called as "Antikörpermangelsyndrom" or antibody deficiency syndromes.[18, 172] However, since there is not only the deficiency of serum immunoglobulins but also the deficiency of cell-mediated immunity, PETERSON and others have designated as "immunologic deficiency diseases".[293] Since then, the terms such as immune-deficiency, immunological deficiency and immunodeficiency have been introduced by various authors.[41, 158, 168]

Immunodeficiency syndromes are clinically characterized by repeated infections, resulted from deficiencies of the body immune response. As pointed out by HITZIG,[150] the susceptibility to infections in this disorder is characterized by the fact that infectious episodes occur repeatedly and also in different sites.

Immunodeficiency is pathogenetically divided into the primary and the secondary form. Primary immunodeficiency includes the hereditary and idiopathic disorders resulted from a failure to produce the effectors of the immune response. Secondary immunodeficiency states may result from hypercatabolism, or various disorders involving the immune organs and tissues (Table 72).

Immunodeficiency is also classified into the following three groups: 1) a failure to produce both humoral antibodies and sensitized lymphocytes, 2) a failure to produce humoral antibodies, and 3) a failure to produce sensitized lymphocytes.

A failure of the immune response may occur in different steps of its process, as shown below.[41]

Disorders at the afferent limb of immunity: Immunodeficiency results from disorders in localization, phagocytosis, recognition and processing of various antigen molecules. That is, the antigenic informations do not reach to m-RNA of the antibody-producing cells. A representative clinical abnormality of this type includes WISKOTT-ALDRICH syndrome. Experimentally the disorder has been encountered in irradiated mice, where antigen

Table 71   Classification of primary immunodeficiency disorders (recommended by World Health Organization).

Type	Suggested cellular defect		
	B-cells	T-cells	Stem cells
Infantile X-linked agammaglobulinemia	+		
Selective immunoglobulin deficiency (IgA)	+ (some)		
Transient hypogammaglobulinemia of infancy	+		
X-linked immunodeficiency with hyper-IgM	+	?	
Thymic hypoplasia (pharyngeal pouch syndrome, DiGeorge's syndrome)		+	
Episodic lymphopenia with lymphocytotoxin		+	
Immunodeficiency with or without hyperimmuno-globulinemia	+	+ (sometimes)	
Immunodeficiency with ataxia telangiectasia	+	+	
Immunodeficiency with thrombocytopenia and eczema (Wiskott-Aldrich syndrome)	+	+	
Immunodeficiency with thymoma	+	+	
Immunodeficiency with short-limbed dwarfism	+	+	
Immunodeficiency with generalized hematopoietic hypoplasia	+	+	+
Severe combined immunodeficiency,			
autosomal recessive	+	+	+
X-linked	+	+	+
sporadic	+	+	+
Variable immunodeficiency (common, largely unclassified)	+	+ (sometimes)	+

Table 72   Classification of immunodeficiency syndromes.

I.   Primary immunodeficiency syndrome
    A.   Deficiency of the humoral immune mechanism, without any apparent deficiency of the cellular immune mechanism.
        A-1.   With the decreased serum immunoglobulin levels.
        A-2.   Without the decreased serum immunoglobulin levels, but with lacking of a specific antibody response.
    B.   Deficiency of the cellular immune mechanism.
        B-1.   Combined with the decreased serum immunoglobulin levels.
        B-2.   Not combined with the decreased serum immunoglobulin levels.
II.   Secondary immunodeficiency syndrome

processing in macrophages is deficient, resulting in failure to respond against *Shigella* antigens.[41)]

Disorders at the efferent limb of immunity: In this disorder, the effectors of the immune response, themselves, are deficient or injured.   Clinically, agammaglobulinemia of Bruton's type results apparently from a lack of the immunoglobulin-producing cells, while Di George syndrome results from a lack of the effector organ involving the cell-mediated immunity.

At any rate, the extraordinary variability of immunological abnormalities presents a difficulty in classification of immunodeficiencies at the present.   No satisfactory classification nor nomenclature is available, and WHO recommends simply to list up clearcut immunodeficiency states as in Table 71.[341,408)]

The author classifies the immunodeficiency syndrome roughly as in Table 72 mainly for clinical diagnostic approaches. For the patients with repeated infections, the representative tests for assessing the immune status, as mentioned later, are performed to see if there is any deficiency in the humoral antibody and/or the cell-mediated response. At the next steps, an individual type of immunodeficiencies may be diagnosed, considering also clinical symptoms and heredity.

Laboratory Tests Necessary for Diagnosis of the Immunodeficiencies

As described previously, clinical symptoms are extremely important for diagnosis of immunodeficiency. In addition to characteristic clinical symptoms and signs, the following laboratory examinations are useful for assessment of immunologic abnormalities.

1. Routine Hematological Tests

Routine blood cell counts are useful, particularly thrombocytopenia and lymphopenia. Eosinophilia is frequently associated with a deficiency of cell-mediated immunity. On bone marrow examination, a lack of plasma cells is a very important finding.

2. Quantitation of Serum Immunoglobulins

Paper electrophoresis and immunoelectrophoresis are performed. On filter-paper or cellulose acetate electrophoresis, notable changes on the serum protein pattern include a marked decrease in the γ fraction and an abnormal modal mobility of the γ fraction. Only IgG is the one which influences the amount of the γ fraction. Alterations in IgA and IgM cannot be clearly observed on cellulose acetate electrophoresis. Therefore, immunoelectrophoretic analysis of serum proteins has to be done routinely in order to screen any abnormality involving serum immunoglobulins. However, especially in infancy, IgA and IgM are physiologically low in their serum concentrations, and their lines of precipitation are not observed frequently even in normal states. Therefore, immunochemical quantitation of serum immunoglobulins is essential for diagnosis of immunodeficiency states.

For immunochemical quantitation of serum immunoglobulins, single radial immunodiffusion method of MANCINI has been widely used,[232] and also immuno-electrodiffusion of FREEMAN & SMITH.[100] Because of the use of different standards, analytical results may vary significantly among different laboratories. Therefore, a careful interpretation is necessary for clinical diagnosis.

3. Metabolic Studies of the Immunoglobulins

In order to exclude hypercatabolic immunodeficiency, metabolic studies of immunoglobulins may become necessary. In hypercatabolic states, the half-life of immunoglobulins is shortened while it is prolonged in primary immunodeficiency.

4. Qualitative Analysis of the Serum Immunoglobulins

Serum immunoglobulins are composed of innumerable numbers of different antibody globulins. Even though the serum immunoglobulin levels are normal, a certain antibody globulin may not be present. Therefore, various serological procedures are used to measure antibody activities in serum.

At first, it is easy to titrate the "natural" antibodies in serum. For examples, A and B iso-hemagglutinins, heteroagglutinins, heterolysins, anti-streptolysin and bactericidins against *Escherichia coli* are useful for screening, if present.

Anti-A and anti-B isohemagglutins of the complete type belong to IgM, and they are extremely useful on routine screening. They are detected in full-term newborn babies in approximately 60%. As shown in Fig. 249, their titers differ in different ages. Therefore, the titer of less than 1 : 4 is significantly low in infancy, while the titer of less than 1 : 16 is significant in childhood.

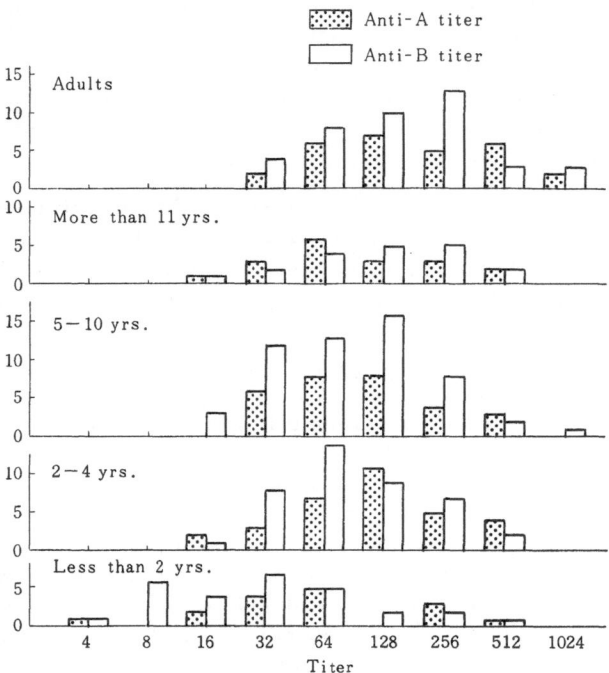

Fig. 249   Distribution of the hemagglutinin titers
in apparently healthy individuals.

Anti-streptolysin O (ASO) titer is also useful in some instances.   As shown in Fig. 250, the titer of less than 12 Todd units may be significantly low in puberty and adulthood, thus suggesting the presence of immunodeficiency.

Also used are the tests for antibody formation following active immunization with diphtheria, tetanus and pertussis (DPT) vaccines, inactivated virus vaccines, bacterial polysaccharides, hemagglutinogens, and others.

5.   Functional Studies of the Lymphoid Tissues

They include delayed-type skin reactions (following tuberculin, histoplasmin, *Candida* antigens, trichophytin, etc.), measurements of lymphoblastic transformation, mitotic index and H^3-thymidine uptake on *in vitro* lymphocyte culture following non-specific (PHA, streptolysin S, etc.) and specific stimulations (streptolysin O, viral antigens, etc.), and skin homograft rejection.   Also, a skin test with sensitization to 2, 4-dinitrochlorobenzene (DNCB) may be useful.

These tests are important for confirming the presence of a failure in the cell-mediated immunity.

6.   Pathological Studies

Either by biopsy or by necropsy, histopathological examinations of the bone marrow, the lymph nodes, the thymus and probably the tonsils are quite useful to confirm the diagnosis of immunodeficiencies.

Characteristics of the Agammaglobulinemic Serum Protein Electrophoretic Pattern

On electrophoresis using various supporting media, the three abnormal patterns are

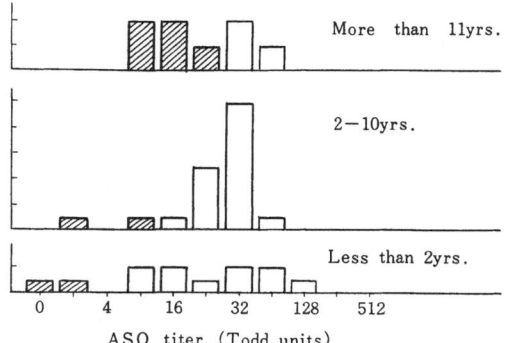

ASO titer (Todd units)

Fig. 250 Relation between the hemagglutinin titer and the ASO titer.
Among the cases which showed the ASO titer of less than 12 x, the distribution of the anti-A and anti-B titers was plotted. The shaded columns indicate the cases which showed low hemagglutinin titers. In the patients of more than 11 years of age, the low ASO titer usually parallels to the low hemagglutinin titer.

observed in immunodeficiency states, as explained below.   Without clinical complications, the total serum protein concentration is usually within normal limits, but may be slightly decreased in agammaglobulinemia.   The serum protein electrophoretic fractions other than the γ fraction should be within normal limits, unless acute infection is associated with.   When active infectious lesions are present, the electrophoretic pattern of agammaglobulinemic type is combined with acute phase response pattern.

1.   Marked Decrease or Lack of the γ Fraction

Only when the serum IgG is altered, the γ fraction shows significant changes.   Particularly in cases with animmunoglobulinemia, no γ fraction band is recognized.   On filter-paper electrophoresis, because of significant albumin tailing, the γ fraction of serum proteins does not become zero even though IgG is completely lacking in serum (Fig. 251).   However, it is easily suspected because the γ fraction band does not extend from the point of serum application towards the cathodal end.   On cellulose acetate or agar gel electrophoresis, a complete deficiency of IgG results in a complete lack of the γ fraction.

2.   Deep γ_1-Zone

With a significant decrease in IgA or IgM, the γ_1 zone becomes extremely empty on agar gel or cellulose acetate electrophoresis.   On filter-paper electrophoresis, however, the finding is not clear.

3.   Changes in the Modal Mobility of the γ Fraction

The modal mobility of the γ fraction band may be calculated as shown in Fig. 252 on cellulose acetate electrophoresis.   In normal adults, with the routine electrophoretic condition, the modal mobility of the γ fraction band ranges from 1.46 to 1.64.   In infants and children, also, the similar values are obtained (Fig. 253).   In some children, a significantly decreased modal mobility has been recognized (Fig. 254).   Some of the representative cases showing abnormal modal mobility of the γ fraction are listed in Table 73 (Fig. 255).   No clinical significance of these findings has been apparent up to date. Abnormal modal mobility of the γ fraction may indicate abnormal distribution of the clonal populations of IgG-producing cells.   It is interesting that Hong and Good reported on oligoclonicity of the IgG-globulins in the cases with hypogammaglobulinemia.[157]

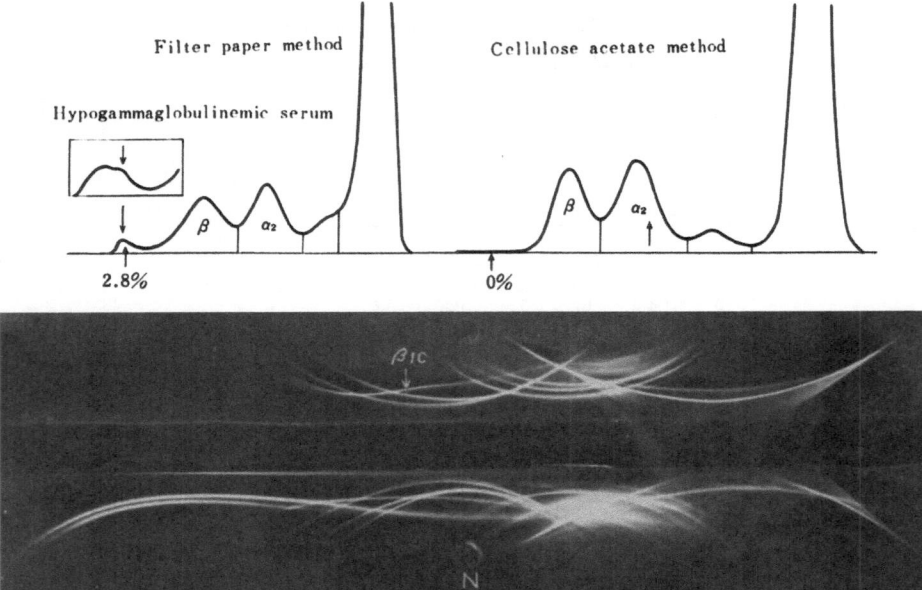

Fig. 251   Serum protein electrophoretic pattern of hypogammaglobulinemic type.

T. K., 6 y.o., male, agammaglobulinemia (BRUTON's type?)

No consanguineous marriage.   No familial occurrence noted.   Developed high fever in many occasions, and suffered repeatedly of tonsillitis, otitis media and suppurative meningitis.

Serum total protein was 6.0 g/100 ml.   On cellulose acetate electrophoresis (right upper), the albumin fraction was 68.1% or 4.1 g/100 ml, and the γ fraction was 0%.   On filter paper electrophoresis (upper left), however, the γ fraction was 2.8% or 0.16 g/100 ml. Agammaglobulinemic pattern is distinctly different from the hypogammaglobulinemic pattern.

On immunoelectrophoresis, IgG, IgA and IgM were not demonstrated (upper well).   Using the single radial diffusion method, IgA and IgM were completely lacking, and IgG was less than 0.01 g/100 ml. Blood group A, no anti-B identified.   No antibody response against vaccination of diphtheria, pertussis, tetanus and measles.   No plasma cell in the bone marrow.

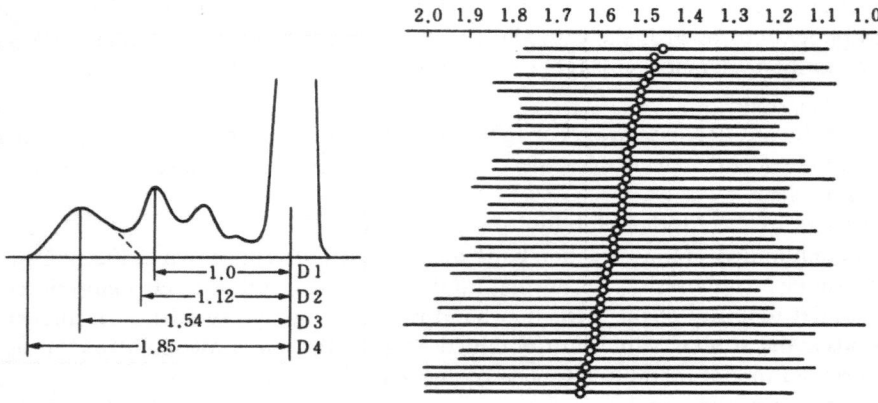

Fig. 252   Relative mobility of the γ fraction calculated from the cellulose acetate electrophoretic patterns.

The relative mobility of the γ fraction was calculated as shown in the left diagram. On the right, the white dots indicate the modal mobility (D_3) and each solid line indicates the distance between D_2 and D_4.

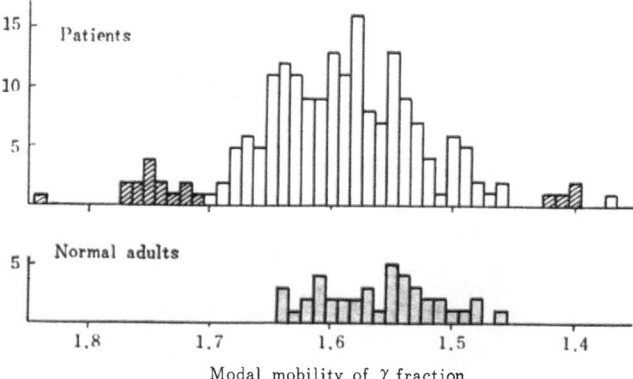

Fig. 253 Distribution of the modal mobility of the γ fraction separated by the cellulose acetate electrophoresis.
The patients were of the ones having either hypogammaglobulinemia or increased susceptibility of infections, including 32 adults and 191 children. In comparison to normal adults, the mode of the frequency distribution seems to be slightly smaller. Five cases with smaller modal mobility and 15 cases with larger modal mobility were separated for clinical studies.

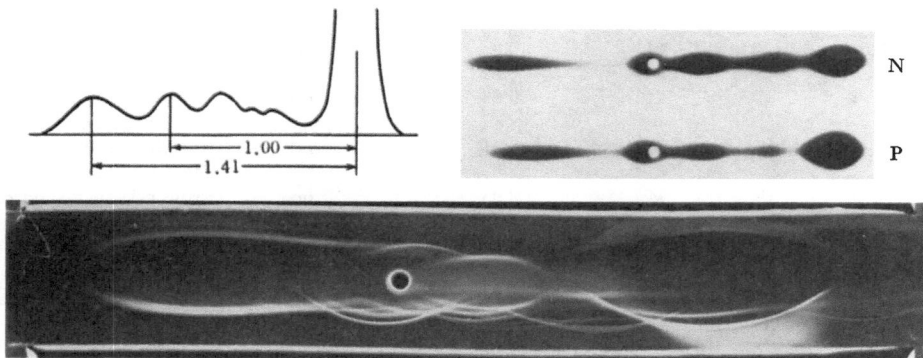

Fig. 254 Serum protein electrophoretic patterns showing faster modal mobility of γ fraction. T. M., 10 y.o., male, STURGE-WEBER syndrome.
Serum total protein was 7.4 g/100 ml. The electrophoretic patterns are almost normal, except that the modal mobility of the γ fraction was less than the normal range. The IgG precipitin line is not particularly deformed. IgM was within normal limits, but IgA was decreased markedly.

Primary Immunodeficiencies

Primary immunodeficiencies occur in various forms, and as stated previously no satisfactory classification has yet been available. Therefore, various types of primary immunodeficiencies are conveniently divided into two major groups: disorders with and without significant abnormalities in the cell-mediated immunity.

Primary immunodeficiencies are frequently associated with various types of malignancies, particularly in WISKOTT–ALDRICH syndrome.[107, 408] Also associated are autoimmune diseases and M-proteinemias.[27, 103] Therefore, reticular malignancy, immunodeficiency, M-proteinemia and autoimmune diseases seem to be intimately related to each other, probably based on the genetic abnormality of the reticular tissues.

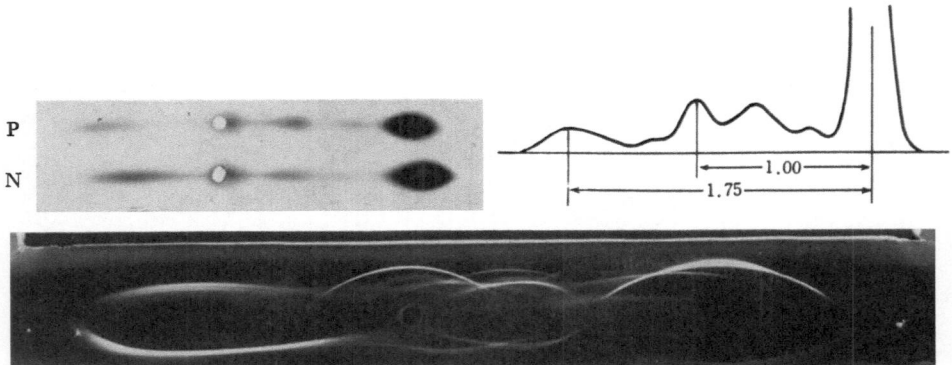

Fig. 255 Serum protein electrophoretic patterns showing slower modal mobility of γ fraction. S. N., 3 y.o., male, Down's syndrome and tetralogy of FALLOT. The cellulose acetate electrophoretic patterns were almost normal except for slight decrease of the γ fraction. The modal mobility of the γ fraction was slower.

Table 73 Cases with abnormal modal mobility of the serum γ-fraction.

I. The modal mobility of less than 1.45

Case	Age	Sex	Serum level of γ-fraction (g/100 ml)	Susceptibility for infections	Diagnosis
M. T.	4	M	0.8	yes	
H. T.	5	M	0.82	yes	
T. M.	7	M	0.63	no	Non-progressive myopathy
Y. O.	9	M	0.71	no	Auto-intoxication Thymic enlargement
T. M.	19	M	0.93	no	STURGE-WEBER syndrome

M. T. and H. T.: siblings suggested to have immuno-deficiency.

II. The modal mobility of more than 1.70

Case	Age	Sex	Serum level of γ-fraction (g/100 ml)	Susceptibility for infections	Diagnosis
H. D.	7 mo	M	0.56	no	?
M. A.	10 mo	F	0.68	yes	Vaccinia
S. I.	1	M	0.51	?	Polyneuritis?
M. I.	3	F	0.30	no	Cold agglutinin 4× < ASO 12× < Isohemagglutinin 4×
S. N.	3	M	0.50	no	Down syndrome
T. M.	3.5	M	0.52	yes	
M. Y.	7.7	M	0.71	no	Thrombocytopenia
K. Y.	10	M	0.65	yes	Immunodeficiency in sister
A. K.	10	M	0.59	yes	IgM deficiency

Other 6 boys with asthmatic bronchitis.

Primary Immunodeficiencies without Significant Deficiency of the Cell-Mediated Immune Response.

1. Infantile Sex-linked Animmunoglobulinemia or Agammaglobulinemia

This type of immunodeficiency has been variously called as BRUTON's disease, sex-linked recessive type congenital agammaglobulinemia, and congenital agammaglobulinemia of BRUTON type.[19,121,172,299] About 20 cases of this disorder have been reported in Japan (Fig. 250), and it has been encountered in frequently.

1) Clinical findings: It is only observed in males, taking a sex-linked recessive form. It usually develops clinical symptoms in 6–9 months after birth, and rarely after 3 years of age. Repeated bacterial infections are characteristically recognized.

2) Hematological findings: Peripheral lymphocytes are normal, but the plasma cells are lacking in bone marrow.

3) Immunological findings: All major classes of serum immunoglobulins are extremely low, and the humoral antibody response is either completely lacking or extremely poor. The cell-mediated immune response is normal.

4) Pathological findings: The thymus is normal. The lymph nodes show a lack of germinal centers, and normal paracortical region in most cases.

2. Dysgammaglobulinemias

WHO designates this disorder as "selective immunoglobulin deficiency". In this disorder, not all of the major classes of immunoglobulins are deficient, but a single or two classes of immunoglobulins are deficient. The nomenclature on dysgammaglobulinemias is quite variable among different investigators. Table 74 shows the classifications of dysgammaglobulinemias reported by HOBBS[153] and JANEWAY et al.[172] Many investigators in this field are reluctant to admit dysgammaglobulinemia as an independent clinical entity, except for selective deficiency of IgA.

Table 74 Classification and frequency of dysgammaglobulinemias.

by JANEWAY	by HOBBS	IgG	IgA	IgM	Cases studied by HOBBS*	Cases studied by the author**
Type II	Type I	N	↓	↓	6	8
Type I	Type II	↓	↓	↑, N	0	2
	Type III	↓	N	N	1	3
Type III	Type IV	N	↓	N	24	16
	Type V	N	N	↓	23	11
	Type VI	N	N	N	3	5**
	Type VII	↓	↑	↓	1	1

* Numbers of patients found among 11,000 studied.
** Numbers of patients found among 200 selected immunoelectrophoretically.
   Immunoglobulin levels are measured by single radial immuno-diffusion technique. All cases of Type VI showed a marked decrease in the isohemagglutinin titers.

Selective IgA Deficiency:

1) Clinical findings: No definite form of inheritance is established, but it sometimes takes an autosomal recessive form.[339] Frequently it is associated with bronchitis, sinusitis, malabsorption syndrome and other infectious diseases. However, it is encountered infrequently even in healthy individuals.[16,67,311,405]

2) Hematological findings: Normal peripheral lymphocytes are seen. With fluorescent antibody techniques, the IgA-producing cells are entirely absent or decreased markedly in the bone marrow and the mucosa.

3) Immunological findings: IgA is deficient in both serum and various secretions, but other classes of immunoglobulins are generally normal. Frequently, IgM tends to be increased in various secretions (Fig. 256) (Table 75). No deficiency is noted in the cell-mediated immune response and also in the specific antibody response involving the immunoglobulins other than IgA.

4) Pathological findings: The lymph nodes maintain their normal structure.

Table 75   Serum concentrations of immunoglobulins in the cases of isolated IgA deficiency.

Case	Age	Sex	TP	IgG	IgM	IgA	Clinical findings
F. W.	2.3	M	7.0	760	84	7.7	Bronchial asthma, low IgA in mother
T. K.	2.1	M	6.8	840	80	4.6	Asthmatic bronchitis
Y. O.	3.7	M	7.2	1140	144	11.8	Asthmatic bronchitis, anti-A 8×
T. A.	8	M	6.8	720	108	8.8	Reapeated angina, otitis media, rheumatic arthritis

TP: serum total protein concentration in g/100 ml.
      Immunoglobulins: mg/100 ml.

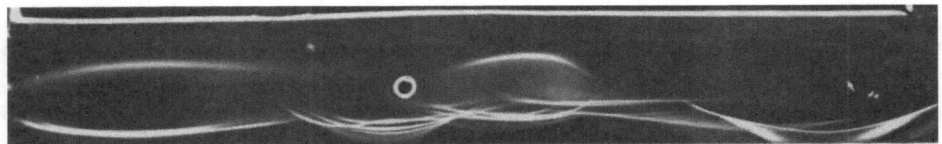

Fig. 256   Immunoelectrophoretic pattern of the serum proteins in isolated IgA deficiency F. W., 2 y.o., male, bronchial asthma.
Suffered from right pleuritis at the age of 1 year and 6 months. Since then, developed frequently asthmatic attacks.
The serum total protein was 7.0 g/100 ml, and the cellulose acetate electrophoresis resulted in the albumin fraction 66%, α_1 5%, α_2 14%, β 9% and γ 6%. On immunoelectrophoresis, IgA was not demonstrated. Immunochemical quantitation resulted in IgG 760 mg/100 ml, IgM 84 mg/100 ml, and IgA less than 7 mg/100 ml. Anti-B was 64 x.
His mother had slightly decreased serum IgA, being 91 mg/100 ml, but anti-B titer was 512 x.

3.   Transient Hypogammaglobulinemia of Infancy

It is also called as "slow starter", and it develops its characteristic clinical symptoms in 2–6 months after birth and usually it improves significantly by 9–15 months of age.[172, 405]

1)   Clinical findings: No definite form of inheritance is known. Clinically, repeated infections due to pyogenic bacteria are recognized transiently.

2)   Hematological findings: Normal peripheral lymphocytes are seen, and the bone marrow may show a decreased number of the plasma cells.

3)   Immunological findings: IgG is particularly decreased in serum, and IgM and IgA may be decreased simultaneously (Fig. 257). Specific antibody response is deficient, but the cell-mediated immune response seems to be not significantly deficient.

4)   Pathological findings: The germinal centers are rarely detected in the lymph nodes.

4.   Non-sex-linked Primary Immunodeficiency with Variable Onset and Expression

Included in this group are "acquired primary agammaglobulinemia", some cases of dysgammaglobulinemia and "sporadic type congenital agammaglobulinemia" and other syndromes, because of lack of sufficient information.

1)   Clinical findings: In some cases the disorder is inherited as a autosomal recessive trait. Repeated infections due to pyogenic organisms are observed frequently, and viral or fungal infections are also seen occasionally. It is frequently associated with autoimmune diseases, reticular malignancies and occasionally with amyloidosis.

2)   Hematological findings: Peripheral lymphocytes are normal, and diminution of the plasma cells may not be apparent in the bone marrow.

3)   Immunological findings: Serum immunoglobulin deficiency may occur in vari-

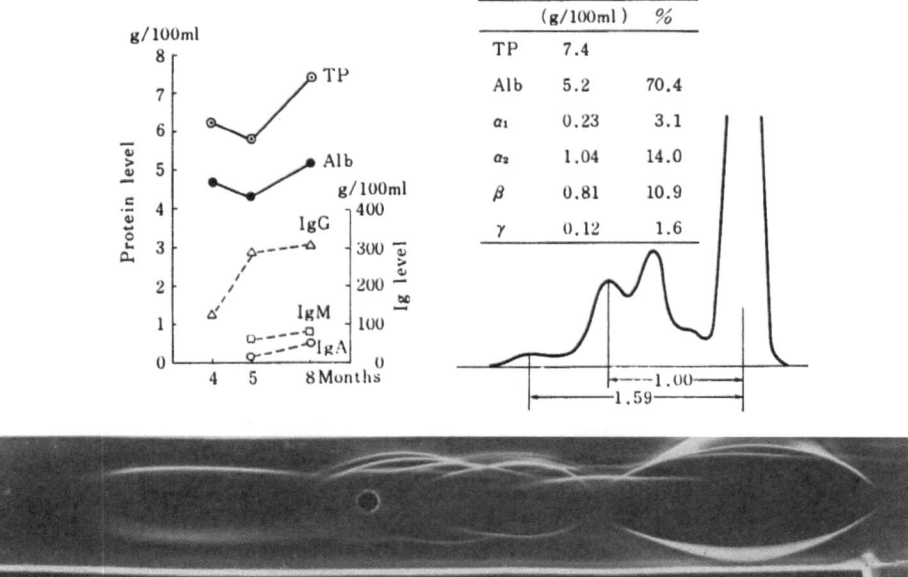

	(g/100ml)	%
TP	7.4	
Alb	5.2	70.4
α_1	0.23	3.1
α_2	1.04	14.0
β	0.81	10.9
γ	0.12	1.6

Fig. 257 Serum protein electrophoretic patterns in transient hypogammaglobulinemia of infancy. N. S., 8 m.o., male, poor development.
Admitted with poor development and poor appetite at the age of 4 months. At that time, the cellulose acetate pattern of the serum proteins showed the γ fraction to be 0%. However, later it increased to 2.5%, and IgG was shown to be 300 mg/100 ml. The isohemagglutinin titer was still low, the anti-A being only 4 x. The modal mobility of the γ fraction was normal. On immunoelectrophoresis, IgA and IgM were not demonstrated clearly.

able degrees and expressions. All three major classes of immunoglobulins may or may not be involved (Fig. 258). Specific antibody response against various antigens is decreased, but the cell-mediated immune response may not be disturbed significantly.

4) Pathological findings: The thymus seems to be normal, but no detail analysis has been available in many instances. The lymph nodes usually fail to show the germinal center, and demonstrate hypoplasia of the paracortical region. Occasionally it is accompanied with reticulosis and hypertrophy of the tonsils.

5. Deficiency of Specific Antibody Response without Hypoimmunoglobulinemia

There are a few cases where no antibody response is recognized against certain antigenic stimulation despite of normal or increased serum immunoglobulin concentration.[172, 224] Although an exact mechanism is not known, the condition is quite similar to immunological paralysis frequently demonstrated experimentally. In addition, the similar phenomenon is also encountered clinically on aging, where decreased antibody response is usually noted even though the serum immunoglobulin concentration tends to be increased.

The condition has been named as specific immunologic unresponsiveness by JANEWAY et al.[172], while serum globulin dyscrasia by LeGRIPPO.[224] A case has been followed up in the author's laboratory with this deficiency (Fig. 259).

Primary Immunodeficiencies always Associated with Significant Deficiency of the Cell-Mediated Immune Response

1. Immunodeficiency with Thymoma

This type of immunodeficiency has been called as agammaglobulinemia with thymoma

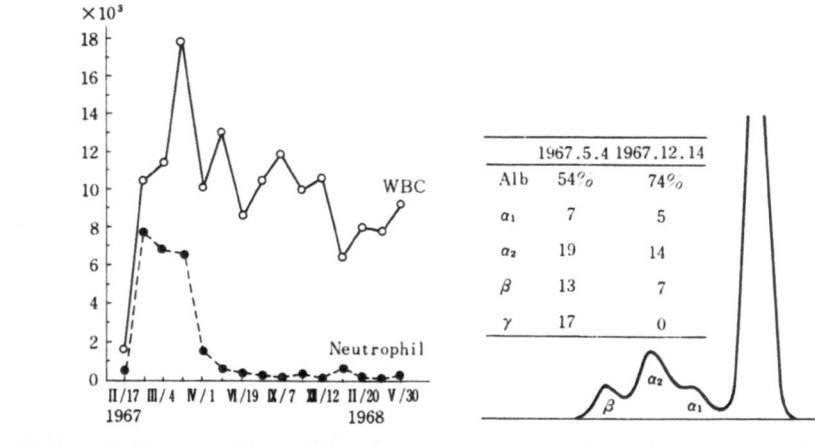

	1967.5.4	1967.12.14
Alb	54%	74%
α_1	7	5
α_2	19	14
β	13	7
γ	17	0

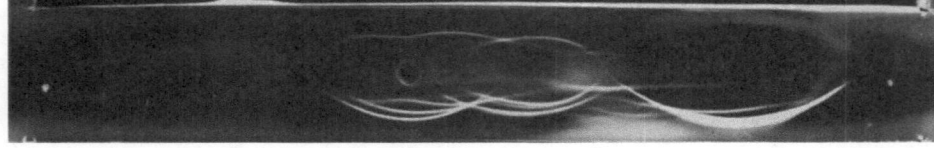

Fig. 258 Serum protein patterns in impotent neutrophil syndrome.
M. O., 2 y.o., male, pulmonary abscess.
Developed pulmonary abscess and pyothorax in February, 1967.
The serum proteins showed dysgammaglobulinemia Type 1, and
complicated with granulocytopenia. The plasma cells were de-
creased in the bone marrow.

Immunoglobulins in serum

IgG	55 mg/100 ml, markedly decreased
IgA	14 mg/100 ml, decreased
IgM	63 mg/100 ml, normal

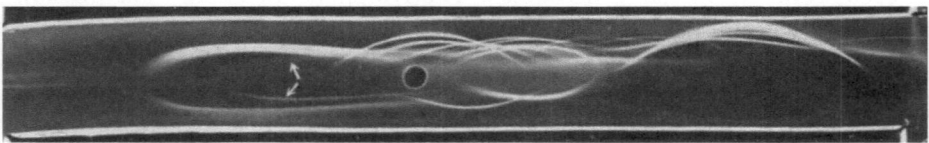

Fig. 259 Immunoelectrophoretic pattern of the serum proteins in specific immunologic unresponsive-
ness.
M. S., 2 y.o., male, asthmatic bronchitis.
Suffered from repeated episodes of upper respiratory infection and asthmatic bronchitis.   Tuberculin
reaction remained negative, although BCG was given.   Blood group A, anti-B 4 x.
The cellulose acetate electrophoretic pattern of the serum proteins is normal.   On immunoelectro-
phoresis, IgM was prominently recognized (arrow).   IgG 1090 mg/100 ml, IgA 190 mg/100 ml, and
IgM 222 mg/100 ml.

or GOOD syndrome.   It was previously included in primary acquired agammaglobulinemia, but it has been recognized as a clinical entity because the association of agammaglobulinemia with thymoma seems to be closely related.[122, 339]   However, the exact
mechanism on the association of two conditions has not be well understood.

1) Clinical findings: A hereditary transmission has been suggested, but no
definite relation has been understood.   Frequently encountered is the repeated infec-

tion with pyogenic bacteria, but a decreased resistance is also noted against viral and fungal infections.

2) Hematological findings: It is often associated with marked lymphopenia, and also with eosinopenia and decreased plasma cells in bone marrow. Occasionally, pure red cell aplasia is noted.

3) Immunological findings: All major classes of immunoglobulins are decreased in serum, and antibody response is markedly decreased. Cell-mediated immune response is also decreased.

4) Pathological findings: Thymoma is noted, most frequently of epithelial cell type. Lymph nodes show a lack or marked decrease of the germinal center, and also hypoplastic paracortical region.

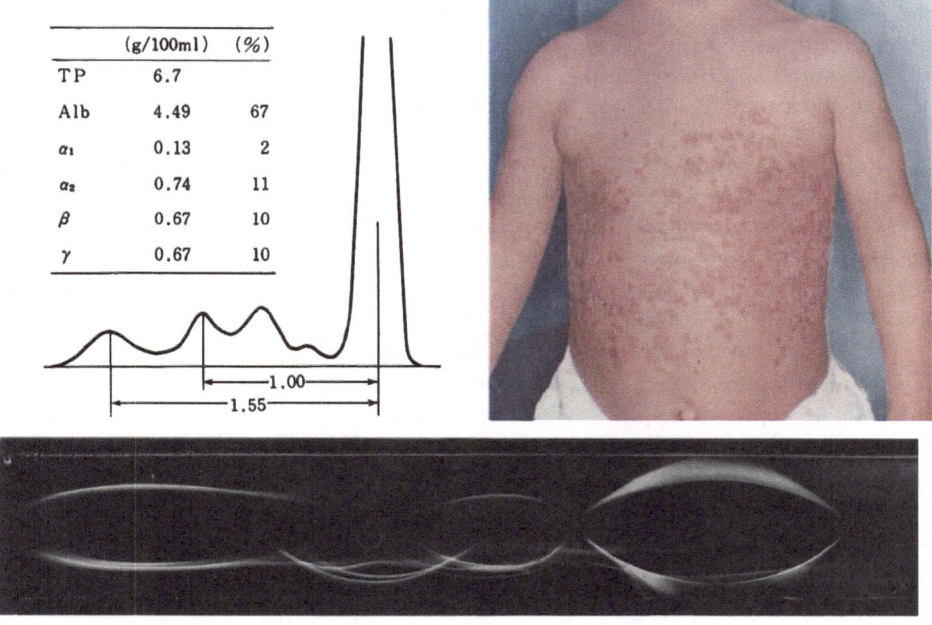

	(g/100ml)	(%)
T P	6.7	
Alb	4.49	67
α_1	0.13	2
α_2	0.74	11
β	0.67	10
γ	0.67	10

Fig. 260 Serum protein patterns in immunodeficiency with thrombocytopenia and eczema (Wis-kott-Aldrich syndrome).

H. A., 1 y. 11 m.o., male, Wiskott-Aldrich syndrome.

Delivered at full term. Suffered from upper respiratory infection repeatedly. Generalized eczematous dermatitis appeared episodically. Admitted with purpura and subcutaneous ecchymosis.

Hb 12.7 g/100 ml, platelet 4,500/mm³, bleeding time more than 7 min., poor clot retraction, positive Rumpel-Leede test. Decreased numbers of megakaryocytes and plasma cells in the bone marrow. The serum protein electrophoretic pattern was normal. The serum immunoglobulin levels and isohemagglutinin titers are listed separately.

The serum protein electrophoretic pattern was normal. The serum immunoglobulin levels and iso-hemagglutinin titers are listed separately.

Tuberculin test became positive one month after BCG was given. The antibody response against typhoid vaccine was extremely poor. Lymphoblastogenesis of the lymphocyte after addition of PHA was normal.

	IgG mg/100 ml	IgA mg/100 ml	IgM mg/100 ml
patient	637(N)	72(N)	36(N)
mother	985(N)	210(N)	25(↓)
father	1040(N)	180(N)	150(N)
sister	870(N)	130(N)	50(N)

	Hemagglutinins	
patient (B group)	anti-A	0×
mother (O group)	anti-A	64×
	anti-B	128×
father (B group)	anti-A	128×
sister (B group)	anti-A	32×

2. Immunodeficiency with Thrombocytopenia and Eczema

The condition was first reported by WISKOTT in 1937, and was recognized as a distinct clinical syndrome by ALDRICH in 1954.   Therefore, it is frequently called also as WISKOTT–ALDRICH syndrome.[3, 41, 65, 199, 363]   More than 60 cases have been reported throughout the world, and only a few cases have been recognized in Japan.

1) Clinical findings: A sex-linked hereditary transmission is recognized, and it has been encountered on in males. In addition to repeated infections, thrombocytopenia and severe eczema occur repeatedly (Fig. 260).

2) Hematological findings: Lymphopenia is usually encountered, but the plasma cells are normal in bone marrow.   Episodes of marked thrombocytopenia are repeatedly seen, and also a decrease in megakaryocytes is noted.   No platelet antibody is detected.

3) Immunological findings: No significant deficiency of the serum immunoglobulins is seen, but occasionally a decreased IgM or an increased IgA may be seen. Antibody response against various polysaccharide antigens is selectively deficient, and it is known to be an abnormality in the afferent limb of immunity.   Normal isoagglutinin titer is extremely low, and frequently no isoagglutinin is detected (Fig. 260). Cell-mediated immunity is also deficient.

4) Pathological findings: The thymus is normal.   The germinal centers are usually present in lymph nodes, and lymphocytes are depleted markedly in the paracortical region.   Reticular malignancy is occasionally associated with this condition.

3. Immunodeficiency with Ataxia-Telangiectasia

In 1941, the first report was made by Mrs. LOUIS-BAR, and it has been called also as Mrs. LOUIS-BAR syndrome.[85, 217, 294]   More than 150 cases have been reported throughout the world, and more than 15 cases have been reported in Japan.

1) Clinical findings: An autosomal recessive hereditary transmission is established. In addition to repeated respiratory tract infections, progressive cerebellar ataxia and telangiectasia of conjunctiva and skin are recognized (Fig. 261). These clinical symptoms may appear later during the childhood, and the differentiation from isolated IgA deficiency may be difficult in early stages.

2) Hematological findings: Lymphocytes and plasma cells may be decreased significantly, and eosinophilia may be seen frequently.

3) Immunological findings: IgA deficiency is most frequently associated with this syndrome, but the serum immunoglobulins may be within normal limits in some cases. Cell-mediated immune response is always deficient.

4) Pathological findings: The thymus is markedly hypoplastic or completely lacking.   Its cortico-medullary distinction is not clear, showing marked depletion of small lymphocytes. HASSALL's corpuscles are entirely lacking, and its histological appearance is quite similar to that of the fetal thymus.   The germinal centers are present in lymph nodes, and the paracortical region shows marked lymphocytic depletion.   It is frequently associated with reticular malignancy.

4. Immunodeficiency with Short-limbed Dwarfism

The first case was reported by McKUSICK and CROSS in 1966,[245] and several cases have been reported.[108, 227]

1) Clinical findings: It seems to be transmitted in an autosomal recessive form, but some sporadic cases have been also recognized. In addition to repeated infections, short-limbed dwarfism due to skeletal disorder is always associated clinically. Ectodermal dysplasia characterized with absence of hair and eyebrows may be noted.

2) Hematological findings: Lymphopenia is a significant finding, frequently as-

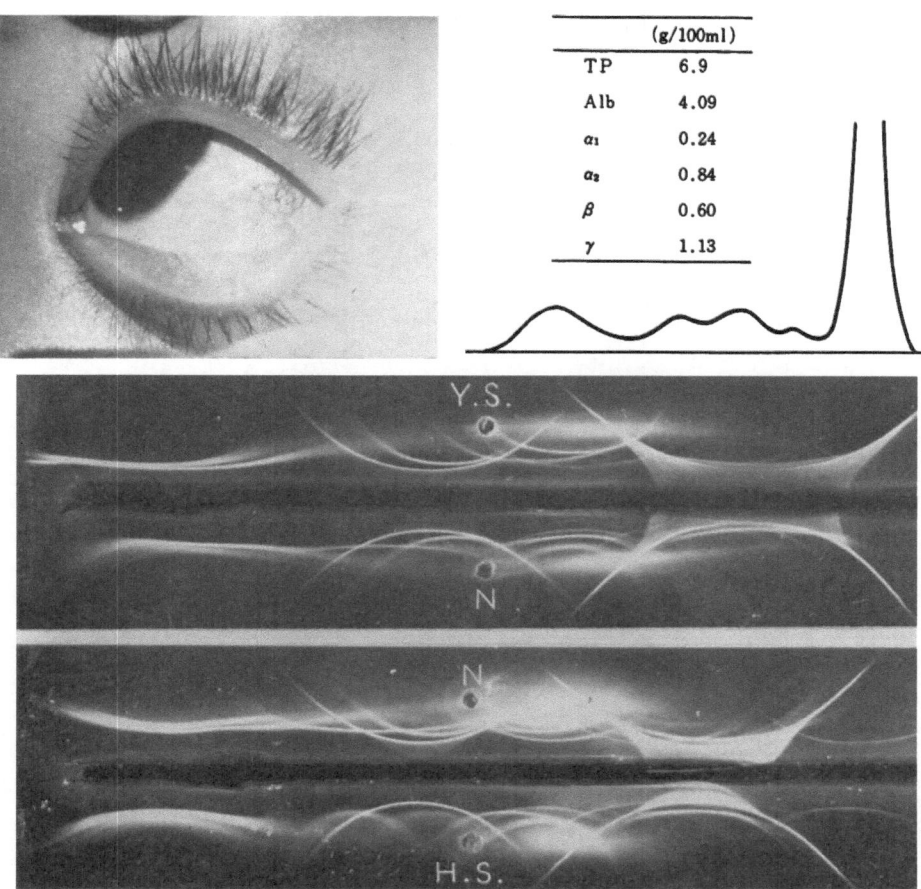

	(g/100ml)
TP	6.9
Alb	4.09
α_1	0.24
α_2	0.84
β	0.60
γ	1.13

Fig. 261 Serum protein patterns in immunodeficiency with ataxia-telangiectasia.
Y. S., 11 y.o., female, Ataxia telangiectasia.
No familial occurrence.
Suffered from otitis media at the age of 2 weeks, pneumonia at the age of 7 years. Said to have cerebral palsy at the age of 2 years, and developed progressive cerebellar ataxia from the age of 7 years. Telangiectasia of the sclerae was found at the age of 3 years.
On cellulose acetate electrophoresis, the serum proteins were normal. On immunoelectrophoresis, a complete deficiency of IgA was noted (IgG 1.21 g/100 ml, IgM 0.1 g/100 ml). The patient's mother (H. S.) also showed a lack of IgA line, although no clinical symptoms were seen.

sociated with eosinophilia. Neutrophilia may occur. The plasma cells may decrease significantly.

3) Immunological findings: Cell-mediated immune response is always decreased, while the serum immunoglobulins are deficient or normal or elevated.

4) Pathological findings: The thymus is of fetal type, showing the absence of HASSALL's corpuscles, lymphocytic depletion and a lack of differentiation into the cortex and medulla. Lymph nodes show lymphocytic depletion and absence of germinal centers.

5. Primary Lymphopenic Immunodeficiency

In this category included are various forms of immunodeficiencies and it is also called as GITLIN's syndrome.[110, 322]

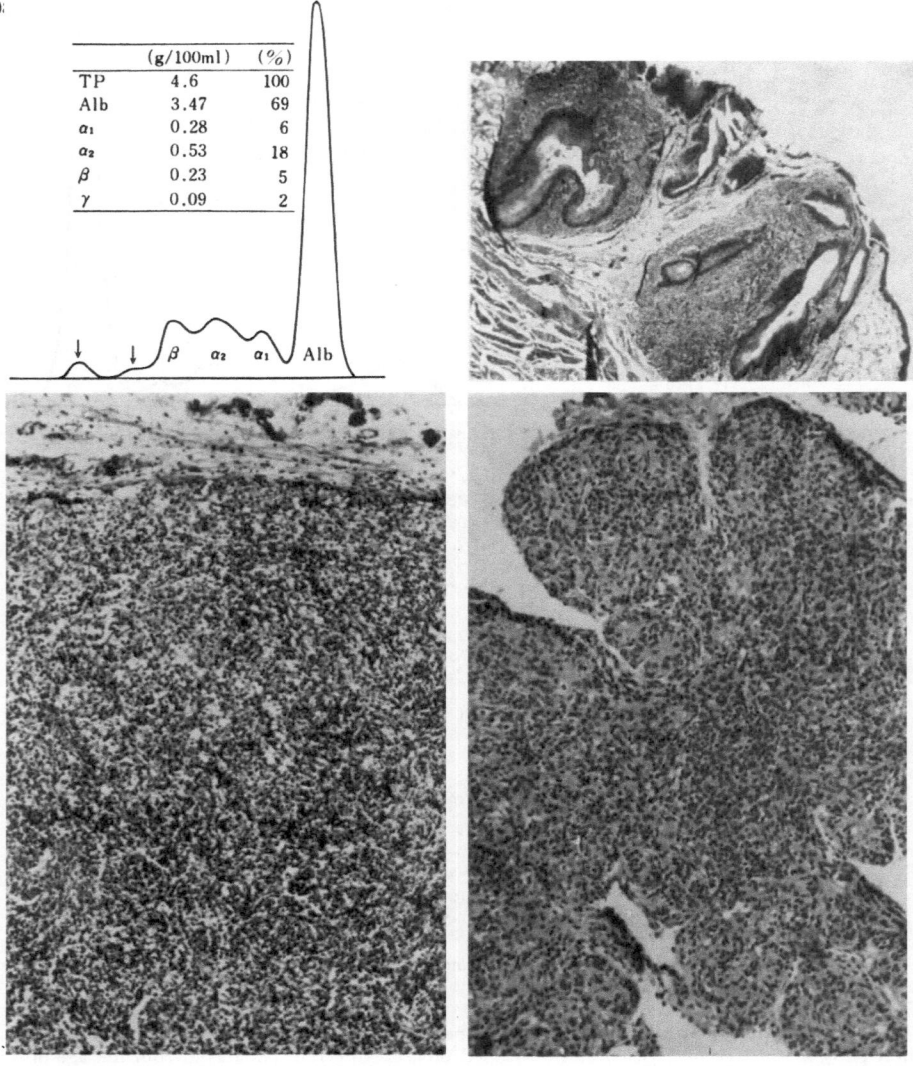

	(g/100ml)	(%)
TP	4.6	100
Alb	3.47	69
α_1	0.28	6
α_2	0.53	18
β	0.23	5
γ	0.09	2

Fig. 262 a Serum protein electrophoretic patterns in thymic alymphoplasia with M-proteinemia. M. M., 7 m.o., female, thymic alymphoplasia.

Suffered from repeated upper respiratory infections and pyoderma in the past. Admitted with watery diarrhea, fever and myoclonic seizure. Died of bilateral pneumonia 5 days after the admission. *Pseudomonas aeruginosa* cultured from blood and femoral abscess. Blood group O with no isohemagglutinin detected. ASO 100 Todd units, CRP test 12+, RA test (−) WBC 4,900–7,900/mm³ with lymphopenia (800–900/mm³) and eosinophilia (over 50%).

At autopsy, marked hypoplasia of the thymus and lymphoid tissues was recognized. The thymus weighed only less than 1 g, showing the histological features of the fetal type (lower right). The lymph nodes (lower left) showed absence of the follicles and marked depletion of the small lymphocytes. Also noted were bilateral hemorrhagic pneumonia and bilateral adrenal hemorrhage. On cellulose acetate electrophoresis, the serum γ fraction was markedly decreased, demonstrating two distinct minor spikes (arrows). The palatine tonsils (upper left) are markedly hypoplastic, showing absence of the lymph follicles and marked depletion of small lymphocytes. Many plasma cells are recognized in the lymphoid tissue which is quite similar to that of the lymph nodes. On immunoelectrophoresis, IgG-K type and IgA-K type M-proteins were clearly identified. The serum IgG was markedly decreased, measuring 146 mg/100 ml on single radial immunodiffusion method, and the serum IgA was slightly increased for this age, measuring 250 mg/100 ml. IgM showed a minor-M-bow on immunoelectrophoresis, and measured 23 mg/100 ml, but the monoclonicity of the IgM was not confirmed.

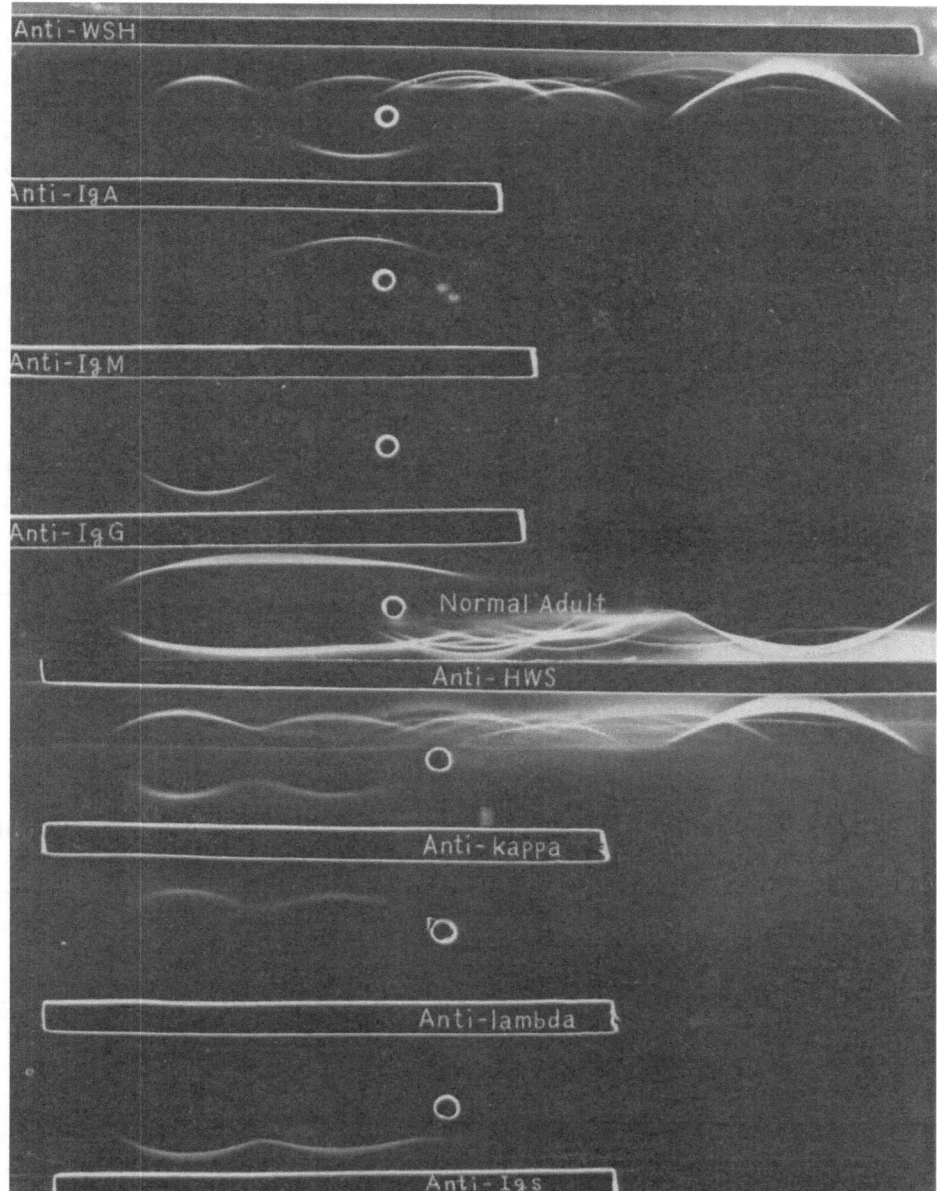

Fig. 262 b Immunoelectrophoretic Serum Protein Patterns.

1) Clinical findings: It is transmitted either in sex-linked or autosomal recessive form in the most cases, but no definite hereditary transmission may be recognized. Many cases may die during infancy of fungal or viral infections.

2) Hematological findings: Peripheral blood shows a varying degree of lymphopenia. Plasma cells may be decreased in the bone marrow.

3) Immunological findings: Serum immunoglobulins are decreased in a varying degree and also in various combinations (Fig. 262). Antibody response may be disturbed only against certain antigenic stimulations. Cell-mediated immune response is always decreased significantly.

4) Pathological findings: The thymus is hypoplastic, showing lymphocytic depletion and absence of HASSALL's corpuscles (Fig. 262). In general, small lymphocytes are depleted in the lymphoid tissues, and only small aggregates of lymphocytes may be recognized in the lymph nodes and spleen.

6. Autosomal Recessive Alymphocytic Agammaglobulinemia

In 1950, GLANZMANN and RINIKER reported the first case under the diagnosis "essentielle Lymphozytophthise", and the condition has been variously called as Swiss type agammaglobulinemia, alymphocytosis, hereditäre lymph-plasmocytäre Dysgenesie, familiäre Lymphopenie, thymic alymphoplasia, thymic dysplasia, hereditary thymic aplasia, etc.[151, 339, 380]

Recently WHO has recommended to classify it as "autosomal recessive severe combined immunodeficiency". Included in severe combined immunodeficiency are also the sex-linked form and the sporadic form. Therefore, some of the cases named as GITLIN's syndrome may be included in this category. The term thymic alymphoplasia has been used interchangeably with Swiss type agammaglobulinemia by some investigators. However, it generally means the hereditary condition resulted from fetal maldevelopment of the thymus.

1) Clinical findings: It is transmitted in an autosomal recessive form, and appears in both sexes. In general, it starts within 3 months after birth with repeated severe infections, and usually the patients die within 2 years.

2) Hematological findings: Marked lymphopenia is almost always recognized, and the plasma cells are not seen in the bone marrow. Lymphocytes show normal morphology, but they do not respond to phytohemagglutinin.

3) Immunological findings: Three major classes of immunoglobulins are markedly decreased. Isohemagglutinins are absent, and no antibody response is lacking against various antigenic stimulations. Cell-mediated immune response is always decreased significantly, and no delayed type hypersensitive reaction is recognized against various types of allergens.

4) Pathological findings: The thymus is markedly hypoplastic, showing absence of HASSALL's corpuscles and marked depletion of lymphocytes. Lymph nodes show marked depletion of lymphocytes and absence of plasma cells.

7. Autosomal Recessive Lymphopenia with Normal Immunoglobulins

In 1964, NEZELOF et al. reported the first case, and it has been frequently called as NEZELOF's syndrome.[111, 268, 339] Included in this category is "episodic lymphopenia with lymphocytotoxin".[408] The abnormality may be summarized as selective thymic deficiency, characterized by normal antibody response and decreased cell-mediated immune response.

1) Clinical findings: It usually is transmitted in an autosomal recessive form. Repeated fungal or viral infections are characteristically noted, and the patients may die within 2 years after birth.

2) Hematological findings: Lymphopenia is recognized, but plasma cells are present normally in the bone marrow.

3) Immunological findings: Serum immunoglobulin levels are normal. Cell-mediated immune response is always significantly decreased.

4) Pathological findings: The thymus shows absence of HASSALL's corpuscles and

marked depletion of lymphocytes. Lymph nodes show usually marked depletion of lymphocytes, but may maintain some germinal centers.

8. Thymic Aplasia or Hypoplasia

In 1965, DI GEORGE reported the first case, and the condition is frequently called as DI GEORGE syndrome or pharyngeal pouch syndrome.[74, 339] The condition is due to failure of the normal derivatives of the third and fourth pharyngeal pouches to develop.

1) Clinical findings: No hereditary transmission has been recognized. The patients usually die of repeated fungal or viral infections during infancy. Because of the combined mal-development of the parathyroid glands, hypocalcemic tetany is recognized. Various types of cardio-vascular malformations may be associated with immunodeficiency state.

2) Hematological findings: No significant changes are seen in plasma cells, but progressive lymphopenia may be recognized.

3) Immunological findings: Serum immunoglobulin levels are within normal limits, but antibody response seems to be disordered. Cell-mediated immune response is always severely deficient.

4) Pathological findings: Usually the thymus is completely lacking. Lymph nodes show marked depletion of lymphocytes in the paracortical region, usually with normally developed germinal centers.

9. Reticular Dysgenesis

The conditon was first described in 1959 by de VAAL et al. in male twins dying of sepsis. Leukocytes are totally absent in the peripheral blood.[71] It is one of the most fulminant immunodeficiencies, and no detail immunological investigation is available because the patients die within ten days after birth. Pathologically, the thymus is markedly hypoplastic, and the lymphoid organs such as lymph nodes, tonsils and PEYER's patches are not recognized macroscopically.

Secondary Immunodeficiencies

Hypogammaglobulinemia is associated with various clinical conditions as listed in Table 76. Among those, reticular malignancies and immunosuppressive therapy are the two most important conditions causing immunodeficiency state.

Table 76 Classification of secondary immunodeficiency syndromes.

1. Plasma protein abnormality of the malnutritional type
2. Plasma protein abnormality of the protein-losing type
   a. Nephrotic syndrome
   b. Protein-losing gastroenteropathy
   c. Exudative dermatopathy
3. Increased endogenous catabolism of IgG (hereditary ?)
4. Neoplastic lesions of the lymphoid tissue
   a. Acute leukemia
   b. Chronic lymphatic leukemia
   c. Malignant lymphomas (lymphosarcoma, reticulum cell sarcoma)
   d. HODGKIN's disease
   e. Myeloma, macroglobulinemia Waldenström, etc.
5. Iatrogenic inhibition of the lymphoid tissue
   a. Administration of immuno-suppresive drugs
   b. Irradiation

Table 77 Frequency of bacterial infections in the patients with malignant lymphomas and leukemias. (from MILLER).

Diagnosis	Cases	Percentage (%)
Chronic lymphocytic leukemia	9/24	37.2
Lymphosarcoma	11/87	12.6
Reticulum cell sarcoma	14/83	16.9
Acute leukemia	44/91	48.5
Chronic myelocytic leukemia	0/7	0
HODGKIN's disease	18/144	12.5

Table 78 Changes in the serum γ fraction concentration in the patients with malignant lymphomas and leukemias (from MILLER).

Diagnosis	Cases	Decreased	Normal	Increased
Chronic lymphocytic leukemia	104	35.6 %	48.1 %	16.3 %
Lymphosarcoma	47	14.9	61.7	23.4
Reticulum cell sarcoma	37	10.8	64.9	24.3
Acute leukemia	42	7.2	47.6	45.2
Chronic myelocytic leukemia	26	0	69.2	30.8
HODGKIN's disease	69	2.9	63.8	33.3

The results obtained with filter-paper electrophoresis.

1. Immunodeficiency States Associated with Lymphoreticular Malignancies

Increased susceptibility to infections: Patients with either lymphoma or leukemia very often are complicated with bacterial infections as well. The frequency can be seen on Table 77 which shows acute leukemia as the highest, and chronic lymphatic leukemia next.[251] However, there are practically no infections suffered in chronic myelogenous leukemia. Both malignant lymphoma and HODGKIN's disease stand in the middle. Pneumonia usually occurs in chronic lymphatic leukemia, while septicemia, stomatitis, and dermatitis are often found in acute leukemia. In HODGKIN's disease, fungal and viral infections are apt to be accompanied with.

Decreased serum immunoglobulin levels: In malignant lymphoma, leukemia and HODGKIN's disease, the serum concentration of immunoglobulins varies among different patients (Table 78).[251] Hypogammaglobulinemia is observed most frequently in chronic lymphatic leukemia, followed in frequency by lymphosarcoma and reticulum cell sarcoma. It is relatively rare that hypogammaglobulinemia is encountered in acute leukemia and HODGKIN's disease. Further, the association of hypogammaglobulinemia with chronic myelocytic leukemia is hardly observed. On immunoelectrophoresis no constant changes are recognized in IgG, IgA and IgM.[120] Association of hypogammaglobulinemia in reticular malignancies indicates usually a poor prognosis.[252, 390]

According to our experiences once the γ fraction decreases significantly, it does not increase again even if clinically improved (Fig. 263, 264). When hyperimmunoglobulinemia is associated with these diseases, the γ fraction is frequently decreased by immunosuppressive agents. However, immunosuppressive therapy alone may not cause significant hypogammaglobulinemia.[251]

Decreased immune response: According to MILLER,[251] decreased humoral antibody response is most frequently observed in lymphoproliferative disorders. But in myeloproliferative disorders, this tendency is not prominent. A delayed type allergic reaction is demonstrated in 90% of acute leukemia, and in 80.6% of chronic lymphatic leukemia, but, despite this, in HODGKIN's disease, it is traced only in 25%.[251] Thus, there appears

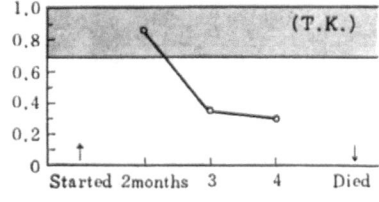

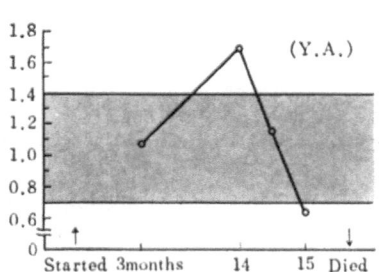

Fig. 263 Serum γ fraction levels in malignant lymphoma.

T. K., 66 y.o., male, reticulum cell sarcoma.
Died in 4 months after the clinical symptoms started, having a fulminating course. Marked decrease in the γ fraction level,

Y. A., 4 y.o., male, HODGKIN's disease
Died in one year and 3 months after the clinical symptoms started. Along with the decrease in the γ fraction, the anti-A titer became lower; 128 × →64 × →32 × →8 ×.

J. T., 40 y.o., male, reticulum cell sarcoma.
Died in 3 years and 9 months after the clinical symptoms started, having a long clinical course.

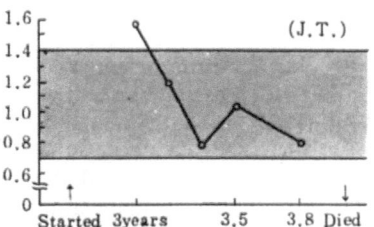

	V/2 (g/100ml)	VII/12 (g/100ml)
TP	6.7	7.2
Alb	4.0	4.75
α₁	0.44	0.33
α₂	1.48	0.94
β	0.61	0.89
γ	0.18	0.28

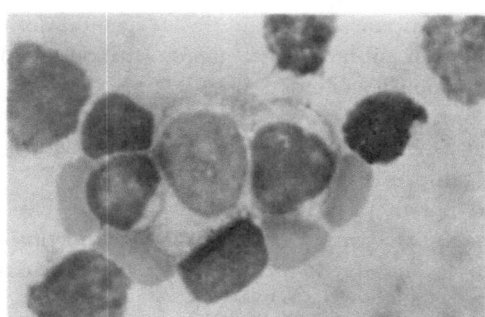

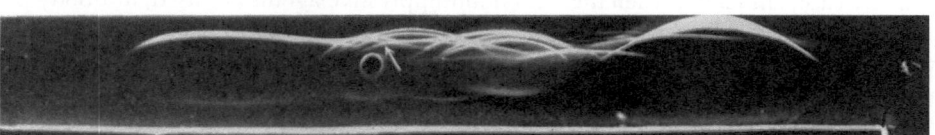

Fig. 264 Serum protein electrophoretic patterns in acute lymphatic leukemia.
K. W., 3 y.o., male, acute lymphatic leukemia.
Admitted with fever and purpura. Marked anemia (Hb 3.9 g/100 ml), and thrombocytopenia with prolonged bleeding time. WBC 2400/mm³. Many peroxydase-negative lymphoblasts found in the peripheral blood and bone marrow (upper left).
On cellulose acetate electrophoresis, the serum protein pattern was that of the acute phase response type, complicated with hypogammaglobulinemia (upper right). IgG, IgA and IgM were all decreased. With the administration of corticosteroids, the leukemic cells decreased rapidly, but the γ fraction level remained to be low. IgG, 460 mg/100 ml; IgA, 56 mg/100 ml; and IgM 43 mg/100 ml.

Table 79　Various immuno-suppressive factors (from TSUBURAYA).

1. Anti-lymphocyte serum (ALS)	6. Pyrimidine antagonists
2. X-irradiation	5-Fluorouracil
3. Adrenocorticoids	5-Bromodeoxyuridine
4. Alkylating agents	7. Folic acid antagonists
Nitrogen mustard	Amethopterin
Cyclophosphamide (Endoxan®),	8. Antibiotics
Busulfan (Myleran®)	Actinomycins
5. Purine antagonists	Mitomycin C
6-Mercaptopurine (6-MP)	9. Anti-inflammatory agents, EACA, chloroquine, etc.
6-thioguanine	
Azathioprine (Immuran®)	

to be a strong tendency of accompanying hypogammaglobulinemia in chronic lymphatic leukemia, and in particular, the humoral antibody response is considered to be deficient. In contrast, with HODGKIN's disease, as with other granulomatous diseases such as sarcoidosis, the cell-mediated immune mechanism seems to be deficient predoninantly.[251]

　2.　Immunodeficiency due to Immunosuppressive Agents

　Recently, various immunosuppressive agents, have been discovered, and are utilized clinically for different purposes (Table 79).[30, 386]　Although most widely used, the immunosuppresive mechanism of adrenocorticosteroids is entirely unknown.　However, a large dosis of steroids results in lymphopenia in such experimental animals as rabbits, rats and mice, causing marked depression of the primary and secondary immune reactions.[88, 161]　Man, monkey and guinea-pig are relatively resistant to a large dosis of steroids, but if extremely large doses of adrenocorticosteroids are given, antibody response may be depressed.[161, 246]

　With X-ray irradiation or administration of alkylating agents, cell division is suppressed to an extreme degree, and as a result, antibody response is impaired.　If its fatal doses are given, all cells are completely destroyed with their subfatal doses given, antibody response will eventually restored, after their administration is ceased.[161]

　With the use of such antimetabolites as purine antagonists, pyrimidine antagonists, or folic acid antagonists, the synthesis of either DNA or RNA is inhibited, and as a result, both cell division and protein synthesis are suppressed.[161]

　Since actinomycins C and D inhibit RNA-polymerase, RNA synthesis is suppressed.[119] On the other hand, anti-lymphocyte serum seems to suppress the activation and proliferation of lymphocytes induced by antigen stimulation, and to prevent antigenic informations to reach the antibody-producing cells.[80]

　In any case, clinically, when these immunosuppressive agents are used, antibody production is suppressed to varying degrees.　However, the effect to serum immunoglobulin levels is variable among different agents.　For example, in nephrotic syndrome, marked decrease in the serum immunoglobulin level may be seen during the administration of adrenocorticosteroids, but it will be recovered after the steroid administration is discontinued (Fig. 142).

　In such reticular malignancies as leukemia and malignant lymphoma accompanying hyperimmunoglobulinemia, the administration of immunosuppressive agents will frequently cause a significant diminution of the serum immunoglobulins.　However, the immunosuppressive therapy alone is not likely to cause hypogammaglobulinemia, and the latter condition seems to be resulted from progressive replacement of the immunoglobulin-producing tissues by neoplastic cell proliferation (Fig. 263).

Chapter 26

Hyperlipoproteinemia

CLASSIFICATION OF THE HYPERLIPOPROTEINEMIA

Almost all of the serum lipids are bound to lipoproteins, and thus hyperlipidemia accompanies always hyperlipoproteinemia. Since there is a lack of unification in the terminology used to express serum lipid abnormalities the author will here define some of the terms to be used subsequently.

1) Hyperlipidemia: Not withstanding changes in the composition of serum lipids, it is a situation which represents an increase in the total serum lipids as a whole. Bearing practically the same meaning, the term "hyperlipemia' " is used at times as well. However, since the word hyperlipemia is used at times to indicate the conditions where the serum appears lactescent.

2) Hypercholesterolemia, hyperglyceridemia, hyperphospholipidemia: Such terms are used to indicate an increase in either one of the serum lipids.

Any abnormality of lipid metabolism in the body is included in "lipopathy" or "lipidosis." Since there is a close relation between hyperlipidemia and lipid metabolic abnormalities, the representative disorders in lipopathy are listed in Table 80.[410]

Hyperlipidemia has been classified as follows:

1) Primary hyperlipidemia
   a. Hypercholesterolemia
   b. Hyperglyceridemia
2) Secondary hyperlipidemia
   a. Hypercholesterolemia
   b. Hyperglyceridemia

However, it is an impossible task to clearly divide hyperlipidemia into two groups, namely, hypercholesterolemia and hyperglyceridemia. SANDOR[319] rather divides them into (1) hyperlipidemia with hyperglyceridemia; and (2) hyperlipidemia without hyperglyceridemia.

On the other hand, according to FREDRICKSON and others,[99] hyperlipoproteinemia has been divided into 5 major groups as seen in Table 81.

In addition AHRENS would further classify as follows: fat-induced or exogenous hyperlipemia brought on by an excessive lipid intake and its poor processing in the body; and, carbohydrate-induced hyperlipemia, caused by an excessive intake of carbohydrates. However, even though "endogenous hyperlipemia" and "carbohydrate-induced hyperlipemia" are used very often in the same context, they do not necessarily bear the same meaning.[99]

Table 80   Classification of representative lipopathies (from WILLIAMS and GLOMSET).

I. **Excess Intracellular Lipid (hyperlipexia)**
   A. Triglyceridoses
      1. Adipose cells
         a. General: obesity
         b. Local: lipomatosis, insulin lipohypertrophy
      2. Liver cells ("fatty liver")
   B. Sphingolipidoses
      1. Gangliosidosis: infantile amaurotic family idiocy, gargoylism
      2. Sphingomyelinosis: NIEMANN-PICK disease
      3. Cerebrosidosis: GAUCHER's disease
   C. Cholesterolipidoses
      1. Xanthomatosis with hyperlipidemia
         a. Xanthoma tuberosum: hypercholesterolemia and hyperglyceridemia
         b. Xanthoma planum: predominantly hypercholesterolemia
         c. Tendon xanthoma: predominantly hypercholesterolemia
         d. Xanthoma eruptivum: predominantly hyperglyceridemia
      2. Xanthomatosis without usually hypercholesterolemia
         a. Xanthoma disseminatum
         b. Histiocytosis X: LETTERER-SIWE disease, HAND-SCHÜLLER-CHRISTIAN disease,

II. **Excess Intracellular and Eosinophilic Granuloma or Extracellular Lipid**
   A. Cholesterolipidoses
      1. Angiopathies
         a. Atherosclerosis
         b. Arteriosclerosis
         c. Microangiopathy (diabetic)
      2. Biliary diseases
         a. Cholesterolosis
         b. Gall stones
      3. Nephrosis
      4. Cholesterolosis with inflammatoin, neoplasia, trauma

III. **Hyperlipidemias**
   A. Hyperglyceridemia
   B. Hypercholesterolemia
   C. Hyperphospholipidemia
   D. Hyperketonemia

IV. **Hypolipidemias**

V. **Deficient Intracellular Lipid**
   A. Undernutrition
      1. Starvation
      2. Diabetes
      3. Hypermetabolism
      4. Steatorrhea

VI. **Miscellaneous**
   WHIPPLE's disease, acanthosis, sclerema neonatorum, etc.

Table 81   Classification of hyperlipoproteinemias (from FREDRICKSON et al.).

1. Type I   hyperlipoproteinemia: increased chylomicron
2. Type IIa   hyperlipoproteinemia: increased β-lipoprotein fraction
3. Type IIb   hyperlipoproteinemia: increased both β-lipoprotein and pre-β lipoprotein factions.
4. Type III   hyperlipoproteinemia: broadly increased β-lipoprotein fractions
5. Type IV   hyperlipoproteinemia: increased pre-β lipoprotein fraction
6. Type V   hyperlipoproteinemia: increased both chylomicron and pre-β lipoprotein fractions.

SERUM PROTEIN CHANGES IN HYPERLIPOPROTEINEMIA

Serum Protein Changes in the Increased Serum High-Density Lipoprotein

Hyperlipoproteinemia is chiefly due to significant increase in the low-density lipo-proteins. The alterations of the high-density lipoprotein or α_1-lipoprotein may not be significant quantitatively. However, in such situations as toxemia of pregnancy or during the 3rd trimester of pregnancy where a great increase of the α_1-lipoprotein is encountered, a distinct prealbumin band may be recognized, particularly when the stored serum is used.

Serum Protein Changes in the Increased Serum Low-Density Lipoproteins

LDL is a rather unique component, and its mobility changes under varying analytical conditions.

1. The Total Serum Protein Concentration

Since the refractive index of lipoproteins is especially high, the total serum protein concentration measured by refractometry may be falsely high in hyperlipoproteinemia. It is not infrequent that the value obtained by refractometry is as much as 2 g/100 ml higher than the value obtained through the biuret method.

2. Changes in Tiselius Electrophoretic Patterns

LDL is migrated mostly within the β-fraction. With the Tiselius electrophoresis, the relative concentration of each protein fraction is obtained by the Schlieren optical system, and thus the increase of the lipoproteins tends to be exaggerated.

3. Changes in the Filter-Paper Electrophoretic Patterns

On filter-paper electrophoresis, most of the LDL is migrated in the β fraction, and hyperlipoproteinemia may cause a significant increase of the β fraction (Fig. 265). How-ever, the increase of the β fraction may not be so great as in the Tiselius pattern, since the protein moiety of the LDL occupies a relatively small portion of their molecules.

Chylomicrons are adsorbed on the filter-paper, and they remain at the application point. Therefore, hyperchylomicronemia may show a relatively narrow band at the application point (Fig. 92). Lipoproteins are easily denatured on storage, and thus the hyperlipo-proteinemic serum which has been frozen may show a sharp band of the denatured lipo-proteins at the application point (Fig. 93).

4. Changes in Cellulose Acetate Electrophoretic Patterns

The low-density lipoproteins migrate in various zones, depending on the types of the cellulose acetate strips to be used. For example, they migrate around the α_2 zone with the Separax and Oxoid membranes, while at the β zone with the Cellogel, Sepraphor–III and Selecta membranes. By using the Millipore membrane, they even migrate at the β_2 zone. The β-lipoprotein is frequently separated as a characteristic sharp protein band (Fig. 266), and this is particularly prominent with the Separax membrane. The char-acteristic sharp band is frequently refered as Lp-peak or Lp-band. The Lp-band, as a rule, does not occupy a certain position, but changes its mobility between the α_2 and β fractions (Fig. 266). Its mobility varies not only among different patients, but also is dependent on its serum level. That is, the more its serum concentration, the slower the mobility.

By this method, chylomicrons seem to migrate around the application point, but at least a portion of it may migrate in different zones.

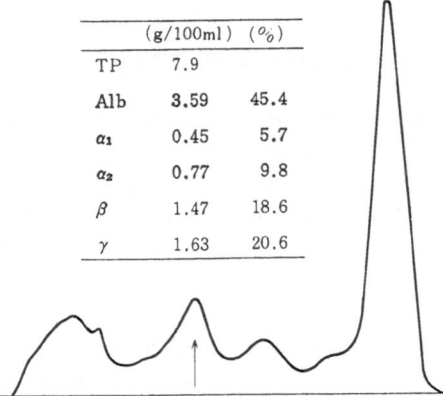

	(g/100ml)	(%)
TP	7.9	
Alb	3.59	45.4
α_1	0.45	5.7
α_2	0.77	9.8
β	1.47	18.6
γ	1.63	20.6

Fig. 265   Serum protein electrophoretic pattern in hyperlipoproteinemia.
T. S., 9 y.o., female, familial hyperlipoproteinemia, Type II.
Admitted with xanthomatosis. Serum total cholesterol was 800 mg/100 ml, but the serum was transparent. No increase in the serum triglyceride level. Her father and sister were also involved.
The filter paper electrophoretic pattern showed a marked increase of the β fraction, due to a marked increase in the serum LDL. The serum total protein was measured with the micro-biuret method.
Filter paper pattern of the serum lipoproteins, showing Type II hyper lipoproteinemia.

5.   Agar Gel Electrophoretic Patterns

The LDL migrate similarly as in the cellulose acetate electrophoresis. As its serum concentration increases, its mobility slows down (Fig. 266). Chylomicrons remain at the application point.

LDL reacts with agaropectin, forming white precipitates. Therefore, agar gel is not used for electrophoretic or immunoelectrophoretic analyses of the LDL.

6.   Changes in the Agarose Gel Electrophoretic Patterns

Since agarose is the purified agar devoid of agaropectin, the LDL do not give rise to non-specific precipitation in agarose gel. The mobility of the LDL in agarose gel is about the same as in agar gel. Agarose gel, therefore, is used for analyzing the LDL instead of agar gel.

7.   Changes in Thin-Layer Filtration Patterns

With the gel filtration technique using Sephadex G–200, the LDL are included in the M-spot while the HDL is in the G-spot (Fig. 267).

8.   Changes in Other Analytical Methods

β–L test is based on the specific immunological precipitation, using the anti-β lipoproteins. In hyperlipoproteinemia, its precipitate increases significantly. When pre-β lipoprotein or VLDL is increased markedly, white floccula may be recognized.

Lipoproteins have high intrinsic viscosity, and serum relative viscosity frequently increases in hyperlipoproteinemia. However, it does not exceed 3.0, while its normal values ranges from 1.6–1.9.

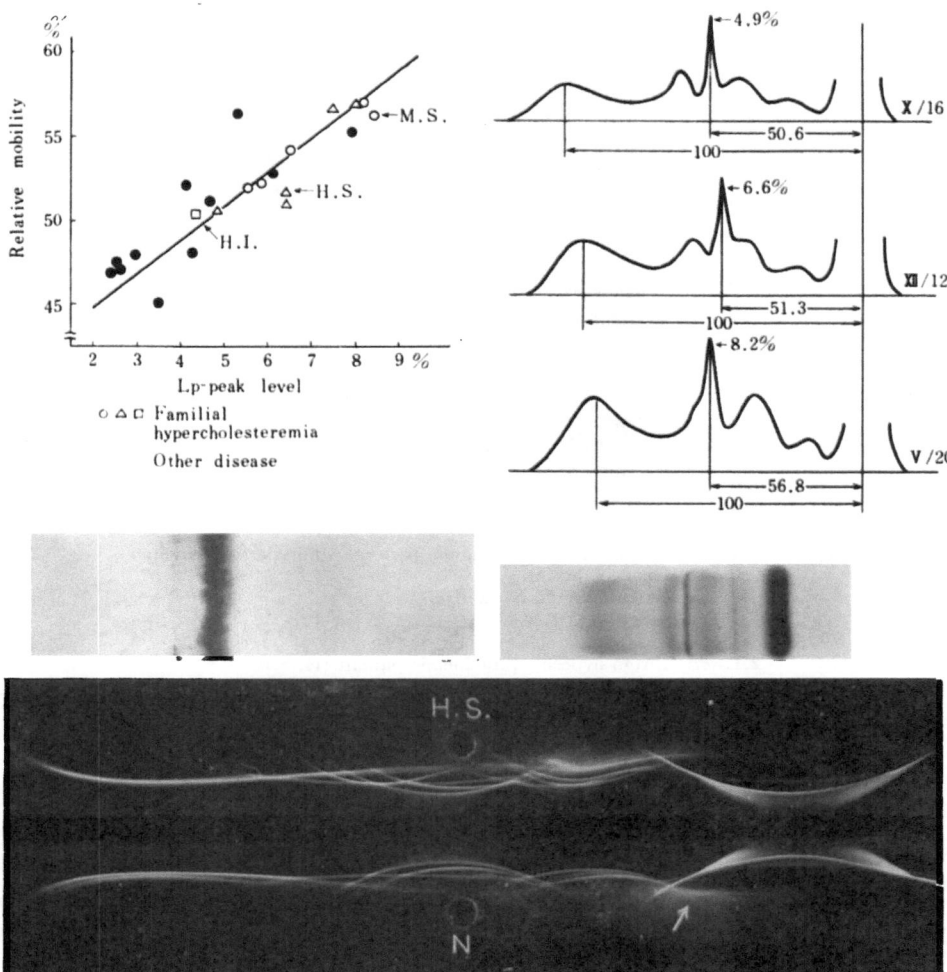

Fig. 266 Serum protein electrophoretic patterns in hyperlipoproteinemia.
H. S., 9 y.o., female, familial hyperlipoproteinemia, Type II.
Complained of multiple xanthomatosis since the age of 6 years. The serum total protein was 8.1 g/100 ml, measured with the micro-Biuret method, and 9.7 g/100 ml with refractometry. On cellulose acetate electrophoresis, a characteristic linear band of Lp was clearly seen. The Lp-band of the low-density lipoprotein tends to migrate slower with increasing its serum concentration (upper left). On agar gel immunoelectrophoresis, the LDL line was vaguely recognized because of its interaction with agar gel (arrow). The serum relative viscosity was slightly increased, being 2.41.

Serum lipid composition (mg/100 ml)

	Total lipids	Triglyceride	Total cholesterol
Normal	500–700	40–110	150–200
H.S.	1465	84	742
M.S.	1485	71	723

M.S.: 11 y.o., sister of H. S.

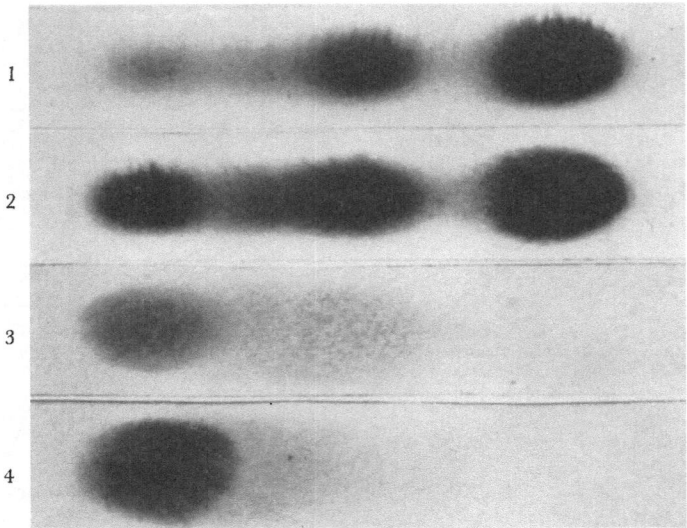

Fig. 267 Thin-layer gel filtration patterns in hyperlipoproteinemia (Type II).
1,2: stained with Amido Black 10 B.
3,4: stained with Oil Red O.
1,3: normal serum
2,4: serum from hyperlipoproteinemic patient (H. S.).
The low-density lipoproteins are separated in the M fraction, while
the high-density lipoproteins are in the G fraction.

DIFFERENTIATION OF HYPERLIPOPROTEINEMIC
SERUM PROTEIN ELECTROPHORETIC PATTERN

Differentiation from the Changes Resulted from *in vitro* Hemolysis

In hemolysis, the hemoglobin in the erythrocytes is freed into the serum, and the hemolyzed serum contains various amounts of hemoglobin. Therefore, on electrophoretic
fractionation of the hemolyzed serum, the free hemoglobin migrates at the β zone, causing
a relative increase of the β fraction. In addition, even in slight hemolysis, the hemoglobin-
haptoglobin complex formed migrates between the α_2 and β fractions, resulting a poor
separation of both the α_2 and β fractions. The hemoglobin can be easily identified on
the electrophoretic strips, because of its characteristic red color.

Differentiation from the β Type M-Proteins

When the M-protein bands are separated as a very narrow band, they can be easily
differentiated from the hyperlipoproteinemic pattern. However, they frequently are
difficult to be differentiated, and the immunoelectrophoretic analysis has to be done.

REPRESENTATIVE DISEASES ASSOCIATED
WITH HYPERLIPOPROTEINEMIA

Primary Hyperlipoproteinemia

These disorders are sometimes called as essential or hereditary hyperlipoproteinemia,
but they are not necessarily of the hereditary type. Therefore, the author prefers to call
them as primary hyperlipoproteinemia.

Table 82 Classification and diagnostic criteria for familial hyperlipoproteinemias (from FREDRICKSON et al.).

A. Changes in serum lipoproteins and lipids.

Types	Electrophoretic findings	Ultracentrifugal findings*				Serum lipids**		
		Chyl	VLDL	LDL	HDL	Chol	PL	TG
I	Chyl. increased	↑	—	—	—	↑	↑	↑↑
II	β increased	—	—,↑	↑	—	↑↑	↑↑	—,↑
III	broad β	—	↑	↑	↓	↑↑	↑↑	↑
IV	pre-β increased	—	↑	—	↓	↑	↑	↑↑
V	Chyl. pre-β increased	↑	↑	↑	↓	↑	↑↑	↑↑

B. Clinical findings.

Types	PHLA***	Glucose tolerance	Diet induction		Frequency	Xanthoma	Serum appearance
			Fat	Sugar			
I	↓	—	+	—	rare	+	milky
II	—	—	—	—	often	+	clear
III	—	abnormal	—	+	often	+	turbid
IV	—	abnormal	—	+	often	—	turbid
V	—	abnormal	+	+	rare	+	milky

↑: increased, ↑↑: markedly increased, ↓: decreased, +: present or induced.
—: normal, not present, or not induced.
  * Chyl.: chylomicron, VLDL: very low density lipoprotein, LDL: low density lipoprotein,
    HDL: high density lipoprotein.
 ** Chol: total cholesterol, PL: phospholipids, TG: triglycerides
*** PHLA: post-heparin lipolytic activity (post-heparin lipoprotein lipase)

Principally, working with an improved filter-paper electrophoretic technique,[99] familial hyperlipoproteinemia has been classified into five different types[99]. Their characteristics are summarized in Table 82.

Secondary Hyperlipoproteinemia

The pathogenesis of atherosclerosis is not certain, but it is included here for convenience sake.

1. A marked increase of serum triglyceride, usually associated with hypercholesterolemia in varying degrees.

Taking FREDRICKSON's classification, the lipoprotein pattern of Type IV is seen most frequently.   In pancreatitis, the patterns of Type I and Type V may be encountered occasionally.

   1)   Dietary or postprandial hyperlipoproteinemic state
   2)   Diabetes mellitus
   3)   Pancreatitis
   4)   Alchoholic intake and intoxication
   5)   VON GEIRKE's disease
   6)   Idiopathic hypercalcemia
   7)   M-proteinemia (increased pathological immunoglobulin, especially IgA type)
   8)   Pregnancy and progesteone administration
   9)   Gout
   10) NIEMANN–PICK's disease, GAUCHER's disease

2. A marked increase of serum cholesterol, not associated with significnt hypergly-ceridemia. Again, taking FREDRICKSON's classification, the lipoprotein pattern of Type II is seen most frequently. The pattern of Type IV may be encountered occasionally.

    1) Hypothyroidism (myxedema)

    2) Obstructive jaundice, biliary liver cirrhosis

3. Mixed type hyperlipidemia, accompanying a marked increase in both triglyceride and cholesterol. In FREDRICKSON's classification, they show the lipoprotein patterns of Types II, III, and IV. In arteriosclerosis, they are encountered in the order III, II, IV.

    1) Nephrotic syndrome

    2) Atherosclerosis

Chapter 27

Plasma Protein Changes in Pregnant and Fetal Periods

PLASMA PROTEIN CHANGES IN PREGNANCY

Pregnancy itself is a physiological phenomenon. From the metabolic standpoint, however, not only are there tremendously unique changes to be seen, but very often various pathological conditions are complicated with pregnancy. Therefore, it will be discussed separately in this chapter.

Serum Protein Electrophoretic Pattern in Normal Pregnancy

The filter-paper electrophoretic fractionation of serum proteins is shown in Fig. 268. Among the most significant changes are the decrease in the albumin fraction and the increase in the β fraction. These changes vary in different stages of pregnancy, and our results are shown in Fig. 269.

The decrease of the albumin fraction becomes marked during pregnancy, but it once again has tendency to increase slightly at the last month. When an attempt is made to examine the α_1 and α_2 fractions separately, it is difficult to see any constant change, but looking at both fractions together, then there is a slight tendency to increase. The β fraction increases markedly, particularly so during the last trimester of pregnancy, reaching a level of roughly 150% of normal. Besides this, the increase of the β fraction parallel to the serum total cholesterol level, while the β-lipoproteins are presumed to increase. The γ fraction shows a slight tendency to increase.

As far as the total serum protein concentration is concerned, it decreases significantly, reflecting the decrease of the serum albumin. It becomes especially low between the 4th and 5th months, and they reach their lowest ebb during the 7th and 8th months. Subsquent to that period, because of a marked increase in both the α and β fractions, the total protein concentration during the last trimester gradually shows a tendency to rise.

Plasma Protein Changes in Normal Pregnancy and Their Physiological Background

As can be seen in Fig. 269, the serum albumin concentration begins to decrease from the first trimester of pregnancy, and this tendency becomes very evident during the third trimester. However, in the last month of pregnancy, it once again shows an increase, and returns to the level seen in the 7th month. The alteration of serum albumin is closely related to the increased circulating plasma volume. In normal pregnancy, from the first trimester, the circulating plasma volume gradually increases, and by the 9th month, it reaches about 130% of that to be seen in non-pregnant women.[77,159] However, in the last month of pregnancy, the circulating plasma volume tends to become less. Further, as regards albumin metabolism, the intravascular albumin shows little difference from non-pregnant women, and its synthetic and catabolic rates are the same. How-

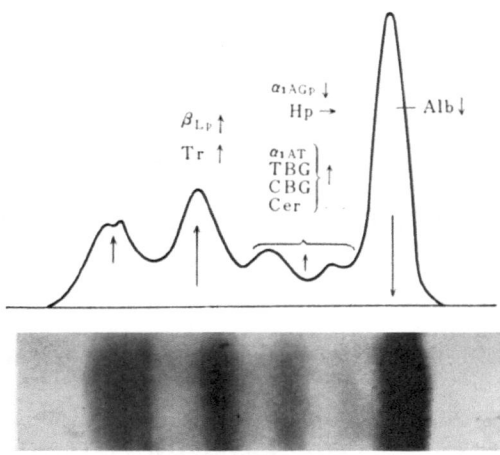

Fig. 268 Serum protein electrophoretic pattern in normal pregnancy (9 m.o.).
The serum total protein was 6.8 g/100 ml, alb, 2.5 g/100 ml; α_1, 0.5 g/100 ml; α_2, 0.7 g/100 ml; β, 1.6 g/100 ml; γ, 1.6 g/100 ml.   On filter paper electrophoresis, noted were a decrease in the albumin fraction, a moderate increase in the β fraction and slight increase in the γ fraction.

ever, if we look at the bodily distribution of albumin, 53% of the total exchangeable albumin is distributed intravascularly, while 47% in non-pregnant women.   In contrast, the extravascular albumin is decreased.[211]

    2.   Estrogen-Sensitive Proteins

    From the very first stage of a normal pregnancy, both estrogen and pregnanediol secretions gradually increase, and it maintains to be high during the second and third trimesters.   However, prior to delivery, the estrogen level suddenly comes down to the normal.[223]   Along with the increased estrogen secretion, distinct alterations in the serum proteins can be recognized.

    Estrogen causes the increased serum levels of ceruloplasmin, thyroxine-binding globulin (TBG) and transcortin.

    During pregnancy, ceruloplasmin increases twice or three times of the normal level, and it is thought to be due to the increased synthesis in the liver, apparently stimulated by estrogen.[213,236]   At the same time, there is recognized an increase of both free and protein-bound copper in the serum (Table 83).

    An increase both in thyroxine-binding globulin (TBG) and protein-bound iodine (PBI) has been also pointed out.[76, 231]   Despite the increase shown in both the basal metabolic rate and PBI, the ^{131}I-thyroid uptake is normal and there are no toxic symptoms suggesting hyperthyroidism.   The Triosorb test shows a rather low uptake (Fig. 209).   These results

Table 83 Serum concentrations of copper and ceruloplasmin in normal pregnancy (from MARKOWITZ).

	Cases	Serum copper (mg/L)			Serum Celuloplasmin (mg/100 ml)
		Total	Free Cu	Bound Cu	
Non-pregnant	10	1.08	0.05	1.03	34
Pregnant	10	2.57	0.29	2.28	84

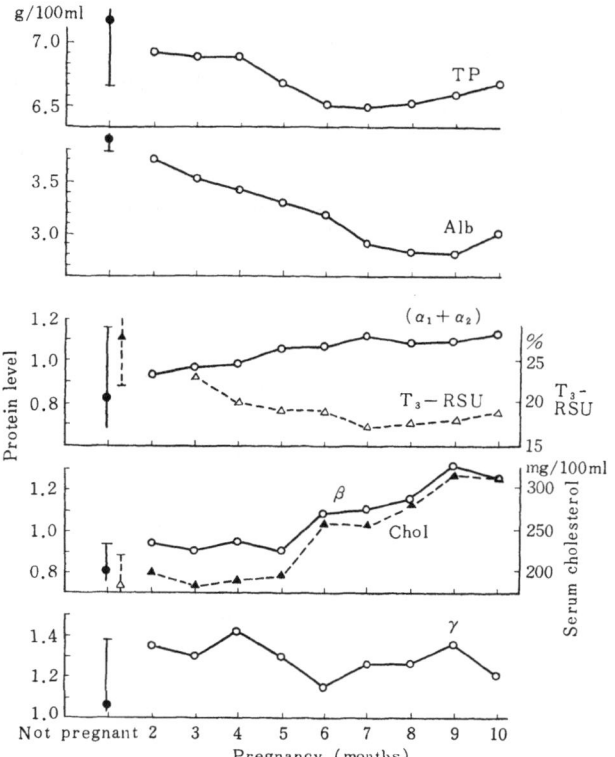

Fig. 269 Serum total protein and serum protein fractions in pregnancy.

The average values obtained from normal pregnant women were plotted. The fractional values were measured with the filter paper electrophoresis. Triosorb test (T3-RSU) and serum total cholesterol values were also described.

can be explained by the increase of TBG in serum. By the same relationship, transcortin is found to increase as well.[67,308,323]

On the other hand, both α_1-acid glycoprotein and haptoglobin are thought to be decreased on hypersecretion and administration of estrogen.[211,214,273] During pregnancy, α_1-acid glycoprotein decreases to 72% (30–125%) of that found in non-pregnant woman.[211]

The serum haptoglobin level is thought to be influenced by the increase of the circulating plasma volume, as well as estrogen secretion. During normal pregnancy, it shows no significant alteration.

3. Acute Phase Reactants

α_1-antitrypsin increases to twice of the normal level, and α_2-macroglobulin increases also.[211] Furthermore, in the third trimester of pregnancy, C-reactive protein appears to be detected, and is said to increase during labor.[343] Again, as stated previously, the serum haptoglobin level remains almost within the normal limits although the circulating plasma volume is significantly increased. Therefore, haptoglobin, too, must keep the increased synthesis during pregnancy. Although α_1-acid glycoprotein shows a significant decrease, the acute phase reactants are generally increased. Whether the increase of these acute phase reactants is influenced by estrogen or other hormones remains to be studied further.

Plasma fibrinogen plays an important role in blood coagulation, and in various inflammations, it markedly increases as the α_2-globulins. The author has observed this same tendency, and thinks that fibrinogen, much like α-glycoproteins, is one of the acute phase reactants. Even during pregnancy, in the same fashion as α_1-antitrypsin and α_2-macroglobulin, it gradually increases, and by the 9th month its serum level reaches 1.5–2 times the level of the normal adult. At the last month of pregnancy, there is a rather abrupt decrease almost to the normal level, but the reason for this has not yet been uncovered (Fig. 270).[189]

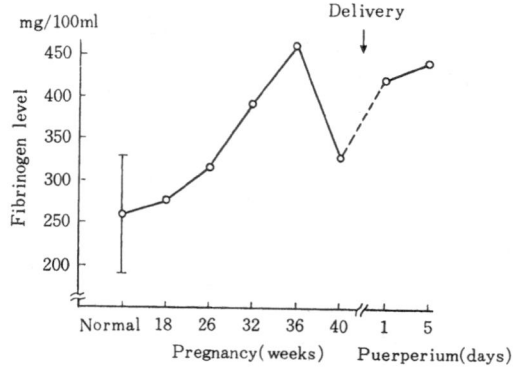

Fig. 270 Plasma fibrinogen levels in pregenancy and puerperium (from KINCH).

4. Transferrin

During pregnancy, both the red blood cell count and hematocrit value are lowered. From the 8th month on, they rise again gradually, but never quite reach the normal. This is generally termed "physiological anemia of pregnancy." Even though there are still many doubtful points about the causes and diagnostic criteria, it is thought generally to be at least due to a relative lack of iron. Therefore, the serum iron shows a gradual decrease and the total iron binding capacity of serum (TIBC) increases. As a result, serum iron saturation becomes gradually decreased.[302] Also, an increase in the serum transferrin level has been observed from the very early stage of pregnancy. During the third trimester, it increases markedly, reaching about 1.5–2 times that of the normal adult level.[192,210] The mechanism of causing the increased transferrin level is uncertain. However, considering the facts that its metabolic pool is increased and its half-life is prolonged, the increased transferrin might represent a defence mechanism to preserve the serum iron binding capacity.

5. Lipoproteins

Serum lipids have been known to increase during pregnancy.[315,397] The serum total lipids increase 1.5 times that of the normal. The total cholesterol and phospholipid are increased. Because of a marked increase in the phospholipid, Chol./PL ratio is lowered (Fig. 269, 271). The serum lipoproteins are also increased during pregnancy, and in the third trimester β-lipoproteins are particularly increased.[315,319,397] As shown in Fig. 271, in the first trimester, the α-lipoprotein fraction and β-cholesterol concentration tend to increase, while the β-lipoprotein fraction and β-cholesterol concentration show a decrease.

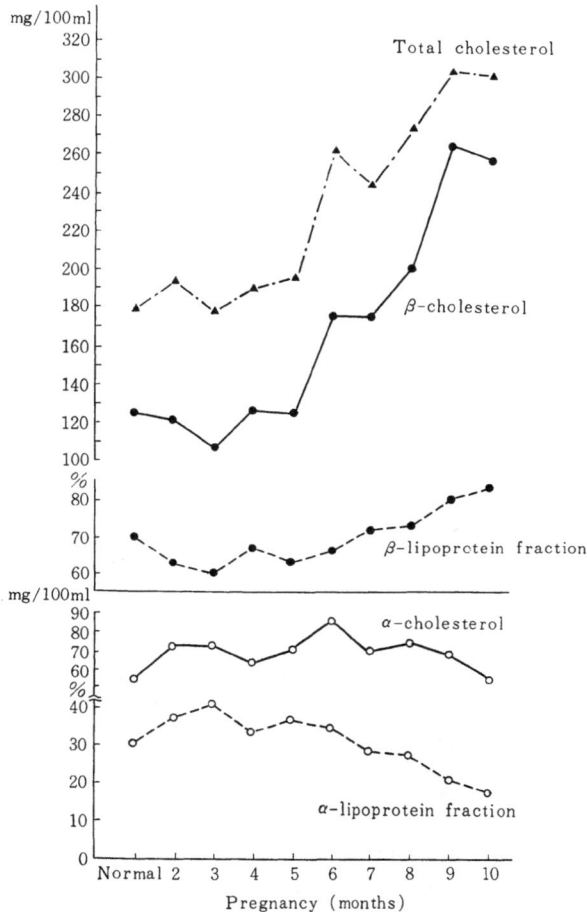

Fig. 270 Serum total cholesterol and lipoprotein fractions in pregnancy.
Lipoprotein fractions were measured by the filter paper electrophoretic method of SWAHN. Fractional cholesterol values were calculated roughly from the total cholesterol level and the lipoprotein fraction level.

However, from the second trimester on, the β-lipoproteins and the β-cholesterol begin to increase and become markedly increased by the last stage of pregnancy. On the other hand, the α-cholesterol concentration remains to be increased in a minor degree, while the relative percentage of the α-lipoprotein fraction decreases because of a marked increase in the β-lipoprotein fraction. At the last month of pregnancy, both α and β-cholesterol concentrations tend to decrease significantly.

The mechanism for these serum lipid changes during pregnancy has not yet been entirely cleared. However, the lipoprotein is one of the so-called "estrogen-sensitive proteins", and is said to be influenced by estrogen or other hormones such as the thyroid or male sex hormones.[221] With the administration of estrogen, the serum cholesterol becomes decreased, α-lipoprotein increased, and β-lipoproteins decreased. Consequently, at least, the changes during the first trimester of pregnancy can be explained by the increase

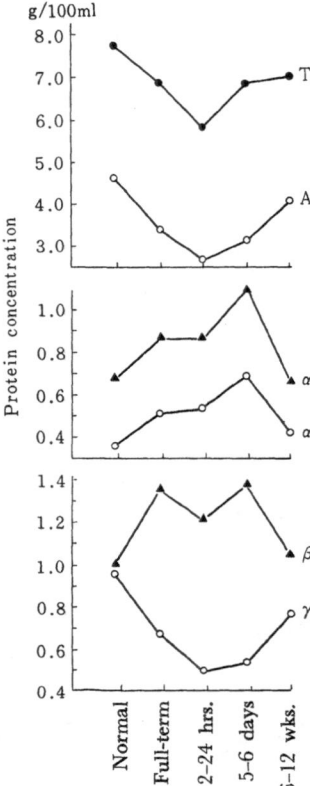

Fig. 272 Serum protein electrophoretic values in puerperium (from Mack).

of estrogen secretion.   During the third trimester, however, it is thought that the estrogen secretion increase helps to maintain the slight increase of α-lipoprotein, but that the marked increase of β-lipoproteins is brought about by still another different mechanisms. Some investigators interpret the β-lipoprotein increase as a possible secondary phenomenon due to increased carbohydrate metabolism during pregnancy.

Plasma Protein Changes in Normal Puerperium

MACK, BOIRON and others have made rather detailed reports on the changes in the serum proteins during normal puerperium.[277]   Within several hours after delivery, the total serum protein as well as all protein fractions begin to decrease, and this can be attributed to bleeding during delivery.   However, the serum levels of the α fractions are relatively well preserved.   Subsequent to this, the circulating plasma volume gradually increases, and returns to practically the normal level in about a month.   Along with this, the serum concentrations of protein fractions also begin to rise.   Within two or three months, the α-globulins, lipoproteins and transferrin likewise return to normal.   However, the albumin and γ fractions still fail to show a complete return to the normal.   Fibrinogen, as during pregnancy, shows the same tendency as the α-globulins (Fig. 272).

Plasma Protein Changes in the Complications Associated with Pregnancy

A normal pregnancy is in itself a physiological phenomenon, and as mentioned before, is thought to be a state of "physiological instability" accompanying tremendous changes

Table 84   Classification of toxemia of pregnancy.

The first group   Diseases not specific for pregnancy
    1.   Hypertension
        1)   Benign (Essential hypertension)
        2)   Malignant
      2.   Renal diseases
        1)   Nephrosclerosis
        2)   Acute and chronic nephritis
        3)   Acute and chronic nephrosis
        4)   Other serious renal diseases (cf. pyelonephritis)
The second group   Renal diseases specific for pregnancy
    1.   Preeclampsia
        1)   Mild
        2)   Severe
    2.   **Eclampsia**
        1)   with convulsion
        2)   without convulsion
The third group   Hyperemesis gravidarum
The fourth group   Unclassified

Table 85   Renal lesions associated with preeclampsia and eclampsia (from ALLEN).

  I.   Renal lesions of preeclampsia and eclampsia antedating pregnancy
    1.   **Chronic glomerulonephritis** (most common)
    2.   Chronic pyelonephritis
    3.   Benign and/or malignant nephrosclerosis
    4.   Various anomalies of the kidney with reduction of renal reserve
  II.   Renal lesions of preeclampsia and eclampsia initiated during pregnancy
    1.   Bilateral cortical necrosis
    2.   **Acute diffuse membranous glomerulonephritis** (common)
    3.   Acute proliferative, exudative or necrotizing glomerulonephritis
  III.   Renal sequelae of preeclampsia and eclampsia
    1.   Chronic glomerulonephritis of various forms
    2.   Benign and/or malignant nephrosclerosis

within the maternal body.   During pregnancy, a great influence is exerted on any existing disorders present in the maternal body, and many complications are likely to occur. From the stand point of the serum protein abnormalities, it would be convenient to divide into three major groups; namely, (1) toxemia of pregnancy, (2) acquired hypofibrinogenemia, and (3) other complications.

1.   Toxemia of Pregnancy

The etiology for toxemia of pregnancy is still unknown.   It is called "a disease of theories" and there are numerous ways of classifying it.   However, the classification shown in Table 84 may be convenient for understanding their plasma protein abnormalities.

In the first group, the serum protein changes are quite similar to those seen in various kidney diseases which are described elsewhere.

In pre-eclampsia and eclampsia, the main pathology exists in the kidneys, and ALLEN recognized various renal lesions, as shown in Table 85.[5]   Therefore, in preeclampsia and eclampsia, the plasma protein abnormalities seen are quite similar to those seen in chronic glomerulonephritis.   However, the decrease of the albumin fraction and the increase of the β fraction are more marked in this case.[319,333]   It is said that hypoproteinemia is encountered in approximately 10% of pregnant nephrites but in one third of the patients with toxemia of pregnancy.[319]

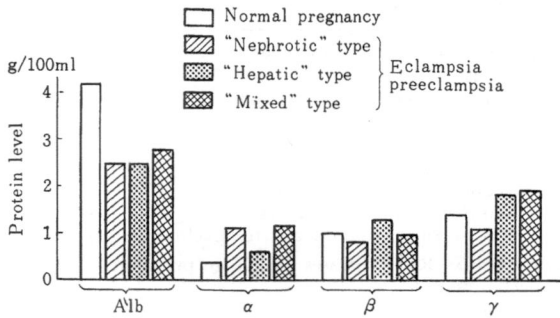

Fig. 273 Various types of the serum protein abnormalities in preeclampsia and eclampsia (from FRIEDBERG).

Table 86 A/G ratio of urinary and serum proteins seen in pregnant women (from LORINCZ et al.).

Types	A/G ratio	
	Urinary protein	Serum protein
Physiological proteinuria	3.55 (2.67–4.56)	0.96 (0.64–1.32)
Chronic hypertension	1.53 (1.31–1.87)	0.62 (0.46–0.87)
Chronic glomerulonephritis	1.08 (0.72–2.15)	0.82 (0.32–1.51)
Preeclampsia	1.10 (0.48–2.81)	0.57 (0.46–0.76)
Eclampsia	0.57 (0.21–1.24)	0.73 (0.50–1.03)

FRIEDBERG classified the serum protein changes in preeclampsia and eclampsia into the following three types: 1) "nephrotic" type, 2) "hepatic" type, and 3) "mixed" type (Fig. 273).[102]   In each of these types, there is a marked decrease of the albumin fraction. In the "nephrotic" type, the α fraction mainly increases.   In the "hepatic" type, there is a marked increase mainly of the γ and β fractions.   Varying degrees of liver damage are known to occur during the third trimester of pregnancy,[102] but remarkable increase of the γ fraction seen in the "hepatic" type may be brought to come from the difference of renal rather than the hepatic changes.

Recent investigations suggest that auto-immunity is involved in toxemia of pregnancy.[319]   Therefore, it is possible that in the "hepatic" and "mixed" types which are accompanied by the increase of the γ fraction, there might exist some contribution of auto-immunity.   Also, according to the report of LORINCZ et al,[226] there is an increase of the serum α_2 fraction in preeclamptic patients, a decrease of the γ fraction as well as an increase of the α_2 fraction is seen, being similar to the nephrotic syndrome.[226]   As a rule, urinary protein loss in the patients with toxemia of pregnancy is more abundant than in the nephritic patients with pregnancy, amounting to 3–10 g daily.[226]   In toxemia of pregnancy, the A/G ratio of urinary proteins is characteristically low (Table 86).[102]   In hyperemesis gravidarum, the renal lesions are not associated with, and hypoalbuminemia is recognized in association with slight increase of both α_2 and γ fractions.   Such changes are thought to be due to poor nutritional intake and hepatic injury.

2.   The Complications Other than the Toxemia of Pregnancy.

Depending upon the disorders combined with pregnancy, it is only natural to expect differences in the serum protein changes.   During pregnancy, because of a marked increase in α_1 fraction (α_1-antitrypsin, TBG, transcortin, etc.), should complications such

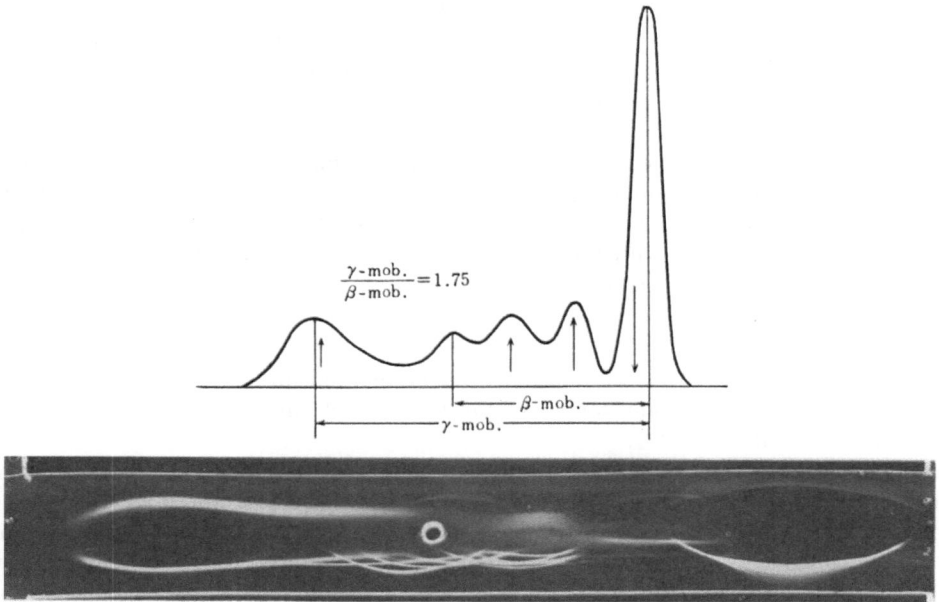

$$\frac{\gamma\text{-mob.}}{\beta\text{-mob.}} = 1.75$$

Fig. 274 Serum protein electrophoretic patterns in complicated pregnancy.
24 y.o. m female, artificial abortion and panperitonitis.
Artificial abortion at the 6th month of pregnancy, followed by hysterectomy because of continuous bleeding. Died of panperitonitis and hypofibrinogenemia, 9 days after the surgery.
On cellulose acetate electrophoresis of the serum proteins, TP was 7.0 g/100 ml, Alb 3.08 g/100 ml (↓), α_1 0.91 g/100 ml (↑↑), α_2 0.84 g/100 ml (↑), β 0.70 g/100 ml (normal), and γ 1.47 g/100 ml (↑). On immunoelectrophoresis, the IgG line is somewhat deformed. The modal mobility of the γ faction was increased, being 1.75.

as acute stress or acute inflammation be present, there is added an increase of α_1-acid glycoprotein, and the α_1 fraction becomes higher than the α_2 fraction (Fig. 274). Hypofibrinogenemia is discussed in detail, elsewhere.

PLASMA PROTEIN CHANGES IN FETAL AND NEONATAL PERIODS

Fetal growth depends on the maternal body, and the serum protein changes are closely related to both fetal and maternal metabolic conditions. Therefore, it will be treated in this chapter together with pregnancy. The neonatal metabolism is already independent from the maternal one, but it is certainly dependent much on the fetal one. Therefore, it will be rather convenient to treat both periods, the fetal and the neonatal, together.

Serum Protein Electrophoretic Pattern of Normal Fetus

Because of difficulty in obtaining serum samples during the fetal period, their results are still insufficient. With practically all researches, results have been obtained from aborted fetuses at various stages, and conclusions presumed from them.[31,32,259,319,332)

The total serum protein concentration increases gradually during fetal growth. At about the 2nd month after conception, it is only 1.55 g/100 ml, but up to the time of delivery, it reaches to about 5.5 g/100 ml.

The electrophoretic patterns of serum proteins are pretty much the same as those of the maternal (Fig. 275).   During the fetal growth the α_1, α_2 and β fractions show only a minimal tendency to increase, and the significant increase of the total serum protein concentration is chiefly dependent on a rapid increase in the albumin and γ fractions.

Characteristically recognized in fetal serum is the fast-moving α_1-globulin, which is to be discussed later.

Since the mobility of this feto-specific protein is faster than the α_1 fraction of the adult, it is identified between the albumin and α_1 fractions as a separate band, if it is present in a high concentration during the early stage of the fetal life.   Ordinarily, the feto-specific protein band cannot be identified with the use of cellulose acetate electrophoresis, in the fetus delivered later than the 6th month of pregnancy.

Plasma Proteins in Normal Fetus

The plasma proteins of the normal fetus can be divided into two groups; (1) the feto-specific proteins, or fetoproteins found only in the fetal plasma, and (2) adult proteins.
1.   Fetal proteins

Among the fetoproteins, one of the most well studied components is the α-fetoprotein. Its existence was first clearly established by BERGSTRAND and CZAR, and it is easily identified immunochemically because they have distinct immunological antigenicity different from the adult proteins.[31,32]   Its relative electrophoretic mobility is between the albumin and α_1-globulin fractions, and it has been called with various synonyms, such as α-fetoprotein, α_1-fetoprotein, fetal α-globulin, α_f-globulin, α_1-fetoglobulin, α_1-fetospecific protein, etc.   The molecular weight of the α-fetoprotein is 64,400, the sedimentation constant 4.5S, the nitrogen content 14.4%.[300]   The carbohydrate content of this protein is relatively small, and the protein-bound hexose is contained only in about 2.3%.[32]

The α-fetoprotein is found in a relatively high concentration during the early stages of fetal life, being the highest around the fifth month.   However, it gradually decreases during fetal growth, and will not be detected at birth (Fig. 275, 276).   Biologically, the

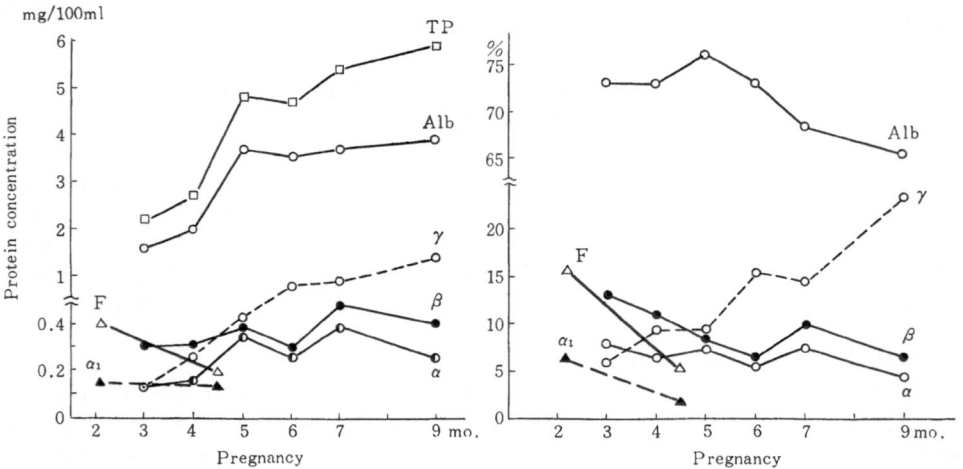

Fig. 275 Serum protein electrophoretic patterns in fetus.
The thin solid lines were made from the data reported by MOORE et al.   (TISELIUS method, pH 7.7, phosphate buffer), and the thick solid lines by BERGSTRAND et al.   (filter paper method).   With the TISELIUS method, the fetal protein (F) was included in the albumin fraction, while it was included in the α_1 fraction with the filter paper method.

α-fetoprotein resembles the 'fetuin' present in fetuses of cows and other animals in many points, but immunologically, there seems to be no common antigenicity.[319,332]

The α-fetoprotein is not found in the serum of the normal adult, but recently it has been found in most of the patients with primary hepatoma and embryonal testicular tumors. Also found is a small amount of it occasionally in the patients with fulminant hepatitis where regeneration of the liver cells is particularly prominent.

2. Adult Proteins

Putting aside the above discussed fetoproteins, approximately 85% of the serum proteins is occupied by the adult proteins during the first stages of fetal life, and almost 100% during the last stage.

Origin: The plasma proteins found in fetal plasma originate from either one of the following two; (1) those synthesized by the fetus itself; and (2) those transferred from the maternal body. Only the IgG-globulin is transfered from the maternal circulation in man. All the others are apparently synthesized by the fetus itself.[319,332] However, there are reports that they are synthesized in the placenta and amnion, but a more detailed study of this is required to prove it.

Times of the first appearance: According to SCHEIDEGGER et al.,[324] immunologically identified, within 8 weeks after conception, at least 5 different adult serum proteins including prealbumin, albumin, transferrin and two α-globulins (probably α_1-antitrypsin and α_2-macroglobulin). As fetal life progresses, the immunoelectrophoretic patterns become complex, but in the full-term baby practically all of the serum protein components can be detected.[332] Nevertheless, their serum concentration varies greatly as listed in Table 87.

Table 87 Approximate concentrations of plasma proteins in neonates delivered at full term (summarized from SCHULTZE and HEREMANS).

A. With the serum concentration similar to that seen in adults.
   1) Albumin
   2) α_1-Antitrypsin (one half of the maternal concentration)
   3) α_2-Macroglobulin (slightly lower than the maternal)
   4) Transferrin (slightly lower than the maternal)
   5) Fibrinogen (slightly lower than the maternal)
   6) IgG-globulin (the same as the maternal)
B. With the serum concentration significantly lower than the adult level.
   1) α_1-Lipoprotein
   2) α_1-Acid glycoprotein (approximately one third of the maternal)
   3) Ceruloplasmin (approximately one third of the maternal)
   4) Low-density lipoproteins
   5) β_{1C}-Globulin and other complement components
C. With the very low serum concentration, not detectable on immunoelectrophoresis.
   1) Haptoglobin (not detected in 90 %)
   2) IgA-globulin
   3) IgM-globulin

An especially detailed investigation has been done on the immunoglobulins. The results obtained in early days were obtained by electrophoretic techniques. Therefore, the γ fraction or the IgG-globulin was studied (Fig. 275). IgG-globulin rapidly increases as the fetal age advances. At the time of delivery, its serum level is almost the same as that of the maternal serum (Fig. 277). As a matter of fact, the fetus itself does not synthesize

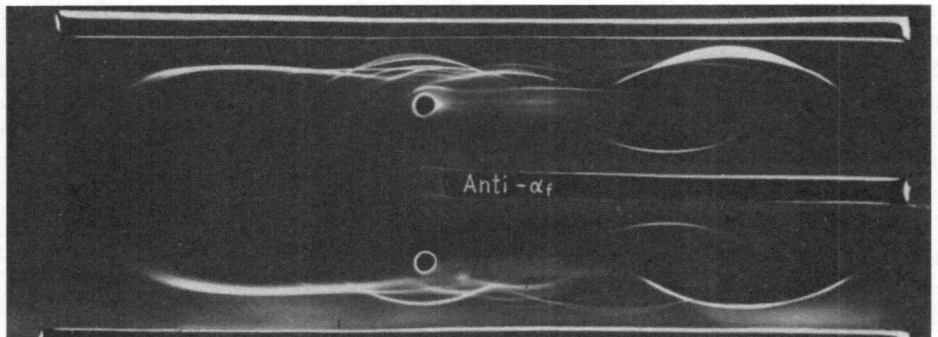

Fig. 276 Immunoelectrophoretic patterns of the α-fetoprotein.
With the sera taken from the patients with primary hepatoma, the symmetric arcs of the α-fetoprotein are seen betweenthe albumin and antitrypsin lines.

IgG, and it is transfered from the maternal body through the placenta. As shown in Fig. 277, IgG and IgM cannot be demonstrated in fetal serum on immunoelectrophoresis, but only a trace amount of them may be detected by using more sensitive analytical methods.[332] Serum IgM is said to amount approximately 20% of the adult level at birth.[418]

Various antibodies are demonstrated in the cord blood. SCHULTZE and HEREMANS listed them as shown in Table 88, concluding as follows[332]: 1) the IgG antibodies having almost the same antibody titers as the maternal blood, 2) the IgA and IgM antibodies not demonstrable or having very low titers, if any, 3) some of the antibodies (against staphylococci, *N. diphtheriae*, streptococci, etc.) occasionally having the serum titer higher than the maternal blood.

The plasma protein components listed as Group A in Table 87 are those which may become physiologically important during the late fetal life or immediately after delivery. For example, albumin is necessary for maintaining colloid osmotic pressure. Transferrin is needed for iron metabolism and gas exchange, fibrinogen for blood coagulation and IgG for immunological resistance. Sufficient activity of all these is called for as soon as the fetus is exposed to the outside world after birth.

On the other hand, the plasma protein components included in Group B and C show the serum concentrations significantly less than that of the maternal, and they are synthesized in a smaller amount than in the adults. However, it is not adequate to assume that their synthetic capacity is insufficient in the fetal and neonatal periods. As will be explained later, even some of the plasma proteins listed as Groups B and C (including lipoproteins, haptoglobin, etc) may increase their serum concentrations tremendously within from several hours to several days after birth. Therefore, it may be safe to suppose that those protein components are not so much needed physiologically during fetal life as in the adults.

Molecular immaturity: As indicated above, the adult serum protein components found during the fetal period, both biologically and immunologically, seem to possess the same characteristics as those found in adult serum. However, there have been reports pointing out molecular immaturity.

According to FINE et al.,[89] the mobility of fetal haptoglobin is rapid, and it forms an atypical precipitin line with the anti-haptoglobin rabbit immune serum. Furthermore, fetal haptoglobin is said not to bind hemoglobin, and to have molecular immaturity. Also, the neonatal haptoglobin is said to be catabolized faster.[178]

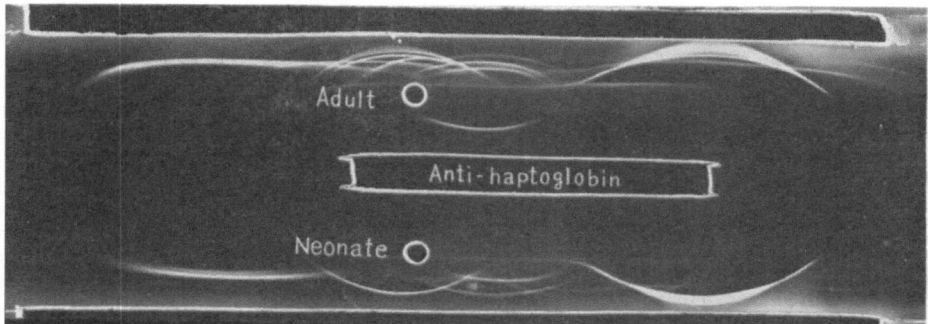

Fig. 277 Immunoelectrophoretic patterns of neonatal serum proteins.
The serum IgG level is almost the same as that of normal adult. IgA and IgM are not seen.
Haptoglobin and hemopexin are also not demonstrated.

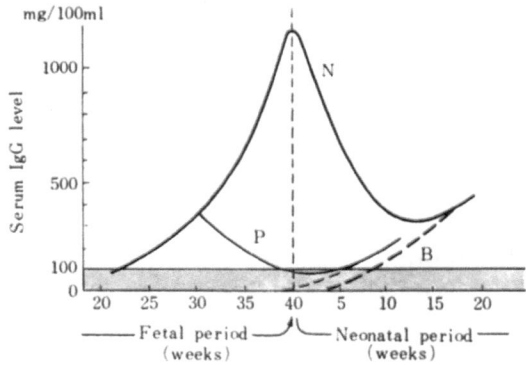

Fig. 278 Serum IgG levels during the fetal and neonatal periods
(from Hobbs and Davis).
Curve N indicates the maternal IgG level in the serum of the
babies. Curve B indicates the serum level of the IgG which is
synthesized by the neonates. The shaded area indicates the so-
called dangerous low level (less than 100 mg/100 ml) reported by
British Medical Research Council. Curve P indicates the serum
IgG levels in premature babies born at the end of the 30th week,
and it is apparent that the IgG level becomes very low around
10–15 weeks after birth.

From about 1959, Miyoshi and his co-workers have emphasized the presence of fetal
type human albumin (Alb. F),[256] and the following characteristics are found. First, the
fetal type human albumin is difficult to be digested by either bacterial protease or trypsin; second, it is difficult to be denatured with 1N NaOH; third, it is difficult to be adsorbed hydroxylappatite and DEAE cellulose; fourth, it has the isomerization equilibrium
different from that of adult albumin; fifth, no differences are to be found in their antigenicity; sixth, after birth, the fetal albumin (Alb. F) changes to adult albumin (Alb. A);
finally, it is found in a high concentration in the patients suffering from familial fetal
hemoglobinemia (familial persistence of fetal hemoglobin) or thalassemia.

Table 88 Antibodies found in cord blood (summarized from SCHULTZE and HEREMANS).

Antibodies	Comparative levels in mother and child		
	Approximately equal	Lower levels in cord blood	Absent in new born
A. Antibodies against human constituents			
1) Anti-erythrocyte antibodies			
Complete anti-A, anti-B		O	
Incomplete anti-A, anti-B			
Anti-Rh, M, S, Fyᵃ	O		
Anti-Kell, Leᵃ, Leᵇ			
Incomplete autoantibodies			
2) Anti-leukocyte antibodies	O		
3) Anti-platelet antibodies		O	
4) Antinuclear antibodies		O	
5) Rheumatoid factor (IgM)			O
B. Antibodies against infective agents			
1) Neutralizing antibodies	O		
(diphtheria, scarlet fever, tetanus, streptolysin, staphylolysin, hyaluronidase)			
2) Antibacterial agglutinins (IgG, IgA, IgM)	O		
(*H. pertussis, H. influenzae,*	O		
Salm. typhi (O)		O	
E. coli, Shig. dysenteriae, Salm. typhi (H))		O (?)	O
3) Antiviral antibodies (IgG)	O		
(measles, herpes simplex, poliomyelitis, Japanese B encephalitis, vaccinia, coxsackie, influenza, mumps, APC virus, rickettsia)			
4) Antibodies against larger organisms			
Syphilis (IgG, IgA, IgM)		O	
Toxoplasmosis	O		
Fungal diseases	O		
5) Miscellaneous antibodies			
Anti-cow milk (IgG)	O		
Anti-sheep erythrocyte agglutinins (IgG, IgA, IgM)		O	
C. Antibodies against allergens			
Skin-sensitizing reagins			O
Blocking antibodies	O		

Plasma Proteins in Normal Newborn Infants

As shown in Table 87, the full-term baby usually carries all the adult plasma proteins at birth, and during the neonatal period, some of the plasma proteins show tremendous alterations,[319,332] as summarized in Table 89.

The most notable change is recognized in the serum γ fraction. In the cord blood, the serum γ fraction amounts to 1.1–1.2 g/100 ml, but it drops rapidly down to about 0.4 g/100 ml within several weeks after birth (Fig. 276). The reason for this is a rapid destruction of IgG molecules of the maternal origin.[150,319,332,418]

The albumin, α and β fractions show an increase. Both the high-density and low-density lipoproteins show a relatively low level at birth, but within 3 or 4 days after birth, show a transient rapid increase. The reason for this seems to be that during the fetal period carbohydrates and amino acids are the main energy sources, while after birth the lipids in the maternal milk become the important source.[332] From a few hours after

Table 89 Changes of the plasma proteins during neonatal period.

I.   Not significantly changed
  1) Prealbumin
  2) α_1-Antitrypsin
  3) α_1-Acid glycoprotein (gradually increasing during childhood)
  4) α_2-Macroglobulin (significantly increased in infancy)
  5) Fibrinogen

II. Rapidly decreased
  1) IgG-globulin
  2) Transferrin

III. Gradually increased
  1) Ceruloplasmin
  2) β_{1c}-Globulin and other complement components
  3) IgA-globulin

IV. Rapidly increased
  1) Albumin
  2) α- and β-Lipoproteins
  3) Haptoglobin
  4) IgM-globulin

birth and within the first 2 or 3 days, serum haptoglobin also shows a rapid increase. This seems to prepare physiologically for post-natal hemolysis.

No significant alteration can be recognized with prealbumin, α_1-acid glycoprotein, α_2-macroglobulin and fibrinogen. Ceruloplasmin and complement components tend to increase gradually.

Synthesis of the immunoglobulins during neonatal period increases slowly. IgM, however, increases rapidly for the first 10 days after birth, suggesting the primary immune response against the new environment.

Plasma Protein Abnormalities in Newborn Infants

Concerning this situation, there have been no detalied reports so far. It is impossible therefore to give an outline on the subject. However, they could be divided into the following three groups; (1) those found in premature and post-mature babies; (2) those resulted from the maternal pathologies; and (3) those resulted from the fetal pathologies. In these instances, the serum immunoglobulins have been studied most extensively.

In mature babies, the plasma proteins may show the changes characteristic for their fetal age, as discussed before. The British Medical Research Council holds that the serum IgG concentration of less than 100 mg/100 ml during the neonatal period is quite dangerous. As shown in Fig. 278, Hobbs and Davis[154] consider clinically to be critical when the serum IgG drops below 100 mg/100 ml within 6 months after birth, particularly for the premature babies born before the 32nd week. Even when delivery is on the predicted date, the "small-for-dates" babies and "post-mature babies" are likely to have low IgG-level, when compared with mature babies, and it may cause fatal pneumonia after birth.[417] In multiple pregnancy with uneventful course, there are no significant differences in the plasma proteins.[244,417]

When a pregnant woman suffers from diabetes mellitus, it is said that the serum protein-bound carbohydrates of the fetus is increased, and this is attributed to the rapid growth of the fetus.[349] When the serum γ-globulin in the diabetic mother is decreased, it is said that the baby is likely to develop idiopathic respiratory distress syndrome (I.R.D.S.).[135] However, the total serum protein and albumin concentrations in the mother's blood show no significant relationship to it.

In situations where the fetus itself contracts a variety of diseases, it is natural to result in plasma protein abnormalities.   The problem that occurs most frequently is intra-uterine infections, such disorders as rubella,[355] syphilis,[348] toxoplasmosis,[81] cytomegalic inclusion disease,[242] and the like.   Ordinarily, even with these infectious diseases, the fetal serum IgG level is normal, but IgA is detected clearly on immunoelectrophoresis and IgM is definitely increased.   Considering this fact, even during the fetal period, the immuno-globulins are proved to be synthesized, if necessary.

MATERNO-FETAL TRANSFER OF THE PLASMA PROTEINS

Both during the fetal and neonatal periods, as indicated before, the synthesis of many of the plasma protein components is definitely less than that of the adult, particularly, the immunoglobulins.   During the fetal period, the fetus is protected from the external en-vironment by the maternal body.   However, at the time of delivery or immediately after delivery, it will be suddenly exposed to the new environment.   Therefore, not only during pregnancy, but during the neonatal period as well, certain protective mechanisms are transfered from the mother to the fetus.

Materno-fetal transfer of the plasma proteins may occur through the following three ways; amnion, placenta, and yalk sac.   During the neonatal period, it may occur through breast feeding.   In man, the yalk sac atrophies during pregnancy, and this problem will not be handled here.

Materno-Fetal Transfer through the Amnion

The amniotic fluid stored in the amniotic cavity serves an important role for fetal life. The fetal sucks the fluid, and urinates in it.   Additionally, both water and electrolytes are thought to be transfered through the skin and lungs of the fetus.

There has been little study done on the protein compositions of the amniotic fluid. The total protein concentration is at an average of 0.3 g/100 ml, and represents about 1/20th of the plasma level.   With the use of electrophoretic techniques, the fractional patterns resemble fairly closely those of the maternal serum, but both the γ and a_2 fractions are relatively low.   The high-molecular weight components such as low-density lipopro-teins, fibrinogen, a_2-macroglobulin, haptoglobin, IgA and IgM are not found in the am-niotic fluid or a very little, if any.[332]   A few proteins specific for the amniotic fluid are also identified.[206,317]

These specific proteins are probably derived from the amniotic epithelium, but the greater part of the serum proteins are thought to come from the maternal circulation.[206]

Though the proteins in the amniotic fluid are not likely to pass through the skin and lungs of the fetus, but they are swallowed into the gastrointestinal tract.   However, they are not likely to be transfered into the fetal circulation.[332]

Materno-Fetal Transfer through the Placenta

In man, at least, the materno-fetal transfer of the plasma proteins is not likely to occur through the yalk sac and the amnion.   However, the placenta plays the important role. According to the ways of the materno-fetal protein transfer, the mammals can be divided into three groups.[332]

1)   Group I (non-transmitters): No protein transfer is recognized, and this group includes cow, goat, sheep, and horse.

2)   Group II (IgG-transmitters): Only IgG-immunoglobulin is transfered, and this includes man.

3) Group III (full transmitters): Practially all the plasma proteins are transfered, and this group includes rabbit, rat, mouse, guinea pig, dog and cat.

The variations in the placental transfer among different animals depend partly on the anatomical differences of the placenta itself. As shown in Fig. 278, basically, the vascular system of the mother is separated from the fetal vascular system with the four layers; fetal capillary endothelium, connective tissue stroma (the basement membrane and the interstitium of chorionic villi), trophoblast, and endometrium. The endometrial layer varies among different animals. In man, the endometrium attached to the placenta disappears, and the chorionic villi come into direct contact with the maternal blood stream. The animals in Group I, where the endometrium remains, do not show the materno-fetal transfer of the plama proteins. On the other hand, among the animals in Groups II and III, where the endometrium disappears during pregnancy, the protein transfer becomes possible. The exact mechanism for the selective transfer of certain plasma protein components remains uncertain. However, either biological absorption through pinocytosis or an active transport involving the lysosome system in the chorionic villi seems to participate.[332] It is thought that once syncytial cells are passed through, both the intersitium and fetal capillary endothelium are easily passed.

In man, it is known that only IgG-immunoglobulin is transferred from the mother to the fetus. Immediately after birth, the serum IgG-immunoglobulin in the newborn infant parallels very closely that of the mother's. Also, the hereditary type of Gm factors is likewise in perfect accord with that of the mother's. Finally, in the neonatal period, there is a rapid decrease of the serum immunoglobulin. From recent experiments using ^{131}I-labeled proteins, only IgG-immunoglobulin was found to be transmitted easily through the placenta. But the components such as albumin, a_1-acid-glycoprotein, transferrin, fibrinogen and IgM-immunoglobulin were hardly ever transmitted through the placenta.[114] Both Gc-globulin and haptoglobin also are thought not to be transferred through the placenta.[149]

Materno-Fetal Transfer through Colostrum and the Maternal Milk

As can be seen in Table 90, among those animals in which plasma proteins are not transfered in any way from the mother to the fetus through the placenta, the important antibodies are transmitted through colostrum.[418]

Table 90 Materno-fetal transfer of antibodies in mammals (summarized from YOSHIOKA).

Animals	Materno-fetal transfer through	
	Placenta	Colostrum
Horse, pig, goat, sheep, cow	−	+++ (36 hours)
Cat, dog, weasel	+	++ (10 days)
Rat, mouse	+	++ (16–20 days)
Rabbit, guinea pig	+++	−
Man, monkey	+++	−

−: not transfered +: transfered ++: greatly significant +++: transfered solely

In man, also the existence of antibodies in colostrum cannot be denied.[418] The colostrum contains both IgA-immunoglobulin and IgM-immunoglobulin in significant amounts. However, these immunoglobulins are not likely to transfer directly into fetal circulation,[332,418] but the secretory IgA must play an important role for local immunity.

REFERENCES

1) ABE, M. et al.: Saishin-Igaku (in Japanese) **19**, 836, 1964.
2) AKAI, S. and YOSHIDA, K.: Saishin-Igaku (In Japanese), **23**, 1709, 1968.
3) ALDRICH, R. A., STEINBERG, A. G. and CAMPBELL, D. C.: Pediatrics, **13**, 133, 1954.
4) ALEXANDER, P. and FAIRLEY, G. H.: in "Clinical Aspects of Immunology", ed. by P. G. H. GELL, and R. R. A. COOMBS, Blackwell Sci. Pub. 1968.
5) ALLEN, A. O.: The Kidney, Medical and Surgical Diseases, 2nd Ed, Grune & Stratton, N. Y., 1962.
6) ALLISON, A. C., BLUMBERG, B. S. and REES, A. P.: Nature, **181**, 824, 1958.
7) ALPER, C. A.: Acta Med. Scand. 179 (Suppl. 445): 200, 1966.
8) ANDERSON, A. B. and FERRIMAN, D.: Brit. Med. J., **1**, 277, 1960.
9) ANDERSON, E. T. and VYE, M. V.: Ann. Int. Med., **66**, 141, 1967.
10) ANDERSON, J. R., BUCHANAN, W. W. and GOUDIE, R. B.: Autoimmunity, C. C. Thomas Pub. U.S.A., 1967.
11) ANDERSON, S. B., JARNUM, S., JENSEN, H. and ROSSING, N.: Scand. J. Clin. Lab. Invest., **21**, 42, 1968.
12) ANDERSON, W. A. D.: Pathology, Mosby Co., 1964
13) ANZAI, T., SATO, K., FUKUDA, M. and CARPENTER, C. M.: Proc. Soc. Exp. Biol. Med., **120**, 94, 1965.
14) ARTURSON, G.: Acta Chir. Scand., **120**, 309, 1961.
15) AXELSSON, U. and HÄLLEN, J.: Lancet, **2**, 369, 1965.
16) BACHMANN, R.: Scand. J. Clin. Lab. Invest., **17**, 316, 1965.
17) BALLARD et al.: μ-chain.
18) BARANDUN, S., BUHLER, H. and HASIG, A.: Schweiz. Med. Wschr., **86**, 33, 1956.
19) BARANDUN, S., STAMPFLI, K., SPENGLER, G. and RIVA G.: Helv. Med. Acta, **26**, 163, 1959.
20) BARDEN, J., MULLINAX, F. and WALLER, M.: Arch. Rheum., **10**, 228, 1967.
21) BARNETT, E. V., et al.: Ann Int. Med. **73**, 95, 1970.
22) BARR, D. P., RUSS, E. M. and EDER, H. A.: Am. J. Med., **11**, 480, 1951.
23) BASSEN, F. A. and KORNZWEIG, A. L.: Blood, **5**, 381, 1950.
24) BECK, E. A.: "Fibrinogen" ed. by K. LAKI, Marcel Dekker, Inc., New York, 1968.
25) BECK, E. A., CHARACHE, P. and JACKSON, D. P.: Nature **208**, 143, 1965.
26) BECK, G. E. and DORTA, T.: Helv. Med. Acta, **26**, 764, 1959.
27) BECROFT, D. M. O. and DOUGLAS, R.: Arch. Dis. Childh. **43**, 444, 1968.
28) BEIERWALTES, W. H. and ROBBINS, J.: J. Clin. Invest., **38**, 1683, 1959.
29) BENNHOLD, H.: Verh. deut. Ges. Inner Med., Munich, 1954.
30) BERENBAUM, M. C.: Brit. Med. Bull. **21**, 140, 1965.
31) BERGSTRAND, C. G. and CZAR, B.: Scand. J. Clin. Lab. Invest., **9**, 277, 1957.
32) BERGSTRAND, C. G. and CZAR, B.: Scand. J. Clin. Lab. Invest. **10**, 379, 1958.
33) BERLIN, S., ODEBERG, H. and WEINGART, L.: Acta Med. Scand., **183**, 347, 1968.
34) BERMAN, L. B.: Am. J. Med., **24**, 249, 1958.
35) BERNIER, G. M. and PUTNAM, F. W.: Progr. in Hematology, **4**, 160, 1964.
36) BESSON, P. B. and W. McDERMOTT: Cecil-Loeb Textbook of Medicine, Saunders & Igaku Shoin, Tokyo, 1967.
37) BETSUYAKU, T.: Sapporo Med. J. **22**, 269, 1962.
38) BEVAN, G., TASWELL, H. F., and GLEICH, G. J.: J. A. M. A., **203**, 92, 1967.
39) BING, J.: Acta Med. Scand. Suppl. **76**, 1, 1963.
40) BIRKE, G., LILJEDAHL, S. O'. and WETTERFORS, J.: Acta Chir. Scandin. **118**, 353, 1960.
41) BLAESE, R. M., STROBER, W., BROWN, R. S. and WALDMANN, T. A.: Lancet, **1**, 1056, 1968.
42) BLAESE, R. M., STOBER, W. and WALDMANN, T.: Clin. Res. **15**, 465, 1967.

43) BLAINEY, J. D., BREWER, D. B., HARDWICKE, J. and SOOTHILL, J. F.: Quart. J. Med. **29**, 235, 1960.
44) BLAUMONT, J. L.: C. R. Acad. Sci. Paris, **216**, 4563, 1965.
45) BLOMBÄCK, M., BLOMBÄCK, B., MAMMEN, E. F., and PRASAD A. S.: Nature **218**, 134, 1968.
46) BOLLET, A. J.: Arch. Int. Med. **104**, 152, 1959.
47) BRACHFELD, J. and MYERSON, R. M.: J. A. M. A. **161**, 865, 1956.
48) BRACKENRIDGE, C. J. and CSILLAG, E. R.: Acta Med. Scand. **172**, Suppl. 383, 1962.
49) BRUTON, O. C.: Pediatrics **9**, 722, 1952.
50) BUCKLEY, C. E. et al.: Ann. Int. Med. **64**, 508, 1966.
51) BURCH, R. R., PEARL, M. A. and STERNBERG, W. H.: Ann. Int. Med. **56**, 54, 1962.
52) BURTIN, P. and LOISILLIER, F., BUFFE D., GUILLERM, M. and GLUCKMAN, E.: Cancer **23**, 80, 1969.
53) BUSH, S. T., SWEDLUND, H. A. and GLEICH, G. J.: J. Lab. Clin. Med. **73**, 194, 1969.
54) CAGGIANO, V., CUTTNER, J. and SOLOMON, A.: Blood **30**, 265, 1967.
55) CALCAGNO, P. L. and M. I. RUBIN.: J. Ped. **58**, 686, 1961.
56) CANNON, E. J.: Arch. Int. Med. **101**, 620, 1958.
57) CATCHCART, E. S. and COHEN A. S.: J. Immunol. **96**, 239, 1966.
58) CATCHPOLE, H. R.: Proc. Soc. Exptl. Biol. Med. **75**, 221, 1950.
59) CHANDRASEKHAR, N.: Canad. J. Bioch. Physiol. **39**, 1489, 1961.
60) CHAPTAL, J., JEAN, R., CRASTES DE PAULET, A., CRISTOL, P. and DOSSA, D.: Semaine Hôp. **32**, 3871, 1956.
61) CHOW, B. F.: Ann. N. Y. Acad. Sci. **47**, 297, 1946.
62) CITRIN, Y., STERLING K. and HALSTED, J. A.: New Engl. J. Med. **257**, 906, 1957.
63) CLAMAN, H. N. and MERRILL, D.: J. Lab. Clin. Med. **67**, 850, 1966.
64) CONWAY, E. J.: Physiol. Rev. **37**, 84, 1957.
65) COOPER, M. D., CHASE, H. P., LOWMAN, J. T., KRIVIT, W. and GOOD, R. A.: Am. J. Med. **44**, 499, 1968.
66) COSTEA, N., YAKULIS, V. and HELLER, P.: J. Immunol. **99**, 558, 1967.
67) CRABBE, P. A. and HEREMANS, J. F.: Am. J. Med. **42**, 319, 1967.
68) DAMMACCO, F. and CLAUSEN, J.: Acta Med. Scand. **179**, 755, 1966.
69) DAUGHADAY, W. H., ADLER, R. E., MARIZ, I. K. and RASINSKI, D. C.: J. Clin. Endocrinol. **22**, 704, 1962.
70) DAVIDSOHN, I. and WELLS, B. B.: Clinical Diagnosis by Laboratory Methods, 13th Ed. Saunders Co. 1962.
71) DE VAAL, O. M. and SEYNHAEVE, V.: Lancet **2**, 1123, 1959.
72) DEMEULENAERE, L. and WIEME, R. J.: Am. J. Digest. Dis. **6**, 661, 1961.
73) Denki Eido Gakkai Society of Electrophoresis, Standard Committee: Physico-Chemical-Biology **11**, 351, 1966.
74) DI GEORGE, A. M.: J. Ped. **67**, 907, 1965.
75) DONALDSON, V. H. and EVANS. R. R.: Am. J. Med. **35**, 37, 1963.
76) DOWLING, J. T., FREINKEL, N. and INGBAR, S. H.: J. Clin. Invest. **35**, 1263, 1956.
77) DUHRING, J. L.: Am. J. Med. Sci. **243**, 808, 1962.
78) DURY, A. and TREADWELL, C. R.: J. Clin. Endocrin. **15**, 818, 1955.
79) EDER, H. A., RUSS, E. M., PRITCHETT, P. A. R., WILBER, M. M. and BARR, D. P.: J. Clin. Invest. **34**, 1147, 1955.
80) Editors: Lancet **2**, 661, 1967.
81) EICHENWALD, H. F. and SHINEFELD, H. R.: J. Ped. **63**, 870, 1963.
82) EISEN, N., LITTLE, J. R., OSTERLAND, K. and SIMMS, E. S.: Cold Spring Harbor Symp. 1967.
83) EMMRICH, R.: MÜNCH. Med. Wschr. **98**: 481, 1956.
84) ENGLE, R. L., Jr. and WALLIS, L. A.: Tice-Harvey Practice of Medicine, Vol. 1., W. F. Prior Com. Inc. Hagerstown.
85) EPSTEIN, W. L., FUDENBERG, H. H., REED, W. B., BODER, E. and SEDGWICK, R. P.: Intern. Arch. Allergy **30**, 15, 1966.

86) ERIKSSON, S.: Acta Med. Scand. 177 (Suppl. 432) 1, 1965.
87) FATCH-MOGHADAM, A., WÜRZ, H., OESSER, R. M. and KNEDEL, M.: Dtsch. Med. Wsch. 93, 1965, 1968.
88) FELDMAN, J. D.: Endocrinology 46, 552, 1950.
89) FINE, J. M., IMPERATO, C., BATTISTINI, A. and MORETTI, J.: Nouv. Rev. Frang. Hemat. 1, 72, 1961.
90) FISHER, B.: SCHAER, L. R. and MESSINGER, S.: Am. J. Clin. Path. 40, 291, 1963.
91) FLETCHER, A. P., BIEDERMAN, O., MOORE, D., ALKJAERSIG, N. and SHERRY, S.: J. Clin. Invest. 43, y81, 1964.
92) FLORSHEIM, W. H., DOWLING, J. T., MEISTER, L. and BODFISH, R. E.: Clin. Res. 9; 71, 1961.
93) FORMAN, W. B., RATNOFF, O. D. and BOYER, M. H.: J. Lab. Clin. Med. 72, 455, 1968.
94) FRANGIONE, B., MILSTEIN, C. and FRANKLIN, E. C.: Nature 221, 149, 1969.
95) FRANKLIN, E. C.: Prog. in Immunology, I. 746, Acad. Press, 1971.
96) FRANKLIN, E. G., LOWENSTEIN, J., BIGELOW, B. and MELTZER, M.: Am. J. Med. 37, 332, 1964.
97) FREDRICKSON, D. C.: J. Clin. Invest. 43, 228, 1964.
98) FREDRICKSON, D. S., ALTROCCHI, P. H., AVIOLI, L. V., GOODMAN, D. S. and GOODMAN, H. C.: Ann. Int. Med. 55, 1016, 1961.
99) FREDRICKSON, D. S., LEVY, R. I., and LEES, R. S.: New Engl. J. Med. 276, 273, 1967.
100) FRIED, M. and PUTNAM, F. W.: J. Biol. Chem. 235, 3472, 1960.
101) FREEMAN, T. and SMITH, J.: Biochem. J. 118, 869, 1970.
102) FRIEDBERG, V.: Deut. Med. Wochschr. 76, 798, 1951.
103) FUDENBERG, H. H. Arth. and Rheum. 9, 464, 1966.
104) FULGINITI, V. A., et al.: Lancet 2, 5, 1966.
105) GABL, F. and HUBER, E. G.: Ann. Paed. 202, 81, 1964.
106) GANROT, P. O.: Scand. J. Clin. Lab. Invest. 21, 177, 1968.
107) GATTI, R. A. and GOOD, R. A.: Proc. of 10th Intl. Cancer Congress, Houston, Texas, Man, 1970.
108) GATTI, R. A. et al.: J. Ped. 75, 675, 1969.
109) GELL, P. G. H. and COOMBS, R. R. A.: Clinical Aspects of Immunology, 2nd Ed. Blackwell Sci. Pub. 1968.
110) GILLES, H. M. and HENDRICKSE, R. G.: Brit. Med. J. 2, 27, 1963.
111) GITLIN, D., CORMWELL, D. G., NAKASATO, D., ONCLE, J. L., HUGHES, W. L. Jr. JANEWAY, C. A.: J. Clin. Invest. 37, 172, 1958.
112) GITLIN, D. and CORMWELL, A.: J. Clin. Invest. 35, 706, 1956.
113) GITLIN, D. and CRAIG, J. M.: Pediatrics 32, 517, 1963.
114) GITLIN, D., KUMATE, J., URRUSTI, J. and MORALES, C.: J. Clin. Invest. 43, 1938, 1964.
115) GLEICHMANN, E. and DEICHER, H.: Klin. Wschr. 45, 684, 1967.
116) GLENNER, G. et al.: Biochem. Biophys. Res Commun. 41, 1013, 1287, 1970.
117) GLUECK, H. I., WAYNE, L. and GOLDSMITH, R.: J. Lab. Clin. Med. 59, 40, 1962.
118) GLYNN, L. E.: in "Clinical Aspects of Immunology", ed. by P. G. H. GELL and R. R. A. COOMBS, 2nd Ed. Blackwell Sci. Pub. 1968.
119) GOLBERG, I. H. and RABINOWITZ, M.: Science 136, 315, 1962.
120) GOLDMAN, J. M. and HOBBS, J. R.: Immunology 13, 421, 1967.
121) GOOD, R. A., KELLY, W. D., ROTSTEIN, J. and VARCO, R. L.: Progr. Allergy 6, 187, 1962.
122) GOOD, R. A. and ZAK, S. J.: Pediatrics 18, 109, 1956.
123) GOODMAN, S. I., RODGERSON, D. O. and KAUFFMAN, J.: J. Lab. Clin. Med. 70, 57   1967.
124) GORDON, R. S. Jr., BARTER, F. C. and WALDMANN, T.: Ann. Int. Med. 51, 553, 1959.
125) GOTOFF, S. P., FELLERS, F. X., VAWTER, G. F., JANEWAY, C. A. and ROSEN, F. S.: New Engl. J. Med. 273, 524, 1965.
126) GÖTSCHLICH, E. C. and EDELMAN, G. M.: Proc. N.A.S. 54, 558, 1965.
127) GOKCEN, M.: J. Lab. Clin. Med. 59, 533, 1962.
128) GRABAR, P. and BURTIN, P.: Immunoelectrophoretic Analysis, Elsevier Pub. Co., 1964.
129) GREEN P., and BERGMAN, B.: Canada. Med. Ass. J. 86, 418, 1962.

130) GREY, H. M. and KUNKEL, H. G.: J. Exp. Med. **120**, 253, 1964.

131) HAFERKAMP, O., SCHLETTWEIN-GSELL, D., SCHWICK, H. G. and STÖRIKO, K.: Gerontologia **12**, 30, 1966.

132) HÄLLEN, J.: Acta Med. Scand. **173**, 737, 1963.

133) HÄLLEN, J.: Acta Med. Scand. Suppl. **462**, 1, 1966.

134) HARBOE, M., FURTH, R., SCHUBOTHE, H., LIND, K. and EVANS, R. S.: Scand. J. Haemat. **2**, 259, 1965.

135) HARDIE, G. and KENCH, J. E.: Lancet **1**, 809, 1967.

136) HASSELBACK, R., MARION, R. B. and THOMAS, J. W.: Canada. Med. Ass. J. **88**, 19, 1963.

137) HÄSSIG, A. et al.: Path. Microbiol. **27**, 542, 1964.

138) HAURANI, F. I.: Serum Proteins and the Dysproteinemias, ed. by SUNDERMAN and SUNDERMAN, Lippincott. 1964.

139) HAUROWIZ, F.: Immunochemistry and the Biosynthesis of Antibodies, Intersdience Pub., 1968.

140) HAUT, A., CARTWRIGHT, G. E. and WINTROBE, M. M.: J. Lab. Clin. Med. **63**, 277, 1964.

141) HEILMYER, L.: Acta Haemat. **36**, 40, 1966.

142) HEILMYER, L., KELLER, W., UIVELL, O., KEIDERLING, W., BETKEK., K. WÖHLER, F. and SCHULTZE, H. E.: Deutsch. Med. Wschr. **86**, 1745, 1961.

143) HEILMYER, L., MERKER, H., WETZEL, H. P., KLEMM, D., BURMEISTER, P. and HAAS, R.: Dtsch. Med. Wschr. **90**, 1649, 1965.

144) HEIMER, R. and LEVIN, F.: Ann. N. Y. Acad. Sci. **124**, 879, 1965.

145) HEVÉR, Ö. and VADÁSZ, G.: J. Ped. **67**, 1156, 1965.

146) HEYMANN, W.: J. Ped. **58**, 609, 1961.

147) HIRAI, H.: Physico-Chemical Biology (in Japanese), **10**, 215, 1965.

148) HIRAYAMA, C., FUKUDA, T. and MURAI, N.: Saishin-Igaku, (in Japanese), **23**, 1598, 1968.

149) HIRSCHFELD, J. and LUNELL, N. O.: Nature **196**, 1220, 1962.

150) HITZIG, W. H.: Die Plasmaproteine in der klinischen Medizin, Springer, Berlin, 1963.

151) HITZIG, W. H. and WILLI, H.: Schweiz, Med. Wschr. **91**, 1625, 1961.

152) HOBBS, J. R.: Brit. Med. J. **3**, 699, 1967.

153) HOBBS, J. R.: Lancet i; **110**, 1968.

154) HOBBS, J. R. and DAVIS, J. A.: Lancet **1**, 757, 1967.

155) HOFFMAN, R. G. and WAID, M. E.: Postgrad. Med. **41**, A–20, A–47, A–69, 1967.

156) HOKOMA, Y., COLMAN, M. R., and RILEY, R. F.: J. Immunol. **98**, 521, 1967.

157) HONG, R. and GOOD, R. A.: Science **156**, 1102, 1967.

158) HONG, R., KAY, H. E. M., COOPER, M. D., MEUWISSEN, H., ALLAN, M. J. G., & GOOD, R. A.: Lancet **1**, 503, 1968.

159) HØNGER, P. E.: Scand. J. Clin. Lab. Invest. **21**, 3, 1968.

160) HOWE, C., AVRAMEAS, S., DE VAUX ST Cyr. C., GRABAR, P. and LEE, L. T.: J. Immunol. **91**, 683, 1963.

161) HUMPHREY, J. H.: in "Immunological Diseases", edited by M. SAMTER, Little, Brown and Co. Boston, 1964.

162) HUNTER, C. C., Jr., PIERCE, J. A. and LaBORDE, J. B.: J.A.M.A. **205**, 93, 1968.

163) ICHIBA, K.: Tokyo Jikei Med. Col. Bull., **82**, 951, 1967.

164) IMPERATO, C. and DETTORI, A. G.: Helv. Paed. Acta **13**, 380, 1958.

165) ISRAEL, H. L.: "Immunological Diseases", ed. by M. SAMTER, Little, Brown, Co., 1965.

166) ISRAEL, H. L., PATTERSON, J. D. and SMUKLER, N.: Proc. Intern. Conf. Sarcoid. Acta med. Scand. Suppl. **425**, 40, 1964.

167) IWASAKI, M.: Jap. J. Int. Med. (in Japanese), **55**, 22, 1966.

168) JACOBS, J. C., et al.: Lancet **1**, 499, 1968.

169) JAGER, B. V.: New Engl. J. Med. **266**, 579, 1962.

170) JAHNKE, K., SCHOLTAN, W. and HEINZLER, F.: Helv. Med. Acta **25**, 2, 1958.

171) JAMES, W. P. T. and HAY, A. M.: J. Clin. Invest. **47**, 1958, 1968.

172) JANEWAY, C. A., ROSEN, F. S., MERLER, E. and ALPER, C. A.: The Gamma Golbulins, Little, Brown and Co. Boston, 1967.

173) JANUM, S. and LASSEN, N. A.: Scand. J. Clin. Lab. Invest. **13**, 357, 1961.
174) JENCKS, W. P., JETTON, M. R. and DURRUM, E. L.: Biochem. J. **60**, 205, 1955.
175) JENSEN. H., BRO-JØRGENSEN, K., JARNUM, S., OLESEN, H., and YSSING, M.: Scand. J. Clin. Lab. Invest. **21**, 293, 1968.
176) JOACHIM, G. R., CAMERON, J. S., SCHWARTZ, M., and BECKER, E. L.: J. Clin. Invest. **43**, 2332, 1964.
177) JOHANSSON, S. G. O. and BENNICH, H.: Immunology **13**, 381, 1967.
178) KAHLICH-KOENNER, D. M. and WEIPPL, G.: Wien. Klin. Wochschr. **72**, 674, 1960.
179) KALBFLEISCH, J. M. and BIRD, R. M.: New Engl. J. Med. **263**, 881, 1960.
180) KAWAI, T.: Bull. Univ. Miami School of Med. **16**, 89, 1962.
181) KAWAI, T.: Saishin-Igaku (in Japanese) **19**, 1001, 1964.
182) KAWAI, T.: Physico-Chemical Biology, **10**, 111, 1964.
183) KAWAI, T.: Physico-chemical Biology, **13**, 307, 1968.
184) KAWAI, T. and TADANO, J.: Medicine and Biology, **74**, 251, 1967.
185) KAWAI, T., YAMADA, K. and TAKAHASHI, K.: Physico-Chemical Biology (in Japanese), **11**, 228, 1966.
186) KELLEY, V. C., KIRSCHVINK, J. F., and ELY, R. S.: Am. J. Physiol. **171**, 738, 1952.
187) KILLANDER, J. (Editor): Gamma Globulins, Nobel Symposium 3, Almqvist and Wiksell, Stockholm, 1967.
188) KILLMAN, S., GJØRUP, S. and THAYSEN, J. H.: Acta Med. Scand. **158**, 43, 1957.
189) KINCH, R. A. H.: Am. J. Obst. & Gynec. **71**, 746, 1956.
190) KLEMPERER, P., POLLACK, A. D. and BAEHR, G.: J. A. M. A. **119**, 331, 1942.
191) KOCHWA, S., SMITH, E., BROWNELL, M. and WASSERMAN, L. R.: Biochemistry, **5**, 277, 1966.
192) KONITZER, K.: Z. ges. inn. Med. u. ihre Grenzgebiete **10**, 801, 1955.
193) KORNGOLD, L.: Progr. in Clin. Path. **1**, 340, 1966.
194) KOSAKA, A. et al.: Clinical Immunology (in Japanese) **1**, 57, 1969.
195) KOZURU, M.: Jap. J. Int. Med. (in Japanese), **52**, 629, 1963.
196) KOZURU, M.: Physico-chemical Biology (in Japanese), **12**, 231, 1967.
197) KRAUSS, S. and SOKAL, J. E.: Am. J. Med. **40**, 400, 1966.
198) KREBS, H. A.: Klin. Wochschr. **7**, 584, 1928.
199) KRIVIT, W. and GOOD, R. A.: Am. J. Dis. Child. **97**, 137, 1959.
200) KUNKEL, H. G. and AHRENS, E. H. Jr: J. Clin. Invest. **28**, 1575, 1948.
201) KUSHMER, D. S., et al.: J. Lab. Clin. Med. **47**, 409, 1956.
202) KUTT, H. and McDOWELL, F.: J. Lab. Clin. Med. **59**, 118, 1962.
203) KWAAN, H. C., McFADZEAN, A. J. and COOK, J.: Lancet **1**, 132, 1956.
204) KYLE, R. A., and BAYRD, E. D.: Am. J. Med. **40**, 426, 1966.
205) LAHEY, M. E., CUBLER, C. J., CARTWRIGHT, G. E. and WINTROBE, M. M.: J. Clin. Invest. **32**, 329, 1953.
206) LAMBOTTE, R. and SALMON, J.: Compt. Rend. Soc. Biol. **156**, 530, 1962.
207) LANGE, F.: Acta Med. Scand. Suppl. **176**, 1945.
208) LAUFER, A., Tal. C., and BEHAR, A. J.: Brit. J. Exp. Path. **150**, 1, 1959.
209) LASSER, E. C., FARR, R. S., FUJIMAGARI, T., and TRIPP, W. N.: Am. J. Roent. Rad. Therap. and Nucl. Med. **87**, 338, 1962.
210) LAURELL, C. B.: Acta Physiol. Scand. **14**, Suppl. 46, 1947.
211) LAURELL, C. B.: Scand. J. Clin. Lab. Invest. **21**, 136, 1968.
212) LAURELL, C. B.: Anal. Biochem. **10**, 358, 1965.
213) LAURELL, C. B. and NYMAN, M.: Blood **12**, 493, 1967.
214) LAURELL, C. B. and SKANSE, B.: J. Clin. Endocr. **23**, 214, 1963.
215) LEE, F. I.: Lancet ii, 1043, 1965.
216) LEONHARDT, T.: Acta Med. Scand. Suppo. 416, 1964.
217) LEVEQUE, B., DEBAUCHEZ, C., DESBOIS, J. C. FEINGOLD, J. BARBET, J. & MARIE, J.: Semaine hôp. Paris, **42**, 2709, 1966.
218) LEVY, R. I. and FREDRICKSON, D. S.: Circulation **34** (Suppl. III), 156, 1966.

219) LEVEY, S. and JENNINGS, E. R.: Am. J. Clin. Path. **20**, 1059, 1950.
220) LEWIS, I. A., ROBERTSON, A., MARS, H., and WILLIAMS, G. Jr.: Circulation **34** (Suppl. III), 19, 1966.
211) LINDGREN, F. T. and NICHOLS, A. V.: in "Plasma Proteins" Vol. II, ed. by F. W. Putnam. Academic Press, N. Y., 1960.
222) LINDSTRÖM, F. D. WILLIAMS, R. C., Jr., SWAIM, W. R. and FREIER, O. F.: J. Lab. Clin. Med. **71**, 812, 1968.
223) LLOYD. C. W.: in "Textbook of Endocrinology" ed. by R. H. WILLIAMS, Saunders, Philadelphia, 1963.
224) LO GRIPPO, G. A., WOLFRAM, B. R., and HAYASHI, H.: J. A. M. A. **191**, 97, 1965.
225) LONGSWORTH, L. G., SHEDLOVSKY, T. and MACINNES, D. A.: J. Exp. Med. **70**, 399, 1939.
226) LORINCZ, A. B., McCARTNEY, C. P., POTTEINGER, R. E. and LI, K. H.: Am. J. Obst. & Gynec. **82**, 252, 1961.
227) LUX, S. E. et al.: New Engl. J. Med. **282**, 231, 1970.
228) LYTLE, R. I., ROSENBAUM, M. J., MILLER, L. F. and ROSENTHAL, S.: J. Lab. Clin. Med. **63**, 117, 1964.
229) MACK, C.: The Plasma Proteins in Pregnancy, C. C. Thomas, Springfield, 1956.
230) MACLEOD, C. M. and AVEY, O. T.: J. Exp. Med. **73**, 191, 1941.
231) MAN, E. B. and WHITEHEAD, R. J. Jr.: Clin. Chem. **14**, 1002, 1968.
232) MANCINI, G., CARBONARA, A. O. and HEREMANS, J. F.: Immunochemistry **2**, 235, 1965.
233) MANCINI, R. E., GARBERI, J. C. and DE LA BALZE, P. A.: Rev. Soc. Biol. **27**, 285, 1951.
234) MANNIK, M.: J. Immunol. **99**, 899, 1967.
235) MARCOLONGO, R., CARCASSI, A., FRULLINI, F. BIANCO, G., and BRAVI, A.: Ann. Rheum. Dis. **26**, 412, 1967.
236) MARKOWITZ, J. L., GUBLER, C. J., MAHONEY, J. P., CARTWRIGHT, G. E. and WINTROBE, M. M.: J. Clin. Invest. **34**, 1498, 1955.
237) MARTIN, W. J., MATHIESON, D. R. and EIGLER, J. O. C.: Proc. Staff Meet'g Mayo Clinic **34**, 95, 1959.
238) MARTINEZ-TELLO, F., BRAUN, D., SAWADE, H., und HAFERKAMP, O.: Virchows Arch. Path. Anat. **339**, 337, 349, 1965.
239) MATSUMOTO, I.: J. Jap. S. Dig. Dis. (in Japanese), **65**, 828, 1968.
240) McCALLISTER, B. D., BAYRD, E. D., HARRISON, E. G., Jr., and McGUCKIN, W. F.: Am. J. Med. **43**, 394, 1967.
241) McCALLY, M. and VASICKA, A.: Obst. Gynec. **19**, 359, 1962.
242) McCRACKEN, G. H. Jr. and SHINEFIELD, H. R.: Ped. **36**, 933, 1965.
243) McDONAGH, T. J., GUEFT, B., PYUN, K. and ARIAS, I. M.: Gastroenterology **48**, 642, 1965.
244) McKAY, E., THOM, H. and GRAY, D.: Arch. Dis. Childh. **42**, 264, 1967.
245) McKUSICK, V. A. et al.: Bull Johns Hopkins Hosp. **116**, 285, 1965.
246) McMASTER, P. D. and FRANZL, R. E.: Metabolizm **10**, 990, 1961.
247) MEIERS, H. G.: Dtsch. med. Wschr. **93**, 1795, 1968.
248) MENACHÉ, D.: Thromb. Diath. Haemorth. Suppl. **13**, 173, 1964.
249) METCOFF, J. and JANEWAY, C. A.: J. Ped. **58**, 640, 1961.
250) MILGROM, F., KASUKAWA. R. and CALKINS, E.: J. Immunol. **96**, 245, 1966.
251) MILLER, D. G.: Ann. Int. Med. **57**, 703, 1962.
252) MILLER, D. G.: in "Immunological Diseases" ed. by M. SAMTER, p. 372, Little, Brown & Co. Boston, 1964.
253) MIYOSHI, K.: J. Jap. S. Int. Med. (in Japanese), **57**, 179, 1968.
254) MIYOSHI, K. et al.: Nihon Rinsho (in Japanese), **21**, 489, 1963.
255) MIYOSHI, K. et al.: Clinical Science (in Japanese) **2**, 651, 1966.
256) MIYOSHI, K. and SAIJO, K.: Physico-chemical Biology (in Japanese), **12**, 217, 1967.
257) MONMA, K. and YOSHINO, K.: Shonika-Shinryo (in Japanese). **30**, 593, 1967.
258) MONTGOMERY, D. A., NEILL, D. W. and DOWDLE, E. B. D.: Clin. Sci. **22**, 141, 1962.
259) MOORE, D. H., DUPAN, R. and BUXTON, C. L.: Am. J. Obst. Gynec. **57**, 312, 1949.

260) MOORE, J. G., MENA, P., PERRY, S., GREENBERG, P. M. CHANG, N., CROOKS, W. S.: Am. J. Obst. Gynec. **83**, 1036, 1962.
261) MORSE, J. H.: J. Immunol. **95**, 722, 1965.
262) MURRAY, R. C. and HOFMEISTER, F. J.: Obst. Gynec. Survey **16**, 449, 1961.
263) MUSTACCI P., PETERMAN, M. L. and RALK, J. E.: J. Clin. Endocrin. **14**, 729, 1954.
264) MÜLLER, H. E. and MÜLLER-VON VOIGT, I.: Dtsch. med. Wschr. **93**, 1707, 1968.
265) MUTO, A. et al.: Physico-chemical Biology (in Japanese), **13**, 91, 1968.
266) NAGEL, R. and KATZ, R.: Am. J. Med. Sci. **245**, 198, 1963.
267) NEUFELD, A. H., MORTON, H. S., and HALPENNY, G. W.: Canad. Med. Ass. J. **91**, 374, 1964.
268) NEZELOF, C., JAMMET, M. L., LORTHOLARY, P., LABRUNE, B. and LAMY, M.: Arch. franç. pédiot, **21**, 897, 1964.
269) NICOLOFF, J. T. and DOWILNG, J. T.: Clin. Res. **9**, 74, 1961.
270) NISHIKAWA, M.: J. Jap. S. Dig. Dis. (in Japanese), **63**, 1309, 1966.
271) NORBERG, R.: Acta Med. Scand. Suppl. **425**, 43, 1964.
272) NORBERG, R.: 4th Intern. Conf. on Sarcoidosis, Paris, 1966.
272) NORBERG, R. 4th Intern. Conf. on Sarcoidosis, Paris, 1966.
273) NYMAN, M.: Scand. J. Clin. Lab. Invest. **11**, Suppl. 38, 1959.
274) NYS, A.: Rev. belge Pathol. et Méd. exp. **23**, 329, 1954.
275) OBA, Y., SASAKI, K. and IDA, K.: Ped. (in Japanese) **8**, 1062, 1967.
276) OLIVER, W. J. and SPORZYNSKI, K.: J. Lab. Clin. Med. **58**, 788, 1961.
277) OKABAYASHI, A.: J. Jap. Path. Soc. (in Japanese), **51**, 223, 1962.
278) OSHIMA, K.: J. Jap. Soc. Int. Med. (in Japanese), **57**, 1043, 1968.
279) OSSERMAN, E. F.: Immunopathology, IVth Intern. Symposium, p. 283, Schwabe Co. 1966.
280) OSSERMAN, E. F.: Cecil-Leob Textbook of Medicine, 12th ed. Philadelphia, Saunders, Co. 1967.
281) OSSERMAN, E. F. and FAHEY, J. L.: Am. J. Med. **44**, 256, 1968.
282) OSSERMAN, E. F. and TAKATSUKI, K., Am. J. Med. **37**, 351, 1964.
283) PACHTER, M. R. and HAVERY, G.: Am. J. Clin. Path. **37**, 248, 1962.
284) PARASKEVAS, F., HEREMANS, J. and Waldenström, J.: Acta Med. Scand. **170**, 575, 1961.
285) PARONETTO, F., RUBIN, E., and POPPER, H.: Lab. Invest. **11**, 150, 1962.
286) PASNICK, L. J., BEALL, G. N., VAN ARSDEL, P. P.: Am. J. Med. **33**, 774, 1962.
287) PATTERSON, R., NELSON, V. L. and PRUZANSKY, J. J.: Immunology **9**, 477, 1965.
288) PEARSON, C. M.: Proc. Soc. Exp. Biol. Med. **91**, 95, 1956.
289) PEETOM, F. and KHAMER, E.: Vox Sang. **7**, 298, 1962.
290) PENICK, G. D., ROBERTS, H. R., WEBSTER, W. P. and BRINKHOUS, K. M.: Arch. Path. **66**, 708, 1958.
291) PERILLIE, P. E. and CONN. H. O.: J. A. M. A. **167**, 2186, 1958.
292) PERNIS, B. and PARONETTO, F.: Proc. Soc. Exp. Biol. Med. **110**, 390, 1962.
293) PETERSON, R. D. A., COOPER, M. D. and GOOD, R. A.: Am. J. Med. **38**, 579, 1965.
294) PETERSON, R. D. A., COOPER, M. D., and GOOD, R. A.: Am. J. Med. **41**, 342, 1966.
295) PHILLIPS, G. B., and DODGE, J. T.: J. Lab. Clin. Med. **71**, 629, 1968.
296) POLLAK, U. E., MANDENA, E., DOIG, A. B., MOORE, M. and KARK, R. M.: J. Lab. Clin. Med. **58**, 353, 1961.
297) POMERANTZ, M. and WALDMANN, T. A.: Gastroenterology **45**, 703, 1963.
298) POPPER, H. and SCHAFFNER, F.: Liver Structure and Function, McGraw-Hill, New York, 1957.
299) PORTER, H. M.: Pediatrics **20**, 958, 1957.
300) PURVES, L. R., MacNAB, M., GEDDES, E. W. and BERSOHN, I.: Lancet 1; (No. 7548), **921**, 1968.
301) RAE, A. I., EPSTEIN, W. V., GULYASSY, P. F., and TAN, M.: Ann. Int. Med. **68**, 48, 1968.
302) RATCLIFF, P., SOOTHILL, J. F. and STANWORTH, D. R.: Clin. Chim. Acta **8**, 91, 1963.
303) RATH, C. E., CATON, W., REID, D. E., FINCH, C. A. and CONROY, L.: Surg. Gynec. Obst. **90**, 320, 1950.
304) RIBEIRO, L. P., ABREN, L. A. and ABREN, R.: Clin. Chem. Acta **2**, 252, 1957.

305) RIGGIO, R. R., SCHWARTZ, G. H., STENZEL, K. H. and RUBIN, A. L.: Lancet 1, 1218, 1968.

306) RITZMANM, S. E. and LEVIN, W. C.: Arch. Int. Med. 107, 754, 1961.

307) RIVA, G.: Das Serumeiweißbild, Verlag Hans Huber, Bern u. Stuttgart, 1957.

308) ROBBINS, J.: in "Serum Proteins and the Dysproteinemias" ed. by SUNDERMAN and SUNDER-
MAN, Lippincott, Philadelphia, 1964.

309) ROBBINS, J. and NELSON, J. H.: J. Clin. Invest. 37, 153, 1958.

310) RABE, F. and SALOMON, E.: Deutsch. Arch. Klin. Med., 132, 240, 1920.

311) ROCKEY, J. H., HANSON, L. A., HEREMANS, J. F., and KUNKEL, H. G.: J. Lab. Clin. Med.
63, 205, 1964.

312) ROSEN, S., CORTELL, S., ADNER, M. M., PAPADOPOULOS, N. M., and BARRY, K. G.: Am. J.
Clin. Path. 47, 567, 1967.

313) ROSSING, N.: Scand. J. Clin. Lab. Invest. 21, 26, 1968.

314) ROWE D. S. and FAHEY, J. L.: J. Exp. Med. 121, 171, 1965.

315) RUSS, E. M., EDER, H. A. and BARR, D. P.: J. Clin. Invest. 33, 1662, 1954.

316) SAITO, Y.: J. Jap. Soc. Int. Med. (in Japanese), 56, 441, 1967.

317) SALMON, J., LAMBOTTE, J. R., and SMOLLAR, V.: Arch. Intern. Physiol. Biochemi. 70, 731,
1962.

318) SAMTER, M.: Immunological Diseases, Little, Brown Pub. 1966.

319) SANDOR, G.: Serum Proteins in Health and Disease, Chapman & Hill, Ltd. London, 1966.

320) SANPE, K. et. al.: J. Jap. Soc. Int. Med (in Japanese). 56, 1336, 1967.

321) SCANU, A. M.: Advances in Lipid Research 3, 63, 1965.

322) SCHAFFNER, F., POPPER, H., and DALLA TORRE, M.: Gastroenterology 30, 357, 1956.

323) SCHAUNWHITE, W. R., Jr. and SANBERG, A. A.: J. Clin. Invest. 38, 1290, 1959.

324) SCHEIDEGGER, J. J., Martin, E. and Riotton, G.: Schweiz. Med. Wochschr. 86, 224, 1956.

325) SCHEINBERG, I. H. and STERNLIEB, I.: Gastroenterology 37, 550, 1959.

326) SCHEURLEN, P. G.: Dtsch. med. Wschr. 90, 40, 1965.

327) SCHMID, K., BURKE, J. F., DEBRAY-SACHS, M. and TOKITA, K.: Nature 204, 75, 1964.

328) SCHMITZ-MOORMANN, P.: Virchows Arch. Path. Anat. 339, 45, 1965.

329) SCHNEIDER, A. J.: Pediatrics 26, 973, 1960.

330) SCHOEN, I., GLOVER, S. N. and LABINER, G.: Am. J. Clin. Path. 43, 456, 1965.

331) SCHULTZ, R. T., CALKINS, E. and MILGROM, F.: Am. J. Path. 50, 957, 1967.

332) SCHULTZE, H. E. and HEREMANS, J. F.: Molecular Biology of Human Proteins, Elsevier,
Amsterdam, 1966.

333) SCRIMSHAW, N. S. and ALLING, E. L.: Federation Proc. 8, 68, 1949.

334) SCRIMSHAW, N. S. and BÉHAR, M.: Science 133, 2039, 1961.

335) SEHON, A. H. and GYENES, L.: in "Immunological Diseases", ed. by M. SAMTER, Little,
Brown and Co., 1965.

336) SEHON, A. H., GYENES, L., GORDON, J.: Richter, M. and Rose, B.: J. Clin. Invest. 36, 456,
1957.

337) SEIBERT, F. B., SEIBERT, M. V., Atno, A. J. and Campbell, H. W.: J. Clin. Invest. 26, 90,
1947.

338) SELIGMAN, M. and BASCH, A.: XII Cong. Internat. Soc. Hemat. New York, 1968.

339) SELIGMAN, M., DANNON, F., MIHAESCO, C. and FUDENBERG, H. H.: Am. J. Med. 43, 66,
1967.

340) SELIGMANN, M., DANON, F., HUREZ, D., MIHAESCO, E. and PREUD' HOMME, J.: Science
162, 1396, 1969.

341) SELIGMANN, M., FUDENBERG, H. H., and GOOD, R. A.: Am. J. Med. 45, 817, 1968.

342) SENECAL, J., AUBRY, L., DUPIN, H., DAVIN R. and DARRASSE, F.: Bull. Méd. Afrique Oc-
cidentale franc. (Dakar), 1, 148, 1956.

343) SHETLAR, M. R., PAYNE, R. W., STIDWORTH, G. and MOCK, D. L.: Ann. Int. Med. 51, 1379,
1959.

344) SHETLAR, M. R. SHETLAR, C. L., BRYAN, R. S. and EVERETT, M. R.: Cancer Res. 9, 515,
1949.

345) SHIMIZU, A.: Nihon Univ. Med. J. (in Japanese), 20, 403, 1961.

448   DIAGNOSIS AND PATHOGENESIS OF ABNORMALITIES

346) SIA, R. H. P.: China Med. J. **38**, 35, 1924.
347) SILVERSTEIN, A. M.: Blood **16**, 1338, 1960.
348) SILVERSTEIN, A. M.: Nature **194**, 196, 1962.
349) SIREK, O. V., SIREK, A., and LEIBEL, B. S.: Diabetes **10**, 375, 1961.
350) SMITH, E., KOCHWA, S., and WASSERMAN, L. R.: Am. J. Med. **39**, 35, 1965.
351) SMITH, R. T.: J. Clin. Invest. **36**, 605, 1957.
352) SOLOMON, A. and KUNKEL, H. G.: Am. J. Med. **42**, 958, 1967.
353) SOLOMON, A., WALDMANN, T. A., FAHEY, J. L. and McFARLANE A. S.: J. Clin. Invest. **43**, 103, 1964.
354) SOMER, T.: Acta Med. Scand. 180 (Suppl. 456); **1**, 1966.
355) SOOTHILL, J. F.: Lancet **1**, 1385, 1966.
356) SOOTHILL, J. F. and HENDRICKSE, R. G.: Lancet **2**; 629 (Sept. 23), 1967.
357) SREBNIK, H. H. and NELSON, M. M.: Endocrin. **70**, 723, 1962.
358) STANWORTH, D. R., HUMPHREY, J. H., BENEICH, H., BENEICH, H., and JOHANSSON, S. G. O.: Lancet **2**, 330, 1967.
359) STEFANINI, M. and DAMESHEK, W.: The Hemorrhagic Disorders, Grune & Stratton, Inc. 1955.
360) STEINER, J. W., LANGER, B. and SCHATZ, D. L.: AMA Arch. Path. **70**, 424, 1960.
361) STEINFELD, J. L.: J. Lab. Clin. Med. **55**, 904, 1960.
362) STERNLIEB, I. and SCHEINBERG, H.: JAMA **183**, 747, 1963.
363) STEIHM, E. R. and McINTOSH, R. M.: Clin. Exper. Immunol. **2**, 179, 1959.
364) STOBO, J. D. and TOMASI, T. B., Jr.: J. Clin. Invest. **46**, 1329, 1967.
365) STORIKO, K.: Ztsch. f. gesamte Blutforsch'g **16**, 200, 1968.
366) STOBER, W., WOCHNER, R. D., CARBONE, P. P. and WALDMANN, T. A.: J. Clin. Invest. **46**, 1643, 1967.
367) SUGIMOTO, R. et al.: Saishin-Igaku (in Japanese), **10**, 2144, 1955.
368) SUGIYAMA, K. et al.: J. Jap. Soc. Int. Med. (in Japanese), **57**, 584, 1968.
369) SUNDERMAN, F. W. Jr.: Am. J. Clin. Path. **42**, 1964.
370) SUNDERMAN, F. W. and BOERNER, F.: Normal Values in Clinical Medicine Saunders Co. Philadelphia, 1950.
371) SUNDERMAN, F. W., Jr. and SUNDERMAN, F. W.: Am. J. Clin. Path. **27**, 125, 1957.
372) SVANBORD, A.: Acta Med. Scand. **41**, Suppl. 264, 1951.
373) SWITZER, S.: J. Clin. Invest. **46**, 1855, 1967.
374) TAKASUGI, T.: Naika (in Japanese), **15**, 1031, 1965.
375) TAKATSUKI, K.: Proc. Jap. Clin. Met. Soc. (in Japanese) (II), 95, 1965.
376) TAKEI, H. et al.: Physico-chemical Biology (in Japanese), **13**, 289, 1968.
377) TANAKA, S. and STARR, P.: J. Clin. Endocrinol. **19**, 485, 1959.
378) TERRY, W. and FAHEY, J. L.: Science **146**, 100, 1964.
379) TILLETT, W. B. and FRANCIS, T. O.: J. Exp. Med. **52**, 561, 1930.
380) TOBLER, R. and COTTIER, H.: Helvet. Paediat. Acta. **13**, 313, 1958.
381) TODD, J. S. and PHILLIPS, L. L.: Surgery, Gynec. & Obst. **114**, 333, 1962.
382) TOMASI J. B. and TISDALE, W. A.: Nature **201**, 834, 1964.
383) TÖNDER, O. and HARBOE, M.: Immunology **11**, 361, 1966.
384) TORISU, M. et al.: J. Immunol. **99**, 629, 1967.
385) TURPIN, R., SCHMIDT-JUREAU H. and JERÔME, H.: Compt. rend. Soc. Biol. **144**, 352, 1950.
386) TSUBURA, H.: Nihon-Rinsho (in Japanese), **26**, 1609, 1968.
387) UEDA, H.: J. Jap. Soc. Int. Med. (in Japanese), **42**, 591, 1953.
388) UEDA, Y.: J. Jap. Soc. Int. Med. (in Japanese) **56**, 1250, 1967.
389) UEDA, Y. et al.: Saishin-Igaku (in Japanese), **23**, 1607, 1968.
390) ULTMANN, J. E., FISH, W., OSSERMAN, E. and GELLHORN, A.: Ann. Int. Med. **51**, 501, 1959.
391) UTIYAMA, O.: Proc. Jikei Med. Col., Dept. of Physiology (in Japanese), **4**, 475, 1961.
392) VAERMAN, J. P., JOHNSON, L. B., MANDY, W. and FUNDEBERG, H. H.: J. Lab. Clin. Med. **65**, 18, 1965.

393) VAN BUCHEM, F. S. P., Pol, G., DE GIER, J., BÖTTCHER, C. J. F., and PRIES, C.: Am. J. Med. **40**, 794, 1966.

394) VAUGHAN, J. H.: in "Immunological Diseases", ed. by M. SAMTER, Little, Brown and Co. 1965.

395) VON FELTEN, A., DUCKERT, F., and FRICK, P.: Schweiz. Med. Wsch. **95**, 1453, 1965.

397) VON STUDNITZ, M.: Scand. J. Clin. Lab. Invest. **7**, 324, 1955.

398) WADA, T.: Saishin-Igaku (in Japanese), **19**, 849, 1964.

399) WALDENSTRÖM, J. G.: Monoclonal and polyclonal hyper-gammaglobulinemia, Cambridge Univ. Press, 1968.

400) WALDENSTRÖM, J. G., WINBLAD, S., HÄLLEN, J. and LIUNGMAN, S.: Acta Med. Scand. **176**, 619, 1964.

401) WALDMANN, T. A., GORDON, R. S., Jr. and ROSSE, W.: Am. J. Med. **37**, 960, 1964.

402) WEIMER, H. E., REDILISH MOSHIN, J. R. and NISHIHARA, H.: J. Natl. Cancer Inst. **19**, 409, 1957.

403) WEINBREN, I. TAGGART, P. I., and GLASS, H. I.: Lancet **1**, 512, 1965.

404) WELTON, J., WALKER, S. R., SHARP, G. C., HERSENBERG L. A., WISTAR, R. and CREGER, W. P.: Am. J. Med. **44**, 280, 1968.

405) WEST, C. D., HONG, R. and HOLLAND, N. H.: J. Clin. Invest. **41**, 2054, 1962.

406) WEST, C. D., NORTHWAY, J. D., and DAVIS, N. C.: J. Clin. Invest. **43**, 1507, 1964.

407) WHITTAKER, M.: Am. J. Clin. Path. **50**, 454, 1968.

408) WHO-Bull. Wld. Hlth Org. **45**, 125, 1971.

409) WILLIAMS, R. C. Jr., ERICKSON, J. L., POLESKY, H. F. and SWIN, W. R.: Ann. Int. Med. **67**, 309, 1967.

410) WILLIAMS, R. H. and GLOMSET, J. A.: Textbook of Endocrinology, ed. by WILLIAMS, R. H., 3rd. Ed., Saunders, Philadelphia, 1963.

411) WOLLHEIM, F. A.: Acta Med. Scand. **183**, 473, 1968.

412) WOOD, H. F., McCARTY, M. and SLATER, R. J.: J. Exp. Med. **100**, 71, 1954.

413) WUHRMANN, F. and WUNDERLY, C.: Die Bluteiweisskörper des Menschen, B. Schwabe, Basel, 1952.

414) WUHRMANN, F. and WUNDERLY, C.: The Human Blood Proteins, Grune & Stratton, N. Y. 1960.

415) YAMADA, T.: Kotsu-Igaku (in Japanese), **14**, 369, 469, 1960.

416) YAN, S. H. Y. and FRANKS, J J.: J. Lab. Clin. Med. **72**, 449, 1968.

417) YEUNG, C. Y. and HOBBS, J. R.: Lancet **1**, 1167, 1968.

418) YOSHIOKA, H.: Shoni-Igaku (in Japanese), **1**, (No. 3), 85, 1968.

419) ZELDIS, L. J., ALLING, E. L., McCOORD, A. B. and KULKA, J. P.: J. Exp. Med. **81**, 515, 1945; 82: 157, 1945.

420) ZETTERVALL, O.: Nobel Symp., **3**, 349, 1967.

421) ZETTERVALL, O., SJÖQUIST, J., WALDENSTRÖM, J. and WINBALD, S.: Clin. Exp. Immunol. **7**, 213, 1966.

422) ZLOTNICK, A. and RODMAN G. P.: Proc. Soc. Exp. Biol. Med. **109**, 742, 1962.

Index